Historical Pollution

Francesco Centonze · Stefano Manacorda
Editors

Historical Pollution

Comparative Legal Responses to Environmental Crimes

Springer

Editors
Francesco Centonze
Università Cattolica del Sacro Cuore
Piacenza
Italy

Stefano Manacorda
Università della Campania
Naples
Italy

ISBN 978-3-319-56936-9 ISBN 978-3-319-56937-6 (eBook)
DOI 10.1007/978-3-319-56937-6

Library of Congress Control Number: 2017937920

Printed on acid-free paper

This Springer imprint is published by Springer Nature
The registered company is Springer International Publishing AG
The registered company address is: Gewerbestrasse 11, 6330 Cham, Switzerland

Foreword

The protection of the environment is an extraordinarily interesting topic for a criminal law scholar who is attracted by the issue of the compatibility of the protection of legal assets traditionally included among those protected under criminal law—such as life and human health—with the protection of legal assets—such as sustainability—which have more recently appeared among the interests to be protected, even by means of criminal sanction. These latter appear to be far more nuanced and multifaceted—as in a prism—and therefore, including them in a criminal offense is a more complex process.

There are numerous reasons for this. First of all, we are faced with a legal asset of broad spectrum, which has relatively recently emerged, and whose protection fluctuates between an "ecocentric" perspective aimed at preserving the health of the environment itself and an "anthropocentric" perspective which gives relevance to the dangers for human health in a broad sense. Secondly, this is a field of law which is traditionally characterized by the anticipation of protection through the use of offenses structured as crimes of "abstract" or "concrete" danger. A strong interaction between primary and secondary law is also present in this field, and thus, the criminal sanction, according to the scheme of "crimes based on the violation of a duty" (*Pflichtdelikte*), often integrates areas of life which already have their own legal configuration. Furthermore, we are dealing with a multilayered regulatory framework that includes civil provisions, administrative provisions, and finally, criminal provisions, and which over time has seen the liability of legal persons develop alongside the liability of the individual.

It must not be forgotten that the legislative choices reflect the sensitive nature of a subject matter that requires a difficult balancing between the need to safeguard the environment—with the inevitable effects, as was just mentioned, on legal assets strictly connected to the individual (think, e.g., of human health, but to some extent also of the well-being of future generations)—and the needs set by other rights, such as the freedom of economic initiative.

It is not by chance that, in this field, the theory of allowed risk has developed: We are dealing with activities which are on the one side socially useful, but on the other side inevitably linked to a certain margin of risk.

It is not surprising, therefore, that environmental criminal law has been a testing ground for principles, such as the precautionary principle, that challenge the cornerstones of traditional criminal law (I am thinking of the ascertainment of negligence). And also that it has been the occasion to test the compatibility of the legislation considered in each case with principles such as those of "harm" (*offensività*) and *extrema ratio*.

By looking at the regulatory scenario—I am specifically referring to the Italian one, but I think these considerations can also be extended to other legal systems—a composite reality does emerge, made up of offenses spread between the complementary legislation and the Criminal Code, and characterized above all by a stratification of the criminal protection: from a collective dimension linked to the protection against threats to public safety, to an individual dimension anchored to crimes aimed at protecting life and physical integrity, to a "formal" protection connected to the use of the criminal sanction as a safeguard of the administrative activity of planning and control. Thus, a transition from a "nuclear" criminal law based on the protection of solid legal assets to an "artificial" criminal law oriented to the protection of functions.

If we add to this the progressive and precise obligations to introduce criminal sanctions in this field—emblematic, in this case, are the well-known events that led to the emission of the Directive of 2008, which anticipated the contents today outlined by the Treaty of Lisbon—we immediately understand the reasons for the above-mentioned extraordinary interest for scholars, linked to the fact that new paths are being explored, but also to the circumstance of dealing with a number of issues which are difficult to resolve.

The point of view adopted in the contributions that follow is, if possible, even more stimulating than the one I have tried to roughly outline up to now; but precisely for this reason, it has plenty of unclear elements on which light must be shed.

The entire volume, in fact, addresses the issue of historical pollution, exploring the phenomenon in all its many facets. The contributions not only carry out an analysis of the topic—considering the European and supranational legislation and case law, and comparing different national perspectives—but they also provide a reconstruction of the relevant criminological background. And already at this stage, the critical points that must be faced come to light: first, the need to start from an approach that gives value to the analysis of the cases that have drawn attention to the phenomenon in question (reference is made, among others, to events such as *Love Canal* and *Huelva*, to which the contributions repeatedly return); second, the difficulties of retracing an elusive phenomenon with "invisible" traits that, even according to the collective perception, suffers from the conflict that can arise between the benefits associated with industrial production (in terms, e.g., of occupation) and the dangers to human health and the environment; and third, the obstacles—one thinks of the "time" element—to the development of social awareness of the phenomenon itself.

Right from the start, the topic is addressed using a multidisciplinary approach that puts together the results of an analysis of industrial history—i.e., changes in production processes over time—and the results of an analysis of the geographical dimension—i.e., the areas where the industries and their waste are located—while also taking into account changes in social awareness and environmental legislation.

The guideline for the analysis carried out in this volume, however, is the stipulative definition of historical pollution: This represents the theme common to all the contributions. And here, the methodological approach selected, and widely explained, consists of referring to the contamination of sites and natural resources that began or took place in the past, due to industrial or other related productive activities. The fundamental characteristic of the phenomenon—the relevance of time—immediately comes to light: historical pollution, as highlighted, usually involves the emergence—or the persistence—of the contamination of sites and natural resources long after the pollution has taken place.

Therefore, such a definition encompasses cases of "legacy pollution" (connected to past activities now closed), but could also include cases of "ongoing pollution" that originated in the past, as the environmental harm is a cumulative effect of the overall activity.

The framework for the following analysis is therefore introduced. Its main characteristic, as we are given to understand from the very start of the volume, is the problem of coping with scientific uncertainty, which makes it difficult to prove specific causation every time there is a problem of verifying the link between the conduct (often cumulative) and the specific event. Furthermore, with reference to the ascertainment of negligence, the scientific uncertainty pushes us toward a precautionary approach, shifting the focus from the "preventive-precautionary" nature of the rule allegedly violated, and toward the risk dimension. The result is that "hindsight bias" makes its entrance in the process of ascertainment, attenuating the recognizability of the danger and the foreseeability of the event that concretely occurred.

Even the analysis of the supranational legislative panorama reflects the difficulty for criminal law to present itself as an effective legal tool to combat the phenomenon in question. While it is true, in fact, as anticipated, that the Directive of 2008 gave birth to the EU obligations to introduce criminal offenses for the protection of the environment—a criminalization in line with the harm principle—it is equally true that the national transposal provisions will only apply to facts committed after their entry into force. Such difficulties also arise if we take into account the fundamental criminal law principles of "non-retroactivity," but also "culpability" and *extrema ratio*: The analysis highlights that the cleanup administrative procedures are often a more effective remedy and a more suitable way to balance the conflicting interests—not to mention again the proof of causation issues which are relevant also in this context. Likewise, we must point out how the case law of the European Court of Human Rights adopts a cautious stance when it comes to ascertaining the liability of a State for the violation of ECHR rights (the specific case concerned the violation of Art. 2) in the case of death resulting from historical pollution.

Moreover, the overall picture provided to the reader is particularly interesting as it points out the vitality of the supranational legislative scenario in the field of environmental protection—after all, it is in this scenario that the guiding principles of modern environmental law have come to light ("the polluter must pay", prevention and precaution)—although the solutions to the phenomenon of historical pollution in large part still have to be found.

The central part of the volume is dedicated to an interesting and in-depth analysis of the situation in other legal systems both of civil law and common law. This analysis concerns all fields—in accordance with the guidelines provided from the very beginning—in order to offer a comprehensive picture of the national remedies regarding both the liability of the individual and the liability of legal persons.

In general, there is a meeting point of the lines of development of the strategies in the fight against environmental pollution, with a wide range of solutions which include civil law, administrative law, and criminal law—even here, as mentioned before, with the simultaneous presence of different bodies of legislation.

And it is interesting to note that, overall, there is a strong influence of European legislation, as the national provisions of the Member States in this field appear to be the result of the transposition of EU directives.

Despite the specificity of the national remedies, we can note—even at first glance—the predominant role of administrative law: In fact, it is in this context that, even in the absence of a precise definition of the concept of historical pollution, we come across the topic of "historical contamination": In the Italian experience, this arises with regard to the discipline of "remediation of sites". Undoubtedly, the cleanup procedures and their complex mechanisms represent a strong weapon provided by the various national systems. Criminal law maintains, in this context, a role in the implementation of the mentioned procedures and intervenes to punish conducts which violate the administrative regulations—in Italy, this happens through the numerous misdemeanors provided for by the Environmental Code.

The effectiveness and suitability of criminal law is far more complex when it comes to events which affect life or physical integrity, or having regard to "macro-events" causing disaster or environmental pollution. The above-mentioned problems of proof, and the need to ensure individuals a high level of guarantees when it comes to a severe penalty, imply that the criminal tool often moves along a narrow path: On the one side, the mentioned principles must be correctly observed (and this, however, often leads to the ineffectiveness of criminal law); on the other side, their flexibility must be considered—here, we must recall the observations regarding causation and negligence—and this represents, in turn, a cost too high to bear.

Even on the administrative side, there are critical issues, which concern, for example, the application of the polluter pays principle (in terms of retroactive implementation of the regulations concerning remediation of sites) and the liability of the current "innocent" owner.

Therefore, the topic of historical pollution—with its crosscutting nature—raises multifaceted problems and requires, in the search for solutions, a great equilibrium and a capacity to tip the balance toward the correct position between the opposite needs of efficiency and guarantees.

As it emerges from the analysis, the phenomenon is clearly present in the various States, although with different levels of regulations and sanctions; this has raised the need for uniform legislation, which, instead, still struggles to find its way and has placed scholars before legal issues of high importance (the recent Italian experience regarding the *Eternit* case is an example).

The contributions found in this volume represent a firm guide to navigate a path that has proven to be rugged and to try to find solutions to the numerous legal problems raised by the phenomenon of historical pollution.

Rome, Italy Paola Severino

Acknowledgements

This book is the result of an international research project on the topic of historical pollution. The project would not have been possible without the support of many institutions and the dedication of some extraordinary people.

As editors, we wish to express our warm gratitude to Piergiuseppe Biandrino, General Counselor of Edison, who initially discussed this proposal. Edison S.p.A. offered through an appreciated financial sponsorship its support for this full independent scientific research. Furthermore, we wish to thank Camilla Beria di Argentine, Director of the Centro nazionale di prevenzione e difesa sociale Foundation CNPDS/ISPAC, and Professor Gabrio Forti, Director of the Centro Studi "Federico Stella" sulla Giustizia penale e la Politica criminale, who both enthusiastically agreed to undertake this work.

The overall research began in January 2015 and involved the participation of an international team of prominent academic institutions: Università Cattolica del Sacro Cuore di Milano, University of California, Irvine, Université Panthéon-Sorbonne (Paris I), Queen Mary University of London, Universität Tübingen, Università LUISS Guido Carli, Universidad Castilla-La Mancha, and Seconda Università di Napoli. The group of young researchers deserve special thanks—it has been a great pleasure working with them. The research program, and the present publication, would never come into being without the invaluable support of Johanna Caputi-Mallmann at the secretariat of CNPDS/ISPAC, who assisted the whole group with great patience and competence. We also owe sincere thanks to Ben Young for his careful work in reviewing the manuscript.

Last but not least, we express our appreciation to Springer, our publisher.

Contents

Editors and Contributors

About the Editors

Francesco Centonze is Full Professor of criminal law and business criminal law at the University "Cattolica del Sacro Cuore" Piacenza. He is Director of Rivista Italiana di Medicina Legale e del diritto in campo Sanitario, Coordinator for the Master's program in business criminal law at "Cattolica del Sacro Cuore" University, and Member of Scientific Committee of Centro Studi «Federico Stella» sulla Giustizia Penale e la Politica Criminale—Università Cattolica Milan. He has published extensively in the field of criminal law and business criminal law. He is the author of several books and scientific articles.

Stefano Manacorda is a Full Professor of criminal law at the Università della Campania, Naples, and Visiting Professor at Queen Mary University of London.

He is Vice President of SIDS (Société internationale de Défense sociale), of ARPE (Association de Recherches Pénales Européennes), and of ISPAC (International Scientific and Professional Advisory Council of the United Nations Crime Prevention and Criminal Justice Program). He has published extensively in the fields of European, comparative, and international criminal law. He is the author of more than one hundred scientific articles in several languages and has published several books in the area of European, international, and comparative law.

Contributors

Luis Arroyo Zapatero Universidad de Castilla-La Mancha, Ciudad Real, Spain

Ava Badiee University of California, Irvine, USA

Francesco Centonze Università Cattolica del Sacro Cuore, Piacenza, Italy

Luca d'Ambrosio Collège de France, Paris, France

Joseph F.C. DiMento University of California, Irvine, USA

Elena Fasoli Queen Mary University of London, London, UK

Jordi Gimeno Beviá Universidad de Castilla-La Mancha, Ciudad Real, Spain

Eliana Greco Università Cattolica del Sacro Cuore, Milan, Italy

Katharina Lienert Ludwig-Maximilians-Universität, Munich, Germany

Stefano Manacorda Università della Campania, Naples, Italy

Calogero Miccichè Università Cattolica del Sacro Cuore, Milan, Italy

Valsamis Mitsilegas Queen Mary University of London, London, UK

Lorenzo Natali Department of Sociology and Social Research, University of Milano-Bicocca, Milan, Italy

Alejandro Luis de Pablo Serrano Universidad de Valladolid, Valladolid, Spain

Giuseppe Rotolo Università Cattolica del Sacro Cuore, Milan, Italy

Rossella Sabia LUISS Guido Carli University, Rome, Italy

Paola Severino LUISS Guido Carli University, Rome, Italy

Benedetta Venturato Università Cattolica del Sacro Cuore, Milan, Italy

Donato Vozza Seconda Università di Napoli, Naples, Italy

Chapter 1
Preventing and Sanctioning Historical Pollution Beyond Criminal Law: An Introduction

Francesco Centonze and Stefano Manacorda

Contents

1 Research Objectives and Methodology

The present volume gathers the results of a research project on legal approaches and responses to the phenomenon of *historical pollution*, which may come from criminal law as well as from other branches of law. The research was conducted

F. Centonze
Università Cattolica del Sacro Cuore, Piacenza, Italy
e-mail: francesco.centonze@unicatt.it

S. Manacorda (✉)
Università della Campania, Naples, Italy
e-mail: stefano.manacorda@unicampania.it

F. Centonze and S. Manacorda (eds.), *Historical Pollution*,
DOI 10.1007/978-3-319-56937-6_1

from 2015 to 2016—with the support of private funding—by a group of scholars from different countries and with different academic backgrounds.[1]

As will be discussed further, the expression "historical pollution" denotes

> the pollution of sites and natural resources (intended as the relevant *behaviour*), which started or took place in the past, due to industrial activities or other production-related activities. Historical pollution usually entails the emergence—or the persistence—of contamination of sites and natural resources (meant as the *effect* of such a conduct) a long period of time after the original pollution took place.[2]

Among scholars and in public debate, expressions such as "legacy pollution" (World Bank Group 2012, p. 57), "past pollution" (Monti 2002, p. 11), "historical pollution" (Mieck 1990, pp. 73 ff.), "historical contamination" (Brandon 2013, p. 6), "residual pollution" (Gassner 1993, pp. 3 ff.), "gradual pollution" (Monti 2002, p. 11), "cumulative pollution" (Ori 1993, p. 16), "chronic pollution" (European Commission 1993, pp. 8–9), and "on-going pollution" are often used: all these expressions refer to concepts which partially overlap or otherwise interfere with the notion of historical pollution which forms the subject of the present research.[3]

Indeed, the phenomenon of historical pollution represents the result of the convergence and interaction of a number of different factors: namely, industrial and technological development which has led to the rapid spreading of production activities characterized by a strong impact on the ecosystem; the complex evolution of scientific knowledge relating to the potential harms to the environment and to human life which may come from these activities; and the lack of prompt legislative initiatives appropriately framing the industrial impact on the environment.

The late discovery of historical forms of environmental pollution arises from an increasing awareness, including among legal scholars, of the harmful effects of industrial development together with an increasing enhancement of the victims' perspective (on this issue, also for the relevant references, see Hall 2013a).

Due to the submersion of the phenomena of historical pollution, legal systems today—i.e., several years after the relevant behaviors took place—must face issues that

---

[1]The research team was composed of the professors Francesco Centonze (Università Cattolica del Sacro Cuore di Piacenza), Joseph F.C. DiMento (University of California, Irvine), Gabrio Forti (Università Cattolica del Sacro Cuore di Milano), Geneviève Giudicelli-Dèlage (Université Paris 1 Panthéon-Sorbonne), Stefano Manacorda (Seconda Università degli Studi di Napoli), Valsamis Mitsilegas (Queen Mary University, London), Mauro Renna (Università Cattolica del Sacro Cuore di Milano), Frank Saliger (Ludwig-Maximilians-Universität München), Paola Severino and Antonio Gullo (Università Luiss "Guido Carli", Rome), and Luis Arroyo Zapatero (Universidad de Castilla-La Mancha); and the researchers Ava Badiee, Luca d'Ambrosio, Alejandro Luis de Pablo Serrano, Elena Fasoli, Jordi Gimeno Beviá, Eliana Greco, Katharina Lienert, Calogero Micciché, Lorenzo Natali, Giuseppe Rotolo, Benedetta Venturato, and Donato Vozza. The research activities were coordinated by the National Centre of Prevention and Social Defence (CNPDS).

[2]See in this volume, Chap. 3, Sect. 2.

[3]On the importance of the factor of time in the understanding of this phenomenon, see in this volume, Chap. 2.

are extremely complex and all but easy to solve. Among these are: protecting human health, which is potentially threatened by environmental contamination; ensuring remedial actions and cleanups; satisfying the need for payment of damages to victims and, more generally, reconciling the social conflicts triggered by the emergence of the harmful effects of historical pollution; allocating the costs for these activities (World Bank Group 2012, pp. 57 ff.). Above all, the main issue at stake relates to the identification of the responsible individuals and corporations, as well as the need for judicial procedures with this as their aim. In short, historical pollution represents an extraordinary challenge to the law for many States around the world, which requires resort to a wide interdisciplinary approach in order to be effectively tackled.[4]

The Italian legal system's solution to these issues served as a starting point for the international research we introduce here. In Italy, in effect, historical pollution issues have been addressed mostly by means of punishment and, in particular, by resorting to criminal law sanctions along with administrative ones. This has resulted in a clear failure of regulation: the dramatic social problems which have arisen in the geographical areas of the country affected by historical pollution remain unsolved; cleanup procedures of the most contaminated sites are far from complete and in some cases not yet even begun; public opinion is numbed by the focus in criminal proceedings on acts carried out during the seventies or eighties.

Starting out from the Italian experience, then, this book intends to open up new horizons, looking at the ways in which other countries face the complex issues connected with historical pollution.

The choice regarding the number and type of countries to be considered for the comparison was preceded by a so-called pilot-study. In this preliminary phase, the research team framed the issue of historical pollution and the solutions to it as found within a set of legal systems characterized by an advanced level of industrialization, including the USA, Switzerland, Japan, and Australia. These countries were selected on the basis of a criterion which reflects their criminological similarity, all having a long industrial tradition and present-day historical pollution issues as well. From a first study of these legal systems, it emerged that they have developed approaches and models for the management of historical pollution, but that these differ significantly one from another. The team then conducted additional research on a second set of legal systems from Europe (the UK, Germany, France, and Spain), Asia (China, Japan and India),[5] and Africa (South Africa and Nigeria).[6] This second part of the preliminary analysis was developed on the basis of a case-law-oriented criterion, aimed at assessing the existence of cases concerning historical pollution in the relevant legal systems. Again, the study brought to light a wide range of different results. Generally, historical pollution phenomena tend to

[4]On this aspect, see again Chap. 2.

[5]For China, see Zhao (2013), Zhang and Silverman (2015, p. 33). For Japan, see Itabashi et al. (2015, p. 23).

[6]For South Africa, see among others du Plessis and Kotze (2007), Feris (2012). For Nigeria, see among others Fagbohun (2007).

occur in every industrialized country, often around industrial clusters built in the past and currently located in the neighborhood of highly populated residential areas.

The research team then decided to limit the scope of the research to the legal systems of a selected set of European countries (France, Germany, Italy, the UK, Spain) and to that of the USA, in order to enable a significant comparison between regional experiences and, at the same time, identify differences and similarities within the European area.

In addition, the research combined two different methodological approaches: one based on a comparative analysis of selected issues, the other based on a comparative analysis of the relevant case law. Following the *issue-oriented analysis*, each of the research units was asked to contextualize historical pollution within its own legal system by filling out a questionnaire which provided a common framework for analysis. The questionnaire was formed of sixty questions, organized into sections, concerning the notion of historical pollution, the social and economic context of this phenomenon, and both national and supranational legal solutions offered to it. Researchers answered these questions, drawing on statutes, case law, and the scholars' debate. More importantly, in so doing each unit was encouraged to freely leverage and stress those particular aspects and results which appeared to them as the most relevant to their legal system, so as to prevent excessive standardization in the research, which might have not allowed a comprehensive description of each system to emerge.

Following the *case-law-oriented approach*, national reports highlighted actual judicial cases concerning historical pollution brought before the courts within each of the selected jurisdictions, with the aim of identifying differences and similarities between the solutions found in different countries. Furthermore, following the same approach, the research team elaborated some *hypothetical scenarios*, abstracted from the cases under consideration. These summarized certain factual elements which represent common features of the actual cases occurring across the different countries. National research units within the team then provided a solution to each of the hypothetical scenarios, based on national legislation, case law, and scholars' opinions.[7]

---

[7]More specifically, each research unit was asked to frame the solution provided by each national legal system to the following five hypothetical scenarios of historical pollution:

– Scenario 1: Agent A owns land and contaminates it over a long period of time. No damage nor threats to human health occur as consequences of the pollution.

– Scenario 2: Agent A owns land and contaminates it. The land is subsequently purchased by Agent B. No damage nor threats to human health occur as consequences of the pollution.

– Scenario 3: Agent A owns land and contaminates it. The polluted land is subsequently purchased by Agent B. In the meantime, land contamination results in contamination of groundwater which is regularly used in agriculture.

– Scenario 4: Agent A owns land and contaminates it. The polluted land is subsequently purchased by Agent B. In the meantime, land contamination results in contamination of groundwater which is regularly used in human nutrition.

– Scenario 5: Agent A owns land and contaminates it. The polluted land is subsequently purchased by Agent B. There is evidence of a causal link between the pollution and the death or injury of a number of people.

The protean dimension of the empirical findings, combined with the variety of both the legal systems and the levels of analysis where these findings have been observed, led to complex and articulated results. Indeed, the comparative analysis highlighted that in several countries the main responses to historical pollution are based on a variety of compensatory and reparatory measures belonging to civil or administrative law; criminal law, to the contrary, plays a rather residual and supplementary role, normally limited to situations in which the contamination is recent or still ongoing.

The following sections deal with the main results of the research relating to the empirical dimension of historical pollution (Sect. 2), the complexity of the historical pollution phenomenon in the different social and economic contexts of each of the countries taken into consideration (Sect. 3), and the criminal and non-criminal responses to it provided for by domestic, European, and international law (Sects. 4 and 5). The scientific aim is to frame in a systematic perspective the main models for the prevention and management of historical pollution and to test their functionality (Sect. 6).

2 Historical Pollution: A Conceptual Framework

The development of legal responses to the phenomenon of historical pollution calls for a detailed empirical study. This belief underlies the research conducted by the team and, first and foremost, by Lorenzo Natali.[8]

From a empirical viewpoint, historical pollution is characterized by certain *recurrent features*. First of all, there are *environmental damages in the present which are the final product of industrial pollution behaviors repeated and protracted over time*: the contamination of environmental matrices and its negative effects on human health are, therefore, the result of industrial activities carried out in the past or in any event protracted over time. Reference is made here to cases such as Love Canal, Hinkley, Porto Marghera, Huelva, and Minamata. For the purposes of our study, then, historical pollution must be distinguished from cases where pollution is *generated by a single event, but has historical effects*, i.e., where the damage to the environment is caused by destructive events—disasters connected with ongoing industrial activities which are inscribed into history along with the date, hour, and sometimes even the minute on which they occurred. In effect, notwithstanding certain common features, the two phenomena differ significantly, and therefore need different legal responses (Ligi 2014).

In cases where the pollution is generated by a single event which has concomitant historical effects, the damage is indeed *visible and immediate*. The Seveso case in Italy (July 10, 1976 at 12.37) (Centemeri 2006), the Chernobyl case in Ukraine (April 26, 1986 at 1.23) (Kolbasov 1992), the Bhopal case in India

[8]See Chap. 2.

(December 3, 1984) (Abraham and Abraham 1991), and the Fukushima Daiichi case in Japan (March 11, 2011) (Davies 2011; Osaka 2012; Gayed 2015) are all emblematic of large-scale technology-related disasters, whose negative effects on the environment and on human health will last for a long period of time.

In proper historical pollution cases as presented above, however, the harmful effects on the environment are usually invisible and the damages to human health arising from people's exposure to contaminants come to light only after a long period of time (Walters 2014, p. 142). In the end, one may say that, besides its particular origin, the *invisibility of contamination* also represents a typical feature of the phenomenon of historical pollution (Davies et al. 2014).

Lastly, historical pollution is characterized by the presence of a *distance in time between the polluting behaviors and the emergence of harm*. This is a crucial aspect of the phenomenon, and the temporal factor may vary depending on a number of different factors. Among these are: the sector or the type of activity generating the pollution (industrial activities, waste management, nuclear activity, farming, military activities); the type of contaminants produced or used in the relevant processes (PCD, DDT, etc.); the type of environmental matrices subject to the pollution (soil, groundwater, lakes, coasts, flora, and fauna), etc. (Brandon 2013, pp. 12–16).

In addition to the aforementioned recurrent features, historical pollution is also characterized by the presence of several *variable features*. With respect to the *source of the pollution*, the present volume focuses on cases in which historical pollution is the result of outflows deriving from *industrial activities* carried out by both State and private or privatized companies (Earnhart 2004). However, it is well known that historical pollution may have different sources, such as for instance illegal waste management. Another common source of historical pollution is represented by toxic substances (such as depleted uranium) contained in weapons and munitions released during military training and war (Sanders 2009).

An additional variable feature of historical pollution concerns the type of behaviors which may lead to the occurrence of the phenomenon. In this respect, it is possible to distinguish first between *historical pollution in the strict sense and historical pollution in the broad sense*. Under the first notion, one can properly refer to historical pollution only when the latter is the result of industrial activities which started and ceased in the past and which led to the release into the environment—in the form of industrial waste waters, production waste, air pollutants, etc.—of persistent organic pollutants, chemical substances, oil products, and heavy metals, from which further damages to environmental matrices and possibly to human health have arisen over time (Brandon 2013, p. 199). Under the second and broader notion, historical pollution also includes the result of industrial activities which, although having started in the past, are still ongoing (so-called ongoing pollution).

On the basis of this fundamental difference, further distinctions may be traced according to which historical pollution may represent: (i) the result of a sum of past and present polluting behaviors carried out by one or more individuals (and more often corporations) who have succeeded one another (so-called cumulative historical pollution); (ii) the result of pollution originated by multiple unidentifiable sources—which may be, for instance, of industrial, urban, or agricultural nature (diffuse historical pollution) (Brandon 2013, p. 7); (iii) the result of a combination of past, present, and diffuse polluting behaviors (diffuse and cumulative historical pollution). All of these situations were taken into consideration in this research, which also tries to put in its proper context cases in which the source of the historical pollution was not only diffuse and varying over time, but also now extinguished (so-called orphan sites) (Larson 2013).

Further variable features of historical pollution may be identified with respect to its effects. First of all, environmental damages—which, as mentioned above, represent a recurrent feature in the phenomenon—may arise in different forms: for instance, they may affect one or more environmental matrices, they may have different intensities or characteristics, and they may be circumscribed to a certain area or have a transnational impact. Moreover, environmental damages constitute one of the direct effects of polluting behaviors, but the latter may also produce an indirect effect, namely damage to human health caused by the contamination. Although it is more common that historical contamination should affect residential areas and lands for agricultural use, and thus potentially pose a threat to or harm human health, there are still situations in which pollution affects abandoned and unused lands, far from residential areas, where no threat or harm to human health can arise. The fact that interests of primary importance—such as life and health—may be harmed as an indirect (and possible) consequence of historical damages to the environment remains nonetheless a very important element in the description of this phenomenon. And this appears more clearly if only one considers that whenever any such harm actually occurs, historical pollution takes on a quite different harm potential, which requires different legal responses aimed at addressing the victims' legitimate requests for damages as well as the different punishments at stake (Hall 2013b). In this case, the latency of the harmful effects on people arising from their exposure to toxic substances is added to the invisibility of the contamination (Rall 1972).

In conclusion, pollution may be defined as "historical" when the following conditions are met: (i) the damage to the environment is—wholly or partially—the result of industrial outflows of polluting substances which either took place in the past or were protracted over time; (ii) there is a time lag between the relevant industrial pollution behaviors and the emergence or the ascertainment of the environmental damage; (iii) the contamination process has been invisible for a long period of time. The nature of the environmental damages makes this phenomenon different from that of one single catastrophic polluting event with historical effects, typical of technology-related disasters. This latter aspect, then, triggers the need for specific legal responses, which take into consideration—in its several aspects—the specificity of this phenomenon and of the role in it played by time.

3 The Empirical Dimension: The Variable Geometry of a Global Phenomenon

In the course of the research, the phenomenon we have just described turned out to be deeply connected with each country's economic and industrial development processes.

In outline, it is possible to distinguish between countries with a very long industrial tradition (such as the UK, the USA, France, and Germany), where the development of industry dates back two centuries and where historical pollution first arose; countries with a long industrial tradition (such as Italy), whose internal economy and industry developed especially after World War II and where historical pollution has progressively emerged; and countries where the development of industry is recent (such as Spain) characterized by a relatively new industrial tradition and where the number of cases of historical pollution is still limited.

Starting with the countries with a very long industrial tradition, we can look at the experience of the USA, the UK, France, and Germany, where industrial revolutions have led—earlier than in other parts of the world—to a transformation of the relevant economies from a rural, artisan, and commercial model to an industrial one. The resort to new power sources and the use of modern technologies with a significant impact on the environment led to the spread of textile, steel, chemical, and petrochemical factories. Strong and rapid economic growth followed, which also led to significant changes in the social context, a demographic explosion, and the rise of industrial cities, i.e., industrial areas around which highly populated residential areas have developed. In those centuries, the industrialization process mostly concerned sectors of industry involved in the production of fast-moving consumer goods, which—with low costs and a high environmental impact—secured the highest profits for corporations. Moreover, it was precisely because of this strong industrial development that the need arose—to an extent previously unseen—to handle a large amount of toxic waste, this being both extremely long-lasting as well as dangerous to human health (Wyatt 2016, pp. 1 ff.). As a result of an initial economic development much more focused on the growth of work and capital than on environmental sustainability, urban and industrial areas have seen a gradual contamination of land and groundwater. Furthermore, the onset of health issues in the population prompted the need for regulation both in the industrial and in the environmental sectors. However, any such regulation was fully developed only in recent decades, together with the evolution of a new level of awareness of environmental protection and health issues. In these constantly evolving economic, industrial, social, and legal contexts, a number of cases of historical pollution occurred, to which governments provided effective responses only in the seventies, when social protest movements arose urging the prevention and control of this form of environmental pollution (Gibbs 2011). Over recent decades, prompted by victims' requests as well as by the public's need for reassurance triggered by the fear of new invisible forms of pollution, the governments of these countries framed stricter regulations for the protection of the environment and of human health. As is well known, this is one of the reasons why there has been a progressive decentralization of certain production activities in those countries—mostly emerging economies—where the burden imposed on corporations with respect to environmental-protection-related costs is significantly lower. At the same time, in developed economies the deindustrialization process has given rise to significant economic, social, and cultural issues (Altena 2003), which often go along with relevant historical contamination issues.

Having outlined the general background underlying the phenomenon of historical pollution, we turn now to look at the USA. This is, in effect, one of the most flourishing economies in the world, with significant productive activity—since the beginning of the century—in industries ranging from chemical, pharmaceutical, and petrochemical products to white goods and electronics: all of which are industries with a high impact on the environment. It is no coincidence that cases such as Love Canal—a paradigmatic case that constitutes a point of reference in the illustration of historical pollution—occurred within an economy with a high level of industrialization. The Love Canal site, located in the city of Niagara Falls in the state of New York, was turned over to urbanization after having been used as a landfill for chemical and city waste between 1940 and 1950. Starting from the mid seventies, certain studies began to flag the occurrence of serious diseases among the population. In particular, the number of cancers and genetic malformations rose significantly. People gathered in protest movements, the citizens brought lawsuits for damages, and legal initiatives were taken at a federal level (Newman 2016; Hahn-Baker 2001; Beck 1979).

In the UK, the process which led to industrial development started in the eighteenth century. Indeed, the industrial revolution effectively originated in this country (Wrigley 1990), which left a negative legacy, including historic damage to the environment. It is well known that activities relating to the extraction and processing of raw materials—especially coal, natural gas, and oil—in the steel, engineering, chemicals, and pharmaceuticals industries were developed particularly in the UK. An exemplary case of historical pollution recently came to prominence in Bawtry, in South Yorkshire (England), in relation to industrial activities in the field of coal production. Since 1915 the Bawtry Gas Works company had been producing coal gas and coal by-products. During the course of its life, the company was first transferred between private owners, then came into State ownership, then was again privatized. The gas production activity ceased in 1952, and since then the site had been used only for storage and supply. During the sixties, the site was closed and sold to a construction company, which in the relevant sale agreement was made aware of the fact that the site was contaminated. The site then became a residential area. Only in 2001, when one of the residents discovered a coal tar well, investigations were conducted which ascertained that both the land and the groundwater had been contaminated. Under English law, the area was classified as a special site to be decontaminated and reclaimed. Due to the winding-up of the construction companies and the transfer of the company which had originally produced the contamination, remedial actions were taken by the State and ended in 2006.

Similarly, France was one of the first European countries to see a national industrialization process, developing activities in various sectors including the chemical, automotive, and steel industries, mostly concentrated around the main hubs of the country. However, French policies on industrial development also include farming activities, in the context of which new technologies play a greater role and chemical substances such as pesticides and fertilizers are widely used. Despite the prior development of environmental laws, it was especially from 1970—as in other developed countries—that France was required to address complex historical pollution cases involving both heavy industry and agriculture. One case drawing particular media attention in France relates to the historical pollution caused by the foundry Metaleurop Nord, located in Noyelles Godault, in the region of Nord-Pas de Calais. Metaleurop Nord had been producing lead and zinc since the end of the nineteenth century, but in 2003 it went bankrupt, causing a serious employment crisis (Mazade 2010).

When the site was closed, a 45 km^2 area was found to be contaminated by lead, cadmium, and zinc including the groundwater. This latter aspect had a further impact on the socioeconomic context. Both the site rehabilitation and the payment of damages to victims were ensured by public funds, following the company's bankruptcy.

Germany represents a further powerful economy on a global scale, whose industrial history dates back a long way and was supported by the presence of rich coal deposits in the Ruhr and Saar basins, close to which several industrial hubs arose. Automotive, steel, chemical, and pharmaceutical industries also played a very important role and have a long tradition in the German economy. Here, a number of cases concerning the contamination of land and groundwater occurred in various industrialized areas. Among these, one case appears as particularly emblematic: it concerns the pollution of the river Emscher caused by industrial and city drains. Renamed by the locals the "sewer-river," the Emscher flows through the most important cities of the Ruhr—which is the most industrialized area in Germany. The river has been decontaminated by a public–private partnership and represents one of the most spectacular and costly cases of remedial measures ever.

Among the countries with a long industrial tradition, attention should be drawn to Italy. Between 1950 and 1960—in the aftermath of World War II—the so-called economic boom took place: industrialization developed rapidly, boosting the country's economic and social growth. In a few decades, Italy became one of the economic superpowers of the world and a symbol of wealth and well-being at a global level (Mori 1977). However, this stage was followed by a new one, starting around 1990 and characterized by a gradual economic decline, during which various industrial activities were moved abroad or interrupted. It is in this new context, marked by economic crisis and by an advanced social and institutional awareness of environmental protection needs and human rights, that historical pollution phenomena started coming to light. Damage to the environment emerged, which turned out to have been caused by industrial activities carried out in the past, often interrupted or converted, and affecting different regions of the country at different times. Although a number of cases of historical pollution were brought to the public's attention, the Porto Marghera case appears as the most paradigmatic of this phenomenon. Porto Marghera is an industrial area in the surroundings of Venice, where—since the beginning of the twentieth century—dozens of factories mainly operating in the chemical industry had been located. Over the years, the workers of one of the main petrochemical plants in the area were exposed to Vinyl Chloride Monomer (VCM) and Polyvinyl Chloride (PVC). Moreover, every year hundreds of thousands of tons of toxic smoke were released in the air and more than twenty tons of toxic and carcinogenic compounds drained into the lagoon waters. In 1980, the plants were closed and the production strongly reduced. Starting from 1990, significant proceedings were brought against the factory's top management for having caused damage to workers' health and to the environment (Cerasi 2007).

The third model we have looked at as a point of reference in this research is the Spanish one. Spain is a country in which historical pollution phenomena are—for the moment and with respect to their number—limited. This is due to the fact that industrialization in Spain had been significantly delayed by the political and economic isolation imposed by Franco's 36-year dictatorship. But the fall of the dictatorship and the entry of the country into the European Union led to rapid economic and social growth and, above all, to the development of industrial activities in certain areas, including, for instance, Huelva in Andalusia, in southwestern Spain. From the sixties onwards, this area, used also for agriculture (for example, the farming of strawberries largely for export), has been polluted by the waste from chemical factories, significantly and negatively affecting the health of people living there (Natali 2014). Moreover, "in Spain, the widespread use of asbestos occurred later than in other countries. For this reason, Spain was over twenty years behind countries like Sweden, Denmark, the UK, Holland, and Germany in tackling the problem."[9]

Beyond this general overview, the comparative research has also shown that historical pollution is strictly connected with the specific and local socioeconomic context in which it takes place, so that it occurs in different forms and with different features depending on the country. With respect to the experience of the USA, historical pollution cases are not limited to Love Canal, and we might mention other cases of great importance such as the Valley of the Drums and Hinkley. In France, historical pollution cases have also arisen as the result of the intense, continued, and reiterated use of a large amount of pesticides, fertilizers, and other substances in farming activities. Further cases of historical pollution, as for example the Corby Steelworks case, may be identified in the UK as well. In Germany, land contamination issues have arisen in relation to uranium mining activities in Saxony, before Germany was unified. In Italy, historical pollution cases arise in connection with the petrochemical industry and the processing of heavy metals (Porto Marghera, Taranto, Priolo-Augusta, Gela, Crotone, Piombino, Bagnoli, etc.), the storage and

[9]See in this volume, Chap. 10, Sect. 2.1.

management of waste (Pitelli, Bussi, so called "Terra dei fuochi", etc.) and the presence of abandoned industrial sites (Casale Monferrato, Pioltello, etc.).

In conclusion, historical pollution is a problem of global significance, despite the fact that it occurs in different forms and with different features at a local level. This depends on the fact that it is strongly influenced by the relevant industrial and economic model in which it arises and by a large number of additional factors which matter in particular cases. Despite its many facets, historical pollution is characterized by the presence of polluting behaviors that are protracted over time, especially on the part of heavy industries—once operative in advanced economies, although now mostly operating in emerging countries—whose activity produces often invisible long-term effects on the environment and on human health (Ruggiero and South 2013). Moreover, State reactions to these effects generally come late and are, therefore, not able to prevent and control this phenomenon.

4 Comparing Legal Responses: Remedying Environmental Damage and Compensating Victims

The legal systems compared in this book (the USA, Italy, the UK, Germany, and Spain) are all characterized by the presence of a particularly articulated environmental legislation, which has evolved in recent times, also as an effect of supranational initiatives.[10] However, although these systems present some common features, the relevant legal frameworks differ significantly in each of them, making the elaboration of a complete comparative analysis a very complex task.

With specific reference to the phenomenon of historical pollution, we note the existence today of a wide range of legal instruments aimed at striking a balance between different key interests, such as the protection of the environment, the protection of public health, the need for victims' compensation, safeguarding the industrial sector, and the maintenance of social and economic development. Everywhere, preventive legislation—aimed at lowering the risk of harm to environmental matrices through the enactment of regulation concerning industrial activity—coexists with ex post facto intervention policies meant to ascertain responsibility and remedy the effects of polluting behaviors. With reference to the latter, a distinction is then usually traced between, on one side, the legal instruments drawn up at national, European, and international level aimed at ensuring compensation for damage to victims or the undertaking of remedial actions for the environment; and, on the other side, the legal responses aimed at sanctioning the infringements, which may be of criminal or of administrative nature.

With respect to the first set of legal tools, within the various legal systems taken into consideration, it is necessary to further distinguish between the responses

[10]See in this volume, Chap. 14.

designed to lead to payment for damage to the environment (compensatory measures) and those designed to lead to the undertaking of cleanups of the contaminated sites and of the groundwater (remedial actions). Within the responses to historical pollution, cleanups play a central role, both because they prevent environmental damage from continuing (so-called static historical pollution) and because they prevent further aggravation of the environmental damage by, for example, stopping the expansion of the contamination from land to groundwater and thereby harming the health of the people exposed to it (so-called dynamic historical pollution).

It is a matter of common knowledge that cleanups are particularly complex from a technical viewpoint and very expensive in economic terms. In some cases they cannot lead to the full recovery of the environmental matrix which was damaged, especially when such damage was caused by persisting toxic substances. In legal terms, the identification of the person who should be burdened with the payment of damages and of the costs relating to cleanup remains a key aspect of the complexity of the regulation in this field (World Bank Group 2012; OECD Report 2012).

The analysis carried out in this volume has allowed us to shed some light—albeit in outline—on the systems of compensatory and remedial measures present in the various countries taken into consideration. Everywhere the regulation is very complex, involving sources of varying nature and level. In European (EU) countries, the rules on the prevention and remediation of environmental damage contained in Directive 2004/35/EC shape the relevant domestic legislation.[11] At the federal level in the USA, the CERCLA stands out, combining with the legislation of each individual State.

In brief, the "polluter pays" principle usually applies to environmental damage. However, exceptions to this principle exist, especially with reference to historical pollution. Indeed, both factual and legal scenarios may be subject to change over the time span between the past polluting conduct (historical pollution behaviors) and the ascertainment of the responsibilities for the environmental damage, thus arising, making the attribution of such responsibilities a very complex task.

Factual scenarios appear very complex both in cases of "diffuse" historical pollution—in which the contamination is the result of polluting substances coming from a variety of sources, industrial and urban, continuing over time and impossible to distinguish—and in cases of "cumulative" historical pollution—i.e., when several industries operating in the same area succeed one another. Diffuse pollution makes it problematic to identify the persons to whom responsibility should be attributed, leading to the resort to presumptions before the judge. In cumulative historical pollution cases, domestic laws often resort to the criteria of proportionality and solidarity in order to allocate costs.

The comparative analysis has also shown that the regulations establishing responsibility for environmental damage may change over time, sometimes leading to a retroactive attribution of responsibility, as happened with the CERCLA in the

[11]See again Chap. 14, Sect. 5.2.

USA (Brandon 2013, p. 186). Conversely, Directive 2004/35/EC does not apply retroactively and is effective only within precise time limits, which significantly narrow the possibility to apply it to historical pollution cases.

Generally, it is also a common feature that the cleanup of the contaminated site falls to those who caused the pollution. However, in the time between the relevant conduct and result, changes to the structure of the corporation which caused the pollution, and indeed its dissolution, may also occur: examples of these changes would be business conversions, restructuring, mergers and acquisitions, privatization, succession, and winding-up. In the event that the original polluters cannot be identified, or if successions or extinction of the corporation has taken place, the need arises to recompose the picture of those who were involved over time.[12] Finally, where there are "orphan sites," the costs connected with cleanup procedures are borne by the State, which in turn—in certain systems—draws from funds created via private contributions.[13]

The payment of damages to third parties harmed by the pollution is an issue worth a final mention. These damages include losses of economic value to land suffered by the owners of properties close to contaminated industrial areas, or—as is more frequent—death and injuries as a consequence of diseases contracted by the people living in the neighborhood of contaminated sites. As is well known, in such cases certain international human rights tools (i.e., ECHR, ICHR) may serve as instruments to impose an obligation on States to protect and compensate victims.[14]

The resort to civil law in cases of historical pollution is widespread, with different results. In the UK, for instance, the legal instruments offered by civil law are used as a matter of course to address the victim's claims for damages in cases of historical pollution.[15] Conversely, in Germany "the compensation mechanisms of private law for damages to health and property resulting from contamination sites have failed in cases of historical pollution", leading to the adoption of new rules.[16]

In final summary, with respect to the use of remediation measures it is possible to distinguish, according to the comparative experiences, two main models for the allocation of responsibility. In the first, the responsibility for historical pollution is attributed to private actors, and thus there is a significant expansion of the scope of their liability, therefore reducing the costs borne by the community. In the second, by contrast, the trend is to limit private actors' liability and to collectivize the environmental costs linked to historical pollution. As for victims' compensation, here we observe a trend toward the resort to compensation mechanisms under ordinary civil law, in spite of the limits they present when applied to cases of historical pollution.

[12] See in this volume, Chap. 9, Sect. 5.4; Chap. 12, Sect. 5.3.

[13] On the Superfund Program, see in this volume, Chap. 8, Sect. 4.3.6.

[14] See in this volume, Chap. 13.

[15] See in this volume, Chap. 9, Sect. 7.

[16] See in this volume, Chap. 11, Sect. 7; for a detailed analysis, see Sect. 4.

5 Punishing Historical Pollution: What Role for Criminal Law?

The other side of this research, and certainly the most original one in view of the existing literature on the topic, concerns the legal responses, in terms of applicable sanctions, provided by the various systems which have been compared.

In presenting the outcome of the research on this issue, attention will be drawn to an overview of the sanctioning system as a whole, therefore making reference to sanctions of both administrative and criminal nature (and sometimes of civil nature). There is a common feature to all these sanctions, specifically concerning their afflictive dimension, which justifies the decision to analyze them jointly, despite the irreducible differences between the various models of responsibility.

A first note emerging from the national reports concerns the absence of criminal proceedings for historical pollution behaviors, with the sole and significant exception of Italy. Although all the legal systems considered here provide for criminal offenses aimed at punishing behaviors which threaten the environment, in practice these are not applied to cases of historical pollution, where—as noted in the previous section—the phenomenon is mainly addressed through resort to civil or administrative sanctions.

An exception to this trend—as already remarked—is represented by the Italian legal system, where the resort to criminal sanctions is characterized by the frequent application of criminal offenses originally meant to protect interests different from the environment, such as the offense of so-called unnominated disaster (Art. 434 of the Italian criminal code), that of poisoning (Art. 439 of the Italian criminal code), or the offense of adulteration of water to be used for nutrition (Art. 440 of the Italian criminal code). In the well-known cases of Porto Marghera, Bussi sul Tirino, and Casale Monferrato,[17] the difficulties connected with the use of criminal law in addressing historical pollution have come to light. The passing of time makes it problematic to prove causation with respect to the harmful effects of the contested conduct. Likewise, the identification of the responsible party is hampered when several individuals or corporations have succeeded one another in the management of a site or industrial activity. Due to scientific uncertainty, or the absence of established knowledge at the time in which the relevant behaviors took place, the attribution of fault and the ascertainment of negligence requirements become very difficult. Further problems connected with the evolution of the applicable legal framework—which makes illicit "today" behaviors that in the past were fully legitimate or simply ignored by regulation—as well as issues connected with the statute of limitations are highlighted in the analysis dedicated to the Italian legal system.[18]

[17]See in this volume, Sect. II of the Chap. 5, by B. Venturato and E. Greco, Sect. II.1.

[18]See in this volume Sect. I of the chap. 4 on Italy, by G. Rotolo.

It is worth noting that difficulties in the use of criminal law arise also in connection with the issue of workers' previous exposure to toxic substances, as for example to asbestos. In Italy, the criminal proceedings often engaged in relation to this issue have resulted in dismissals or acquittals, either on procedural grounds or in the merits.[19] In France, the Supreme Court has recently ruled out criminal responsibility in relation to the death of workers exposed to asbestos in the past.[20] In one case in Spain (Unión Naval Case), the parties reached a settlement agreement involving the "compensation of €11 million to some of those affected, and prison for the legal representative of the company and three occupational safety managers."[21]

Faced with such a diversified comparative background, which—with respect to historical pollution—encompasses legal systems characterized by the adoption of non-criminal models of responsibility (the wide majority) as well legal orders where criminal law represents the main and undisputed response (essentially Italy), one may wonder whether a common and unifying feature really exists. Yet such a feature is represented, following a certain number of reports, by the existence of an offense punishing the failure to comply with orders to take remediation measures. These orders may come from the public administration, the judiciary, or from a settlement between the parties. In fact, in certain legal systems the failure to comply with remediation orders is a regulatory offense, in others it is punished with fines of criminal nature, and in still others it is punished with a joint application of fines and imprisonment, as happens in Italy (see Art. 257 of the Consolidated Laws on the Environment and Art. 452-terdecies of the Italian criminal code). The topic is treated by the national reports, albeit in heterogeneous ways depending on the importance of the offense of non-compliance with remediation measures in each order.[22] In some reports, finally, criminal responsibility is dealt with not in connection with the failure to comply with remediation orders itself, but rather with the damage to the environment following any such failure, as for example groundwater contamination resulting in damage to human health.[23]

In conclusion, with respect to the legal systems wherein we have studied the "sanctioning issue," historical pollution is mostly dealt with in merely hypothetical terms—again with the exception of Italy, where there are a significant number of judgments. The analysis carried out in the individual reports shows both the practical and theoretical limits of an approach to historical pollution based on the resort to criminal law.

---

[19]See in this volume Sect. II of the Chap. 5 on Italy, by B. Venturato and E. Greco, Sect. II.1.5.

[20]See, again, Chap. 12, Sect. 1.

[21]See in this volume, Chap. 10, Sect. 2.2.

[22]See, again, Sect. I of the chapter on Italy, by G. Rotolo, Chap. 4, Sect. I.4.

[23]See in this volume, Chap. 12, Sect. 7.2.

6 Legacy Pollution: Lessons from the Past and a Challenge for the Future

The research presented in this volume is intended to start a critical debate on the legal instruments that are available to face the common problem of historical pollution. This is a phenomenon which has tragically emerged in several industrial contexts, and which is likely to emerge in the future in other countries that are currently in the midst of industrial and economic development, and where development is being achieved “at any cost” and to the detriment of the environment and human health.

As is apparent from the complexity of the phenomenological scenarios and from the plurality of legal instruments available, the problem of historical pollution is underlain by a complex balance of a number of interests. Among these interests are the protection of human health, the protection of the environment, and the need to safeguard social, economic, and industrial development. And another important factor is the need to correctly allocate the extremely high economic burdens deriving from the recovery of environmental matrices.

The comparative research has shown that the main goal, in all of the legal systems taken into consideration, must be the identification of legal instruments aimed at remedying environmental damage. Moreover, it is necessary to protect the rights of the victims of historical pollution, so that they may be compensated for the economic and noneconomic losses they suffer.

Conversely, it has also emerged that criminal law, within the systems studied in this volume, makes only a limited contribution in cases of historical pollution. Currently it plays a residual role, sometimes limited to cases of failure to comply with remediation or recovery orders. There is also a resort to criminal law in cases in which death or injury occurred as a consequence of previous polluting behaviors that were protracted over time. However, even in these cases the ascertainment of criminal liability at trial remains very problematic—especially with respect to causation and mens rea—often resulting in the frustration of legitimate requests for justice. Consistent with these considerations, the comparative case law analysis has shown that, apart from certain specific cases, there are no criminal proceedings specifically concerning historical pollution in any of the legal systems we have studied. Only in the Italian experience—a unique system within those analyzed—is it possible to identify a long list of judgments issued by criminal courts on historical pollution cases, many of which have proven to be problematic in the ascertainment of individual and corporate liability and have been criticized by scholars.

In conclusion, the research presented in this volume promises to shed light on an issue of crucial modern relevance but which remains underexplored, and thus to orient the choices that stand to be made within countries, especially as regards the sanctioning system. The potentials as well as the limits that have emerged within the various legal systems here reviewed may serve as a basis for reflection on legal systems in other countries that are currently involved in deep industrialization processes. Furthermore, these other systems may also be considered in the context of a critical review and rethinking of the legislative and judicial choices made in the USA and in Europe, promoting the adoption of balanced solutions to the equivocal legacy that industrialization has bequeathed to present and future generations.

References

Abraham, C., & Abraham, S. (1991). The Bhopal case and the development of environmental law in India. *International and Comparative Law Quarterly, 40*(2), 334–365.

Altena, B., & van der Linden, M. (Eds.). (2003). *De-industrialization: Social, cultural, and political aspects*. Cambridge: Cambridge University Press.

Beck, E. C. (1979). The love canal tragedy. *EPA Journal, 5*(1), 17–20.

Brandon, E. (2013). *Global approaches to site contamination law*. Dordrecht, Heidelberg New York, London: Springer.

Centemeri, L. (2006). Ritorno a Seveso. Il danno ambientale, il suo riconoscimento, la sua riparazione. Milano: Bruno Mondadori Editore.

Cerasi, L. (2007). *Perdonare Marghera: la città del lavoro nella memoria post-industriale*. Milano: Franco Angeli.

Davies, L. L. (2011). Beyond Fukushima: Disasters, nuclear energy, and energy law. *Brigham Young University Law Review, 2011*(6), 1937–1990.

Davies, P., Francis, P., & Wyatt, T. (2014). Taking invisible crimes and social harms seriously. In P. Davies, P. Francis, & T. Wyatt (Eds.), *Invisible crimes and social harms* (pp. 1–25). UK, London: Palgrave.

du Plessis, W, Kotze, L. (2007). Absolving historical polluters from liability through restrictive judicial interpretation: Some thoughts on Bareki No v. Gencor Ltd. In Stellenbosch law review (vol. 18, no. 1, pp. 161–193).

Earnhart, D. (2004). Liability for past environmental contamination and privatization. *Environmental & Resource Economics, 29*(1), 97–122.

European Commission. (1993).Green paper on remedying environmental damage. COM (93) 47 final, 14 May 1993 [COM Document]. www.aei.pitt.edu/950/

Fagbohun, O. (2007). The imperatives of environmental restoration due to oil pollution in Nigeria. *Stellenbosch Law Review, 18*(2), 347–363.

Feris, L. (2012). The public trust doctrine and liability for historic water pollution in South Africa. Law, Environment and Development Journal, 8(1), 1-[iv].

Gassner, U. (1993). Legal approaches to the removal of Altlasten (residual pollution) in the Federal Republic of Germany. *Environmental Claims Journal, 6*(1), 3–14.

Gayed, M. (2015). Cleaning Fukushima: Designing multiparty stakeholder damage mitigation system. *American Journal of Mediation, 8,* 113–138.

Gibbs, L. M. (2011). *Love Canal and the birth of the environmental health movement*. Washington, DC: Island Press.

Gorman, H. S. (2003). Brownfields in historical context. *Environmental Practice, 5*(1), 21–24.

Hahn-Baker, D. (2001). Reflections on love canal. *Buffalo Environmental Law Journal, 8*(2), 225–242.

Hall, M. (2013a) Victims of environmental harms and their role in national and international justice. In R. Walters, D. S. Westerhuis, & T. Wyatt (Eds.), Emerging issues in green criminology. Exploring power, justice and harm. Critical criminological perspectives (pp. 218–241). Basingstoke: Palgrave.

Hall, M. (2013b). Victims of environmental harm. In Rights, recognition and redress under national and international law. Abingdon, Oxon; New York: Routledge.

Itabashi, K., Ida, M., & Ohta, H. (2015). Japan. In Baker & McKenzie (Eds.), Baker & McKenzie's international guide to contaminated land (pp. 22–26). www.lexology.com/library/detail.aspx?g=14731d66-c7bd-438c-9977-31392864b6ec

Kolbasov, O. (1992). Ecological disaster area: The Chernobyl case study. *Boston College Environmental Affairs Law Review, 19*(3), 637–640.

Larson, R. B. (2013). Orphaned pollution. *Arizona State Law Journal, 45,* 991–1032.

Ligi, G. (2014). *Antropologia dei disastri*. Roma-Bari: Laterza.

Mazade, O. (2010). *La reconversion des hommes et des territoires. Le cas Metaleurop*. Paris: L'Harmattan.

Mieck, I. (1990). Reflections on a typology of historical pollution: Complementary conceptions. In P. Brimblecombe & C. Pfister (Eds.), The silent countdown. Essays in European environmental history (pp. 73–80). Berlin, Heidelberg: Springer.

Monti, A. (2002). Environmental risks and insurance. A comparative analysis of the role of insurance in the management of environment-related risks [OECD Report]. www.oecd.org/finance/financial-markets/1939368.pdf

Mori, G. (1977). *Il capitalismo industriale in Italia: processo d'industrializzazione e storia d'Italia*. Roma: Editori Riuniti.

Natali, L. (2014). Green criminology, victimización medioambiental y social harm. El caso de Huelva (España). *Revista Crítica Penal y Poder, 7,* 5–34.

Newman, R. S. (2016). *Love canal: A toxic history from colonial times to the present*. New York: Oxford University Press.

Report, O. E. C. D. (2012). *Liability for environmental damage in Eastern Europe, Caucasus and Central Asia (EECCA): Implementation of good international practices*. Paris: OECD Publishing.

Ori, A. (1993). Soil pollution countermeasures in Japan. *Environmental Claims Journal, 6*(1), 15–26.

Osaka, E. (2012). Corporate liability, government liability, and the Fukushima nuclear disaster. *Pacific Rim Law Policy Journal, 21*(3), 433–460.

Rall, D. P. (1972). The invisible pollution. *The New England Journal of Medicine, 287*(22), 1146–1147.

Ruggiero, V., & South, N. (2013). Toxic state-corporate crimes, neo-liberalism and green criminology: The hazards and legacies of the oil, chemical and mineral industries. *International Journal for Crime, Justice and Social Democracy, 2*(2), 12–26.

Sanders, B. (2009). *The Green Zone. The environmental costs of militarism*. Oakland: AK Press.

Walters, R. (2014). Air pollution and invisible violence. In P. Davies, P. Francis, & T. Wyatt (Eds.), Invisible crimes and social harms (pp. 142–160). UK: Palgrave.

World Bank Group. (2012). *Getting to green—A sourcebook of pollution management policy tools for growth and competitiveness*. Washington: World Bank.

Wrigley, E. A. (1990). *Continuity, chance and change: The character of the industrial revolution in England*. New York: Cambridge University Press.

Wyatt, T. (2016). The brownness of green crimes and harms. In T. Wyatt (Ed.), Hazardous waste and pollution. Detecting and preventing green crimes (pp. 1–7). Switzerland: Springer.

Zhang, D., & Silverman, S. (2015). People's Republic of China. In baker & McKenzie (Ed.), Baker & McKenzie's international guide to contaminated land (pp. 32–36). http://www.lexology.com/library/detail.aspx?g=14731d66-c7bd-438c-9977-31392864b6ec

Zhao, X. (2013). *Developing an appropriate contaminated land regime in China: Lessons learned from the US and UK*. Berlin, Heidelberg: Springer Science & Business Media.

Chapter 2
The Contribution of Green Criminology to the Analysis of Historical Pollution

Lorenzo Natali

The past is never dead. It's not even past.
(Faulkner 1951).

Contents

I am extremely grateful to Fulvio Coltorti, Giovanni Ovi, Veronica Binda, Mario Perugini, Matteo Di Tullio, Alessandro Corda, Juan Diego Pérez Cebada and Lara Ferla for their reflections, which helped in the realization of this work. I would like to express my gratitude to Avi Brisman for his invaluable advice and close reading of the text. My thanks also to Mirella Giulidori for her help in the translation of this contribution.

L. Natali (✉)
Department of Sociology and Social Research, University of Milano-Bicocca, Milan, Italy
e-mail: lorenzo.natali1@unimib.it; lorenzonat78@gmail.com

F. Centonze and S. Manacorda (eds.), *Historical Pollution*,
DOI 10.1007/978-3-319-56937-6_2

1 Towards a Hidden Geography: A Criminological Proposal

Scientific and technological developments have introduced into our late modern societies a new type of vulnerability, which translates into risks likely to lead to serious consequences—often irreparable[1]—for human and natural environments. Environmental disasters such as Minamata,[2] Seveso,[3] Love Canal,[4] Chernobyl,[5] Exxon Valdez, Deepwater Horizon, and Fukushima bear names that reflect what Laura Centemeri (2006, p. 59) defines as a "shared geography of prototypical disasters"—a sort of collective memory of such events—that has led to an acknowledgement of industrial, technological, and environmental risk on a global scale. It is along these paths that social actors have started to become aware of the dramatic consequences of the human–environment relationship.

The aim of this contribution is to map out, from a criminological perspective, some contextual and theoretical scenarios that help to interpret and properly situate the phenomenon of "historical pollution" within wider cultural and scholarly frameworks. Like a lump of naked rock that is worked little by little to uncover a hidden shape, just such an attempt will be made to outline a number of relevant criminological facets of historical pollution. In its narrowest meaning, this expression refers to industrial processes that have ceased but whose effects and impacts—harms and threats to the environment and human health—continue to emerge and to manifest, even dramatically, after many years. On the other hand, according to a wider definition of the phenomenon, the expression may also include activities still in progress, as long as they are waning.[6] More specifically, "historical pollution" denotes

> the pollution of sites and natural resources (intended as the relevant *behaviour*), which started or took place in the past, due to industrial activities or other production-related activities. Historical pollution usually entails the emergence—or the persistence—of

---

[1]On this aspect, see South (1998), Beck (2009 [2007]), Brisman and South (in press). See also Centonze (2004).

[2]The acknowledgement that harm to human health and the environment had been caused by mercury contamination deriving from the chemical plants of Minamata (at the end of the 1950s) marked a true turning point for Japan. From that moment on, efforts to reduce pollution could not be ignored.

[3]See in this volume, Chap. 4.

[4]See in this volume, current chapter, and Chap. 8.

[5]See Galli and Nigro (1987) on the perception of radioactivity following the Chernobyl disaster.

[6]For the purposes of our research, the concept investigated also includes the phenomenon that is named "legacy pollution". The concept of "ongoing pollution" will not be used in opposition to it, but rather will be treated as a type of legacy pollution, identifying past conducts whose effects are still ongoing. Moreover, since the concept may refer to a wide range of hypotheses, it is necessary to limit the extension of the phenomenon on which the research is focused: attention has to be paid only to pollution caused by industrial and other production-related activities. See in this volume, Chap. 1.

contamination of sites and natural resources (meant as the *effect* of such a conduct) a long period of time after the original pollution took place.[7]

If criminology has had difficulties in viewing itself as a "prospective" discipline (Brisman and South 2013b, p. 412), this temporal myopia[8] has also concerned the "historical" depth of its own gaze, normally limiting it to the best-known crimes that do not require a radical rethinking of the horizons of space and time through which we approach crime and its consequences. On the contrary, this criminological contribution could be useful in promoting increased knowledge and theoretical reflexivity about what can be defined as the "historical inheritance" of industrial development.[9] By following the outline and entering the boundaries of what can be defined as a hidden geography, the traditional boundaries of the criminological discipline will be breached, in order to inaugurate connections with and within *green criminology*. Green criminology allows the meeting of a wide range of theoretical orientations aimed at connecting a series of issues of crucial importance for today's world: environmental crimes, harm, and various forms of (in)justice related to the environment, the animal species, and the planet. Without becoming a single unity, these approaches come together within an increasingly broad perspective (see, for example, South 1998; Lynch 1990; Lynch and Stretesky 2003; Walters 2010; White 2011; Sollund 2012; Natali 2013a) that examines the bio-physical and socio-economic consequences of the diverse sources of environmental harm, such as pollution, the deterioration of natural resources, the loss of biodiversity, and climate change (see South et al. 2013, pp. 28–29; Natali 2015a).[10]

With this background, I will focus on some thematic nuclei helpful in observing and defining the phenomenon of historical pollution under the heading of complexity. Keeping away from a deductive approach, I will take a "photograph" which sums up within itself some clues concerning the object under consideration,[11] and remains sensitive to its variegated aspects. I will do this by considering three

[7]See in this volume, Chap. 3, Sect. 2.

[8]On this, see also Brisman (2014b, 2015b).

[9]"Unlike other forms of crime, green crime's effects tend to be persistent and long term. Ordinary crimes or street crimes usually cover short durations in time. […] The time span of green crimes, however, tends to be measured in terms of decades or even centuries since pollutants may last and accumulate in the environment. Once a pollutant is emitted into the environment, the life course of that crime begins and continues until the pollutant is cleaned up, or until it becomes inert" (Jarrell et al. 2013, p. 423).

[10]It can certainly be said that even before the contributions of green criminologists, criminology did deal with the environment, albeit in an indirect way. The object of study was, in these cases, organized crime and corporate crime. For example, one can think of waste management or the construction industry, where the environment is adversely impacted through illegal business patterns and practices, yet the criminological inquiry focused on the illegal acts and omissions, not the ensuing environmental harm. On the various typologies useful to deepen our understanding of environmental crimes and harms, see South et al. (2013). On the value positions that shape green criminology, see Potter (2015).

[11]The research into these "clues" directly recalls the type of inference known as "abduction" (see Verde and Nurra 2009; Ceretti and Natali 2009, pp. 387–395; Eco 1983).

different scenarios of historical pollution: Love Canal (USA), Brescia (Italy), and Huelva (Spain). In describing these particular contexts—just three examples among the various cases of historical pollution—it will be possible to highlight some common features of this phenomenon and, at the same time, to see its global relevance.

Before describing the contextual scenarios, a few methodological points are in order. Every researcher knows full well that the choice of a certain method implies a series of decisions as to which research questions to formulate and pursue, which theories to explore, which data to collect, and how to analyse and interpret the data acquired (Cardano 2011). With regard to the distinctive features of environmental crimes, an eco-global criminological perspective (Heckenberg and White 2013, White and Heckenberg 2014) suggests that we consider: (1) *who* is the victim: humans, non-human animals, an entire ecosystem; (2) *where* the harm takes place: at a local level, globally, or somewhere in between; (3) what is the *context* in which the harm manifests itself: the natural or the built environment—urban or rural; and (4) what is the *temporal frame* within which the consequences of the harm can be analysed: immediate, delayed, or intergenerational. In this regard, the first question to be asked concerns the exact "when" of the environmental phenomenon that is to be researched: *before* the harm happens, *while* the harmful consequences are developing, or *after* the harm has occurred (assuming that there is a "stopping" point).[12] In the case of historical pollution and contamination it seems essential to keep together more temporal perspectives in relation to single cases, matching the analysis of the harmful consequences caused by the pollution of the area over the course of time with the study of environmental harms as they are developing here and now.[13]

Moreover, the meeting between two emerging sensitivities such as green criminology and cultural criminology[14] could prove useful in positioning these phenomena on the cultural and historical level. In the convergence of those two theoretical orientations—a middle ground that Brisman and South (2014) refer to as "green cultural criminology"—one might find those studies that analyse the intersection of crime, environment, culture, and justice, also including the significant dimension of socio-environmental conflicts.[15] As Jeff Ferrell (2013, p. 349) writes:

> By the nature of their subject matter, both green criminology and cultural criminology push against the conventional boundaries of criminology, and so tend to upset the definitional and epistemic order of the discipline. Likewise, both are open to exploring a range of social harms and social consequences, whether these harms are conventionally defined as

[12]On this, see Heckenberg and White (2013, p. 86).

[13]Heckenberg and White (2013, p. 86) write: "If we choose to examine an event that is *a number of years in the past* (e.g. poisoning of waterways over many years), then the historical method can be utilized, drawing upon documents, maps and photographs and site records that facilitate a retrospective analysis of the phenomenon in question."

[14]Since its inception, cultural criminology has called for the development of a form of criminological *verstehen* that is capable of exploring the universes of sense and the emotional processes related to crime and to its control. See Ferrell et al. (2008).

[15]On this topic, see Brisman et al. (2015).

> "criminal," currently left outside the orbit of law and criminality, or even themselves propagated by the criminalization process. At their best, both link their overt substantive concern—environmental harm in the case of green criminology, meaning and representation in the case of cultural criminology—with broader issues of power and inequality. And certainly both attempt to situate their subject matter historically, both in terms of its long-range development and its current residency within the crisis of late modernity and late capitalism.

Historical analysis thus becomes crucial. In this regard, environmental historian Simone Neri Serneri (1990) points out how academic culture has for too long ignored the processes that have over time led to present-day environmental deterioration. In order to develop an appropriate "environmental history" it seems necessary to: (1) pay attention to the relationship between historical and biological time; (2) go beyond a rigidly anthropocentric and modernistic view of the relationship between humans and the environment; and (3) deepen the "history of conscience" that historical actors have with respect to their action on the socio-ecological context, including not only severe threats to the environment and to human health, but also apparently less significant behaviours which, when repeated, produce effects in the long run (Neri Serneri 1990, pp. 891–893). From such a standpoint—and recalling an international conference of economic historians in Milan in 1994 (Saba 1997, p. 466)—Neri Serneri (1990) highlights how the environmental historian should not only analyse and explain the past, but also warn or *alert* us about the future. The respective roles of the environmental historian and the green criminologist find here an important place of convergence.[16] Another point of contact with historical knowledge is found not only with reference to "macroscopic, obvious and calamitous transformations (acute environmental crises)" but also to "gradual and diffused ones, harbingers of equally dangerous alterations for the ecosystem where the human species lives" (Neri Serneri 2005, p. 69).

More importantly, strengthening the historical ties between events that are distant from each other in time (past and present) implies the adoption of a method that is open to the complexity and to the oxymoronic character of such ties; it also means bringing to light a plurality of narratives (Natali 2013b) and saving them from oblivion. It means, ultimately, academic and cultural work on memory—from both a historical and criminological viewpoint—with profound consequences for our present and possible futures. Even from a purely historical perspective—essential to the investigation here described—only an interdisciplinary method can adequately interpret the complexity of the observed phenomena (Saba 1997).

Finally, as the relationship between temporal and spatial analysis refers directly to the dialogue between historians and geographers,[17] it is also essential to decide upon a geographical starting point, in order to catch the particular features of the context considered.[18]

---

[16]On the role of the criminologist, see Forti (2000, p. 318), and Natali (2015a, p. 30).

[17]On this aspect, see Neri Serneri (2005, p. 17); see also Armiero and Barca (2004).

[18]To this end, one can access the online *Environmental Justice Atlas*: http://www.ejolt.org/maps/. This online platform is also useful because environmental crimes and harms of considerable

2 Love Canal: A Prototypical Case of Historical Pollution?

In 1978, the Love Canal disaster resonated nationally in the US media, changing forever the awareness of environmental risks and of contaminated sites (Szasz 1994; see also Rotolo 2016). Love Canal is certainly "one of the most discussed and influential case studies in environmental history" (Blum 2008, p. 148). Its story might be summarized as follows:

> Hooker Chemical dumped large amounts of some very nasty stuff in a hole in Niagara Falls, New York, and then covered it up—literally and figuratively—before selling the property to the local school board for $1. In the 1970s, large amounts of rainfall caused some of the buried waste to leak out of its containers, flow to the surface, and contaminate homes in the area. The oozing waste generated vocal complaints, which led Michael Brown, a Niagara Falls reporter, to investigate the contamination. After reading Brown's articles, Lois Gibbs, a shy young housewife, connected the chemicals to her son's numerous illnesses. Gibbs then single-handedly began a crusade to get the entire neighbourhood relocated, manipulating the media to her advantage. She became a savvy political leader, extolling the dangers the chemicals posed to children and pressing for safety above all else. She gained the attention of the nation and the president, and as a result, the state and federal government agreed to purchase all the homes in the beleaguered neighbourhood. Congress later passed the Superfund legislation to help others in similar situation. (Blum 2008, p. 1)

This description, however, like many others, is just a "standard version of the Love Canal story" that almost inevitably disregards the deep historical context of the place[19] and the complex dimensions of grassroots activism (Blum 2008, p. 2). With respect to the history prior to 1978, it suffices to remember that, since the middle of the 1700s, the Niagara Falls have epitomized an *ambivalence* that represents a common tension of the human–environment interaction: on the one hand, they were admired for their "natural magnificence", which takes one back directly to the experience of the sublime; on the other, they were appreciated for their power-generating potential for various types of industry (Blum 2008, p. 18). In fact, it is really because of this new source of power that "industry sprung up along the banks of the Niagara River in the late 1800s and early 1900s" (Blum 2008, p. 20). It is within this historical and environmental context that the

> Hooker Electrochemical Company (later Hooker Chemical Company), founded by Elon Hooker, began to produce chemicals in the Niagara Falls area. Over the next seventy years, Hooker Chemical became an integral part of the community, providing jobs to residents and tax revenue to the local governments. In 1942, Hooker Chemical obtained the right to dispose of its chemical wastes in the old Love Canal, and it later purchased the property from the city. This development continued the long-standing tradition of exploiting the land of Niagara Falls. (Blum 2008, p. 21)

(Footnote 18 continued)

seriousness (such as historical contamination) may be intertwined with environmental conflicts. On this, see Brisman et al. (2015).

[19]See also Colten and Skinner (1996).

The second aspect normally ignored by "standard" accounts concerns the environmental activism that characterized the citizens' response to the socio-environmental disaster of Love Canal. An in-depth study endowed with a historical sensitivity to context cannot but emphasize the complexities and the "complications of race, gender and class in grassroots activism" (Blum 2008, p. 2). Each social actor involved in that story had a unique, personal perspective and told a "different version of the truth about what really happened" (Blum 2008, p. 3; see also Levine 1982, pp. 1, 193–194). This means to bring to light and recognize the multiplicity of voices and actions that have given form to the interpretation and the social construction of the reality of the disaster and of the following response. The personal histories and the changing definitions of the situation therefore make complex a macro history that could appear clear, straightforward, and compact in a less reflexive analysis.

Focusing on the peculiar aspects that make the case of Love Canal a paradigm for the study of cases of environmental pollution, it is possible to identify some significant points:

- The Love Canal disaster developed over several decades: the impact of leaching chemicals was uncertain and slow in developing, the visible effects limited and difficult to detect (Levine 1982, pp. 1, 14).
- Love Canal may be defined also as a "conceptual event": "[t]he physical phenomena are out there, present in the world. Equally important, so are the social processes whose consequences are very real" (Levine 1982, p. 219). In fact, the physical manifestations of the disaster have been "readily overlooked, ignored, denied, and minimized" (Levine 1982, pp. 1, 170).
- In the Love Canal story, "there was no crisis or disaster until authorities defined it publicly and the event was reported in the world press. Unlike more familiar disasters, with known and uncontrollable moments of impacts, the Love Canal emergency could have been defined earlier or later" (Levine 1982, p. 1).
- Love Canal represents a clear example of the "complex ethical dilemmas we face in balancing human health and well-being against economic costs and benefits" (Levine 1982, pp. 1–2).
- The social awareness of the ongoing, creeping disaster was a lot less clear than told by the "standard versions of the Love Canal story". For example, "[b]y the mid-1970s, only a few old-timers remembered the whole history of the canal that had turned into a schoolyard and open fields. Most of the people there had arrived in the mid-1960s and later, after the school and road construction was complete, and were unaware of the early history of the place" (Levine 1982, p. 13). Furthermore, "the mental process of moving from a condition of ignorance about the chemicals, to belief in the possibility of personal danger from them, to sharing the beliefs of others and joining in activities—some of them unaccustomed ones—with organized residents was rather slow and uneven overall" (Levine 1982, p. 193).

- The temporal dimension therefore assumes a peculiar sociocultural and institutional connotation: "Both individuals and institutions reacted to Love Canal with slowly dawning awareness of the far-reaching consequences of past activities: burying chemical waste and encouraging people to live in homes and send their children to school close to the disposal site. The consequences might or might not have been anticipated by the people who decided long ago to bury and to build, or who allowed these things to be done. Their decisions and actions affected the future, but the individual decision makers were not accountable at the time of their actions. The day of reckoning was deferrable, with the impact and responsibility to be borne by someone else in the distant future" (Levine 1982, p. 2).
- Scientific studies "became part of an emerging interpretation of what happened to and what was done for the people of Love Canal" (Levine 1982, p. 115).[20] The role of science acquires a decisive meaning in a highly complex context, such as the environmental one, that is inevitably open to scientific uncertainty and to conflicting scientific interpretations (see Natali 2015a; Hall 2014, p. 104).

The story of Love Canal, briefly summarized above, is, then, the story of the "loss of innocence" of a community with respect to the environment in which it is situated, but it is also a "success story" because it demonstrates that "ordinary citizens", if organized, can gain power and force corporate and governmental accountability.[21]

Exploring now two other historical–geographical contexts involved in phenomena of historical pollution, it will be possible, on the one hand, to trace the thin lines which relate them to a prototypical disaster such as that of Love Canal; on the other hand, we will see how the peculiar developments that have characterized the "success story" told so far—that is, the social and institutional answers prompted by the "discovery" of the disaster—do not find significant correspondences in the cases that will be described.

[20]"From the very beginning, the definitions of the Love Canal health and environmental problems —where they are, what they are, how serious they are—have varied considerably, depending on who defined the problem, when, to whom, and what they stood to gain or lose from the definition. The words, however, were usually spoken in the language of science" (Levine 1982, p. 168).

[21]A victimological approach becomes extremely useful in this case (see Williams 1996). From a historical–sociological perspective, Levine (1982, pp. 176–177) pinpoints some shared beliefs about the dramatic turns the inhabitant of Love Canal experienced: (1) we are the blameless victims of a disaster; (2) the problems we face are too large for us and thus we need help; (3) we are good citizens and we deserve help from the government; (4) the government can and should help us now; (5) we are being treated unfairly; (6) we must stick together to take care of ourselves; (7) family and community help is not enough for our needs; (8) no one but the government has enough resources for our pressing needs; and (9) we must work together to force the government to provide us that with which we are entitled.

3 Clues About the Phenomenology of Historical Pollution in Italy: Caffaro Industry in Brescia as a Case in Point

It is well known that Italy's changeover to industrialization involved peculiar costs and socio-environmental consequences that emerged in all their complexity only after a long period of time.[22] Compared to other European countries, the process of industrialization that took place in Italy during the twentieth century started slowly and belatedly, making it a "latecomer" country. Its delayed start explains, at least in part, why both legislative regulations and social awareness of environmental issues in Italy arose comparatively late.[23]

The Italian industrialization process was, in many respects, a dramatic and ambivalent transformation (Ginsborg 1989). Particularly after the Second World War, Italy underwent a period of great business upheaval and industrial development: "In a very short time our country was able to conquer the poverty of centuries" (Ferrero 1999, quoted in Coltorti 2011, p. 143). The Italian economic miracle (1952–1963) was characterized by a decisive increase in production and by an important process of accumulation, in the basic economic sectors and in heavy industry. Because of a complex set of factors (which cannot be summarized here), after 1975 what happened was a "marked slowing-down of the role that big businesses performed in the overall development of the Italian economy" (Coltorti 2011, pp. 143–145).

Very briefly, it is possible to describe the specific nature of Italian industrialization as follows: the industrial revolution took place between 1896 and 1914 with the development of the iron and steel industry and mechanical engineering, but it only really took off after the "economic miracle" of the 1950s—first in the industrial triangle in the north-west (Lombardia, Piemonte, Liguria), then in the north-east, and finally in the centre (Amatori 2000, p. 66; Giannetti and Vasta 2003, pp. 9–41, 183; Zamagni 2005, p. 84; De Simone 2014, pp. 286–288); one can observe a territorial and structural polarization (between geographical areas and between large and small industries) (Neri Serneri 2005, p. 37); in the steel, petroleum, and petrochemical industries, businesses were subsidized by the State or belonged to it (e.g. the Istituto per la Ricostruzione Industriale or the oil and gas company Eni); in the 1970s, the big businesses declined; in the 1990s, privatization processes began.

In the Italian context, certain researchers were pioneers in the field, for they moved from a perspective of history of industry to the analysis of specific case studies which were "particularly exemplary of the environmental and territorial impact of the large industries, as they are the source of serious and repeated

[22]On the traffic of toxic waste in Italy, see also an interesting piece of investigative journalism by Pergolizzi (2012).

[23]See Neri Serneri (2005, p. 42). See also in this volume the quadripartite chapter on Italy, by G. Rotolo, B. Venturato, E. Greco, R. Sabia, and C. Micciché. Obviously, there are examples in which latecomers to industrialization address environmental issues earlier in their histories as a result of seeing what has transpired in other locations.

phenomena of industrial and urban pollution" (Neri Serneri 2005, p. 36).[24] More specifically, the scenarios relevant for this investigation can found in three different parts of the country: the industrial triangle of north-western Italy (Lombardia, Piemonte, Liguria), the centre and the regions of the Triveneto (Veneto, Trentino-Alto Adige, Friuli-Venezia Giulia) and, finally, the south and the islands.[25] Within this tri-partition, it is possible to start identifying the "hot spots" of historical pollution in Italy, albeit with a cartographic exercise still in progress. On the basis of the defining premises shared with the research team[26] and the sources that may be employed (e.g. the Istituto Superiore di Sanità and its epidemiological study SENTIERI[27]), we can distinguish three distinct scenarios.

The first context is represented by the petrochemical industries of Porto Marghera,[28] Ravenna, Ferrara,[29] Taranto,[30] Brindisi, Priolo-Augusta (Siracusa, Montedison), Gela, Cagliari (Saras), Crotone (Pertusola Sud)[31]—a far from

[24]See, for example, Ruzzenenti (2001), De Luigi et al. (1995), Poggio (1996).

[25]Ginsborg (1989, p. xii) writes: "this tri-partition seems to me more effective than the traditional North-South division in illuminating the physiognomy of the processes of development that affected [...] the nation."

[26]See in this volume, Chaps. 1 and 3.

[27]It is essential to make reference to the Istituto superiore di sanità (ISS, the Higher Institute for Health). For the past few years, this institution has been concerned with measuring the level of contamination of industrial areas, both used and disused, focusing also on the possible risks for any residents. The project, called SENTIERI (Studio Epidemiologico Nazionale dei Territori e degli Insediamenti Esposti al Rischio da Inquinamento—National epidemiological study of territories and settlements exposed to pollution risks), confirms some correlations between certain substances (e.g., asbestos) and specific pathologies. It was first carried out in 2010 and an account was published in a supplement of *Epidemiologia e Prevenzione*, 34 (5–6), September–December 2010; it was updated in 2011 and can be found in *Epidemiologia e Prevenzione*, 35 (5–6), September–December 2011. The project was the result of a collaboration of CNR and La Sapienza University of Rome with the World Health Organization (WHO). The SENTIERI project assesses the health status of residents of Italian polluted sites through the analysis of mortality for the period 1995–2002, to set priorities in remediation intervention and so prevent environment-related diseases, for a total of roughly six million people in 298 municipalities (twenty-one in northern Italy, eight in the centre, and fifteen in the south). One of the limitations of the study is that when considering landfill sites it only took into account the legal ones, when it is well known that the illegal ones are more harmful, given the total absence of restraint. Only in the case of dumps in the provinces of Naples and Caserta was the study also expanded in this direction; it highlighted a close link between illegal dumps and certain cancers, beside some forms of congenital malformation (Pergolizzi 2012, pp. 156–158). In any case, even with these data, we are still very far from a substantial mapping of the critical environmental issues of our country (Pergolizzi 2012, p. 160).

[28]On this, see Stella (2000, 2003), Centonze (2004), and Perini (2002). See in this volume, section II of the Chap. 5 on Italy, by B. Venturato and E. Greco.

[29]On the chemical plants of Ravenna and Ferrara, see Pergolizzi (2012, pp. 23–24).

[30]On the Taranto case, see Rotolo (2012, pp. 90–103).

[31]In 1930, the petrochemical complex Pertusola Sud (Eni Group) started manufacturing at Crotone (Calabria), bringing (as in many other similar scenarios) work opportunities and wealth on the one hand, and contamination and health hazards on the other. Once the industrial activities were wound down, all that remained were the disastrous consequences of the industrial past.

exhaustive list. But think also of the events related to the reclamation of the Italsider at Bagnoli (Naples) whose productive activity definitively ceased in 1992, or of the industrial area of Porto Torres (Sardinia), or of the Ilva at Genoa-Cornigliano. In Tuscany, the Tioxide at Scarlino (Grosseto)—a British multinational corporation that took over the old complex of Montedison in order to manufacture titanium dioxide—had discharged the residual waste of its manufacturing for years (from 1978 to 1985) (Pergolizzi 2012, p. 163). In addition, one could also think of the scenario of the Bormida Valley, which very clearly illustrates long-term environmental harm and devastation. From the end of the 1800s, at Cengio, in the province of Savona, Acna (Azienda coloranti nazionali e affini) manufactured explosives and nitric acid; in 1990, it came under the control of Enichem and manufactured bon acid, beta naphthol, Schaeffer's acid, and phthalocyanines. On 23 July 1988, the plant was closed down, leaving as a lasting legacy environmental harms that are nearly impossible to remedy. Data gathered by the WHO prove the seriousness of the situation, which has also had direct effects on the health of the resident population.[32]

A second scenario that pinpoints specific phenomena of historical pollution concerns landfill or dump sites, both legal and illegal.[33] One of the best-known cases is that of the Pitelli dump, on the so-called poison hill at La Spezia. The dump started in 1976, and over time, new installations were added which extended it out of all proportion. It was closed down in 1996.[34] In this instance, the connection between the disposal of waste and the ensuing socio-environmental harms presents such extremely high levels of complexity that it is even difficult to identify victims. Another "exemplary" case is represented by the dumps at Bussi on the Tirino (Pescara), discovered in 2007 by the National Forest Rangers (Corpo Forestale dello Stato). The "poison dumps" at Bussi are the most extensive in Europe, with twenty-five hectares of toxic wasteland.[35]

Finally, a different scenario that is part of the possible phenomenology of historical pollution is the so-called brownfield.[36] "Brownfield" is a word coined and used in the USA to describe those sites whose expansion, re-qualification, or reuse are complicated by the possible, or effective, presence of contaminating sources. They are mainly the legacy of processes of industrialization (often carried out at an accelerated pace) that have affected various areas of the planet. The area of Santa Giulia-Montecity (Milan) which is built on what was left of the Montedison and of

[32]See also Pergolizzi (2012, p. 133).

[33]In these contexts, it is important to stress that the often irreparable socio-environmental harms that follow are caused by legal as well as illegal dumps (Pergolizzi 2012, p. 155).

[34]In the Pitelli area, there would also have been buried waste from the factory of the Union Carbide Unisil Spa Termoli (Pergolizzi 2012, p. 141).

[35]See in this volume, section II of the Chap. 5 on Italy.

[36]See also Lynch and Stretesky (2014, p. 164).

the ex-steelworks Radaelli, is only one example,[37] but the same problems of reclamation concern the former area of Falck (a steelworks) and the railway station at Sesto San Giovanni, as well as the SISAS (Società Italiana Serie Acetica Sintetica) at Pioltello, still awaiting reclamation since 1985, or the Eternit factory at Casale Monferrato.[38]

If we move around the map provided by the website *Environmental Justice Atlas*, we find, in Italy, a "hot spot" that gives focus to some profiles that are of interest for this investigation: Caffaro industry in Brescia (Fig. 1).[39]

From the end of the 1930s until 1984, Caffaro industry (Brescia, Italy) produced enormous quantities of polychlorinated biphenyls (PCBs), spilling them, as well as dioxins, into the surrounding environment. After fifty years of pollution, the level of social and environmental harm is particularly serious, so much so that in 2002 the site was included among the fifty-seven sites of national concern identified by the Ministry of the Environment as contaminated and in need of reclamation and clean-up.[40] Research carried out by Paolo Ricci, an epidemiologist of the Azienda Sanitaria Locale of Mantova, in the context of a collaboration between the Istituto

---

[37]The area should have been reclaimed before building the residential neighbourhood, removing the contamination produced by decades of industrial activity. Because of irregularities in the reclamation work, part of the area was confiscated in 2010, including the park and the kindergarten, which were built on mercury- and chloroethylene-contaminated ground (Pergolizzi 2012, p. 147).

[38]Eternit opened its first asbestos production plant in Italy in 1907. Altopiedi (2011, 2013) analyses this case from a criminological perspective. The sociologist explores on the one hand the strategies of denial through which those responsible for the crime facilitate the process of decriminalization of their actions and, on the other, the process of the social construction of victimization. On this, Ruggiero and South (2013, p. 20) write: "The use of asbestos was banned by the European Union in 1999, 101 years after the discovery of its danger. One of the reasons it took so long to ban the substance was because asbestos 'kills slowly' and, in many cases that were ever actually brought to court, lawyers as well as pathologists could easily dismiss the association of asbestos with lethal respiratory conditions". Concerning the Eternit factory in Casale Monferrato, "[p]opular mobilisation and citizen action played a major role in bringing this history of harm into the arena of media news and the courts. A long-running and contentious affair was eventually transformed by popular political protest from a matter officially deemed an 'accident' into a trial […]". On this point, see also South (2015, pp. 18–19); see also in this volume, section II of the Chap. 5 on Italy.

[39]To offer a case in point, suitable for highlighting some sensitive points that arise when speaking of historical pollution, I did not look at contexts in which a present pollution event is added to a past one. The observational field would be too wide and would complicate the definition of some essential aspects that I intend to highlight. On the other hand, it is evident that cases of ongoing pollution, such as that of the Ilva of Taranto, will provide some useful cues because of the presence of analogous dynamics. As for the possibility of generalizing from the results drawn from a case study, Heckenberg and White (2013, p. 96) recall the concept of "naturalistic generalization" described by Melrose (2010). See also Stake and Trumbull (1982).

[40]There are fifty-seven contaminated industrial sites which, because of their dimensions, have been declared of national concern, but according to the ISPRA (Istituto Superiore per la Protezione e la Ricerca Ambientale—the Higher Institute for Environmental Protection and Research), there are more than 15,000 areas in Italy affected by more than a century of industrial history. The best known, as already mentioned, are the iron and steel industry at Bagnoli, the chemical industry at

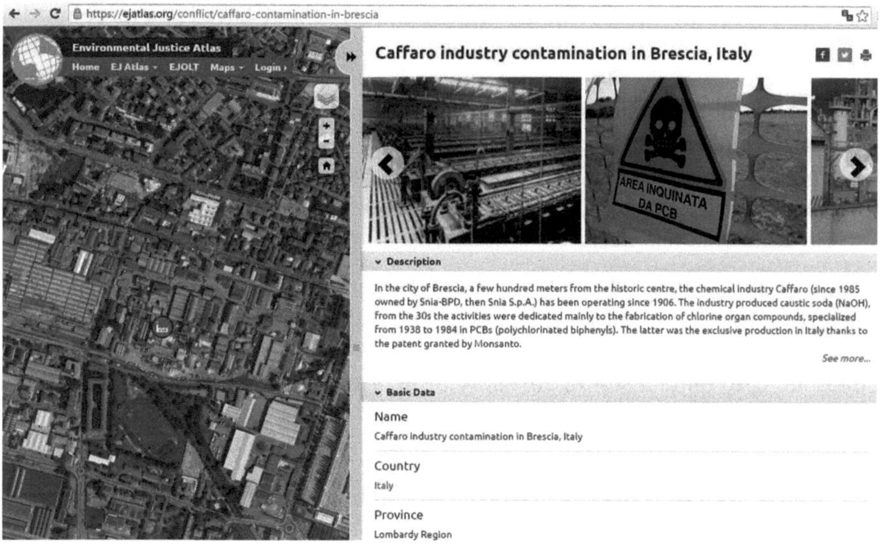

Fig. 1 Caffaro industry in Brescia, Italy. *Source* https://ejatlas.org/conflict/caffaro-contamination-in-brescia (Environmental Justice Atlas)

Superiore di Sanità and Airtum (Associazione Italiana Registri Tumori—National Register of Cancer Cases), highlights the heightened risk to inhabitants of Brescia of contracting various form of cancer (thyroid cancer, liver cancer, breast cancer, lymphoma). As in many other similar cases, however, asserting that certain illnesses are linked to the environmental contamination is anything but straightforward. Proving causation is a huge hurdle in many environmental cases.[41] While there is a Municipality Order forbidding the inhabitants of contaminated areas to walk on any areas not covered in tarmac, the clean-up itself never in fact began, because Caffaro industry went bankrupt.

The points that make this case interesting for a socio-criminological exploration are the following:

(1) The Caffaro is part of the history of the Italian chemical industry (Ruzzenenti 2001) and therefore becomes a tangible example of the present consequences of activities carried out during the historical development of industrialization in Italy.
(2) Production and pollution started eighty years ago and ceased thirty years ago. The effects of contamination continue up to the present.

(Footnote 40 continued)

Marghera, the petrochemical sites at Gela and Priolo, and the asbestos industry at Casale Monferrato.

[41]See also in this volume, Chap. 3.

(3) Today, production of PCBs is forbidden because they have been recognized as powerful carcinogens. When the Caffaro was in operation and producing them, however, their toxicity was not known.[42]
(4) There is greater scientific uncertainty around PCBs compared to other substances already recognized as carcinogenic (e.g. asbestos).
(5) There is still little information about the contamination of the site.
(6) The scenario resembles a bomb that explodes in slow motion, that is, a creeping environmental disaster (see Williams 1996, p. 16; Williams 1998).

All of this echoes similarly serious phenomena found in other countries:[43] oftentimes they are "crimes"—in the broad meaning proposed by green criminology (Brisman 2008, p. 731)—that develop silently and slowly, as creeping disasters, contributing to build a sort of "banality of industrial evil" (Pergolizzi 2012, p. 41).

4 The Case of Huelva, Spain: A Situation of Organized Irresponsibility?

A further scenario useful to highlight certain crucial aspects of the phenomenon of historical pollution is situated in south-west Spain. Huelva is a town seriously polluted by the presence of a massive industrial complex, established at the beginning of the 1960s and composed of a great number of chemical and energy corporations.[44] The plant was constructed close to the town, in what could be described as its "backyard". Figure 2 shows the most critical areas affected by pollution.

Right from the early years of industrial development, life for the inhabitants of Huelva—and the very image of the town itself—started to change dramatically. While material comforts had finally reached them, over time, destructive and

---

[42]In Italy, PCBs were banned in 1983.

[43]In describing a case of PCB waste in Warren County (North Carolina), Jarrell et al. (2013, p. 437) write: "The use of PCBs was significant to the economy and to the expansion of production […]. Over time, PCBs became less important to production because of changes in mining technology and the decreased use of transformers and capacitors. In addition, PCBs were discovered to be highly toxic to the ecosystem, and by the 1960s scientists were reporting that PCBs were a global threat […]. There was considerable political activism surrounding environmental hazards such as PCBs and environmental organizations pushed hard for their ban […]. Eventually, the United States Congress banned PCB production (but not their use) in 1979 […]. The ban of PCB production is not the end of the story, however. While PCBs were no longer critical to manufacturing production and natural resource withdrawals, they still persisted in the environment where they were causing significant harm when released in the form of ecological additions". They add: "While the production of PCBs was not defined as criminal prior to the ban, changes in production and pressure from the public redefined the disposal and use of these chemicals. When the law finally caught up to the harm the chemicals caused it created a market that led to their illegal disposal" (Jarrell et al. 2013, p. 438).

[44]On this case, see also in this volume, Chap. 10.

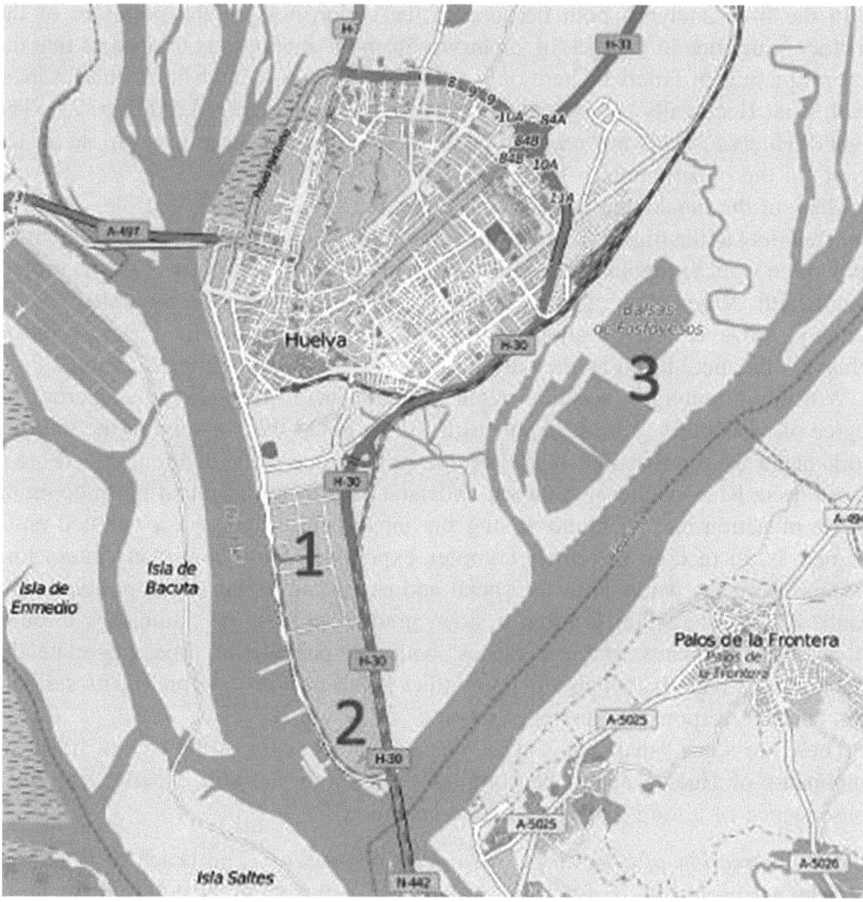

Fig. 2 Huelva (Spain) and its backyard: (*1*) the industrial and chemical plant; (*2*) the beach of Punta del Sebo; (*3*) the so-called *balsas de fosfoyesos*. *Source* https://ejatlas.org/conflict/huelva-industrial-chemical-complex-spain (Environmental Justice Atlas)

irreversible consequences, both for the environment and for people's health, began to take hold and to darken the promise of prosperity brought by the factories.

This critical situation, dominated by the presence of industrial activities that, starting from the 1960s, released their effluent into the Río Tinto, was further aggravated when some of them started discharging the *fosfoyesos* (phosphogypsum)—a waste product of the phosphoric acid production process that contains concentrations of uranium series radionucleides and is stored in piles, the so-called *balsas de fosfoyesos*, extending over an area of 1200 ha, just a few hundred metres from the town (Natali 2010, 2014, 2015a, 2016).[45]

[45]See also Dueñas et al. (2007) and Pérez-López et al. (2007) for a scientific discussion of these aspects.

In the final analysis, both because of the seriousness of the pollution of the territory[46] and due to the lack of an answer from the institutions, it appears that the general picture of Huelva—here only sketchily drawn—would fit, if only at local level, what Beck calls "organized irresponsibility" (Beck 2009 [2007], p. 31). The town of Huelva is truly an "open-air laboratory" where no one is responsible for the results of the experiments.

Many of the inhabitants are convinced that the situation is irreversible, settled in time, and feel at the mercy of powerful actors against whom it is impossible to fight. They often want to regain their living spaces in the local environment and to protect their health. When these factors are weighed against the jobs generated by the industry, however, the question becomes: how is it possible for these people to achieve a balance, faced as they are with such a tragic and unjust dilemma?

Worrying though this scenario might be, it appears to be marked by a surprising degree of invisibility.[47] In this context, it can be noted that, in spite of the untiring work of an association that fights for the gradual recovery of the polluted areas (Mesa de la Ría), neither a common understanding of the problem nor a common course of action can be found among the inhabitants. Living in a polluted environment is, in fact, an extremely complex experience built from many interacting spheres: from the personal to the social and extending as far as the political. The result of this interaction is often a slow, gradual process of "attuning", through which the inhabitants of these places, with the passage of time, negotiate the contaminated reality, though still in conflict about how to interpret it, its seriousness, and the responsibilities related to it.

These are some "snapshots" of the social awareness (or of the lack of it) of the inhabitants of Huelva about the complex drama they are still experiencing, as a consequence of a long history of industrialization:[48]

- *Transformation of a territory and collective memory*. The beach of Punta del Sebo represents the "before" that was "there"—present in the collective memory of all the locals, either through personal experience or through the stories of those living at that time—which no longer exists and stands in stark contrast to what is now there, the reality of the factories. Side by side with these beliefs, by far the most numerous, there are others that challenge the authenticity of the version of reality conjured by this collective memory, going so far as to deny its existence, and labelling it as utopian, given that the area has always been

[46]See also Benach et al. (2004) and Monge-Corella et al. (2008).

[47]See Davies et al. (2014) for a discussion of the link between invisibility and crimes. See also in this volume, Chap. 3.

[48]These "snapshots" are analysed in greater depth in previous works (Natali 2010, 2014, 2015a, 2016), where I explored how the inhabitants of Huelva relate to the "uncomfortable truth" of pollution, noting that they themselves do not always agree on the definition and interpretation of that reality. Starting from a viewpoint in tune with the experiences and the narratives of the inhabitants of Huelva, as actors placed within a well-defined network of symbols, social interactions, practices, and power relationships, I tried to reconstruct and understand what directs their "definition of the situation" of pollution and what might motivate their action or inaction.

polluted because of the mines dating back to Roman times.[49] Capturing these ambiguities in the discourses of the people who live in these contexts is an essential task in order to enter into the symbolic complexity running through the multiple experiences of environmental victimization.[50] These narratives are obviously influenced by discourses circulating in the public sphere through communication media, structured by those who have the power and the means—and not only economic means—to impose or neutralize a certain definition of reality (see Tellechea Rodríguez 2004; Luque 2006).[51] A glaring example is represented by the action of the local press—supported by the chemical industries—that, faced by the mobilization and the fight for the social imagery realized by the Mesa de la Ría, has, on various occasions, denied that the Punta del Sebo was ever an important leisure area for the inhabitants of Huelva (Luque 2006).

- *Work versus health and environment.* The first factories were built in the early 1960s, and within a few years the inhabitants of the area were living in the aftermath of this drastic transformation of the territory. The "flow of pleasure" (Halsey 2006, p. 52) achieved with economic development has long since been overrun and thwarted by real flows of pollution and contamination, as many of the inhabitants of Huelva have realized, leaving them unable to decide whether the plant has been more beneficial or damaging for them. Like many socio-environmental conflicts, the one in Huelva, between those who are for the factories and employment and those who speak up for health and the environment, concerns much more than the single objects of contention (employment, environment, health) and raises important questions regarding who they—the inhabitants of that territory—are (subjectivity), what they can do (power), what they can know (epistemology), and who they might become (desire) (Halsey 2006, p. 4).
- *Environmental perceptions, denial, and time–scape.* The most commonly recurring beliefs concerning the perception of contamination and of its risks fell into the following pattern: "We know about the pollution, but …" This premise is then followed by various explanations to justify inaction in the face of such awareness, including: (1) the mechanisms of denial (Cohen, 2001), which, together with habit, combine to mellow the drama of the reality endured,

[49]See also Ruiz-Ballesteros et al. (2009). The authors develop an interesting study about the transformations occurring in the representations and social perceptions of the environment in certain areas of Andalusia, following the closure of the mines.

[50]On this, see also Waitt (2010, pp. 235–238), and Altopiedi (2011, p. 116).

[51]Starting from a discourse analysis that refers to Michel Foucault's thinking, the geographer Waitt (2010, p. 239) writes: "it is crucial to understand that while discourses may manifest themselves in ways that bring order to social life as rules, maxims, common sense, or the norm, they are always unstable and may be ruptured. Discourse analysis requires remaining alert to such instability, ambiguity, and inconsistency. Well-conducted and thoughtful discourse analysis enables insights into the resilience and rupture of multiple and sometimes conflicting discourses that give meaning to our everyday lives."

allowing people to familiarize themselves with it and adapt to it, making it more acceptable and ever less detectable; (2) the huge expansion of the temporal horizon concerning the contamination phenomena, which, stretching beyond human and industrial times, contributes to evaporating and dispersing the perception of risk (Adam 1998, p. 10); (3) the doubt and the uncertainty of the "reality" of contamination (Auyero and Swistun 2009), which impel people to unceasingly weave the web of a vanishing reality. Moreover, during the unending process of interpreting and defining the reality of pollution, the body becomes for many the main tool for knowledge (Auyero and Swistun 2009) to prove to oneself and to others the rationality of one's convictions about the seriousness of the situation.

- *Differential victimization and the experience of environmental injustice*. In confronting the reality and extent of the correlation between contamination and health problems, a significant narrative emerges in the social accounts of this complex situation, relating to the different forms of victimization that pollution can cause. Some people highlight the equal distribution of the damaging effects among all the inhabitants of Huelva, regardless of proximity to the industrial area; others, instead, trace precise qualitative differences, according to individual state of health, holding that those already carrying diseases were affected more seriously by the damaging effects of pollution (see Williams 1996). If this victimization is perceived as equal or differentiated, the theme of injustice seems to emerge clearly in the reflection that what is happening in Huelva would not have happened in another place—the recurring question being typical of any (collective) victim: "Why us?" "Why right here, in our backyard, and not somewhere else?"

Along these lines, the socio-environmental disaster of Huelva may, indeed, be described as a truly creeping disaster (see Williams 1996), developing in "slow motion" (Auyero and Swistun 2009, p. 4; Natali 2015a, 2016). These expressions are appropriate to highlight the extended temporal dimension and the slow pace—at least with respect to the rhythms of individual biography—with which the harmful and destructive effects have pervaded the environment and the bodies of those who inhabit it. "Green" and cultural perspectives on the processes of environmental victimization prove to be extremely important in showing that experiences of environmental suffering are not simply individual: they are actively created, starting from the position that the residents as a group occupy in the wider social and cultural macrocosm (Auyero and Swistun 2009, p. 159). In particular, focusing on the mechanisms of denial (Cohen 2001; Pulcini 2013) may contribute to understanding silence, apathy, and a range of other possible responses by those who witness daily the destruction of the environment that they inhabit (Williams 1996; Brisman 2012; Natali 2013a; White and Heckenberg 2014, pp. 186–192). In fact, in the face of the dramatic environmental transformations that shake their daily life, environmental victims sometimes learn to accept irreparably altered landscapes, and sometimes they simply "delete" them, as one does with an illness or death (Settis 2010, pp. 73–74; Natali 2015b).

5 Some Relevant Questions Useful for Approaching Historical Pollution

The peculiar phenomena of "historical pollution", only briefly sketched out here, clearly show the traces of an uncomfortable past, impossible to eliminate, its other, dark, side intruding over and over again into the present. Just when everything seems to be finished, the "behind the scenes" of production processes reappears, unexpectedly, back on stage.

After the description of these three scenarios, it should be clear that in order to deal adequately with the toxic legacies left by the industrial past it is necessary to turn our attention to:

(1) changes in production processes over the course of time (industrial history);
(2) areas where the industries and their waste are located (geographical dimension);[52] and
(3) changes in social awareness and in environmental legislation.

A historical–geographical reconstruction is essential because over the course of a century of industrialization, production processes (and the goods being produced) have changed, and with them the residues and the waste, and therefore the various types of pollution and contamination. In other words, through time and in different places, the kinds of contamination have altered according to technological and production changes. To offer just one example, up until the 1950s all chemical production was based on the use of coal, so the characteristics of chemical industry waste depended on the characteristics of coal. These characteristics changed when petroleum took the place of coal as a raw material.

To fully address these complex issues, one would have to retrace the industrial history of the country considered from the beginning of the last century to the present, taking into account the varying cultural and social environments, the different legal contexts, and the multiple social sensitivities that went with them.[53] This huge task falls outside of the specific focus of this criminological exploration. A few, short observations of historical and social character seem to be appropriate, however.

Beginning in the 1800s, the processes of industrialization transformed societies and economies in radical ways. This transformation, which up to a certain point remained gradual, was perceived by those living at that time as "overturning" the world they knew up to then. It was a truly momentous change in the human–

[52]From this viewpoint, "the possibility of capturing within a unitary vision the sense of the problems of space can only come from geography" (Muscarà 1967, p. 16). The *esprit géographique* will prove useful as a "call back to the concrete in the face of the abstract fragmentation of systematic sciences, a call back to unity in the face of the fictitious separation between man and environment, finally a call back to the complexity of inter-dependence in the face of the division of the earth's surface into 'islands' of space" (Muscarà 1967, p. 19).

[53]On this, see Ginsborg (1989); see also Amatori and Colli (1999).

environment relationship (Neri Serneri 2005, pp. 57–58). All at once, industry started to pour the waste and residue of manufacturing processes into the environment, in ever-increasing quantities, "proportionate to a production capacity undergoing continuous expansion" (Neri Serneri 2005, p. 59):

> It consisted largely of solid materials, often stinking, full of toxic minerals or sulphurous residues: they were usually thrown out in heaps upon the ground in large quantities, till they became small hills, and, in time, they released the toxic substances which, mixing with rain and surface waters, soaked into the ground and poured into the waterways. Similarly harmful were the discharges dispersed into the air as fumes and vapours, often acrid, irritating and stinking, most commonly because they contained the sulphurous residues produced by the combustion of coal and the hydrochloric residues of the soda industries, the basic produce of the original chemical industry, used to make bleach and soaps for the textile industry. The picture was completed by liquid discharges which, together with wash waters, poured out into rivers and canals huge quantities of acid and alkaline substances and of toxic metals, among which were arsenic, chromium and silver, widely used in the metallurgical and chemical industries. (Neri Serneri 2005, p. 60)

The introduction of industry into the environment thus developed along particular trajectories:

- There was a dichotomy between "inside" (the factory) and "outside" (the environment from which the natural resources were taken and into which the residues were discharged).
- The legislations that dealt with such questions from the first decades of the 1800s, in France and Germany, had little effect: "Mostly they limited themselves to order that the toxicity to be tolerated was quantified according to the so-called 'state of the technology'—that is of the available technology—accepting, in fact, the existing situation" (Neri Serneri 2005, p. 61).
- The limitations of these countermeasures were mainly due to the following factors: the interests seen as priorities were those of whoever derived an income from the industries (ideological and economic priority of industrial development); scientific and technological knowledge did not perceive the specifically ecological dimension of the problem; the idea prevailed that it was sufficient to *remove* the toxicity of the industries from the towns (moving them out into the country or dispersing the fumes using high smokestacks, for example) to neutralize them and make them harmless.
- The main concerns were about health, hygiene, and aesthetics, not environment and ecology.
- More generally, a "tunnel vision" was prevalent in regard to the paradigm of progress and to a providential image of nature, capable of absorbing the waste of industrial manufacture and infinitely mouldable by man (Neri Serneri 2005, pp. 62, 65).

It is possible to explore and reconstruct such historical stratifications in many contemporary landscapes: in fact, they appear dynamic and heterogeneous, as they imply history and conflict, the material dimension and the structure of feelings and of awareness, the concrete links between cultural forms, power, ideology, and imagination. All these elements "are, literally, *crossings* that leave footprints" (Bonazza 2011, p. 79).[54] In this way, the historical–sociological and geographical approaches can be combined in order to provide a qualitative account of the genealogy of historical pollution in the different scenarios of the planet.

6 An Opening to the Ambiguity of the Observed Phenomena

In addressing historical pollution, one question has silently but unceasingly accompanied our exploration throughout: through what processes does the "gift" (industry) change into "poison", assuming the ambiguous form of a "deadly gift" and thus becoming something disastrous on the social as well as the environmental level?[55]

Different historical times often become part of the collective imagination and memory through the form of the oxymoron: when we reason, in the present, about different periods of our personal and collective history, we often tend to level out the most diverse temporal dimensions in order to reassemble them into a coherent picture, eliminating anything that seems to contradict the linearity of the observed historical period. This is a "normal" mechanism through which we try to order, select, remember, or forget—practicing various forms of denial, whether institutional, social, or personal (Cohen 2001).[56]

Forgetting the many and distinct narratives that developed around our industrial story, however, does not help us to see clearly either the present or the future that lies in wait. On the contrary, it risks further exacerbating any binary opposites and monolithic dilemmas (environment and health versus jobs, for example) that hinder alternative and innovative visions and solutions. To overcome this frozen and one-dimensional narrative, it is important to adopt a "time–scape" perspective that will distinguish, and at the same time connect, the different times—industrial times, juridical times, times tied to social perception, and the times of nature. Such a perspective will avoid the pitfall of easy simplification.

Thus, in approaching historical pollution scenarios, it seems crucial to concentrate our focus on those tensions and frictions that arise between different temporal

[54]The following observations by Bonazza (2011, p. 6), with reference to cultural geography, are also valid in a criminological perspective: "each point is determined by geographical, ideological, political, economic coordinates: this way what I see (or represent) depends on the vantage point from 'where' I see (or represent)."

[55]The gift is also what is dangerous to accept (Mauss 2002, p. 109). In fact, it is never simply free: the donor always expects a return gift. Somewhat predictably, the demonstration of superiority and power expressed by the donor is a counterpoint to the recipients making themselves smaller and more subordinate, especially when the gift is one that cannot be refused. This dependence towards the donor—a dependence, moreover, that gains tenacity over time—permeates many cases of historical pollution.

[56]On this, see also Brisman and South (2015), Kane and Brisman (2013), and Natali (2013b).

fractions ("broken" or "fragmented" time), born of the meeting between a past that enjoyed a sort of environmental naivety and the recent awareness of the serious consequences of environmental contamination—an awareness that has transformed our own reflexivity. We are faced with different times that unfold at different speeds: on the one hand, the speed of industrialization and of its momentum towards an apparently bright future, and, on the other, the speed of today, which has to deal with the residue of that initial impetus.

Regarding these aspects, Neri Serneri (2005, p. 43) speaks of the "incorporation" of nature to underline how the social and natural worlds, even though they are distinct areas, live at the same time within each other, participating in the same dynamics. Even more significant for my discourse is the following:

> The more intense the ways of "incorporation" of nature into social dynamics, the more the ecosystems' assets are modified by them. "Environmental crises" are born of conflicts between the diverse speeds of change of social systems and ecosystems, in particular of those with which the social systems are in more direct contact. (Neri Serneri 2005, p. 44)

It is indeed the intensity, speed, and spatial extension of these processes that marks the historically new character of today: for the first time, industrial technologies have allowed the separation of production processes from the rhythms of nature. The asynchrony between historical and biological times is thus pushed towards catastrophic levels (Neri Serneri 2005, p. 57). In particular, a decisive passage in the modalities of management and exploitation of natural resources was produced by changes in industrial technology—specifically with the development of the mechanical, chemical, and electro-technical industries—during the second industrial revolution (Neri Serneri 2005, p. 39). To confront these temporal frictions, we will need theoretical approaches and legal instruments that are capable of grafting new outlooks onto different profiles of responsibility for the consequences of actions placed in the past, in "history".

To all this, another level of complexity is added by the fact that the expression "historical pollution" contains within it meanings that present a certain amount of ambiguity and uncertainty. In the end, it is possible to say that it is the very phenomenon called "industrialization" that presents a high level of complexity and ambiguity:

> Everybody is prepared to admit that it is a process, that is, a dynamic phenomenon. However, as soon as one tries to find its place, and its sense, within historical times—not to mention a value judgment—the differentiations emerge in an instant and in the full light of day. (Mori 1977, p. 15)

7 What Are the "Discovered" Sensitive Points of a Hidden Geography?

At this point, in my analysis the following question is to be posed and answered: What clues originate from socio-criminological research on historical pollution in a global context? Exploratory research on this phenomenon seems to show some recurrent and common themes.

First, the harmful consequences do not materialize immediately, but only over extended periods of time.[57] When the environmental problems prove to be progressive and accretive rather than being immediately catastrophic, then our ability to perceive and read them fails and is no longer a reliable guide to understanding the phenomenon under observation. In such cases, it is important to employ a complex temporal perspective. As previously outlined, environmental contamination is the result of processes that occur over varying periods of time; it is the outcome of several polluting behaviours repeated through time that may be illegal, but may also be legal so long as they remain within the authorized limits, and yet still harmful to the environment. In the majority of these contexts, even if the polluting behaviours were to be immediately stopped, their harmful effects would still continue to be produced, because of the long periods of latency[58] of the illnesses with which they can be linked.

Second, causation can be extremely difficult to prove, both because of the temporal distance between the behaviour and the production of its harmful consequences (see also Hall 2014, p. 100) and because of the complexity of the social and ecological context representing the background scenario of historical pollution. The consequences may be *visible* today, but their *origin* goes back to a (remote or near) past that is very difficult to appreciate in its reality. Instead of an "origin", we should speak of a *process* which has unfolded itself over time, generating complex reactions and counter-reactions within the concerned environment. Among some of the obstacles on the path to recognition of an environmental harm could be nature's capacity to absorb the harm and reveal the most destructive and visible consequences only after an extended period of time. Moreover, even when the harm is already visible and evident, attributing a cause can be extremely complex—one has only to think of the link between exposure to chemical agents and the illnesses of workers and/or inhabitants of the contaminated areas.[59] Also, some powerful actors may exercise their influence to delay and/or mitigate the law and institutional response (Szasz 1994).

Third, phenomena of historical pollution are often described as "victimless crimes", because the victims are not easily identifiable[60] and sometimes cannot speak for themselves (as is the case with forests and non-human animals). Those who live in contaminated habitats may offer "multiple, confused, and contradictory"

[57]In different contexts, environmental harm occurs quickly, but their effects are felt immediately and for a long time afterwards. An example might be the Deepwater Horizon oil spill.

[58]"Latent" describes something that has not yet clearly manifested itself. In particular, latency is the period that intervenes between, and links, a stimulus (cause/origin) and the response to the stimulus (effect).

[59]On the basis of the work of Slapper and Tombs (1999), Altopiedi (2011, p. 100) underlines how, in corporate crimes (one of the categories to which certain destructive actions on the environmental level can be attributed) the space–time distance between the action and its harmful consequences can be great, entailing "significant implications in terms of awareness and of proof of victimization". See also Cottino (2005).

[60]See Bisschop and Vande Walle (2013, p. 40); see also Geis and Goff (1987, p. xviii).

points of view about it (Auyero and Swistun 2009, p. 65). This “nebulosity” also invests the effect of the contamination upon people’s health and on the natural environment. The ensuing confusion seems linked strictly to different variables: the constitutive uncertainty regarding the toxicity of some substances present in the environment; discursive spheres dominated by those social actors—businesses, the State, mass media, local and national newspapers, local and national authorities—who have the power to achieve an effective work of obfuscation and confusion concerning the issue of contamination; and finally, the relational and physical anchoring of the perception of risk on the part of the residents (Auyero and Swistun 2009; Natali 2010, 2014). Moreover, environmental victimization is anything but a neutral process from the societal point of view: some groups of people suffer more than others when other conditions of social vulnerability are added (White 2013a; see also Brisman 2014b, 2015b). White (2011, p. 110) reminds us that there are differential risks—and therefore, differential victimizations—within the same “at risk” population.[61] In these circumstances, the relevant question will be not only how many people will be harmed but, above all, *who* will be, as well as who will not.[62]

Finally, if it is true that it is extremely difficult to go from the “the mute physical fact of damage” that constitutes environmental harm to its social, political, and juridical salience—a salience understood as the measure of how much an issue truly matters to the social actors involved (Szasz 1994, pp. 30, 40)—it becomes crucial to facilitate the social and political recognition of eventual harms linked to the production process (Natali 2015b).[63] This result may be achieved by shedding light on that “twilight zone” where environmental harms emerge, come to life, as facts not yet in existence in the social and institutional sphere.[64]

8 The Interdisciplinary Relevance of the Process of Definition

Faced with the great complexity that characterizes human–environment interactions, anthropologists, city-planners, criminologists, economists, geographers, historians, lawyers, and sociologists often employ diverging languages and models of

[61]On this aspect, see also Brown (1991) and Bullard (1990).

[62]This is evident in relation to variables such as age, gender, or health. See also Williams (1996), Hall (2013), and Abignente and Scamardella (2013, pp. 66–72).

[63]In this regard, Altopiedi (2011, p. 94) again underlines how is it necessary to carry out an adequate revaluation of the victimological perspective in relation to corporate crime, analysing in detail “careers of victimization”.

[64]Activism from an environmental justice perspective plays a decisive role, in making visible the issues tied to the phenomena described (Szasz 1994, p. 31). Szasz (1994, pp. 165–166) underlines how the grassroots activism of US social environmentalist movements has extended the demographic basis of environmentalism, teaching people who were initially only concerned about the near and immediate threat they perceived to re-frame their problem in a much wider context.

interpretation (Settis 2010, pp. 287–288). There are many scholars, however, who point out that it is really not possible to start from a single disciplinary approach in dealing with the social and political complexity of our relationship with the environment (see Lanzillo 2013, p. v). An interdisciplinary approach—one which refers to a possible communication among disciplines—will also therefore be particularly useful in exploring specific environmental scenarios of criminological relevance (Hall 2014). The expansion of the criminological imagination into the field of green criminology (White 2003) will help to incorporate into our criminological and juridical narrative a past that is still present in its most harmful effects on both the social and environmental levels, with a full awareness that a great part of the socio-environmental harms and threats caused through time is inscribed in historically situated lifestyles and models of production (Szasz 1994).

Criminological research has highlighted the ways in which social acts are labelled "deviant", "harmful", or "criminal" according to the historical period, the geographical location, and the social world under examination. In fact, about the same event, different and sometimes conflicting definitions, rooted in different social and juridical times, may exist (see also Natali 2015a).[65] Recognizing these temporal differences also means articulating a certain value judgment.[66] With specific regard to historical pollution, the polluting activities could have been, *at the time*, completely legal or not subject to any regulation, even if *today* they would be considered illegal (see also Brisman 2015a). The temporal perspective thus takes on a critical relevance in the interplay of multiple, intimately linked levels of reflection: time and responsibility, time and complexity, time and crime are only some of the pairs that can bring new life to a theoretical outlook that is adequately attentive to the environmental challenges of our times.[67] In this context, enhancing the multiple narrative capabilities of the criminological imagination will allow us to keep alive the connections between micro-stories and macro-stories,[68] that is, to keep together the biographical dimension of individual lives and the historical legacy of the great industrial transformations.

---

[65]These aspects are highlighted by critical, cultural, and interactionist approaches to crime and deviance.

[66]According to our shared definition, "historical pollution is characterized by the presence of a *distance in time between the polluting behaviors and the emergence of harm*" (Chap. 1, in this volume).

[67]On this, see Adam (1998, 1999), and Leccardi (2009). Resta (2008, p. 182) writes: "our time: it belongs to us while we belong to it, it binds us and we bind it, it shapes us and we shape it. It defines our expectations and it constructs the space of our experience; it is made of many times that intersect within it and are not measured in the same way. They decompose and re-compose in a cubist picture". From this angle, all systems live off a "contradictory and paradoxical plurality of temporality" (Resta 2008, p. 184).

[68]See also Ferrell (2013, pp. 350–351) and http://www.studiculturali.it/dizionario/lemmi/microstoria.html. On the Love Canal case, Levine (1982, p. 5) highlights the point that "the sociological imagination helps to show us how individual troubles and individual behaviors reveal the larger social world".

The basic idea behind this chapter is that the historical reconstruction of a "complex"[69] event—which we have called "historical pollution"[70]—cannot happen only by way of definition (however, appropriate) but through empirical investigation and socio-historical exploration. Therefore, it seems crucial to promote a "cross-pollination between legal scholarships (and especially scholarship on criminal law) and more sociologically-derived accounts of offending and the criminal justice system in its operational context" (Hall 2014, p. 96).[71] This idea is accompanied by the conviction that when one chooses a perspective it is not because it is the only correct one, the only right way of seeing the situation, but because this way of considering things lets one see things that from a different perspective remain hidden (Becker 1982; see also Natali 2015a). In such fluid scenarios, the exploration of the single cases investigated will throw light upon the main aspects of the defining proposals developed, which in their turn will interact with the empirical scenarios, thus starting an interdisciplinary process of co-construction.[72]

More importantly, the judgment that allows us to define an event as "historical pollution" cannot take place in a vacuum. Like any other process of definition, it develops within a context that is already socially, juridically, and scientifically constructed.[73] A possible definition of "historical pollution" will have to be rooted in the changed environmental sensitivity typical of the time defining it; insofar as it concerns our late modern times, it will come to maturity following the new experiences of destruction and vulnerability linked to human manipulation of the environment. Within the complexity of "post-normal science" horizons (Tallacchini 2005, 2012), it seems crucial to continue taking a stance on the reality status of events as elusive and hazy as environmental events. In this sense, from an eco-global criminological perspective (White 2011, p. 14), considering the issues of environmental (and global) harm does not only mean looking at specific cases of

[69]On the paradigm of complexity, see Morin (1993, 1999).

[70]See above in this chapter; see also in this volume, Chap. 3.

[71]More importantly, "in the field of green criminology other areas of law (especially civil and administrative law) are also integral, and perhaps pivotal. This is largely because to some extent criminal law and criminal justice systems tend to be ill-suited to the specific problems and features of environmental 'offending' and victimisation" (Hall 2014, p. 99).

[72]If it is true, ultimately, that the law always has to take into account the reality that it intends to regulate and within which it chooses its own relevant objects, then in order to accomplish this approach to the "world of facts" it will be necessary, first of all, to go through disciplines which are outside the law, establishing a rapport with fields of knowledge which focus on significant amounts of the world of experience with reference to the knowledge of our times. See also Pulitanò (2006, p. 800).

[73]This is a theoretical position that calls directly upon the "social constructionism" that assumes a peculiar relevance in the environmental field (White 2008, p. 33). In any case, what we "name" socially with the expression "nature" *exceeds* our attempt to define it, and goes beyond the social worlds themselves, with *their* meanings, with *their* times. Therefore, if is true that we can only intervene on a socially constructed level, it must be done always keeping in mind what is beyond it and inevitably escapes us (Natali 2015a).

environmental destruction; it also, and above all, means building new backgrounds for reading the world in which we live (White 2010, pp. 5–6; Natali 2015a). This proposal is essential to an analysis of environmental harms built upon new theoretical perspectives that promote complex thoughts which are in tune with the new (environmental) needs of a changing world (South 1998, pp. 225–226).

These new areas of investigation are opened up to criminology by the green perspective with which our discipline is attending to the disastrous relationship between humans and the environment—including the harmful consequences of behaviours far back in time. Specifically, the importance of turning our attention to the phenomenon of historical pollution and studying it in depth arises from the acknowledgement of the scarcity—in the field of green criminology as well (White 2013b)—of adequate cognitive and learning tools to approach this dynamic reality. Furthermore, from a criminological perspective, the issue of historical pollution fits directly into the patterns set out, through time, by economic-productive processes, and cannot therefore be judged on the basis of traditional notions of crime. This particular point of view challenges the common idea that the "real" issues of crime and society pertain only to the lowest social levels, and instead gives new importance to political economy in the analysis of criminal behaviours (see Greife and Stretesky 2013; Lynch et al. 2013; Jarrell et al. 2013, p. 440; Lynch 2015).

9 Assessing the Perceived Seriousness of Environmental Crime: New Paths in Criminological Research

Studies like the present one are also crucial for exploring the polyhedral nature of the concept of the gravity of the crime, which cannot be reduced to the legal definition, which says very little about the gravity that the behaviours under consideration can in reality assume. To this is added another level of complexity, given that the expressions "environmental crime" and "environmental harm"—from a criminological perspective—include meanings that present a certain ambiguity and vagueness of definition (see also Hall 2014; Potter 2015); hence, the importance of probing the intricate tangle of the social and natural worlds called into question by the empirical dimension of the phenomenon. To this end, it seems essential to adopt a dynamic conception of society, which recognizes the mobility of social values with respect to the norms and sanctions provided: above all in the environmental field, the valuations of legislators and judges do not always necessarily agree with the social conscience of our day.

More generally, an early criminological contribution that tried to "measure" the social perception of the gravity of crimes was that of Sellin and Wolfgang (1964).[74]

[74]In 1964, they published their work *The Measurement of Delinquency*, which led to the creation of the National Survey of Crime Severity (NSCS). That study, as with the numerous studies that followed, highlighted a general (normative) consensus concerning the gravity of crimes,

In Italy, Sellin and Wolfgang's study was repeated in the 1980s by Delogu and Giannini (1982), with a sample of 1600 subjects (students, police, magistrates, ordinary citizens, religious, convicts, members of parliament). A subsequent study on white-collar crime by Wolfgang et al. (1985) also included some scenarios that clearly suggest environmental crimes: for example, a factory that pollutes the water sources of a town, causing the death of twenty people. Studies in the late 1980s and 1990s took into account some socio-environmental consequences of harmful behaviours ("polluting factories causing assorted amounts of harm"), but largely left environmental issues in the background.

Only recently have we witnessed the publication of the first studies focused, in particular, on the social perception of environmental crime (Martín et al. 2013) and of punishment for environmental harms (Shelley et al. 2011).[75] Continuing this trail, we might pose some of the following questions: Are environmental crimes perceived as serious crimes? How are they evaluated, also with respect to "non-environmental" crimes? Is there a (normative) consensus in relation to their gravity, or does the social perception of the gravity vary according to socio-demographic characteristics? More specifically, are social groups more exposed to environmental harms (differential victimization) also more inclined to perceive such crimes as serious, compared to other groups? For example, a person who judges as less serious certain environmental crimes might do so because of a greater sense of safety derived from the possibility of *isolating* themselves from the contaminated places—of neutralizing the danger in a manner akin to environmental sociologist Andrew Szasz's notion of "inverted quarantine" (Szasz 2007). As a possible new path in future criminological research on historical pollution, I suggest taking into account the way social perceptions modulate themselves according to the qualitative peculiarities of the socio-environmental harms; to explore the question of the normative consensus or disagreement on the gravity of crimes involving the relationship between humans and the environment; and to investigate the tendency of the public to support sanctions of a penal type rather than administrative or civil measures of control in relation to the non-desirableness of certain socially and environmentally harmful behaviours. Understanding the ways in which people perceive and evaluate environmental issues of criminological relevance is a crucial means of extending the experience, information, and data

(Footnote 74 continued)

irrespective of ethnicity, gender, age, or social class. In brief, violent crimes were considered more serious than crimes against property, and the latter, in their turn, were seen as more serious than "victimless" crimes. A wider normative consensus corresponded with a higher degree of severity. See also Forti (2000).

[75]To date, any studies that have focused on the social perception of environmental crimes and harms have been primarily quantitative. The main methodological questions addressed by this research have essentially concerned three aspects: the criminal facts to submit to social evaluation, the composition of the samples of the subjects to be interviewed, and the measurement scales to be used.

needed when one has to imagine, plan, and put into practice public and governmental environmental policies fit for the challenges that await us.

In the contemporary horizon of our late modern society, it has become increasingly clear that industry (in its different forms) has generated economic growth and employment as well as, over the same period of time, highly contaminating industrial waste. Considering that there is no eschatological way out of the described dilemmas, the questions then become: What levels of risk and potential consequences are we prepared to tolerate? Which goods are we ready to sacrifice and in what measure? And again: who is responsible for the present situation? How to confront *what is left* of our industrial past? In this chapter, I have suggested that it is not possible to answer these unexpected temporal juxtapositions, which often produce contrasting oxymorons that are difficult to approach, without taking into account a century of industrial history and its ambiguous legacy. Facing the damages produced by this temporal disjunction between "cause" and "effects" is a task that can no longer be postponed. Naturally, if it is indubitable that the recognition of a specific historical pollution and contamination scenario must not impose today's consciousness and knowledge on yesterday's—a gap that we must still bridge—it is also certain that such recognition should not become a way to avoid responsibility, *shifting* it onto others (see also Gaarder 2013).

In conclusion, a criminological approach aimed at carrying out an in-depth analysis of the phenomenon of "historical pollution" will have to tune into and synchronize with the rising importance of these criminal scenarios, until recently largely overlooked. A work of this kind will help to collect the traces of an ignored geography and to imagine policies that take into account its complex physiognomy (see also Franzini Tibaldeo 2013). Only if we can promote a rich and interdisciplinary dialogue between different fields of knowledge (particularly between green criminology,[76] environmental history, and legal scholarship) in order to *make visible* the multiplicity of historical pollution phenomena that emerge at different times and speeds all over the planet, will we be able to see clearly that "the past is never dead" and that, ultimately, "it's not even past" (Faulkner 1951, quoted in Colten and Skinner 1996).[77] These are points and open questions that a policy-oriented criminology, theoretically robust and at the same time empirically grounded, may suggest.[78]

---

[76]On the role of green criminology, see in particular Brisman (2014a, p. 29).

[77]Colten and Skinner (1996, p. xi) write: "Although Faulkner never had to concern himself with the issue of hazardous waste, his observation aptly applies to this topic. For seldom has society had to grapple with a more enduring problem than long-lived deposits of industrial residue."

[78]In terms of criminal policy, it is necessary to abandon the misleading belief that, in order to arrest "the alliances, personal habits and institutional routines that sanction a more generalized mode of environmental decay" (Halsey 2004, p. 837), it would be enough to capture the "environmental criminals". Along these lines, Hall (2014, p. 98) remarks: "the idea that any law (much less criminal law) can or should constitute the *sole* solution to the problems of environmental harm is surely wrong. For legal commentators, the difficulty with a field that is apparently so wide is that it sits uncomfortably with classic doctrinal legal ideals of certainty and predictability". More concretely, "the legal debate may boil down to a basic question of what combination of civil,

References

Abignente, A., & Scamardella, F. (2013, October). Rischio e catastrofe. Il fallimento di scienza ed istituzioni: trovare soluzioni precarie in una vita precaria. *Governare la paura. Journal of interdisciplinary studies. Special Issue. Fear of Nature. The Government of Catastrophe between Human and Social Sciences*, 53–89.

Adam, B. (1998). *Timescapes of modernity: The environment and invisible hazards*. London: Routledge.

Adam, B. (1999). La responsabilità e la dimensione temporale della scienza. In C. Leccardi (Ed.), *Limiti della modernità. Trasformazioni del mondo e della conoscenza*. Roma: Carocci.

Altopiedi, R. (2011). *Un caso di criminalità d'impresa: l'Eternit di Casale Monferrato*. Torino: L'Harmattan Italia.

Altopiedi, R. (2013). Azione collettiva e costruzione della vittimizzazione. Il caso Eternit. *Studi sulla Questione Criminale*, *8*(1), 35–57.

Amatori, F. (Ed.). (2000). *L'impresa: una prospettiva storica*. Milano: Egea.

Amatori, F., & Colli, A. (1999). *Impresa e industria in Italia. Dall'Unità a oggi*. Venezia: Marsilio Editori.

Armiero, M., & Barca, S. (2004). *Storia dell'ambiente. Una introduzione*. Roma: Carocci.

Auyero, J., & Swistun, D. (2009). *Flammable. Environmental suffering in an argentine shantytown*. Oxford, New York: Oxford University Press.

Ayres, I., & Braithwaite, J. (1992). *Responsive regulation: transcending the deregulation debate*. New York: Oxford University Press.

Beck, U. (2009 [2007]). *World at risk*. Cambridge: Polity Press.

Becker, H. (1982). *Art worlds*. Berkeley, Los Angeles, London: University of California Press.

Benach, J., Yasui, Y., Martínez, J. M., Borrell, C., Pasarín, M. I., & Daponte, A. (2004). The geography of the highest mortality areas in Spain: A striking cluster in the southwestern region of the country. *Occupational and Environmental Medicine, 61*(3), 280–281.

Bisschop, L. (2010). Corporate environmental responsibility and criminology. *Crime Law Soc Change, 53,* 349–364.

Bisschop, L., & Vande Walle, G. (2013). Environmental victimisation and conflict resolution: A case study of e-waste. In R. Walters, D.S. Westerhuis, & T. Wyatt (Eds.), *Emerging issues in green criminology. Exploring power, justice and harm*. London: Palgrave Macmillan.

Blum, E. D. (2008). *Love Canal revisited. Race, class, and gender in environmental activism*. Kansas: University Press of Kansas.

Bonazza, A. (2011). *Manuale di geografia culturale*. Roma-Bari: Laterza.

Brisman, A. (2008). Crime-environment relationships and environmental justice. *Seattle Journal for Social Justice, 6*(2), 727–817.

Brisman, A. (2012). The cultural silence of climate change contrarianism. In R. White (Ed.), *Climate change from a criminological perspective*. New York: Springer.

Brisman, A. (2014a). Of theory and meaning in green criminology. *International Journal for Crime, Justice and Social Democracy, 3*(2), 21–34.

Brisman, A. (2014b). The visual acuity of climate change. In P. Davies, P. Francis, & T. Wyatt (Eds.), *Invisible crimes and social harms* (pp. 61–80). Basingstoke, Hampshire, UK: Palgrave Macmillan.

(Footnote 78 continued)

administrative, mediation-based, criminal justice or other legal—versus extra-legal—approaches to the issue of environmental harm will minimize the risk of such harm occurring or reoccurring" (Hall 2014, p. 103). See also Ayres and Braithwaite (1992), Natali (2015a), Bisschop (2010), and Forti (2003, 2007).

Brisman, A. (2015a). Environmental harm as deviance and crime. In E. Goode (Ed.), *Wiley handbook on deviance*. Hoboken, NJ: Wiley.

Brisman, A. (2015b). "Multicolored" green criminology and climate change's achromatopsia. *Contemporary Justice Review, 18*(2), 178–196.

Brisman, A., & South, N. (2013a). Introduction. Horizons, issues and relationships in green criminology. In N. South & A. Brisman (Eds.), *Routledge international handbook of green criminology*. London and New York: Routledge.

Brisman, A., & South, N. (2013b). Conclusion. The planned obsolescence of planet Earth? How green criminology can help us learn from experience and contribute to our future. In N. South & A. Brisman (Eds.), *Routledge international handbook of green criminology*. London and New York: Routledge.

Brisman A., & South N. (2014). *Green cultural criminology. Constructions of environmental harm, consumerism and resistance to ecocide*. London and New York: Routledge.

Brisman, A., & South, N. (2015). State-corporate environmental harms and paradoxical interventions: Thoughts in honour of Stanley Cohen. In R. A. Sollund (Ed.), *Green harms and crimes: Critical criminology in a changing world*. Basingstoke, Hampshire, UK: Palgrave Macmillan.

Brisman, A., & South N. (In Press). Consumer technologies, crime and environmental implications. In M. Maguire & T.J. Holt (Eds.), *The handbook of technology, crime and justice*. London and New York: Routledge.

Brisman, A., South, N., & White, R. (Eds.). (2015). *Environmental crime and social conflict. Contemporary and emerging issues*. Farnham: Ashgate.

Brown, P. (1991). The popular epidemiology approach to toxic waste contamination. In S. Couch & S. Kroll-Smith (Eds.), *Communities at risk: Collective responses to technological hazards*. New York: Peter Lang.

Bullard, R. (1990). *Dumping in Dixie: Race, class, and environmental quality*. Boulder: Westview Press.

Cardano, M. (2011). *Ricerca qualitativa*. Bologna: il Mulino.

Centemeri, L. (2006). *Ritorno a Seveso. Il danno ambientale, il suo riconoscimento, la sua riparazione*. Milano: Mondadori.

Centonze, F. (2004). *La normalità dei disastri tecnologici. Il problema del congedo dal diritto penale*. Milano: Giuffrè.

Ceretti, A., & Natali, L. (2009). *Cosmologie violente. Percorsi di vite criminali*. Milano: Raffaello Cortina.

Cohen, S. (2001). *States of denial: Knowing about atrocities and suffering*. Cambridge: Polity Press.

Colten, C. E., & Skinner, P. N. (1996). *The road to Love Canal: Managing industrial waste before EPA*. Austin: University of Texas Press.

Coltorti, F. (2011). Il ruolo dell'industria: grandi e medie imprese. In L. Paolazzi (Ed.), *Libertà e benessere in Italia. 150 anni di storia unitaria e i traguardi del futuro*. Roma-Bari: Laterza.

Cottino, A. (2005). *Disonesto ma non criminale*. Carocci: La giustizia e i privilegi dei potenti. Roma.

Davies, P., Francis, P., & Wyatt, T. (Eds.). (2014). *Invisible crimes and social harms*. Basingstoke, Hampshire, UK: Palgrave Macmillan.

De Luigi, G., Meyer, E., & Saba, A.F. (1995, January–March). Nasce una coscienza ambientale? La Società italiana dell'alluminio e l'inquinamento della Val Lagarina (1928–1938). *Società e Storia*, 75–110.

De Simone, E. (2014). *Storia economica*. Milano: FrancoAngeli.

Delogu, T., & Giannini, M. C. (1982). *L'indice di criminalità di Sellin e Wolfgang nella teoria generale della misurazione di gravità dei reati*. Milano: Giuffré.

Dueñas, C., Liger, E., Cañete, S., Pérez, M., & Bolívar, J. P. (2007). Exhalation of 222Rn from phosphogypsum piles located at the southwest of Spain. *Journal of Environmental Radioactivity, 95*(2–3), 63–74.

Eco, U. (1983). Corna, zoccoli e scarpe. Alcune ipotesi su tre tipi di abduzione. In U. Eco, & T. A. Sebeok (Eds.), *Il segno dei tre. Holmes, Dupin, Peirce*. Milano: Bompiani.

Faulkner, W. (2012 [1951]). *Requiem for a nun*. New York: Vintage.

Ferrell, J. (2013). Tangled up in green. Cultural criminology and green criminology. In N. South, & A. Brisman (Eds.), *Routledge international handbook of green criminology*. London and New York: Routledge.

Ferrell, J., Hayward, K., & Young, J. (2008). *Cultural criminology: An invitation*. London: Sage.

Ferrero, E. (1999). *La mia storia*. Torino: Centro Studi Piemontesi.

Forti, G. (2000). *L'immane concretezza. Metamorfosi del crimine e controllo penale*. Milano: Raffaello Cortina.

Forti, G. (2003). Tutela ambientale e legalità: prospettive giuridiche e socio-culturali. *Rivista italiana di diritto e procedura penale,* 1353–1377.

Forti, G. (2007). La "chiara luce della verità" e "l'ignoranza del pericolo. Riflessioni penalistiche sul principio di precauzione. *Scritti per Federico Stella*, Napoli: Jovene, I, 573–674.

Franzini Tibaldeo, R. (2013, October). Un'esistenza indivisibile. Complessità, "governance" e responsabilità nell'età globale. *Governare la paura. Journal of Interdisciplinary studies. Special Issue. Fear of nature. The government of catastrophe between human and social sciences*, 192–218.

Gaarder, E. (2013). Evading responsibility for green harm. State-corporate exploitation of race, class, and gender inequality. In N. South & A. Brisman (Eds.), *Routledge international handbook of green criminology*. London and New York: Routledge.

Galli, I., & Nigro, G. (1987, September). The social representation of radioactivity among Italian children. *Social Science Information*, *26*(3), 535–549.

Geis, G., & Goff, C. (1987). Presentazione. In E. Sutherland (Ed.), *Il crimine dei colletti bianchi: la versione integrale*. Milano: Giuffrè.

Giannetti, R., & Vasta, M. (2003). *L'impresa italiana nel Novecento*. Bologna: il Mulino.

Ginsborg, P. (1989). *Storia d'Italia dal dopoguerra a oggi*. Torino: Einaudi.

Greife, M. B., & Stretesky, P. B. (2013). Crude laws. Treadmill of production and state variations in civil and criminal liability for oil discharges in navigable waters. In N. South & A. Brisman (Eds.), *Routledge international handbook of green criminology*. London and New York: Routledge.

Hall, M. (2013). Victims of environmental harms and their role in national and international justice. In R. Walters, D. S. Westerhuis, & T. Wyatt (Eds.), *Emerging issues in green criminology. Exploring power, justice and harm*. London: Palgrave Macmillan.

Hall, M. (2014). The roles and use of law in green criminology. *International Journal for Crime, Justice and Social Democracy, 3*(2), 96–109.

Halsey, M. (2004). Against "green" criminology. *British Journal of Criminology, 44*(6), 833–853.

Halsey, M. (2006). *Deleuze and environmental damage: Violence of the text*. Aldershot, Burlington: Ashgate.

Heckenberg, D., & White, R. (2013). Innovative approaches to researching environmental crime. In N. South & A. Brisman (Eds.), *Routledge international handbook of green criminology*. London and New York: Routledge.

Jarrell, M. L., Lynch, M. J., & Stretesky, P. B. (2013). Green criminology and green victimization. In B. Arrigo & H. Bersot (Eds.), *Routledge handbook of international crime and justice studies*. London and New York: Routledge.

Kane, S. C., & Brisman, A. (2013). Technological drift and green machines: A cultural analysis of the *prius paradox*. *CRIMSOC: The Journal of Social Criminology*, Green Criminology Issue, Autumn, 104–133.

Lanzillo, M. L. (2013, October). Presentazione. Governare la catastrofe fra scienze umane e scienze sociali. *Governare la paura. Journal of Interdisciplinary Studies. Special Issue. Fear of Nature. The government of catastrophe between human and social sciences*, I–XIII.

Leccardi, C. (2009). *Sociologie del tempo. Soggetti e tempo nella società dell'accelerazione*. Roma-Bari: Laterza.

Levine, A. G. (1982). *Love Canal: Science, politics, and people*. Lexington, Massachusetts: Lexington Books.
Luque, E. (2006). *Mobilising memories, evidence and futures: Disentangling huelva from chemical industry*. First draft discussion paper at ECPR Nicosia Joint Sessions (pp. 25–30) April 2006.
Lynch, M. J. (1990). The greening of criminology: A perspective for the 1990s. *The Critical Criminologist, 2,* 11–12.
Lynch, M. J. (2015). Green criminology and social justice: A reexamination of the Lynemouth plant closing and the political economic causes of environmental and social injustice. *Critical Sociology*, 1–16.
Lynch, M. J., Long, M. A., Barrett, K. L., & Stretesky, P. B. (2013). Why green criminology and political economy matter in the analysis of global ecological harms. *British Journal of Criminology, 53,* 997–1016.
Lynch, M. J., & Stretesky, P. B. (2003). The meaning of green: Contrasting criminological perspectives. *Theoretical Criminology, 7*(2), 217–238.
Lynch, M. J., & Stretesky, P. B. (2014). *Exploring green criminology. Toward a green criminological revolution*. Farnham, Burlington: Ashgate.
Martín, A. M., Hernández, B., Hess, S., Ruiz, C., & Alonso, I. (2013). The relationship between moral judgments and causal explanations of everyday environmental crimes. *Social Justice Research, 26*(3), 320–342.
Mauss, M. (2002 [1923–1924]), *Saggio sul dono. Forma e motivo dello scambio nelle società arcaiche*. Torino: Einaudi 2002.
Melrose, S. (2010). Naturalistic generalization. In A. J. Mills, G. Durepos, & E. Wiebe (Eds.), *Encyclopedia of case study research*. Thousand Oaks: Sage.
Monge-Corella, S., García-Pérez, J., Aragonés, N., Pollán, M., Pérez-Gómez, B., & López-Abente, G. (2008). Lung cancer mortality in towns near paper, pulp and board industries in Spain: A point source pollution study. *BMC Public Health, 8,* 288.
Mori, G. (1977). *Il capitalismo industriale in Italia. Processo d'industrializzazione e storie d'Italia*. Roma: Editori Riuniti.
Morin, E. (1993). *Introduzione al pensiero complesso*. Milano: Sperling & Kupfer.
Morin, E. (1999). *Seven complex lessons in education for the future*. Paris: Unesco Publishing.
Muscarà, C. (1967). *La geografia dello sviluppo*. Milano: Edizioni di Comunità.
Natali, L. (2010). The big grey elephants in the backyard of Huelva, Spain. In R. White (Ed.), *Global environmental harm. Criminological perspectives*. Devon: Willan Publishing.
Natali, L. (2013a). The contemporary horizon of green criminology. In N. South & A. Brisman (Eds.), *Routledge international handbook of green criminology*. London and New York: Routledge.
Natali, L. (2013b). Exploring environmental activism: A visual qualitative approach from an eco-global and green-cultural criminological perspective. *CRIMSOC: the Journal of Social Criminology. Special Issue: "Green Criminology"*, Autumn, 64–100.
Natali, L. (2014). Green criminology, victimización medioambiental y social harm. El caso de Huelva (España). *Revista Crítica Penal y Poder, 7,* 5–34.
Natali, L. (2015a). *Green criminology. Prospettive emergenti sui crimini ambientali*. Torino: Giappichelli.
Natali, L. (2015b). A critical gaze on environmental victimization. In R. Sollund (Ed.), *Green harms and crimes. Critical criminology in a changing world* (pp. 63–78). UK: Palgrave MacMillan.
Natali, L. (2016). *A visual approach for green criminology. Exploring the social perception of environmental harm*. UK: Palgrave MacMillan.
Neri Serneri, S. (1990). Storia, ambiente e società industriale. *Rassegna di studi tedeschi in "Società e storia", 50,* 891–893.
Neri Serneri, S. (2005). *Incorporare la natura. Storie ambientali del Novecento*. Roma: Carocci.
Pérez-López, R., Alvarez-Valero, A., & Nieto, J. M. (2007). Changes in mobility of toxic elements during the production of phosphoric acid in fertilizer industry of Huelva (SW Spain) and

environmental impact of phosphogypsum wastes. *Journal of Hazardous Materials, 148*(3), 745–750.
Pergolizzi, A. (2012). *Toxicitaly. Ecomafie e capitalismo: gli affari sporchi all'ombra del progresso*. Roma: Castelvecchi.
Perini, C. (2002, July–December). Rischio tecnologico e responsabilità penale. Una lettura criminologica del caso Seveso e del caso Marghera. *Rassegna italiana di criminologia, XIII*(3–4), 389–412.
Poggio, P. P. (Ed.). (1996). *Una storia ad alto rischio. L'Acna e la Valle Bormida*. Torino: Edizioni Gruppo Abele.
Potter, G. R. (2015). Justifying "green" criminology: Values and "taking sides" in an ecologically informed social science. In M. Cowburn, M. Duggan, A. Robinson, & P. Senior (Eds.), *Values in criminology and community justice*. Bristol: Policy Press.
Pulcini, E. (2013). *Care of the world: Fear, responsibility and justice in the global age*. New York: Springer.
Pulitanò, D. (2006). Il diritto penale tra vincoli di realtà e sapere scientifico. *Rivista Italiana di Diritto e Procedura Penale, 49*(3), 795–825.
Resta, E. (2008). *Diritto vivente*. Roma-Bari: Laterza.
Rotolo, G. (2012). *Tutela penale dell'ambiente e conoscibilità del precetto*. Milano: EDUCatt.
Rotolo, G. (2016). Modelli "dinamici" di tutela dell'ambiente e responsabilità penale: problemi e prospettive. *Jus,* 1–38.
Ruggiero, V., & South, N. (2013). Toxic State-corporate crimes, neo-liberalism and green criminology: The hazards and legacies of the oil, chemical and mineral industries. *International Journal for Crime, Justice and Social Democracy, 2*(2), 12–26.
Ruiz-Ballesteros, E., Valcuende, J. M., Quintero, V., Cortes, J. A., & Rubio, E. (2009). Naturalizing the environment. Perceptual frames, senses and resistance. *Journal of Material Culture, 14*(2), 147–167.
Ruzzenenti, M. (2001). *Un secolo di cloro e Pbc. Storia delle industrie Caffaro di Brescia*. Milano: Jaca Book.
Saba, A. F. (1997). La pollution en Val Lagarina (1928–1938): une nouvelle voie pour l'histoire de l'environnement? *Histoire, économie et société, 16*(3), 463–470.
Sellin, T., & Wolfgang, M. (1964). *The measurement of delinquency*. Montclair: Patterson Smith.
Settis, S. (2010). *Paesaggio Costituzione Cemento. La battaglia per l'ambiente contro il degrado civile*. Torino: Einaudi.
Shelley, T. O., Chiricos, T., & Gertz, M. (2011). What about the environment? Assessing the perceived seriousness of environmental crime. *International Journal of Comparative and Applied Criminal Justice, 35*(4), 307–325.
Slapper, G., & Tombs, S. (1999). *Corporate crime*. London: Longman.
Sollund, R. A. (2012). Introduction. In R. Ellefsen, R. A. Sollund, & G. Larsen (Eds.), *Eco-global crimes: Contemporary problems and future challenges*. Farnham, Burlington: Ashgate.
South, N. (1998). A green field for criminology? A proposal for a perspective. *Theoretical Criminology, 2*(2), 211–234.
South, N. (2015). Green criminology and brown crime: Despoliation, disposal and de-manufacturing in global resource industries. In T. Wyatt (Ed.), *Hazardous waste and pollution. Detecting and preventing green crimes*. Cham: Springer.
South, N., Brisman, & A., Beirne, P. (2013). A guide to a green criminology. In N. South & A. Brisman (Eds.), *Routledge international handbook of green criminology*. London and New York: Routledge.
Stake, R. E., & Trumbull, D. J. (1982). Naturalistic generalization. *Review Journal of Philosophy and social science, 7,* 1–12.
Stella, F. (2000). *Leggi scientifiche e spiegazione causale nel diritto penale*. Milano: Giuffrè.
Stella, F. (2003). *Giustizia e modernità. La protezione dell'innocente e la tutela delle vittime*. Milano: Giuffrè.
Szasz, A. (1994). *Ecopopulism. Toxic waste and the movement for environmental justice*. Minneapolis: University of Minnesota Press.

Szasz, A. (2007). *Shopping our way to safety. How we changed from protecting the environment to protecting ourselves*. Minneapolis: University of Minnesota Press.

Tallacchini, M. (2005). Scienza, politica e diritto: il linguaggio della co-produzione. *Rivista di filosofia del diritto, I, 2,* 313–336.

Tallacchini, M. (2012). Scienza e diritto. Prospettive di co-produzione. *Sociologia del diritto, XXXII, 1,* 75–106.

Tellechea Rodríguez, J. M. (2004). El Conflicto 'Info-ambiental': El Caso de Huelva información (Agosto 1983–Diciembre 1992). *Ambitos, 11–12*(1–2), 319–340.

Verde, A., & Nurra, A. (2009). Criminal Profiling as a plotting activity based on abductive processes. *International Journal of Offender Therapy and Comparative Criminology, 54*(5), 829–849.

Waitt, G. (2010). Doing Foucauldian discourse analysis—Revealing social realities. In I. Hay (Ed.), *Qualitative research methods in human geography*. New York: Oxford University Press, Oxford.

Walters, R. (2010). Toxic atmospheres: Air pollution, trade and the politics of regulation. *Critical Criminology, 18*, 307–323.

White, R. (2003). Environmental issues and the criminological imagination. *Theoretical Criminology, 7,* 483–506.

White, R. (2008). *Crimes against nature: Environmental criminology and ecological justice*. London: Willan Publishing.

White, R. (2010). Globalisation and environmental harm. In R. White (Ed.), *Global environmental harm. Criminological perspectives*. Cullompton: Willan Publishing.

White, R. (2011). *Transnational environmental crime: Toward an eco-global criminology*. London and New York: Routledge.

White, R. (2013a). The conceptual contours of green criminology. In R. Walters, D. S. Westerhuis, & T. Wyatt (Eds.), *Emerging issues in green criminology. Exploring power, justice and harm*. London: Palgrave Macmillan.

White, R. (2013b). Resource extraction leaves something behind: Environmental justice and mining. *International Journal for Crime and Justice, 2*(1), 50–64.

White, R., & Heckenberg, D. (2014). *Green criminology. An introduction to the study of environmental harm*. London and New York: Routledge.

Williams, C. (1996). An environmental victimology. *Social Justice, 23,* 16–40.

Williams, C. (Ed.). (1998). *Environmental victims. New risks, new injustice*. London: Earthscan.

Wolfgang, M. E., Figlio, R. M., Tracy, P. E., & Singer, S. I. (1985). *The national survey of crime severity*. Washington, DC: Government Printing Office.

Zamagni, V. (2005). *Introduzione alla storia economica d'Italia*, Bologna: il Mulino.

Chapter 3
Historical Pollution: In Search of a Legal Definition

Giuseppe Rotolo

Contents

1 Necessity for a *New* Definition

It is not easy to frame concepts and provide clear definitions concerning environmental crime and green criminology (Herbig and Joubert 2009, pp. 51 ff.; Gibbs et al. 2010, pp. 24 f.; Natali 2013). The latter term—green criminology—is widely considered "the most familiar and suggestive," and "serves well as the most

G. Rotolo (✉)
Università Cattolica del Sacro Cuore, Milan, Italy
e-mail: giuseppe.rotolo@unicatt.it

F. Centonze and S. Manacorda (eds.), *Historical Pollution*,
DOI 10.1007/978-3-319-56937-6_3

comprehensive umbrella" (Ruggiero and South 2010, p. 247).[1] Yet, alongside the rather recent upsurge in criminological interest on environmental issues—as a specific subject of study for that discipline—several denominations have been coined, sometimes even competing with each other; in addition, different perspectives and approaches have been used to address the phenomenon and to focus on some of its particular features (Gibbs et al. 2010, pp. 25 ff.).

There is also a crucial question concerning the lexicon used to define the area of interest, as the terminology used requires a precise rendering of the terms "environment" and "environmental harm." Even the use of the term "green" has been contested, suggesting its inadequacy to express the multiplicity of issues ascribable to the same area of interest (Halsey 2004, pp. 833 ff.)

Little surprise, then, that studies in this field are still in progress, with all the research outcomes to be considered only partial elements of the overall figure, much like single tiles of a mosaic. Moreover, our knowledge in this area is to be considered in a state of flux, as all improvements are premised on the awareness previously gained and, at the same time, are bases for further development.

The definition of the phenomenon—as well as the whole area of interest—is dependent on the particular field in which the research is being carried out and, in addition, is liable to change in view of the results that issue from it. This is especially the case in environmental studies since there is a huge range of issues and perspectives concerning the general matter of analysis, and no single or unitary denomination can encompass them all.[2] It is easy to understand, therefore, why so much attention has been paid to the necessity of "naming" further types or aspects of environmental harms, so as to provide a suitable framework for a developing body of empirical and legal knowledge (White 2008, pp. 112 ff.).

On this basis, the present essay aims at introducing the basic elements that characterize the phenomenon of historical pollution, together with key-related problems. Specific attention will be paid to the temporal dimension in framing a suitable definition, since this is a recurring theme across different approaches (White 2008, pp. 100 ff.).

In particular, we will consider the role of time with respect to the dynamics of pollution: specifically, we note that the consequences of polluting conduct are not

[1]Criminological studies are here presented inasmuch as they provide elements that are useful in justifying, supporting, and discussing definitions of "historical pollution." For a more integrated analysis of the criminological contexts pertinent to our topic, which is also complementary to the present essay, see Chap. 2 in this volume.

[2]As is well known, the definition of "green crime" also postulates the philosophical framework which is used to understand the nexus between humans and their environment: in this respect we may distinguish *anthropocentric*, *biocentric*, and *ecocentric* perspectives (Halsey and White 2009). Moreover, a *legalistic* and a *socio-legal* approach may be differentiated to address this concept (Gibbs et al. 2010, pp. 125 ff.). Hence, each definition has both a philosophical and a conceptual aspect, depending on the particular view that is adopted. Different approaches—like ecofeminism, environmental racism, and the red-green movement—can be distinguished within the overall environmental justice movement (Lynch and Stretesky 2009, pp. 89 ff.).

only immediate, but also persistent, and reveal their environmental impact only in the longer term. This can be illustrated by comparing the features of accidents that cause immediate environmental harm with those consistent with long-term pollution. As Raymond Michalowski has argued:

> the effects of pollution are seldom immediately obvious. There are occasional environmental disasters such as the smog inversion in New York City in 1973, which took 400 lives. For the most part, however, the effects of pollution are subtle and cumulative. Individuals may become sick, feel nauseated, or develop cancer, but this is generally perceived as a failure of their own body, not the consequence of living in a polluted environment. Additionally, because the pollution of air and water leading to illness and other costs is most often the cumulative result of the action of a multiplicity of industries, some at great distances from the affected city or a region, it is difficult to identify a guilty party. (Michalowski 1985, p. 341)

We immediately note that the range of harmful consequences includes not only damage or threats to natural resources, but also personal injuries—and even deaths—since contamination can cause harm to human health as a secondary effect.[3] Moreover, in many of the relevant cases, harms to ecosystems and to natural resources do not emerge immediately, but only after an extended period of latency, due to the main features of the specific ecological context or of the pollutant.

In discussing the causation of environmental harms, it is important to make reference to studies that describe ecosystems as complex adaptive systems (Levin 1998; Holling 2001; Holland 2006, pp. 1, 7 ff.).[4] Given their nature as "self-organized systems," ecosystems consist in adaptive processes which maintain natural resources over time.[5] Thus, in brief, sustainability (understood also in terms of sustainable development) is preserved as long as those processes' adaptive capability is maintained in the face of external factors.

The period of time required to recognize harmful effects (both for the environment and human health) depends on the type of contaminant and can vary from one case to another, since the dose–response relationship—which is fundamental for modern toxicology—differs due to the properties of the contaminant (Rowell 2012, pp. 985 ff., 1003 ff.).

These, along with further aspects, will be discussed in more depth below; first, however, we will examine another aspect of Michalowski's argument, namely the part that begins to draw connections between such issues and the relevant legal tools. The difficulty of finding appropriate legal remedies for situations of pollution

---

[3]The point is widely confirmed by the analyses presented in the chapters of this volume dedicated to domestic legal systems.

[4]Of course the theme of complex adaptive systems is enormous and cannot properly be addressed in this context. Here attention will be paid only to those profiles that contribute to highlighting the constitutive features of the phenomenon under investigation. For a more extensive investigation of such studies, see the seminal works by Holland (1995) and Levin (1999).

[5]Such an approach expresses a specific view about the "complexity" of living systems. Although in a different manner, the theme of complexity is discussed as a particular feature of historical pollution in this volume by L. Natali in Chap. 2.

and latent environmental harm may easily be appreciated. White, indeed, goes further, making reference to the variation in the legal system itself.

> Another issue that is worth pursuing in greater depth is how the processes of criminalisation and decriminalisation impact upon perceptions and responses to environmental harm. While the naming exercise just mentioned implies an expansion of existing and potential harms in relation to the criminological agenda, there are countervailing processes at work as well. For example, the decriminalisation of environmental harm occurs when governments change legislation to allow previously illegal activities to now proceed. (White 2008, p. 113)

In addition to questions regarding the regulation and even criminalization of environmental harms—which is a very relevant topic for studies on environmental crimes (Hall 2015, pp. 8 ff.; Hall 2014, pp. 97 ff.)—a different issue is pertinent here as well. Although they share a number of basic features and are all related within the perspective of legal enforcement, the use of different models or concepts of pollution could entail differences in the effectiveness of legal tools. Indeed, as the present research will discuss, the temporal category raises significant concerns related to law enforcement in cases of historical pollution, due to the lack of legal regimes tailored to its basic features, and the inconsistency of the legal categories which are applied retrospectively to facts that may be separated by significant periods of time.

In particular, certain critical aspects have been pointed out on the basis of the observation that typical features of environmental crimes (i.e. that the offender is usually not an individual but a corporation, and victims may be communities or large groups of people) do not perfectly fit with the criminal law system, and to some extent may even impede the application of the relevant criminal provisions (Du Rées 2001).

> For example, it is often difficult in such cases to ascribe direct (or even indirect) causation between the actions/inactions of a specific party (or state accused of bringing about environmental harm) and the undesirable outcomes themselves. This can be particularly difficult given the typically longer timescale of environmentally destructive activities and their consequent impacts compared with most traditional forms of criminality. (Hall 2014, p. 100)

In conceiving of the chronological framework, it is important to distinguish the moment in which the behaviour is realized from that when the consequences (finally or completely) arise. Looking at this scenario as a series of snapshots arranged so as to create a time-lapse film, many further relevant details may emerge: changes may occur in the pertinent legal regime; the social context may change; and the processes through which people come to see themselves as victims may undergo evolution, along with improvements in community awareness concerning the negative effects produced by polluting activities. There is also a chance that spatial concerns could emerge, since given the long timeframe over which the consequences are produced, the contamination could spread over a wider area than that immediately affected by the polluting activity.

All these elements are to be understood as subsidiary and complementary to the pivotal feature of the phenomenon at stake, which consists in the emergence—or the persistence—of natural resources contamination a long time after the polluting conduct took place. Nevertheless, those features are also fundamental to describing the object of the present research and to trace the boundaries of the definition, which is necessarily stipulative, of the term "historical pollution" .

In the next section, we immediately introduce such a definition and then critically discuss it. It is hoped that by exploring lines of argument which may call into question its validity, we may indeed contribute to establishing it. To anticipate what follows, we will say that the concept of historical pollution, rather than being formal, in fact contains substantial ambiguity, as it neither arises from normative acts or provisions nor can be explicitly inferred via interpretative activities. Anyhow, inasmuch as it is useful to trace the emergence of particular problems at the intersection of the empirical description of the phenomenon and the characterization of the pertinent legal framework, such a definition will be tested and—hopefully—verified below.

2 Introducing the Concept of "Historical Pollution"

We begin by proposing a partial definition of the phenomenon which emphasizes its main feature—the temporal—and on this basis exploring its further implications. Thus, we propose that "historical pollution" stands for the pollution of sites and natural resources (intended as the relevant *behaviour*), which started or took place in the past, due to industrial activities or other production-related activities. Historical pollution usually entails the emergence—or the persistence—of contamination of sites and natural resources (meant as the *effect* of such a conduct) a long period of time after the original pollution took place.

Already at first sight many aspects of this appear questionable: the particular context of the production of environmental harm; the distinction between static or dynamic frameworks, since the activity could be considered interrupted or still ongoing; the focus on the polluting act, in order to clearly distinguish it from the harmful consequences which occur at a significant temporal distance. However, the crucial thing to discuss in more depth is the concept of pollution itself, the nature of which is extremely complex, as the seminal anthropological studies of Mary Douglas have shown (Douglas 1992). Indeed, the meaning of "pollution" can hardly be defined in a way that is certain and unambiguous; rather, it is used to address many different situations that share a common element: it is a form of behaviour that has harmful effects in various, and variegated, contexts (not only environmental, but also—among other spheres—social, human, and even in a certain sense religious) (Nagle 2009). Furthermore, the connotation of "pollution" as pre-eminently environmental has to be considered only a quite recent acquisition (Nagle 2009, pp. 14 ff.); that is, environmental pollution is only a particular and focused type of pollution, narrower than the concept has traditionally been, and to which greater relevance has been attributed since the twentieth century.

Even focusing on the limited field of *environmental pollution*, the concept manifests ambiguity. Different lenses—at least two—can be used to view the concept:[6] one that centres on the constitutive features of the polluting act; another that focuses on the consequences, and in particular on the environmental harm that has been defined "the pivotal concern of much of environmental law" (Lin 2006, p. 984).[7] Moreover, these conceptualizations are not separate, but both contribute to the same complex framework, inasmuch as pollution cannot be defined in itself, but only in connection with the harmful effects caused thereby. As Cassuto has contended:

> Pollutants do not exist outside of systems; pollution presupposes a system to pollute. Identifying pollutants involves determining that a foreign presence and potential source of harm exists within the system. Deciding that a substance is a pollutant requires two potentially problematic steps: designating the system's boundaries and defining harm. (Cassuto 2004, p. 107)

Thus, pollution depends on the context and has to be causally related to a sort of environmental harm. Yet the latter concept also resists objective definition: "[r] ather, 'harm' is a normative concept that reflects underlying social judgments about the good and the bad" (Lin 2006, p. 901).

"Pollution" therefore requires normative definitions (Cassuto 2004, p. 107) and is dependent on the social values (Lin 2006, p. 900)[8] that emerge as significant in a temporally circumscribed context; thus, it is precisely at the intersection of these issues that all the problems related to the concept of historical pollution arise. Given that the act of pollution and the emergence of its effects (also understood as the clear perception of such consequences) can be temporally separated, a great deal of social change could occur over the intervening period—together with the evolution of the pertinent legal regime—which can have serious implications concerning the allocation of responsibilities and the enforcement of remediation activities.

Thus, according to the proposed definition, the basic feature of historical pollution consists in the chronological distance between the beginning of the relevant behaviour (pollution) and the point of observation (usually that of the judgement, in a legal perspective). It is also usually characterized by a chronological distance between the conduct (pollution) and the associated consequences (contamination of sites and natural resources), in accordance with the idea that relevant effects can

[6]The distinction between the two streams of argument (the polluting act, and the environmental harm understood as a consequence for natural resources) derives from Nagle's definitional categories (2009, pp. 29 ff.) which are as follows: (i) the comprehensive definition, according to which everything that is added to the environment pollutes; (ii) the definition which aims at pinpointing lists of polluters and pollutants; (iii) the definition which centres on the polluting effects. In other words, the aim here is to coordinate the relevant considerations around two basic analytical factors of the polluting dynamic: the act and its consequences.

[7]Taking as an example the definition presented in the chapter on the legal system of the UK (England and Wales), this pertains primarily to the "contaminated land regime" and lays emphasis on the effects of the conduct rather than on the polluting activity (regarding which, location in time has no particular relevance): see Chap. 9, Sect. 3 in this volume.

[8]The author argues that, because of the dependence on social values, "harm" means different things for different people, consistent with their differing perspectives (Lin 2006, p. 900).

consist in harms and threats to environmental resources and human health (as a secondary consequence of environmental contamination). In order to validate this definition, the following subsections will critically discuss several key points.

2.1 Pollution and Contamination: Synonymous or not?

The first theme is terminological. Is "historical pollution" an adequate label with which to address the phenomenon and the issues in question? If it is, why? And if not, are there more appropriate definitions available?

In fact there are many other terms that are roughly synonymous with the term "pollution," and applied in the same general contexts. As Nagle points out, "corruption," "impurity," "contamination," "uncleanness," "defilement," and "profanation" express concepts akin to that conveyed by "pollution" (Nagle 2009, p. 45).[9] Focusing on the legal context of environmental protection, "pollution" and "contamination" are of course the most frequently encountered in material situations in which harm is caused to ecological resources. And in discussions of themes and legal issues concerning site contamination for which there is a significant temporal dimension, the literature evidences the use of terminology rather similar to "historical pollution"[10] but which inflects the term "contamination": these terms are "historical contamination" (Brandon 2013, p. 6, p. 17, p. 27; Vegter and Kasamas 2011, p. 1025)[11] and "historic contamination."[12] Indeed, these terms have been frequently used indiscriminately and synonymously.[13]

A fundamental question therefore must be addressed before we can fix on the best terminology: namely, is there a difference between "pollution" and "contamination"? The answer that follows takes into account the needs which the proposed

---

[9]Although the author specifies that [n]one of pollution's synonyms [...] captures the same kind of concern about harms occurring in a shared environment.

[10]Critically discussing the connections between the legal regime for insolvency and the environmental one, Mamutse and Fogleman (2013) use the term "historic pollution" to address a phenomenon analogous to what here is called "historical pollution." Although from varying angles, "historic pollution" recurs in several other legal studies (Taylor 1999; O'Quinn 2000).

As will be discussed in more depth in the next subsection, exactly the same term—"historical pollution" —was used by Mieck; and even if the perspective is significantly different, it may well be useful to our inquiry. Mieck uses the term to refer to the different models of "pollution" which have been historically prevalent over different epochs (Mieck 1990).

[11]Vegter and Kasamas refer to historical pollution as "contamination due to past polluting activities, which is still present in soil and groundwater."

[12]While the terms "pollution" and "contamination" are frequently used interchangeably, Larson (2013, pp. 1012, 1014) refers particularly to "historic contamination."

[13]Similarly, the terms "pollutant" and "contaminant" are also deemed synonymous.

definition must fulfil, along with lexical aspects which are of primary relevance; our conclusions, however, will not be entirely immune to ambiguity.

As Nagle has remarked, although many other terms are used in this context, none of them—and this includes "contamination"—addresses the specific concerns or issues in the way that "pollution" does.

> Pollution is similar to and yet distinct from contamination, defilement, poisoned, and other terms. Contamination is broader than pollution. As a federal appeals court remarked in the context of construing a contractual indemnity provision, "all pollution is contamination, but not all contamination is pollution." Thus, rivers, lakes and other bodies of water can be characterized as contaminated (because of the presence of the offending materials) or as polluted (because of the introduction of those materials into that aquatic environment). (Nagle 2009, p. 46)

The argument implies that *contamination* can to some extent be independent from human behaviour that is explicitly intended to pollute environmental resources, or in any case which has this outcome. Rather, in the alternative case contamination overlaps with pollution, that is, when it is provoked by human conduct. So the terms coincide only if the contamination is caused by pollution. Thus:

> pollution suggests something slightly different from any of these synonymous terms. The distinctiveness of the idea of pollution rests in its dual suggestion of an unwanted outside agent and a general environment that the agent enters. (Nagle 2009, p. 47)

Nagle's opinion thus seems to support the lexical choice we are defending in this chapter. Indeed, "pollution" connotes activities—that is to say, human behaviour—that lead to harms or threats of harm to environmental resources, whereas "contamination" better describes scenarios which can be deemed—although not necessarily—a consequence of pollution. Furthermore, whereas "contamination" conveys a static picture of a given ecological context, the concept of "pollution" connotes something more dynamic, encompassing causes and processes that have significant harmful effects.[14]

Thus it serves us best to draw a clear distinction between "pollution"—as behaviour or conduct—and "contamination"—as effect—since this fits the fundamental issue on which we focus in the present research, namely the chronological separation of the polluting conduct from the emergence of harmful consequences

[14]What has here been inferred on the basis of Nagle's statements, also seems validated by his own etymological approach, according to which he recalls that "pollution" comes from the Latin verb "polluere," which means "to soil" or "to defile"; furthermore, according to Webster's first American dictionary of the English Language, the first meaning—of five overall—attributed to "pollution" is "the act of polluting" (Nagle 2009, p. 7).

(contamination).[15] For this reason, we adopt this terminology in what follows, although this does not preclude differing lexical usages on occasion.[16]

2.2 Epochs and Contexts of "Historical Pollution"

Two further aspects will be discussed together, since they are deeply related. Regarding historical pollution, the use of the adjective "historical" calls for some comment inasmuch as it can evoke diverse semantic attitudes besides the ordinary one. Thus far we have been using "historical" to highlight the chronological distance between the dangerous activity that has polluted the environment and the emergence of perception of harm; yet the adjective could also refer to the particular epoch in which the phenomenon at stake takes place.

Indeed, exactly the same term—"historical pollution" —has been applied by the historian Ilja Mieck to address the evolution of several forms of pollution that, since they are typical of the pertinent epoch, have been named "historical." Consistent with such a perspective, he describes six heterogeneous and complementary scenarios, recurring to the French denominations in order to name them: *pollution microbienne*, *pollution artisanale*, *pollution industrielle*, *pollution fondamentale*, *pollution foncière*, and *pollution accidentelle* (Mieck 1990).

Clearly the path Mieck traces in his studies is significantly different from that followed in this essay. Yet a point of connection emerges between these two perspectives, since the rationale behind Mieck's research might play a heuristic role in analysing the concept of historical pollution as it is here understood.

Actually, the idea that each historical epoch has to deal with a particular form of pollution that is expressive of social, economic, and cultural conditions[17] can be

---

[15]Consistent with this distinction, see also the definition provided on the FAO webpage (http://www.fao.org/docrep/x5624e/x5624e04.htm#1.1) concerning marine pollution. The terminological difference between "pollution" and "contamination" is explicated by quoting the Joint Group of Experts on the Scientific Aspects of Marine Pollution (GESAMP), which defines pollution as "the introduction by man, directly or indirectly, of substances or energy into the marine environment (including estuaries) resulting in such deleterious effects as harm to living resources, hazards to human health, hindrance to marine activities including fishing, impairment of quality for use of sea water and reduction of amenities." Meanwhile, "contamination on the other hand is the presence of elevated concentrations of substances in the environment above the natural background level for the area and for the organism."

[16]As examples of different uses of the same terminology, consider the Italian administrative regime, according to which cases of pollution recur when the contamination has exceeded not only the "threshold concentrations of contamination" (CSC) but also the "threshold concentrations risk" (CSR). It can be inferred that "pollution" and "contamination" are qualitatively different concepts: pollution refers to more severe contamination that poses a *real* risk for natural resources, while contamination in itself produces only a *potential* risk. See in this volume one of the chapters dedicated to the Italian national context: that authored by C. Miccichè, Chap.7.

[17]Furthermore, such an approach also appears to be consistent with the "timescape" perspective, presented by Natali (2013, pp. 76 ff.).

helpful in order to comprehend why "historical pollution" has been used in the context of industrial activity.[18] At least for Western countries, there is already universal comprehension that the major sources of long-term pollution[19] are various industrial activities, although the specifics might vary within geographical areas (Brandon 2013, pp. 12 ff.).[20]

> In North America, such activities include the operation of factories, mines, smelters, electrical power plants and other production facilities, and harbours. Other common sources of contamination are above-ground and underground fuel storage tanks, fuel pipelines, dry-cleaning facilities, military.
>
> [...]
>
> Europe reflects a similar trend to North America and Australasia, with typical contaminating activities including industrial production and commercial services (chemicals and heavy metals), town gas manufacturing, oil processing, municipal waste treatment and disposal, industrial waste treatment and disposal, power plants, storage, transport spills on land, mining (such as disposal of tailings, acid mine drainage and use of catalytic reagents) and military sites (ammunition, fuel and chemical usage and storage). In Eastern Europe, major sources of site contamination have been military installations, nuclear reactors and storage of hazardous chemicals, resulting in a legacy of contamination which will take decades or longer to remediate, at great cost to taxpayers. (Brandon 2013, pp. 12–13)

According to this perspective on the issues at stake, among the wide range of activities that could cause air, soil, or water pollution, several forms are especially relevant since they express the characteristic causal dynamics of environmental harms that are due to industrial production cycles. Regarding the branches of industry that are most directly implicated in pollution (electrical, chemical, and gas and oil), it is noteworthy that the long-term incidence of particular pollutants (such as dioxins[21])

---

[18]Of course the argument applies in general terms, although each domestic context can present particularities. For instance, the analysis of the French legal system shows how cases of historical pollution also concern further fields, such as agriculture and the military: see Chap. 12, Sect. 2 in this volume.

[19]The point is not intended to deny the findings of studies on so-called second generation environmental harms. In brief, environmental degradation might arise as the cumulative effect of even very common activities; many sources might contribute to such pollution, although not particularly significant in themselves nor clearly pinpointed; and finally, such behaviours might be viewed as either legal or illegal. These considerations will be presented in more depth in the next subsection.

[20]This aspect remains uncontested also for developing countries, although the different overall economic context plays a significant role in differentiating them. As Brandon notices, "[t]he predominant sources of contamination in developing countries are also industrial activities, particularly in the form of untreated industrial or chemical waste disposal. Groundwater contamination is particularly widespread in developing countries" (Brandon 2013, p. 13).

[21]Consider for instance the Seveso Case, presented in this volume in two of the chapters focussing on the Italian context (see Chap. 4, Sect. 2; see also Chap. 5, Sect. 1.1). Different types of harmful consequences may be observed in distinguishing short-term and long-term effects of exposure to dioxins: besides the environmental harms caused, the long-term health implications (in terms of increasing rates of diseases, pathologies, and even of mortality for the community directly exposed) emerge as more significant effects than in the short term (Bertazzi et al. 1998; Pesatori et al. 2003).

is deeply related to those types of activity (Martin 1985); moreover, the occurrence of particular forms of pollution, such as groundwater pollution, could be explained by reference to the consequences caused over time by activities subsidiary to such production—in particular, waste management and disposal (Buresh 1986; Ng 1989; Gash 2002).

2.3 "Historical Pollution" as an Evolutionary Pattern: Static and Dynamic Frameworks in Comparison

As already highlighted, distinguishing the moment when the dangerous or harmful activity was carried out, and the moment of the emergence of the effects, could evoke the evaluation of two snapshots of the same story, capturing chronologically distant situations. Thus from such a perspective, historical pollution can be understood both as a *new* and at the same time an *old* phenomenon, because it has to be deemed both as actual and as past, since it concerns the emergence and perception of the consequences of distal facts. Moreover, although rooted in past epochs, the awareness of such pollution is quite a recent achievement, and a deep analytical understanding of its key features is still lacking.

Nevertheless, a further and complementary approach to such issues is also feasible, one which would consider the same facts not merely in "static" terms, as constituted by two static images, but also as a dynamic and evolutionary pattern that links those moments. In adopting such a "timescape" attitude towards environmental concerns, "it is necessary to look at time not as another 'object' along with other objects of analysis, but as a perspective through which to observe the social from a rarely considered angle" (Natali 2013, p. 77). According to such a point of view, historical pollution presents a number of different profiles: statically understood, we can compare the two different epochs, along with the social, economic, and cultural context; conversely, deemed as an evolutionary pattern, the entire dynamic process that over time has produced harmful consequences for the environmental resources becomes relevant.

Drawing a distinction between these two analytical frameworks has heuristic benefits, inasmuch as the evaluation of each perspective allows representative issues and concerns to emerge. For the static theoretical framework, the spotlight is directed at issues characterized by the chronological separation of the two moments—such as, for instance, problems concerning brownfields.[22] In fact, the issue of heritage

[22]For a more detailed analysis of the international legal approach addressing this issue, see Chap. 14, Sect. 6 in this volume. For our purpose, it is useful to recall the CABERNET (Concerted Action on Brownfields and Economic Regeneration) definition of "brownfield," as recalled by Oliver et al. (2005): "Brownfield: sites that have been affected by the former uses of the site and surrounding land; are derelict and underused; may have real or perceived contamination problems; are mainly in developed urban areas; and require intervention to bring them back to beneficial use."

industrial sites which are neglected or underused reveals two faces of the same entity: structures once thought to be economically and socially useful are now considered useless and even dangerous, precisely because of the concomitant environmental risks that stand in the way of redevelopment. Of course, these opposite attitudes towards the same object also encompass the different cultural approaches that characterize different historical contexts, along with the growing awareness of the environmental and health implications of such activities.

Turning to the dynamic approach to historical pollution, the processes of harm causation, which are often latent and thus "invisible," become relevant. As already underlined, the environment itself has to be considered a complex system possessing adaptive dynamics.

> These systems are self-organized, and a small set of critical processes create and maintain this self-organization. ("Self-organization" is a term that characterizes the development of complex adaptive systems, in which multiple outcomes typically are possible depending on accidents of history […].) […] These processes establish a persistent template upon which a host of other variables exercise their influence. Such "subsidiary" variables or factors can be interesting, relevant, and important, but they exist at the whim of the critical controlling factors or variables. If sustainability means anything, it has to do with the small set of critical self-organized variables and the transformations that can occur in them during the evolutionary process of societal development. (Holling 2001, p. 391)

Harmful impacts on natural resources, then, could depend on unbalanced and maladaptive evolution in a complex system—which hypothetically might be natural or extraneous to any pollution—or on the introduction of toxic external factors that negatively affect that evolutionary process over time. Conversely, the particular nature of pollution is given by the activity that causes contamination, via the influence of a material factor which is external to the environmental context, recapped in the point by Cassuto already mentioned above (Cassuto 2004, p. 107).

Accordingly, a further aspect that contributes to the general uncertainty pertaining to the issues under examination is the concept of "pollutant" itself (Nagle 2009, pp. 32 ff.). Defining this term introduces relevant ambiguities—as it is significantly extensible[23]—and requires that we consider the external context and evaluate the specific harm that is caused. Moreover, according to modern toxicology, the relationship between exposure to a particular kind of substance and the related effects (dose–response relationship) is not always linear (Rowell 2012, pp. 1003 ff.). So each pollutant presents specific features concerning its propensity to cause harmful consequences—also understood in terms of damage to health, consistent with the toxicological approach—and this clearly contributes to the

[23]As Nagle (2009, pp. 32 f.) remarks with respect to water pollution: "Three lines of decided cases confirm that the CWA regards nearly everything added to water as a pollutant. The first line of cases holds that fish can be pollutants in some circumstances. A second line of cases questions whether the intentional application of chemicals—by farmers, mosquito control districts, and individual homeowners—are pollutants. The third line of cases considers whether water itself can be a pollutant when someone combines two bodies of water containing different pollutants or different amounts of those pollutants."

uncertainty mentioned above. In brief: "pollution may be (1) always harmful, (2) always harmless, (3) harmless at low doses or exposure but harmful at higher doses or exposure, (4) beneficial at low doses or exposure but harmful at high doses or exposure, or (5) beneficial to some but harmful to others" (Nagle 2013, p. 33).

Of course the temporal category also has a significant role in the causal dynamics unleashed by pollutants, as the category of "persistent organic pollutants" (POPs) explicitly reveals. According to the Stockholm Convention on Persistent Organic Pollutants, adopted 22 May 2001 and entered into force in 2004, POPs "possess toxic properties, resist degradation, bioaccumulate and are transported, through air, water and migratory species, across international boundaries and deposited far from their place of release, where they accumulate in terrestrial and aquatic ecosystem."[24]

It is clear, then, that such pollutants involve both diachronic and spatial aspects, while they activate the dynamics of harm production. POPs therefore represent a theme of interest from the perspective of a discussion of historical pollution, notably since the Stockholm Convention requires the management of industrial sites in accordance with precautionary principle while aiming at phasing out the listed pollutants (Brandon 2013, pp. 56 f.).

Moving beyond the issues raised by POPs, the key chronological and spatial aspects noted above are exceptionally significant in connection with the dynamic approach to historical pollution. As illustrated by the excerpt from Michalowski presented above, damage could occur as a cumulative effect of multiple episodes of pollution perpetrated over time: in this respect, contamination coincides with accretion (Mueller 1996, p. 21), that is to say that the overall harmful consequence depends on the accumulation of less significant effects. Thus the impact on environmental resources and on human health could be seen as an accretion of multiple fragmented and less relevant impacts. An awareness of cumulative effects is therefore fundamental to our present investigation.

> For example, this could refer to the way in which dioxins accumulate in fish flesh over time. It could also refer to the cumulative impact of multiple sources of pollution as in cases where there are a high number of factories in one area (such as places along the US–Mexican border). Diseases linked to asbestos poisoning may surface many years after first exposure, and this, too, provides another example of long-term effects of environmental harm. Persistent use of pesticides in particular geographical areas may also have unforeseen consequences for local wildlife, including the development of new diseases among endemic animal species. (White 2008, p. 101)

As regards our present theoretical investigation, the implications of such an awareness will be that we should prefer a conceptualization of historical pollution as not necessarily embracing only cases of so-called legacy pollution [25] (contamination inherited from discontinued activities), but also include cases of ongoing

[24]For a more detailed analysis of the Convention, see Chap. 14, Sect. 4.1 in this volume.

[25]According to the World Bank document "Targeting Legacy Pollution," "Legacy pollution refers to any pollution that remains from past activities where there is no immediately responsible party who can be held liable for the pollution and compelled to carry out remediation."

pollution, provided it originated in the past and the environmental harm has been demonstrated to be a cumulative effect of the overall activity. Of course, in the latter case, the overall pollution can be considered as fragmented where there are reasons to distinguish different chronological segments, for instance, because the type of industrial activity or the concerned pollutant have varied over time; although it could be tricky to circumscribe the specific impact of each portion, even if the temporal phases may be clearly and exactly identified. Furthermore, the effect itself can also be considered dynamic rather than static, meaning that the harmful impact can increase over time or cause secondary consequences.

A concrete example can help us better comprehend the foregoing remarks. Within a 2014 Italian criminal proceeding[26], there was discussion of whether to consider the concrete case a hypothesis of historical pollution. In brief, the case concerned the polluting activity of a refinery which, along with failures in waste management, had led to the release of toxic substances and thus over time to severe soil contamination of the refinery site, and subsequently—as a dynamic consequence—the contamination of groundwater and of the ecosystem, including the Po river. Notwithstanding, those facts impacted on natural resources which were already affected by contamination inherited from the past. Moreover, a report by the corporation that had managed the plant in 2001 identified the main causes of the overall contamination as Second World War bombing and industrial activity carried out until 1983—when, it asserted, a different corporation had been in charge of that refinery. To the contrary, however, the judgement made clear that the same legal entity had owned the activity before 1983, subsequently, and currently, although under different formal denominations; hence the report was deemed inadequate on this point. Formally understood, the decision did not acknowledge the defendant's request to consider this a case of historical pollution and thus to treat the current owner as a third party who was not responsible for the contamination; despite the judgement on this point, and according to the methodology followed in this research, the case can be deemed an example of historical pollution, as it raises legal and social concerns that are consistent with the long-term emergence of environmental harm, as the basic feature of the category under examination.[27]

[26]Tribunale di Cremona, 18 July 2014, giud. Salvini. The decision and the case are discussed in this volume: see Chap. 4, Sects. 4 and 6.2; Chap. 5, Sect. 1.4.

[27]More recently, another Italian criminal judgement has addressed issues concerning the definition of—literally translated—historical contamination: Corte di Assise di Alessandria, 14 December 2015. In this case the defendants affirmed that the material facts were consistent with the category of historical pollution, as they were merely the current owners and did not contribute to polluting natural resources. The decision denied this on the grounds that the contamination could not be considered definitely past and the harmful consequences were still current. Leaving aside the particular judgement on responsibilities, the case seems rather consistent with the definition of historical pollution here proposed, which is aware of the dynamic nature of the phenomenon, and does not require that the contamination be past and inactive.

Moreover, the spatial implications can also vary over time. Think of the case of groundwater pollution, for instance (Boyd 2015, pp. 331 f.):[28] assuming industrial activity or even waste dumping as the main source of pollution, the contamination of the soil originally occurs where the plant or landfill lies; but the subsequent impact on natural resources can be more extended, in terms of surface water and then groundwater contamination. Thus, while the direct effect of polluting activity is to be understood as circumscribed, the indirect or subsequent ones are correlatively more extended, consisting in ecosystem degradation. In addition, as the dimension of contamination varies over time, it cannot be excluded that further sources of pollution could contribute to the overall harmful impact on natural resources.

Along the path so far explored in our discussion of historical pollution, and in pursuit of a reliable definition, complexity emerges as the main trait of the phenomenon.[29] Indeed, in that respect the categories traditionally used to address the basic features of environmental harm to some extent overlap, and thus we note the consistent presence of uncertainty in our definitions. When we come to consider "static" and "dynamic" concerns within the chronological and spatial categories, our description of the phenomenon becomes more articulated, since both "pollution" and "contamination" are conceivable as static (circumscribed in time and space) or dynamic (prolonged and extended). Putting these variables together, we can sketch a complex picture which encompasses a wide range of hypotheses.

Yet this recognition of the static and the dynamic aspects to our theoretical framework does not clear up every ambiguity, nor does it solve every question: although the basics of historical pollution are precisely definable, its conceptual borders remain uncertain. Indeed, it is not clear under which conditions a case can be confidently described as an example of historical pollution, even though its characteristics may be conceptually clear; how long, we may ask, must the period of latent causation of environmental harm be in order to consider a hypothesis of historical pollution?

In accord with the methodology followed in this research and consistent with the stipulative definition proposed, the present exploration will not attempt to frame an all-encompassing explanation which will provide answers to questions such as these. Rather, we suggest that the shared problems and concerns to which different hypotheses of historical pollution give rise could reveal the pertinence of any particular case to the category. Hence, addressing the traditional alternative between the legal-procedural and socio-legal approaches to the description of environmental harm (Halsey and White 2009, p. 27), the following sections discuss the legal concerns bound up with the notion of historical pollution, as well as the related social perceptions, on the assumption that those profiles not only represent significant aspects of the phenomenon under examination, but also reveal the factors that underpin the consistency of material facts with the wider category.

---

[28]For comparison, consider the Bussi sul Tirino case and the Cremona case, as presented in this volume: Chap. 4, Sects. 4 and 6.2; Chap. 5, Sects. 1.2 and 1.4.

[29]On this aspect, see Chap. 2 in this volume.

3 The Legal Concerns of Historical Pollution: Coping with Uncertainty

The inconsistency between the enforcement of environmental crimes and the principles of legal certainty and predictability has long been pointed out; and the prolonged time between the relevant behaviour and the emergence of harmful consequences in cases of historical pollution emphasizes this point. When it comes to criminal enforcement, allocating responsibilities is made even more difficult when handling cases that unfold over such a long timescale (Hall 2014, p. 98). However, legal implications in this sector are even more complex still, as different issues arise in both streams: prospective and retrospective. Regulation of polluting —and, thus, potentially harmful—activities has a *prospective* attitude when it is couched in terms of preventing damage; conversely, the *retrospective* attitude emerges as paramount along with the need for allocating costs and responsibilities for past events (Brandon 2013, p. 6). In both directions (prospective, towards the future, or retrospective, evaluating the past), uncertainty is exacerbated because of the temporal dimension, and as a consequence these profiles may intersect and even overlap, as will be discussed here.

Uncertainty concerning the future requires that regulation of historical pollution should be consistent with the precautionary principle. Actually, the phenomenon in question—considered as a unitary concept, although susceptible to different labels according to the various models so far considered—can be closely related to the precautionary approach because of significant elements that connect both readings. Indeed, the above-mentioned principle aims at preventing environmental harm (Pardy 2002, p. 65) and, although many forms can be differentiated—principally "weak" and "strong" versions-, its fundamental aim is to cope with uncertainty (Sunstein 2005, p. 4).[30] Accordingly, as already mentioned, while uncertainty is characteristic of environmental matters, historical pollution is notable for the additional component of uncertainty it brings due to the timescale in the causation of harm.

In discussing the links between historical pollution and the precautionary principle, we also highlight the risks of misinterpretation or even debasement of that principle. While precautionary measures are tools oriented to the future, the risk nevertheless arises that a precautionary logic could inspire retrospective decisions about the allocation of responsibilities, as will be discussed below.

---

[30]As is well known, the precautionary principle is prominent within the public international law on the environment, as formalized first with the Rio Declaration on Environment and Development 1992; in particular, Principle 15 states: "In order to protect the environment, the precautionary approach shall be widely applied by States according to their capabilities. Where there are threats of serious or irreversible damage, lack of full scientific certainty shall not be used as a reason for postponing cost-effective measures to prevent environmental degradation." For further discussion see DiMento (2003, pp. 18 ff., 129 ff.); additional aspects are also addressed in the present volume: see Chap. 14, Sect. 2.2.

> As a preventive measure, the precautionary principle generally has prospective application. The principle can be applied to the process of developing rules and deciding on proposed developments because these are prospective decisions. They are decisions about what should be allowed to occur in the future. The precautionary concept can be applied by legislative and administrative decision makers to answer the question "What should we do tomorrow?" (Pardy 2002, p. 65)

Conversely, in considering the precautionary principle as a criterion to evaluate past events, further questions are suggested:

> Is applying the principle in this way a good idea? Should it be used as a standard to assess the past actions of particular persons, so as to judge whether they have acted legally? Is it appropriate to use the principle to answer the question, not "What should we do tomorrow?" but "What should you have done yesterday?" and, depending on the answer, impose liability? (Ibid.)

The answers to questions of this sort will be presented in the following sections.[31] At this point, it is necessary to discuss why the adoption of a precautionary approach in the retrospective evaluation of material facts could be fostered in cases of historical pollution, being aware that this would imply defining more severe standards of duty, when assessing responsibilities. In brief, the reason consists in the particular uncertainty rooted in the long-term dynamic, which is not amenable to effective enforcement of the law (Milhollin 1979).

The answer to the present issue should be nuanced, since—for instance—many aspects will depend on the particularities of the legal system in question, given the radical differences between civil and common law systems. Furthermore, the effectiveness of legal enforcement is also subordinate to the pertinent legal regime (administrative, civil, or criminal), although it can be anticipated that criminal enforcement emphasizes the critical concerns, because of the basic principles of that discipline.

Below we present the first significant critical issues pertaining to long-term liability; subsequently, our attention will focus on particular themes concerning the difficulties in giving the relative evidence, namely the proof of causation and the allocation of responsibilities, especially in respect of foreseeability and the related concept of negligence. In general, a common theme in the following remarks is the overall absence of tailor-made legal tools with which to address cases of historical pollution and cope with long-term issues (Milhollin 1979, pp. 1 ff.), together with the inadequacy of traditional remedies in these cases.[32]

---

[31]See more in detail Sect. 3.3. Pardy distinguishes between civil law and criminal law: whether or not the use of the precautionary principle as a retrospective criterion to assess past conduct is acceptable for civil law, it is forbidden in criminal law, since in that case the assessment rationale would assume the form of absolute liability (see Pardy 2002, pp. 66, 70 ff.).

[32]Among those considered in national contexts, perhaps only the United States' Environmental Response, Compensation, and Liability Act (CERCLA) shows a particular attitude to handling issues of this sort, as also acknowledged by Brandon (2013, p. 29). See Chap. 8, Sect. 4.1.4 ff in this volume.

3.1 Obstacles to Legal Enforcement

Several critical aspects appear typical from a long-term perspective, although we consider different models of liability, distinct legal regimes, and various remedies oriented to compensation, recovery, or remediation.[33] Some of those critical aspects are exacerbated when criminal sanctions are used to tackle episodes of historical pollution, as this option could imply the debasement of rules and indeed of fundamental principles. Attention will here be paid first to more general issues and then to particular aspects which are emphasized in respect of criminal enforcement.

The first issue arising in this context could be considered consistent with the "polluter-pays principle,"[34] since according to it only those who are responsible for or have at least contributed to causing the contamination should bear the costs of remediation, although in cases of historical pollution that precept could be hard to follow (de Sadeleer 2015). As is well known, "polluter pays" arises as a consistent principle both in domestic (Brandon 2013, p. 200) and international legal systems.[35] Essentially, the principle is premised on an economic rationale concerning the allocation of costs that pertain to environmental recovery and aims at ensuring "that parties responsible for the contamination of a site, instead of the taxpayer, assume liability for its remediation wherever possible" (Brandon 2013, p. 200). However, it represents a legal ideal, since it is not always possible to guarantee that the party directly responsible for the contamination will bear the costs. Several problems can emerge when the owner is different from the polluter—that is, when the contaminated site has meanwhile been transferred. Thus, by interpreting "owner" as not equivalent to "polluter", "those advising purchasers should take particular care to ascertain the historical use of the site even where former pollution is only considered to be a remote possibility" (Warren 1993, p. 8).

The latter argument introduces the critical topic of the polluter-pays principle in cases of historical pollution. Indeed, as already outlined, throughout the period of latency many activities could be overlaid, such that the environmental degradation has to be considered a cumulative effect. Where many sources coincide in this manner, ascertaining the responsible parties is particularly tricky, given the difficulty of determining each particular contribution and its effects. Moreover, the sources of pollution are not always pinpointed with certainty (point-source

[33]Indeed, the range of remedies in cases of historical pollution has to be understood as wide and not restricted to punitive and criminal sanctions. The analysis of the UK legal system shows how criminal enforcement is extremely residual, while remediation within the clean-up regime—also incentivizing voluntary actions—is the ordinary remedy: Chap. 9, Sect. 1.

[34]The pertinence of the principle at stake as regards remedial activities is demonstrated by European Directive 2004/35 on environmental liability concerning the prevention and remedying of environmental damage.

[35]Concerning the international relevance of such a principle see, Chap. 14, Sect. 2.1 in this volume.

pollution), and sometimes clearly distinguishing the sources is not even achievable (non-point-source pollution), and this exacerbates the present difficulties yet further.

Although the precise individuation of the responsible party is always difficult, a more radical problem is that sometimes such identification is not even possible in principle—for instance where the party no longer exists because of bankruptcy (Milhollin 1979, p. 8), and present production activity has been abandoned. These would be cases of so-called orphaned pollution.

> Orphaned pollution is persistent contamination of natural resources from sources for which no party can be held financially liable for clean-up costs. (Larson 2013, p. 993)[36]

Under such circumstances, remediation can be promoted only by incentivizing enterprises[37] to engage voluntarily in socially beneficial projects (Larson 2013, p. 994). In addition, in many cases it cannot be guaranteed that the harmful activities were in fact illegal: assuming that there is a delay in coming to consider a given environmental topic as a matter of legal interest, the polluting conduct could have occurred before any regulation was in force. And there is yet a further possibility: that although all contributed or still contribute to bring about a relevant harm as cumulative consequence, each individual fact could elude contestation by not having exceeded the permitted limits introduced through pertinent regulations. In other words, it can be generally assumed that "environmental crime by accretion is committed by legitimate enterprises which, for the most part, act more or less lawfully on the margin of ill-defined permissibles" (Mueller 1996, p. 21).

Consistent with the criminological argument that environmental crime has to be conceived as an example of corporate and white-collar crime (Herbig and Joubert 2009, pp. 57 f.; Simon 2009, pp. 103 ff.; Tombs 2008, pp. 18 ff.), another profile emerges concerning the structural features of the corporations involved. Actually, when the harmful activity is carried out by a holding, it is often difficult and sometimes impossible to distinguish between what has to be addressed to the parent corporation and what to subsidiaries. And there is always a risk that in the allocation of responsibilities, certain acts perpetrated by the parent will be directed to elude liability (Milhollin 1979, pp. 8 f.).

Of course, these problems are not restricted to issues concerning remediation; they also pertain eminently to the difficulties in assigning responsibility. In this respect, another aspect emerges, which is typical of historical pollution in view of the prolonged timeframe that separates the polluting conduct and the harmful effect: namely, the statute of limitation (Milhollin 1979, pp. 11 f.). Certainly, a detailed and complete analysis of this issue cannot be provided here since its application is

[36]The article presents the case of abandoned mines.

[37]As has been pointed out, "for-profit firms are best suited to address orphaned pollution. This is particularly true of for-profit firms specializing in natural resource development, like energy companies and mining companies. These firms have the necessary resources and expertise to effectively remediate orphaned pollution. However, legal and regulatory obstacles, including permitting requirements and strict liability 'polluter pays' statutes often prevent effective voluntary engagement by for-profit firms in environmental remediation" (Larson 2013, p. 994).

significantly different depending on whether the legal framework is shaped consistent with the "injury rule"—such that the limitation period starts to run when the offence occurs—or with the "discovery rule"—the limitation period starts to run when the offence is discovered. Accordingly, the rationale behind the provision could be oriented so as to protect the environment or human health from damage or dangerous effects. The moment from which the limitation period starts to run varies depending on which of these options is considered. What is noteworthy is that, because of the latency in harm causation together with the postponement of social perception, effective enforcement often cannot be provided in cases of historical pollution because the limitation period expires before the proceedings are activated, or even before the definition of the trial.

3.2 *Proof of Causation*

As has been anticipated above, determining criminal responsibility is one of the most crucial factors in the analysis of historical pollution, given that the punitive pattern suggests seeking legal remedy in criminal law, and in view of the role played by legal proof in the allocation of responsibilities. It is thus not hard to see how the complexity which is characteristic of historical pollution may influence judicial scrutiny.

Of course, uncertainty plays a significant role on the terrain of legal proof as well. And the first area that is relevant here is the *proof of causation*, as it requires we ascertain the material dynamics that have caused (or have concurred in causing) the contamination. And in fact it is extremely difficult to prove a causal relationship between polluting activities and environmental harm or personal injury, the latter being understood as a secondary consequence of the contamination. A note recently published in the *Harvard Law Review* (2015, p. 2265) recaps how hard it is to handle matters of this sort in a judicial context:

> In some environmental cases, as in some toxic tort cases, demonstrating but-for causation can be nearly impossible, even when the plaintiff has undoubtedly been injured. In global warming and other similar cases, multiple entities may emit a pollutant with no feasible means to determine whether any individual entity's emissions created the complained-of injury. This may be true even when a court has accepted that the result of the accumulated emissions has created the injury. In such cases, the Court has considered whether the defendants' conduct has made a "meaningful contribution" to the pollution. However, the extent of what suffices to show a meaningful contribution remains unclear.

Of course, the specific aspects of the way environmental law is enforced in a particular domestic legal system must be taken into account, in order to avoid misunderstandings that could arise in comparing different legal regimes. And the knot that is most difficult to unravel is the causal nexus. After all, as has been clearly pointed out, theoretical issues are also often involved: whether we are considering the relatively homogeneous class of environmental harms suits, or toxic torts cases (Note 2015, pp. 2256 ff.), in both scenarios courts need to corroborate both general and specific causation, the latter concerning the proof of particular effects rather than the general relationship between a causal factor and the harmful consequence.[38] Specific causation is especially difficult to ascertain in cases of environmental contamination:

> For specific causation, as in the multiple-emitter environmental cases [...], it may be impossible in toxic tort cases to determine which specific entity is responsible for the exposure that led to the plaintiffs' harms. (Note 2015, pp. 2261 f.)

The point clearly illustrates a crucial feature of historical pollution. Within the prolonged period of latency that separates what is assumed to be the causal factor from the occurrence of the effect, many additional sources could concur to determine the same harmful consequence, so that precisely distinguishing "the" cause of contamination is nearly impossible.

In order to better understand the issue at stake, albeit from a different perspective that is consistent with the arguments considered so far, our attention should be drawn to the diachronic and spatial profiles of pollution, which evoke the issues that emerge from studies on what have been called "second-generation" environmental problems:[39] namely, that a significant proportion of environmental harms derives not from large point sources, but from multiple and even small—often not pinpointed—sources.

> Many of the remaining environmental problems are not caused by large, industrial point sources, however. Instead, these problems are the result of numerous small, diffuse, non-point sources. For convenience, these sources are described in this essay as second generation sources, and the problems they cause are described as second generation problems. Examples of second generation problems include: urban and agricultural runoff;

[38]The case Smaltini v. Italy, decided by the ECHR (application n. 43961/09), demonstrates how difficult it is to prove a causal link where pollution is claimed to be the cause of personal injuries—consistent with the historical pollution framework. Ms. Smaltini was diagnosed with acute myeloid leukaemia and then brought proceedings against the manager of the huge steel plant (Ilva) located in her city (Taranto), assuming that its emissions, prolonged over time, led to the occurrence of her pathology. But the proceedings were discontinued, the judge having taken into account the experts' opinion that the incidence of the particular type of leukaemia among people of the same age as Ms. Smaltini was no higher in the local area than in other regions. Thus, the causal link could not be proved. Subsequently Ms. Smaltini laid a complaint before the European Court of Human Rights regarding the violation of the right to life (Art. 2), as in the domestic legal system the protection of her basic interests had not been assured, the proceedings having been discontinued. The ECHR rejected her application as ill-founded, because it was not demonstrated scientifically that there was a causal relation between the pathology and the polluting emissions. In December 2012 Ms. Smaltini died. More recently, new proceedings have been activated before the ECHR, stemming from a complaint by 182 citizens concerning the violation of their right to life and health. This case is discussed in the present volume by D. Vozza, Chap. 14, Sect. 3.3.3.

[39]According to Vandenbergh's opinion, the term "second generation" indicates those sources of pollution or those problems that are still relevant despite the legal tools introduced and enhanced during the 1970s and 1980s in the USA. While those laws focused "on [a] relatively small number of large industrial polluters," the "remaining 'second generation' sources are often not large point sources, but numerous, small, diffuse non-point sources, such as the many contributors to urban and agricultural runoff into waterbodies" (Vandenbergh 2001, p. 191).

> air pollution problems caused by emissions from numerous small businesses and from increases in the number and the use of motor vehicles; increases in waste generation by individuals, and many sources of global warming gases. Examples of second generation sources include the 25,000 to 35,000 dry cleaning facilities in the United States. The EPA has concluded that their cumulative impact is significant. (Vandenbergh 2001, p. 196)

It is not hard to comprehend how, in urban or rural contexts as well as industrialized ones,[40] many factors may contribute to the occurrence of harmful environmental impacts, and the point is particularly relevant from a long-term perspective. Moreover, it is not only the small size of the polluting activities which causes difficulty in ascertaining a causal link when the pollution is diffuse and the sources non-point, but also the fact that often such micro-emissions are not regulated and may even correspond to entirely normal activities (Vandenbergh 2004).[41]

In short, for all the reasons evaluated so far, in any *ex post* claim concerning environmental damage or individual disease due to pollution, proof of causation would be extremely challenging. Indeed, when a causal link between the exposure to a toxic substance and the emergence of a particular harmful consequence could be deemed as ascertained in general, proving causation for the specific, concrete, and individualized effect would nevertheless be nearly impossible (Farber 2007, pp. 1636 ff.; Percival 2010, pp. 42 f.).[42] The tendency in some domestic regulations concerning compensation for environmental damage to overcome the "causation conundrum" by shifting or relaxing the burden of proof is thus understandable—although not justifiable (Percival 2010, pp. 42 ff.).

---

[40]With regard to this argument it is pertinent to recall that in connection with their locational aspects, environmental issues have been divided into "brown," "green," and "white." As White has pointed out: "*Brown* issues tend to be defined in terms of urban life and pollution (e.g., air quality); *green* issues mainly relate to wilderness areas and conservation matters (e.g., logging practices); and *white* issues refer to science laboratories and the impact of new technologies (e.g., genetically modified organisms)" (White 2008, p. 98).

[41]This kind of argument is pointed out explicitly in the following: "When the harm occurs many years after the act, proof of causation can be difficult. The source of groundwater contaminants, for example, becomes more speculative if the spread of leachate over a period of time has been affected by natural processes, or if more than one pollutor has been operating in the area. In the case of cancers, such as leukemia, it is undisputed that the disease is linked statistically to radiation, but the plaintiff may have been exposed to more than one source over the period of latency, and the defendant can always say that it is impossible, in a given case, to know whether this cancer was due to radiation or 'natural' forces" (Milhollin 1979, p. 6).

[42]The same is relevant in cases of exposure to asbestos, where proof of causation can be satisfied only when it has caused "signature injuries," that is injuries "uniquely tied to exposure to it." Otherwise, issues related to exposure to asbestos are emblematic of the "causation conundrum": "in a modem world awash in pollutants from multiple sources, the difficulty of proving causal injury has made common law liability too crude a vehicle to compensate those exposed to environmental hazards" (Percival 2010, p. 42). Moreover, as the etiopathogenetic dynamic activated by asbestos is extremely long and latent, a significant period ordinarily separates exposure from the disease occurrence, so that cases concerning liability for such exposure can be considered consistent with issues that are typical for historical pollution.

3.3 The Mental State Requirements

In litigation, even if proof of causation can be established, there remains another critical obstacle to the allocation of responsibilities in cases of historical pollution: the complexity of the causal dynamics, along with the long-term chronological dimension, would often impede the foreseeability of the consequences that an activity of a certain type could cause. This is a particular dimension of the "uncertainty" already noted above.

Assuming that hazardous activities are rarely intentionally directed at causing harm (Farber 2007, p. 1641), the more complex and articulated the causal dynamic is, the less foreseeable and predictable are the outcomes. Such an issue is of great concern as regards legal enforcement, especially criminal, since both civil and common law legal regimes require proof of *mens rea*. Thus, if the harm has not been wilfully caused, a proof of negligence will be required in order to hold the defendant liable.[43] In the US environmental regulatory system, for instance, the amendments introduced during the 1970s and 1980s upgraded misdemeanours to felonies and simultaneously modified the mental state requirements.

> When Congress amended the environmental laws in the 1980 s to make environmental crimes felonies, it also changed the mental state requirements for Clean Water Act prosecutions from "willfully or negligently" for misdemeanors to "knowingly" for felonies and "negligently" for misdemeanors. By eliminating the willfulness requirement and substituting a knowledge requirement, Congress provided consistency among the criminal provisions of the major environmental statutes and aligned the mental state requirements for environmental crimes with the requirements for many other federal criminal statutes. (Uhlmann 2009, p. 1235)

Hence, if the standard of "knowingly" means (although this is not uncontested) "knowledge of all no jurisdictional facts" (Uhlmann 2009, p. 1238), the complex and chronologically extended material context in cases of historical pollution would generally exclude complete awareness of the "fact." Moreover, the effect would not be foreseeable either, so that a definition of a duty of care consistent with negligence could be precluded as well.

Given these obstacles, it could be inferred that criminal enforcement would be particularly ineffective in addressing historical pollution. Conversely, despite those considerations, courts may be influenced to overcome difficulties of this kind even if this would imply the misinterpretation of relevant rules—notably, here, *mens rea*—and the infringement of fundamental principles on which the criminal system is premised. Such a scenario would be possible and indeed even predictable if the aim of assuring adequate enforcement in particularly severe hypotheses of environmental contamination was allowed to prevail over a consistent interpretation of principles and legal rules.[44]

---

[43]The basics of the domestic legal systems addressed in the present research are set out in the relevant chapters in this volume.

[44]The criminal courts' attitude towards cases of historical pollution in Italy can be considered an example of the adaptive interpretation here described: see Chap. 4 in this volume.

Two different schemes concerning the judicial evaluation of the allocation of responsibilities are relevant: the first pertains to the cognitive attitude of judgement, the second to the definition of standards of care. However, both share the same context, as their occurrence is fostered by the retrospective evaluation of the facts, consistent with the proof of the mental state.

Regarding the first issue, the evaluation of negligence introduces the risk of systematic bias occurring: specifically, there is a risk of judgements being affected by hindsight bias (Fischhoff 1975, pp. 288 ff., 1982, pp. 335 ff.; Fischhoff and Beith 1975, pp. 1 ff.), which is understood as follows:

> In hindsight, people consistently exaggerate what could have been anticipated in foresight. They not only tend to view what has happened as having been inevitable but also to view it as having appeared "relatively inevitable" before it happened. People believe that others should have been able to anticipate events much better than was actually the case. (Fischhoff 1982, p. 341)

In other words, given that hindsight bias could also influence judicial decisions (Casper et al. 1989; Rachlinski 1998; LaBine and LaBine 1996), in respect of a determination of negligence it could also come to influence the *ex post* definition of duties or standards of care. That is, the danger is that legal standards will be based on retrospective determinations of expected outcomes—although they originally were not foreseeable—inflected by hindsight bias, rather than appreciating what *ex ante* should properly be required.[45]

Moreover—moving to the second issue—the difficulties in determining liability in those cases could be overcome by defining more severe standards of care

[45]The experiment conducted by Kamin and Rachlinski had noteworthy results. Consistent with the tort case petition of Kinsman Transit Co. (1964), the authors submitted the same case to different groups of students, who were required to simulate an administrative hearing (in foresight) and a mock trial. In particular, "[i]n the foresight condition, participants learned that a city had constructed a drawbridge and needed to determine whether the risk of a flood warranted maintaining a bridge operator during the winter when the bridge was not in use. Hiring the operator would serve as a precaution. The operator would monitor weather conditions and raise the bridge if the river threatened to flood. The foresight condition asked participants without outcome knowledge to decide whether a flood was sufficiently probable for the city to appropriate funds for the operator" (Kamin and Rachlinski 1995, p. 93). Conversely, "[t]he hindsight manipulations contained the same background facts. The story continued, however, stating that the city had decided not to hire the operator. During the first winter of the bridge's existence, debris lodged under it. This resulted in a flood that could have been prevented had an operator been hired. The flood damaged a neighbouring bakery, whose owner then sued the city. Participants in the hindsight condition were instructed to hold the city liable if the flood was sufficiently probable that the city should have hired the operator to prevent it. The second hindsight condition added a debiasing manipulation in which the judge instructed participants to recognize the influential effects of hindsight and to consider alternative outcomes as had the city in foresight" (ibid.). The results demonstrated the significant incidence of hindsight bias, showing how "a good faith effort to determine a reasonable level of precautions in foresight may receive harsh judgment when viewed in hindsight" (Kamin and Rachlinski 1995, p. 99); on the contrary, no influence was played by debiasing techniques. Indeed, only 6 participants out of 25 (24%) in the foresight perspective decided to hire the operator, while conversely 29 participants out of 51 (56.9%) among those who participated in the two combined hindsight conditions affirmed that the operator should have been hired.

consistent with the precautionary principle, as has already been sketched. However, this would imply placing all the burden of uncertainty on the individuals.

Such an interpretation is impermissible as regards criminal liability, since this requires culpability and moral blameworthiness, whereas civil liability aims at compensation (Pardy 2002, pp. 77 f.). Indeed, defining standards of care consistent with the precautionary principle would imply the modification of the legal scheme in favor of unacceptable forms of absolute liability, so as to override culpability issues. Moreover, to some extent it implies "passing the buck"[46] of uncertainty to those who have neither the particular competence nor the knowledge to handle it which is also unacceptable.

> To create a strict liability offence with a vague statutory provision, and then to require precautionary action on the part of an accused, is essentially to say to private enterprises, "We do not really know what the rule is. We are giving you the job of making the judgement about when the standard might be breached. If you get it wrong, you are on your own. We will hold you criminally responsible for an action that only in retrospect has proved to be contrary to the provision. (Pardy 2002, p. 76)

4 Historical Pollution and Victimology

Social concerns as well as legal issues could contribute to defining cases of historical pollution, although here too we find structural uncertainties and ambiguities. Indeed, while in general the process of becoming a victim of environmental pollution often has a long-term character and involves a relevant mass of people, these features seem further emphasized in cases of historical pollution. Where harms—understood both as environmental contamination and as personal injury—are invisible, victims cannot gain awareness of their personal condition for considerable time.

> Environmental victimization is complex in that it can be direct or indirect, cause immediate harm and long-lasting harm, it can be individual or collective, it may evolve routine practices or "accidents," and the victims include the environment, humans, and animals. Often, environmental crime victims may not even know that they are victims as the harm does not always manifest immediately and it is challenging to determine who caused the harm. (Jarrell et al. 2013, p. 431)

It is exactly this argument that gives the phenomenon at stake its kaleidoscopic character, showing the complex interrelation between the multiple profiles reviewed

[46]Pardy entitles one section of his essay "Passing the buck," and points out: "If the precautionary principle is to be applied to prosecutions for violation of this provision, private citizens and commercial operations will be required to do what government cannot or will not do itself, which is to determine and identify the circumstances that fall on either side of the line. The regulator thus avoids applying the precautionary principle to itself, and instead foists it upon parties with fewer investigative resources" (Pardy 2002, pp. 76–77).

above. Given the evolution of cultural contexts,[47] changes occur over time in the social perception of activities which are initially understood as a source of economic improvement, which later come to be seen as the cause of contamination.[48] Moreover, awareness about the harmful implications of activities once deemed useful could depend on the emergence of diseases or health injuries that, given their indirect—that is, not immediate—relation to pollution, are ordinarily not instantly perceivable. And finally, the interruption of industrial activities carried on over a long period—due to the activation of legal proceedings or to other factors—could have a harsh economic impact, along with the consequent social conflict. To complete such a background, technological and scientific improvements also stand out as factors that cause relevant changes in the social perception of the phenomena at stake.

Taking our first steps on the dialectical path here traced, health injuries of course constitute the most clearly perceivable profile of victimization. Indeed, "certainly compared to the most subtle categories of environmental harm [...], 'health impacts' tend to be relatively obvious (or at least become so over time) and are usually scientifically verifiable" (Hall 2013, p. 28). Yet to some extent paradoxically, these features seem to be denied in cases of historical pollution. Precisely because the pertinent phenomenon consists in the prolonged time that separates the hazardous activities and the emergence of harmful consequences, the awareness of health impacts is also delayed. Moreover, pathologies caused by personal exposure to toxins often themselves are characterized by a significantly long etiopathogenesis;[49] this further contributes to postponing the moment when full awareness of the harmful implications arises. In cases of cumulative pollution, furthermore, since the harmful activity has been prolonged over time, the health impact could be progressive rather than instantaneous. Thus, a social perception of a general harm to health, before the cause becomes fully understood, could also cause "psychological victimization"—here referring to the social harm approach, according to which "harm" encompasses not only physical and financial/economic but also emotional and psychological harms (Hillyard and Tombs 2004, pp. 19–20). Consider the case of a neighbourhood that, due to historical exposure to toxins, registers an increased

[47]Saha and Mohai explore the temporal concerns regarding political choices in the siting of waste facilities, where environmental justice claims backed up by legislative action apparently led to greater exposure to noxious substances by more vulnerable groups. As the authors remark, "[o]ur findings support the argument that siting increasingly has followed the path of least resistance as a result of unprecedented growth in public environmental concern and citizen action. Institutional factors also are likely to have contributed to the historical patterns" (Saha and Mohai 2009, p. 338). "As state and federal agencies assumed responsibility for approving siting proposals of industry, legislatively mandated permitting processes have provided new political opportunities for public involvement, both administrative and judicial. Distributional politics appear to have prevailed such that those segments of the population with fewer political, organizational, and technical resources have borne a disproportionate share of the society's environmental burden" (ibid.).

[48]See the distinction between "gift" and "poison" as symbolizing two opposite attitudes towards the same element, due to the evolution over time of the social perception, in Chap. 2, Sect. 6.

[49]In this respect the case of exposure to asbestos is emblematic: see Tombs (2008, p. 24).

incidence of a particular pathology: those not yet affected by the toxins could still be considered victimized, since they live with the fear of getting sick.

When it comes to scientific verifiability, in cases of historical pollution it is necessary to distinguish two different profiles: that concerning identification of the injury, and that pertaining to the definition of causes. Although the former is generally verifiable, uncertainties may still persist concerning causes: and in the long term, as we have already made clear, many different factors may have concurred to determine the injury and the causal dynamics may be extremely complex.

For all these reasons, it is understandable why environmental crimes are frequently considered "invisible crimes" (Davies et al. 2014, p. 14; Herbig and Joubert 2006, p. 59); consistent with this, its victims are often invisible too, and this is a further factor that could obstruct legal enforcement (Spapens 2014). Indeed, several characteristics which have been identified as typical of environmental victimization seem to overlap with those of invisible crimes, so that—rather than simply sharing common aspects—the former category can be considered a clear example of the latter. In particular, concerning environmental victimization, it has been noted that:

> The victims are not always aware of the fact that they have been victimized.
>
> The victimization is often delayed, with the victim becoming aware of the victimization much later.
>
> Victims are not sure about who victimized them or who exactly is responsible.
>
> The victimization is often serious not so much because any individual victim was seriously affected, but because numerous victims were affected by the crime.
>
> Victimization can often include repeat offences. (Hall 2013, p. 26)[50]

Most of those features have been already introduced above, in terms of the characteristics of historical pollution: thus here we need only to observe that the particular characteristics of environmental victimization make legal enforcement even more difficult still. Harm, as we have noted, is often directed to more than a few individuals and indeed involves entire classes of people, as also environmental justice studies have widely pointed out.[51]

Such forms of victimization can lead to even more complex social conflicts. Consider the case of a large plant which over time has assured a high rate of employment for the community, but which runs the risk of being closed (or has already been shut down) due to a seizure determined in a proceeding on the environmental impact of its activity, or because of administrative measures aimed at complying with emission standards and minimizing environmental risks (Davies 2014; Rotolo 2016). Clearly, while fostering protection of the relevant interests—i.e. the environment—competing social values could be compromised. As has been noted concerning cases like these:

---

[50]The author here follows the characterization of environmental victimization used in Skinnider (2011).

[51]For the emergence of studies centred on the incidence of crime rates in local areas, considering issues of victimization and crime prevention, see Evans and Fraser (2003, pp. 81 ff).

> [S]uch harms impact substantially on the lives of their victims and the communities in which they occur, and, in turn, these injustices impact heavily on the work of social, health, welfare and criminal justice agencies and other regulatory bodies. It is not a crime to close an industrial plant, yet doing so in an area whose economic wealth is generated almost exclusively from a single industry causes significant further harms and losses that impact substantially in terms of costs to individuals, families and communities. (Davies 2014, pp. 308 f.)

In other words, invisible harms—understood as direct and even in terms of secondary or indirect victimization—are extremely relevant: not only because those who are victimized (the whole community, in our case) are exposed to various and different typologies of harm (that is to say harms concerning different aspects); but also in terms of possible "re-victimization," given the difficulties of adequate redress in the judicial context (Davies 2014, p. 309).

5 Concluding Remarks

The present chapter sought to introduce the concept of historical pollution and provide a consistent theoretical framework for it which could be useful for further analyses and discussions. In view of the complexity of the phenomenon—arguably its fundamental trait—the definition was first presented and only then subject to critical analysis. Our attention has generally been directed at the contexts of industrial activity and waste disposal linked to productive cycles, since those are the fields in which relevant cases arise in the countries we are considering.

A number of key problems still resist clarification or resolution. Crucial legal and social concerns have here been treated not only as illustrations of the phenomenon in question, but also as heuristic criteria designed to contribute in providing a better definition. These elements could help us identify relevant material cases, in other words those bearing the characteristic features of our category.

Several concepts familiar from environmental studies have here been recalled and applied to the concept of historical pollution, since they allow us to understand the typical dynamics of harm production in such cases. We have also considered the phenomenon of historical pollution from both a static and a dynamic perspective.

Looked at as static, themes concerning legacy pollution or brownfield redevelopment come to relevance. In this respect, we are led to consider the temporal distance between the harmful activity and the moment when issues concerning the contamination come to be addressed (whether for redress, compensation, or allocation of responsibilities).

Looked at as dynamic, other concepts have also been adduced as relevant to such a perspective, in order to comprehend and to indicate the basics of causal processes, which can determine the occurrence of contamination over time: notably point source vs. non-point sources, and cumulative and diffuse pollution.

Ongoing pollution has not been considered as a category which is opposed to historical pollution, since the same dynamics of harm causation along with other basic features can recur when the polluting activities began in the past, although they are not concluded, or even if the contamination is not wholly inactive. And groundwater pollution has been presented as a concrete example of historical pollution which focuses on the long-term material processes that produce harms to environmental resources.

The aim, then, has been to provide a map which allows us to explore this "new" concept—which is at the same time an "old" phenomenon—despite the outstanding aspects of uncertainty and ambiguity. Of course, the present suggestions, including the illustrative references to domestic legal systems, are not intended to anticipate what will be more properly discussed in later chapters; rather, they are vehicles to expound general arguments concerning the definition here presented. Besides local specificities, some common elements do indeed exist; for instance, in many cases we have registered a lack of pertinent regulation that is tailor-made for historical pollution, or a lack of a consistent formal definition,[52] and even of pertinent terminology.[53] These absences can, of course, be blamed on the delay in introducing legal tools to tackle environmental crime;[54] in addition, they are due to the peculiarities of historical pollution, which centre on its "invisibility," and the necessarily complex nature of any adequate legal regime. For those reasons, it is easy to see how difficult legal enforcement could be.

Nevertheless, one particular legal regime deserves some more comment. Within the US legal system, the Comprehensive Environmental Response, Compensation, and Liability Act (CERCLA) suggests itself as a system of regulation which seems more equal to the task of coping with these matters.[55] It is noteworthy that, on one hand, its introduction was a reaction to some severe trigger cases, including that of Love Canal (Brandon 2013, pp. 108 ff.); and, on the other, that in some respects this regime has been considered a model for successive regulation, including that introduced through EU law (Vanheusden 2007, p. 568).

The themes we have presented here, along with the concept thus introduced, will receive more in-depth analysis in the chapters that follow: the present essay aimed only to break the path and to move a few initial steps into the territory thereby revealed.

[52]Consider among the others the case of Spain: Chap. 10, Sect. 3.1.

[53]This is the case for France, as presented in this volume: see Chap. 12, Sect. 2. Conversely, in Germany the concept of *Altlasten* seems to express the relevant issues: see Chap. 11, Sect. 1.1.1.

[54]The surge in interest in environmental issues can be generally located in the 1970 and 1980s.

[55]See Chap. 8, Sect. 4.1.4.

References

Bertazzi, P. A., Bernucci, I., Brambilla, G., Consonni, D., & Pesatori, A. C. (1998). The Seveso studies on early and long-term effects of dioxin exposure: A review. *Environmental Health Perspectives, 106,* 625–633.

Boyd, C. E. (2015). *Water quality: An introduction*. Dordrecht, Heidelberg, New York, London: Springer.

Brandon, E. (2013). *Global approaches to site contamination law*. Dordrecht, Heidelberg, New York, London: Springer.

Buresh, J. C. (1986). State and federal land use regulation: An application to groundwater and nonpoint source pollution control. *The Yale Law Journal, 95,* 1433–1458.

Casper, J. D., Benedict, K., & Perry, J. L. (1989). Juror decision making, attitudes, and the hindsight bias. *Law and Human Behavior, 13,* 291–310.

Cassuto, D. N. (2004). The law of words: Standing, environment, and other contested terms. *Harvard Law Review, 28,* 79–128.

Davies, P. A. (2014). Green crime and victimization: Tension between social and environmental justice. *Theoretical Criminology, 18,* 300–316.

Davies, P., Francis, P., & Wyatt, T. (2014). Taking invisible crimes and social harms seriously. In P. Davies, P. Francis, & T. Wyatt (Eds.), *Invisible crimes and social harms* (pp. 1–25). New York: Palgrave Macmillan.

de Sadeleer, N. (2015). Case note. Preliminary reference on environmental liability and the polluter pays principle: Case C-534/13, Fipa. *RECIEL, 24,* 232–237.

DiMento, J. F. C. (2003). *The global environment and international law*. Austin: University of Texas Press.

Douglas, M. (1992) [1966]. *Purity and danger: An analysis of the concepts of pollution and taboo*. London, New York: Routledge.

Du Rées, H. (2001). Can criminal law protect the environment? *Journal of Scandinavian Studies in Criminology and Crime Prevention*, 2, 109–126. Reprinted in R. White (Ed.). (2009). *Environmental crime: A reader*, 638–655. Cullompton: Willan Publishing.

Evans, K., & Fraser, P. (2003). Communities and victimisation. In P. Davies, P. Francis, & V. Jupp (Eds.), *Victimisation: Theory, research and policy* (pp. 80–100). Houndmills, New York: Palgrave Macmillan.

Farber, D. A. (2007). Basic compensation for victims of climate change. *University of Pennsylvania Law Review, 155,* 1605–1656.

Fischhoff, B. (1975). Hindsight ≠ foresight: The effect of outcome knowledge on judgment under uncertainty. *Journal of Experimental Psychology: Human Perception and Performance*, 1, 288 ff.

Fischhoff, B. (1982). For those condemned to study the past: Heuristics and biases in hindsight. In D. Kahneman, P. Slovic, & A. Tversky (Eds.), *Judgment under uncertainty: Heuristics and biases* (pp. 335–352). Cambridge: Cambridge University Press.

Fischhoff, B., & Beith, R. (1975). "I knew it would happen"—Remembered probabilities of once future things. *Organizational Behavior and Human Performance*, 13, 1 ff.

Gash, J. (2002). Beyond Erin Brockovich and a civil action: Should strict products liability be the next frontier for water contamination lawsuits? *Washington University Law Quarterly, 80,* 51–125.

Gibbs, C., Gore, M. L., McGarrell, E. F., & Rivers, L., III. (2010). Introducing conservation criminology. Towards interdisciplinary scholarship on environmental crimes and risks. *The British Journal of Criminology, 50,* 124–144.

Hall, M. (2013). *Victims of environmental harm. Rights, recognition and redress under national and international law*. Abingdon, New York: Routledge.

Hall, M. (2014). The roles and use of law in green criminology. *International Journal for Crime, Justice and Social Democracy, 3*(2), 97–110.

Hall, M. (2015). *Exploring green crime. Introducing the legal, social and criminological contexts of environmental harm*. London, New York: Palgrave MacMillan.

Halsey, M. (2004). Against "green" criminology. *The* British *Journal of* Criminology, *44*, 833–853.
Halsey, M., & White, R. (2009). Crime, ecophilosophy and environmental harm. In R. White (Ed.), *Environmental Crime: A Reader* (pp. 27–49). Cullompton: Willan Publishing.
Herbig F. J. W., & Joubert S. J. (2006). Criminological semantics: conservation criminology—Vision or vagary? *Acta Criminologica, 19*, 88–103. Reprinted in R. White (Ed.). (2009), *Environmental crime: A reader*, pp 50–62. Cullompton: Willan Publishing.
Hillyard, P., & Tombs, S. (2004). Beyond criminology? In P. Hillyard, C. Pantazis, S. Tombs, & D. Gordon (Eds.). *Beyond criminology: Taking harm seriously*. London, Ann Arbor: Pluto Press.
Holland, J. H. (1995). *Hidden order: How adaptation builds complexity*. Redwood City: Addison-Wesley.
Holland, J. H. (2006). Studying complex adaptive systems. *The Journal of Systems Science and Complexity, 19,* 1–8.
Holling, C. S. (2001). Understanding the complexity of economic, ecological and social systems. *Ecosystems, 4,* 390–405.
Jarrell, M. L., Lynch, M. J., & Stretesky, P. B. (2013). Green criminology and green victimization. In N. South & A. Brisaman (Eds.), *Routledge international handbook of green criminology* (pp. 423–444). London, New York: Routledge.
Kamin, K. A., & Rachlinski, J. J. (1995). Ex post ≠ ex ante. Determining liability in hindsight. *Law and Human Behavior, 19,* 89–104.
LaBine, S. J., & LaBine, G. (1996). Determinations of negligence and the hindsight bias. *Law and Human Behaviour, 20,* 501–516.
Larson, R. B. (2013). Orphaned pollution. *Arizona State Law Journal, 45*, 991–1032.
Levin, S. A. (1998). Ecosystems and the biosphere as complex adaptive systems. *Ecosystems, 1,* 431–436.
Levin, S. A. (1999). *Fragile dominion*. Cambrige, MA: Perseos Books.
Lin, A. C. (2006). The unifying role of harm in environmental law. *The Wisconsin Law Review*, 897–984.
Lynch, M. J., & Stretesky, P. B. (2009). The meaning of green: Contrasting criminological perspectives. *Theoretical Criminology, 7*, 217–238. Reprinted in R. White (Ed.) (2009). *Environmental Crime: A Reader*, 84–102. Cullompton: Willan Publishing.
Mamutse, B., & Fogleman, V. (2013). Environmental claims and insolvent companies: The contrasting approaches of the United Kingdom and the United States. *The British Journal of American Legal Studies, 2,* 578–634.
Martin, T. J. (1985). Long-term liability for hazardous waste-induced injury in Missouri: Latent harm sufferers beware. *Washington University Journal of Urban and Contemporary Law, 28,* 299–343.
Michalowski, R. J. (1985). *Order, law and crime: An introduction to criminology*. New York: McGraw-Hill Inc.
Mieck, I. (1990). Reflections on a typology of historical pollution: Complementary conceptions. In P. Brimblecombe & C. Pfister (Eds.), *The silent countdown: Essays in European environmental history* (pp. 73–80). Berlin, Heidelberg: Springer-Verlag.
Milhollin, G. (1979). Long-term liability for environmental harm. *The University of Pittsburgh Law Review, 41,* 1–26.
Mueller, G. O. (1996). An essay on environmental criminality. In S. M. Edwards, T. D. Edwards, & C. B. Fields (Eds.), *Environmental crime and criminality. Theoretical and practical issues* (pp. 3–33). New York, London: Garland Publishing.
Nagle, J. C. (2009). The idea of pollution. *The UC Davis Law Review, 43,* 1–78.
Nagle, J. C. (2013). Good pollution. *The University of Chicago Law Review, 79,* 31–34.
Natali, L. (2013). The contemporary horizon of green criminology. In N. South & A. Brisaman (Eds.), *Routledge international handbook of green criminology* (pp. 71–82). London, New York: Routledge.

Ng, L. (1989). A DRASTIC approach to controlling groundwater pollution. *The Yale Law Journal, 98,* 773–791.
Note. (2015). Causation in environmental law: Lessons from toxic torts. *Harvard Law Review, 128*, 2256–2277.
O'Quinn, J. C. (2000). Not-so-strict liability: A foreseeability test for *Rylands v. Fletcher* and other lessons from *Cambridge Water Co. V. Eastern Counties Leather PLC. The Harvard Environmental Law Review, 24,* 287–313.
Oliver, L., Ferber, U., Grimski, D., Millar, K., & Nathanail, P. (2005). The scale and nature of European brownfields. *CABERNET 2005: International conference on managing urban land LQM Ltd,* Nottingham, UK, Belfast, Northern Ireland, UK.
Pardy, B. (2002). Applying the precautionary principle to private persons: Should it affect civil and criminal liability. *Les Cahiers de Droit, 43,* 63–78.
Percival, R. V. (2010). Liability for environmental harm and emerging global environmental law. *The Maryland Journal of International Law, 25,* 37–63.
Pesatori, A. C., Consonni, D., Bacchetti, S., Zocchetti, C., Bonzini, M., Baccarelli, A., et al. (2003). Short- and long-term morbidity and mortality in the population exposed to dioxin after the "Seveso Accident". *Industrial Health, 41,* 127–138.
Rachlinski, J. J. (1998). A positive psychological theory of judging in hindsight. *The University of Chicago Law Review, 65,* 571–625.
Rotolo, G. (2016). Deliberative democracy and environmental law enforcement. In T. Spapens, R. White, & W. Huisman (Eds.), *Environmental crime in transnational context: Global issues in green enforcement and criminology* (pp. 174–192). Abingdon, New York: Routledge.
Rowell, A. (2012). Allocating pollution. *The University of Chicago Law Review, 79,* 985–1049.
Ruggiero, V., & South, N. (2010). Critical criminology and crimes against the environment. *Critical Criminology, 18,* 245–250.
Saha, R., & Mohai, P. (2009). Historical context and hazardous waste facility siting: Understanding temporal patterns in Michigan. *Social Problems, 52*, 618–648. Reprinted in R. White (ed.). (2009), *Environmental crime: A reader*, 313–349. Cullompton: Willan Publishing.
Simon D.R. (2009). Corporate environmental crimes and social inequality: new directions for environmental justice research. In R. White (Ed.), *Environmental Crime: A Reader* (pp. 103–115). Cullompton: Willan Publishing.
Skinnider, E. (2011). *Victims of environmental crime: Mapping the issues*. Vancouver: The International Centre for Criminal Law Reform and Criminal Justice Policy.
Spapens, T. (2014). Invisible victim: The problem of policing environmental crime. In T. Spapens, R. White, & M. Kluin (Eds.), *Environmental crimes and its victims: Perspectives within green criminology* (pp. 221–236). Farnham, Burlington: Ashgate.
Sunstein, C. (2005). *Laws of fear: Beyond the precautionary principle*. Cambridge: Cambridge University Press.
Taylor, R. J. (1999). A process to resolve environmental legacy liabilities: Unlocking assets to pay the costs of historic pollution. *Environmental Claims Journal, 11,* 27–46.
Tombs, S. (2008). Corporations and health and safety. In J. Minkes & L. Minkes (Eds.), *Corporate and white-collar crime*. Los Angeles, London, New Delhi, Singapore: Sage.
Uhlmann, D. M. (2009). Environmental crime comes of age: The evolution of criminal enforcement in the environmental regulatory scheme. *The Utah Law Review, 4,* 1223–1252.
Vandenbergh, M. P. (2001). The social meaning of environmental command and control. *The Virginia Environmental Law Journal, 20*, 191–219.
Vandenbergh, M. P. (2004). From smokestack to SUV: The individual as regulated entity in the new era of environmental law. *Vanderbilt Law Review, 57,* 515–628.
Vanheusden, B. (2007). Brownfield redevelopment in the European Union. *Boston College Environmental Affairs Law Review, 34,* 559–575.
Vegter, J. J., & Kasamas, H. (2011). Bringing sustainable management of contaminated sites into practice—The role of policy and regulations. In F. A. Swartjes (Ed.), *Dealing with*

contaminated sites: From theory towards practical application (pp. 1017–1053). Dordrecht, Heidelberg, New York, London: Springer.

Warren, M. H. (1993). Historical pollution: A European perspective. *Environmental Management and Health, 4,* 7–8.

White, R. (2008). *Crimes against nature: Environmental criminology and ecological justice.* Cullompton, Portland: Willan.

World Bank. *Targeting legacy pollution.* Guidance notes on tools for pollution management. (http://siteresources.worldbank.org/INTRANETENVIRONMENT/Resources/244351-1279901011064/GuidanceNoteonLegacyPollution.pdf)

Chapter 4
Historical Pollution and the Prominence of Criminal Law Enforcement in Italy

Giuseppe Rotolo

Contents

1 The Phenomenon of Historical Pollution in Italy

The phenomenon of historical pollution must be considered rather new in the Italian experience. Although rooted in past economic epochs, and connected especially with industrial and technological development, it is only recently that the consequences of such pollution have begun to emerge in terms of environmental harm and, secondarily, threats to or even impairment of personal interests (health, life). Consequently, neither social awareness of the phenomenon itself, nor full and focused legal regulation, has yet completely emerged.

It is the latter point which is our present concern: here we discuss the relevance that the legal system bestows on issues pertaining to historical pollution, paying

G. Rotolo (✉)
Università Cattolica del Sacro Cuore, Milan, Italy
e-mail: giuseppe.rotolo@unicatt.it

F. Centonze and S. Manacorda (eds.), *Historical Pollution*,
DOI 10.1007/978-3-319-56937-6_4

attention both to regulation and to legal enforcement. To anticipate the argument that follows, we may remark at the outset that there is, at present, no formal or explicit regulation of the phenomenon of historical pollution in Italy. As discussed below, only Article 242 of the *Testo Unico Ambiente* (TUA; Legislative Decree of 3 April 2006, n. 152, also known as *Codice ambientale*) refers to historical contamination at all, in the context of providing a legal regime for the activation of remediation programmes. Conversely, however, legal enforcement appears to be extremely severe, extending very frequently into the realm of criminal liability.

In this respect, some critical points come to light. Although the only legal provision related to historical pollution pertains to administrative regulation—in particular to the definition of remediation programmes—in practical terms the legal tool used to respond to infractions lies in a different field, namely criminal law, which is the most severe. Of course, the reason for this is premised on the perceived need for effective enforcement due to the gravity of the harms to basic interests produced in cases of historical pollution. Yet although comprehensible in some ways, the preference for criminal enforcement transgresses an important legal limit: namely, the absence of crimes that have a structure tailored to the particular features of historical pollution.

We thus arrive at the critical subject for analysis here: namely, that in pursuit of greater effectiveness—at least in theory—the strategy that has been followed debases basic elements of legal theory, and even fundamental principles concerning criminal law.

Given the prominence assigned to criminal enforcement, this Chapter and Chap. 5 of this volume focus on this topic, with analysis of administrative regulation postponed to Chap. 7. Chapter 6 dwells on aspects of corporate liability, and the positioning of this topic after the chapters on criminal law and before that on administrative issues is entirely intentional. In the Italian system, corporate liability—although labelled "administrative," and not neglecting the idea that legal persons should also be involved in crime prevention—depends on the occurrence of a crime, and in fact holds a punitive attitude.

The analysis of criminal law is divided into two chapters. After a general presentation of the phenomenon (including an introduction to the social, economic, and historical background to historical pollution), the present chapter describes the relevant features of the discipline, and discusses the problems that arise with the strategy premised on criminal enforcement. To emphasize the core concerns, aspects from the pertinent case law are recalled and critically discussed. Chapter 5 (authored by Venturato and Greco) begins by setting out the cases which, because of their representativeness, have been identified either as leading cases or as benchmarks. Given the complexity of the cases selected, the presentation is focused on the aspects that are directly relevant to historical pollution, and to the decision finally taken. Also recalling further judicial decisions, the last section of Chap. 5 recaps the legal tools pertinent to each of the five scenarios taken to be relevant for capturing the peculiarities of the phenomenon of historical pollution.

2 The "Invisibility" in the *Seveso* Case: What Can We Learn About Historical Pollution?

On Saturday 10 July 1976, shared social awareness about the environmental effects of industrial activities in Italy received a huge and unwelcome boost. A few minutes after noon, a severe accident occurred in the ICMESA chemical plant (Industrie Chimiche Meda Società Azionaria), located in Meda, a small city in the rich industrial area of Lombardy called Brianza.[1]

From the perspective of environmental harm, the chemical composition of the released cloud is particularly significant, since it contained—among other chemicals—tetrachlorodibenzodioxin (TCDD). The exposure to the dioxin had deleterious effects on the nearby population (Pozzo 2008, pp. 6 ff.): alimentary restrictions were enforced because of suspected contamination of vegetables cultivated in the surrounding areas, and there were cases of skin disease, in particular chloracne. It was also alleged that an increase in the rate of abortions was registered, due to the fear of genetic malformation. Economic effects were also felt, since many activities in the affected area were forced to close. However, the most immediate consequence was the evacuation of a considerable part of the population who lived close to the plant, which was imposed as soon as it became clear that the aerial release contained dioxin (Cemeteri 2006, pp. 29 ff.).

At first sight, the *Seveso* case *cannot* be considered a case of historical pollution. Many aspects speak in favour of this: the consequences were plainly and immediately perceivable, remediation activities were implemented promptly, and legal proceedings were quickly underway (Pozzo 2008, pp. 1 ff.). At the institutional level, furthermore, awareness of the need for more extensive and appropriate environmental protection strategies was heightened in the aftermath of the accident, and not only at the local level: the European Commission adopted Directive 82/501/EEC, referred to as the Seveso Directive, precisely in order to control the risks associated with industrial production, with particular reference to the impact of toxic chemicals on ecosystems and on populations exposed to the consequences of such accidents.[2]

---

[1]In brief, the accident unfolded as follows: a chemical reaction led to a rapid rise in temperature of a particular chemical process, and consequently a relief valve failed in one of the reactors. The resultant toxic release extended forty metres into the air and scattered across the surrounding area, affecting the populations of neighbourhoods in the vicinity. The disaster is referred to as the *Seveso case* due to the name of the municipality whose population suffered the most severe consequences, but other communities were impaired by the toxic cloud as well (Meda, Desio, Cesano Maderno, Barlassina, Bovisio-Masciago). For a detailed description of the accident, see Pozzo (2008, pp. 5 ff).

[2]The link between the Seveso accident and the implementation of the Directive can be inferred from the short presentation available on the institutional European Commission website: "From disasters to success. In Europe, the catastrophic accident in the Italian town of Seveso in 1976 prompted the adoption of legislation on the prevention and control of such accidents. The so-called Seveso Directive (Directive 82/501/EEC) was later amended in view of the lessons learned from later accidents such as Bhopal, Toulouse or Enschede resulting in the Seveso-II (Directive

Nevertheless, a number of lessons can be drawn from this case in order to assist in the study of the phenomenon of historical pollution, related in particular to *invisibility* as a peculiar feature of environmental harm (Davies et al. 2014, p. 14).[3] As a matter of fact, regarding the Seveso disaster, it has remained *unknown* how long and to what extent those effects might persist, how large is the number of victims, and what kind of indirect consequences might be traced back to the accident. In addition, in 1976 the impact of dioxins on human health was not clear, due to the lack of studies, although their toxicity for the ecosystem had already been ascertained (Cemeteri 2006, pp. 26 ff.). And a further aspect of such invisibility can be observed in the social sphere. Public consciousness about the meaning of the accident remained muted, arguably because an animated public response to such an event depends on three different aspects: "a distinguishable actor or a guilty party, a specific victim or set of victims, and a recognisable injury to that victim or group thereof" (Brisman 2014, p. 62, recalling Michalowski 1985, p. 341).

Remarkably, a study on the social impact of the accident thirty years later shows that the affected population was able neither to build a common shared memory of the incident, nor to completely overcome the stigma commonly attached to contaminated people (Stefanizzi et al. 2008, pp. 55 ff.). Even today, it is sometimes difficult for those who have personally experienced the disaster to talk about it, while the new generation lacks a clear and complete awareness about what the Seveso event stands for.

The difficulty in facing up the problem, or even in recognizing it and addressing it rationally, was manifest right after the accident, when several workers in effect denied that it had occurred, in the context of evading calls to publicly assign blame for what had happened, perhaps afraid of losing their positions at ICMESA (Corona 2015, p. 84).[4] Yet the continuing discomfort in recalling the episode and addressing the ongoing consequences does indeed have an objective reason, namely the

(Footnote 2 continued)

96/82/EC). In 2012 Seveso-III (Directive 2012/18/EU) was adopted, taking into account, among others, the changes in the Union legislation on the classification of chemicals and increased rights for citizens to access information and justice. It replaces the previous Seveso-II Directive. The Directive now applies to more than 10,000 industrial establishments in the European Union where dangerous substances are used or stored in large quantities, mainly in the chemical, petrochemical, logistics and metal refining sectors. Considering the very high rate of industrialisation in the European Union the Seveso Directive has contributed to achieving a low frequency of major accidents. The Directive is widely considered as a benchmark for industrial accident policy and has been a role model for legislation in many countries world-wide" ("Industrial Accidents," website of the European Commission, environment subsection: http://ec.europa.eu/environment/seveso/).

[3]For a deeper analysis, see Chap. 3 in this volume.

[4]Indeed, the conflict of basic interests, namely the protection of the environment or human health on the one hand, and the need to increase or preserve employment on the other, is a frequent point of issue in this field, and emerges in more recent episodes of relevant environmental pollution. In fact, the same dynamic seems evident also in the current Italian case of Taranto, where an iron and steel plant and its CEOs have been criminally charged for the long-term pollution of the ecosystem. In terms of social consequences, it is easy to see how the population has been divided into two groups: those who want the production to close in order to protect the health of the

invisibility of the dioxin itself (Stefanizzi et al. 2008, pp. 70 ff.). While immediately after the accident there was a kind of social stigmatization of the people assumed to be contaminated, the personal perception of the effects has remained undefined over the longer term—and the point is indicative of a particular dimension of the social process of the erasure of memory, although a public consensus about the real effect of the consequences of exposure to dioxins has in the meantime been reached.[5]

The *Seveso* case has to be included in the range of industrial accidents (Perrow 1999; Centonze 2004) because of the way the disaster unfolded. Indeed, it cannot be considered as an example of historical pollution, inasmuch as the latter refers to the emergence or persistence of contamination in sites and natural resources a long time after the polluting activity itself has taken place—consistent with the formal definition adopted in this research.[6] Further, the aspects already referred to regarding the latency of the harmful consequences and the difficulties for the community in acquiring complete awareness of them, also contribute to the analysis of the core features of the phenomenon under consideration.

3 Understanding the Phenomenon of Historical Pollution: Social, Economic, and Historic Background

In order to fully comprehend the differences between industrial disasters and cases of historical pollution, the basic narratives of the Seveso disaster can be compared with the *Porto Marghera Petrochemical Plant* case,[7] which can be viewed as the very first episode of historical pollution to gain public and institutional attention in Italy. There are many similarities between them: in both cases, the industrial activities in question were based on chemical processes related to the chlorine cycle; the harmful consequences pertained both to the environment and to human health, and impacted a wide geographical area (Perini 2002, pp. 391–392). Yet

(Footnote 4 continued)

neighbouring community, and those who want the industrial activity to continue in order to preserve local employment (Rotolo 2014, pp. 397 ff.).

[5]In the clinical field the incidence of toxicity has also been studied and verified in the long term. In particular, the carcinogenic nature of dioxins has been confirmed: the cohort study on the population of Seveso massively exposed to TCDD outlined "an excess risk of lymphatic and hematopoietic neoplasms" together with "an elevated risk of breast cancer" (Pesatori et al. 2009, p. 9). Moreover, polychlorinated dibenzo-p-dioxin (PCDD) has been included in the range of permanent organic pollutants (POPs) provided by the Stockholm Convention (Council Decision 2006/507/EC of 14 October 2004), in order to circumscribe the production and release of toxic substances which certainly harm human health and the environment. In brief, 2,3,7,8-tetrachlorodibenzo-*p*-dioxin (TCDD), which is the toxic chemical released through the accident that occurred in Seveso, can be considered "the most toxic congener in the family of polychlorinated dibenzo-dioxins, PCDD" (Pesatori et al. 2009, p. 2).

[6]See Chap. 3 in this volume.

[7]See Chap. 5, Sect. 1.1 of this volume.

there are important differences as well: while the Seveso disaster was an accident for which the material aspects of the case have been clearly identified, the Marghera case has been uncovered through a jurisdictional approach, which has attributed the contamination to a long (and slow) process of industrial activity, specifically as a cumulative effect of that kind of pollution (Perini 2002, p. 392; Vallini 2015, pp. 25 ff.). Shifting attention to the legal aspects, such a difference obviously plays a significant role in the assessment of responsibility because of the complexity of the historical evolution of the facts in the second case: we may remark, for instance, that the judges in the *Porto Marghera* case had to scrutinize the positions of twenty-eight people who had succeeded each other in the relevant management positions since the mid-1960s, and thus could be considered potentially responsible parties (Perini 2002, p. 396; Vallini 2015, p. 27).

In the Italian context, a number of cases are now arising of environmental pollution which may be described as "historical." The situation can be regarded as a long-term outcome of the process of industrialization, which started in the late nineteenth century but accelerated considerably after the Second World War (Corona 2015). In fact, since the second half of the last century a number of important urban and industrial areas have emerged and have had significant impacts on the local ecosystems, impacts which are particularly appreciable from a long-term perspective (Corona 2015, pp. 63 ff.), and the correspondence of some of sites with prominent cases of "historical pollution" seems noteworthy (e.g. Mestre-Marghera, Taranto).[8] Also relevant is the range of industrial sectors which have developed in this period: the chemical, petrochemical, steel, and iron industries (Corona 2015, p. 65). In other words, it is precisely those industries whose production processes involve pollutants with the potential for long-term hazard which have proliferated.

Moreover, the development of industry—itself a prominent source of pollution—has to be seen also as a basis for further activities that are dangerous for the environment: not only the subsidiary consequences of such production, that is the need for waste dumping, but also the associated urbanization itself and the availability of more advanced, often chemical components, or of mechanical support for agriculture (Patrono 1980, p. 12).

At the beginning of Italy's industrial development—which can be placed at the end of the nineteenth century—no effort at all was invested in public strategies for environmental protection: originally the political and economic plan only cared

[8]The *Taranto* case has not been considered among the class of historical pollution cases, although it has several of its basic features. Rather, it can be interpreted as a case of cumulative pollution, mainly due to the activity of a steel and iron plant which started in the early 1960s; other sources of contamination can be identified in the petrochemical activities related to the refinery which operates in that area, and the commercial harbour. Nevertheless, we refrain here from discussing this case along with the class of historical pollution cases, for two reasons: first, the industrial activity is currently ongoing; second, the absence of judgments on responsibility for the contamination (the ruling so far pertained only to preventive seizures). For a complete description of the case and of the respective judicial proceedings, see Ruga Riva (2015c).

about the potential impact on human health (Corona 2015, pp. 50 ff.). In combination with the assumption that industrial development in the second half of the twentieth century happened in areas already contaminated through previous exploitation of natural resources, this observation helps us grasp the emergence of historical pollution. To some extent, it is plausible that the new dangerous activities added their impact to contamination inherited from the past.[9]

Consistent with that process of historical evolution, even the social perception of industrialization's impact on natural resources emerged later (Corona 2015, pp. 80 ff.), the delay being related precisely to the *invisibility* of environmental pollution. And as a consequence, the legal regulation of industrial activity has long been insufficient or indeed totally lacking, this being a period during which industries were free from bounds or restrictions. The first legislation to have regulated industrial activity was the Law of 31 December 1962, n. 1860, on the pacific use of nuclear energy; another example of the sectoral approach to the issue was the Law of 13 July 1966, n. 615, also known as the *Legge Anti-smog*. It was only with the law commonly known as *Legge Merli*—the Law of 10 May 1976, n. 319, on the prevention of water pollution—together with the variations introduced with the *Legge Merli bis*, that comprehensive regulation of the field came into effect (Patrono 1980, pp. 16 ff.). The evolution of administrative regulations on waste dumping followed a similar dynamic: the Law of 20 March 1941, n. 366, can be seen as the first legislation on urban waste management, while it was with the Decree of the President of the Republic of 10 September 1982, n. 915 and the Law of 9 November 1988, n. 475, that toxic and industrial waste disposal was regulated. And it was not until the introduction of the Legislative Decree of 9 February 1997, n. 22 (*Decreto Ronchi*), that more comprehensive regulation developed.

Returning briefly to the Seveso disaster, a clear impression of the concrete effects of such a lack of regulation can easily be attained. Before the accident occurred, other episodes—albeit less severe—had happened, but legal enforcement could not be implemented because of the lack of a legal framework (Pozzo 2008, pp. 3 ff.). It is clear, then, that activities which had no immediate consequences, but did have long-term impacts on natural resources and human health, were able to proliferate in the absence of effective and consistent ex ante control. Hence, the previous analysis seems to anticipate an issue which is relevant also with regard to criminal enforcement in cases of historical pollution: with the law apparently unable to address such a delicate issue, a more proactive attitude has been adopted by the criminal courts (*supplenza giudiziaria*), although this could require the use of analogy—which is forbidden in the criminal field—or even lead to violations of fundamental constitutional principles.

[9]For a historical analysis of one of the most relevant cases of pollution carried out over almost an entire century, see Ruzzenenti (2001), who makes free use of the term *inquinamento storico* (lit.: *historical pollution*) (pp. 522 ff.).

4 The Concept of Historical Pollution: An Overview of the Legal Framework

The preceding section evoked the complexity of the notion of historical pollution. Yet indeed the issue has to be considered even more problematic still, since no explicit and formal definition of the phenomenon is available under Italian law. The term *inquinamento storico* (historical pollution), however, recurs frequently. At first, the term attracted the attention only of scholars; later, it began to appear in judgments addressing the phenomenon. From the perspective of the different legal areas concerned, we may note that attention to historical pollution has shifted from the administrative field—consistent with a regulatory approach—to the criminal, the latter representing the severest form of liability that is frequently activated to tackle this sort of environmental harm.

The term *inquinamento storico* has been used by scholars and courts mainly in connection with material situations characterized by the emergence of environmental harms a long time after the dangerous activities occurred, and sometimes also those inherited from past activities conducted in the same area. Of course, there are many particular scenarios falling within this general category,[10] and it is possible to draw out the legal framework related to historical pollution by reviewing the relevant legal references, and analysing how this concept rose to the attention of scholars and courts. The purpose is to take into account the relevant cases and material contexts that have been addressed using the concept of historical pollution and, at the same time, to trace a heuristic path to arrive at a clear conceptualization of that term.

The most valuable legal reference in addressing the phenomenon is Article 242 of the *Testo Unico Ambiente*. The provision, which has administrative relevance, is entitled *Remediation of contaminated sites*. Here we look only at the aspects of the provision that are relevant for understanding the legal framework.[11] In particular, Article 242 sets out the procedure that the polluter has to follow in case of a risk of site impairment, and refers to "historical contamination that could still pose a risk of aggravation of the situation of contamination."[12]

The most pertinent legal issue consists in the allocation of responsibility and the consequent costs for remediation when it is not clear who the polluter was, or when the current owner of the site has not played any role with respect to the polluting

[10]Different scenarios and the legal enforcement that pertains are described by Venturato and Greco in Chap. 5 of this volume. See also Chap. 3 in this volume.

[11]Further aspects are presented in detail in Chap. 7 of this volume, see in particular Sect. 3.2.

[12]The complete wording of the first paragraph of the provision follows: "Upon the occurrence of an event that has the potential to contaminate the site, the polluter implements within twenty-four hours the necessary measures to prevent and immediately notifies pursuant to and in the manner provided for in Article 304, paragraph 2. The same procedure applies to the act of discovery of historical contamination that could still pose a risk of aggravation of the situation of contamination."

activity.[13] In this respect, the second paragraph of the same provision refers to the particular dynamic of the production of harm, when the contamination is not the effect of a single event but of several, and it is possible to retrace the history of the pollution that has occurred in the relevant area: "In the event that the pollution is not attributable to a single event, the parameters to be evaluated should be identified, case by case, based on the history of the site and the activities carried out there in time." The wording thus seems to confirm that historical pollution can overlap with cases of cumulative pollution.

In discussing such a regulation or commenting on the administrative judgments, scholars have sometimes used the term *inquinamento storico* (literally, "historical pollution") to express the idea that the contamination (meaning the harmful consequence of previous pollution) emerges a significant period of time after the activity (Prati 2003, pp. 159 ff.; Prati 2012, p. 89b; Panni 2007, pp. 844 ff.). The term *contaminazioni storiche* (literally: *historical contamination*) has also been used, in connection with civil liability (Salanitro 2011, pp. 19, 23).

As already sketched, criminal scholars have begun to refer to the concept of historical pollution and indeed to explicitly use that terminology (Micheletti 2007, pp. 372 ff.; Di Landro 2014, pp. 7, 11). In fact, at the intersection between administrative regulation and criminal enforcement lies a misdemeanour entitled *Sites remediation* (Art. 257 TUA),[14] which pertains to contamination that has

---

[13]See the judgment of the local administrative court of the Italian region Lombardia, *TAR Lombardia*, Milan, s. I, 19 April 2007, n. 1913, which emphasizes the requirement of the identity between the party who polluted the site and the one who currently holds it, and also underlines a key chronological aspect: the application of the rule cannot be extended to cases of pollution that occurred when the regulation did not exist, except if the parties who polluted and who currently own the site are one and the same. The judgment is commented on by Vanetti (2007, pp. 1081 ff).

[14]The provision was modified through the Law of 22 May 2015, n. 68. The complete text of the misdemeanour, in its earlier version, is as follows:

1. Whoever causes pollution of the soil, subsoil, surface water or groundwater with concentrations exceeding threshold limit values is liable to the penalty of imprisonment from six months to one year or to a fine from 2600 to 26,000 Euro, if he does not carry out remediation according to the project approved by authority in charge in proceedings under Articles 242 and following. In case of failure in making notifications under article 242, the offender shall be punished with a term of imprisonment from three months to one year or with a fine from 1000 to 26,000 Euro.
2. A term of imprisonment from one year to two years and a fine from 5200 to 52,000 Euro are also applied if the pollution is caused by dangerous substances.
3. In the judgment of conviction for the offense under paragraphs 1 and 2, or in the judgment under article 444 of the Criminal Procedure Code, the benefit of the suspended sentence may be subjected to execution of emergency, remediation and environmental restoration.
4. The compliance with the projects approved under Articles 242 and following is a condition for exemption from punishment for environmental crimes covered by other laws with regard to the same event and the same conduct of pollution referred in paragraph 1.

The new legal regime does not modify the crime structure: the modification only involves aspects that concern the relationship between the misdemeanour and the felony under Art. 452-terdecies, *Omitted remediation*, which the same law introduced. Thus, Art. 257 TUA has been

exceeded a risk threshold and has a diachronic structure. Two different behaviours are criminalized: the omission of the required reporting, which would allow the procedure to begin, and non-compliance with the remediation plan approved by the legitimate authority, under the already mentioned Article 242 TUA. In this regard, two different moments—chronologically distant—are significant: the pollution (i.e. the activity) from which the contamination derived, and the contamination itself. When the contamination emerges, the duty to report arises; in a further and subsequent moment, when the remediation plan is approved it has to be considered compulsory and the responsible party is called upon to comply with it.

It is thus easy to imagine how hard it is to allocate responsibility in cases where significant contamination arises after a considerable period of latency.[15] In particular, the problems concern a situation when the current owner or operator was neither in charge of nor engaged in the polluting activity. Of course, consistent with basic principles regulating the Italian criminal system—according to which criminal responsibility has to be personal and requires culpability—only the person who infringed the law can be called upon to remediate the damage, and—in the case of non-compliance with the plan—be charged. Thus a current owner who did not pollute is generally not compelled to remediate, but is covered under the duty of reporting—which is relevant from the criminal perspective under Article 257 TUA —when the contamination is considered dynamic, that is when it can have harmful effects on further sites (Micheletti 2007, pp. 376–377). Moreover, if the historical pollution has been caused by activities which were undertaken before the approval of the first legal regime covering remediation procedures (Ministerial Decree of 25 October 1999, n. 471) and is considered *static*, since the contamination does not introduce any further risk, no one should be held as the responsible party: neither the current owner, because of the reasons already discussed, nor the original polluter, due to the lack of regulation concerning the duty of remediation at the time when the polluting activity took place (Micheletti 2007, p. 376).[16]

More recently, several criminal judgments have explicitly addressed the pertinent cases as episodes of historical pollution.[17] The innovative aspect in these cases is the explicit use of this term for facts that are not consistent with the legal

(Footnote 14 continued)

modified at paragraph 1 and at paragraph 4: respectively, by introducing a rule according to which it does not apply when a more severe crime comes to relevance, and by specifying the extension of the exemption of relevance only to pertinent misdemeanours (instead of crimes).

[15]See the case described in Sect. II of this chapter, Chap. 5, Sect. 1.6.

[16]According to Art. 2.1 of the Italian Criminal Code: “No one may be punished for an act which did not constitute an offence according to the law in force at the time it was committed.” This quotation, as well as further norms reported in this essay, is drawn from the Italian Criminal Code, as translated by Wise (2005).

[17]*Tribunale di Cremona*, 18 July 2014, Judge. Salvini, p. 121, p. 134, p. 340; *Corte d'Assise di Chieti*, 19 December 2014 (ud.), n. 2, p. 127, p. 128, p. 170, p. 171. Both cases to which the judgments refer are discussed by B. Venturato and E. Greco in Chap. 5 of this volume, respectively, Sects. 1.2 and 1.4.

provision already mentioned, since they are characterized by the emergence of more severe harms to ecosystems, natural resources, and even human health. Thus, this linguistic choice has to be seen as rather innovative in the criminal field. Conversely, the cases that fall within such a "new" category recall traditional and consolidated phenomena: environmental or even health disasters, where the harm is the consequence of prolonged and latent dynamics.

Focusing specifically on the innovative judicial cases, the first we will look at (the *Cremona Petrochemical Plant* case) concerns a refinery which released toxic substances (partly because of the failure to repair the drainage system). This led to groundwater contamination and also a danger of explosions, due to the presence of hydrocarbons underground as a consequence of past toxic emissions. The second noteworthy case (the *Bussi sul Tirino* case), with regard to which the court used the term *historical pollution*, has a clear similarity with the first. According to the prosecutor's charge, the polluting activity of a chemical plant caused groundwater contamination through the protracted release of toxic substances. Moreover, there was a danger to human health given that local neighbourhoods drew on that water for their personal needs. Nevertheless, in comparing these two episodes and the consequent dynamics of criminal enforcement, a very significant difference immediately emerges: while in the former case the responsible parties were sentenced as guilty, in the latter the court declared an acquittal judgment. Both judgments, though, are not final.[18]

5 The Criminal Enforcement of Historical Pollution

The last point is indicative of a problem concerning the effectiveness of criminal enforcement as regards the phenomenon of historical pollution. Indeed, at first sight it seems that the Italian criminal justice system simply cannot give an answer to the question of justice posed by the emergence of environmental harms after a long period of latency: not only because of the lack of a specific definition of the phenomenon or of tailor-made regulation, but also due to the peculiar

[18]In both cases, the first instance judgement has been appealed and the relevant Courts of Appeal have issued their decisions (whose full opinion is not yet available). From publicly available information, it would appear that, in the *Cremona Petrochemical Plant case* the Court of Appeal (*Corte d'Assise d'Appello di Brescia*) confirmed the conviction of one of the defendants (the company's CEO) while it acquitted all the others. The Court also confirmed the first instance sentencing to pay damages arising from the contamination to the civil claimants. With respect to the *Bussi sul Tirino case*, the Court of Appeal (*Corte d'Assise d'Appello dell'Aquila*) partially overruled the first instance judgement—which had acquitted all of the defendants—convicting ten out nineteen of them in relation to the unintentional disaster allegations, although the relevant convictions were then declared subject to pardon (*indulto*). The Court also sentenced the convicted defendants to pay damages to the civil claimants for an overall amount of approximately 4 million euro. Further details on these recent developments may be found in the next Chapter authored by B. Venturato and E. Greco at sections 1.2 and 1.4.

characteristics of the enforcement of environmental crimes in Italy. It is thus necessary to show how the criminal enforcement of the phenomenon at issue has been shaped: to this end we will review the legal provisions that have been deemed relevant in these cases (albeit interpreted extensively by judicial courts). Such a review has three principal segments.

(a) Although with some delays, the legal regime to protect natural resources has been progressively consolidated in what is currently known as the "environmental code," namely the aforementioned Legislative Decree of 3 April 2006, n. 152. The criminal range of this law consists of several crimes, for the most part misdemeanours, premised on the administrative regulation, since they depend on the violation or absence of a compulsory permit, the infringement of particular prescriptions, or on exceeding of emissions thresholds defined by a particular competent authority or in a legal provision (Ruga Riva 2013, pp. 13 ff.). Among these, the provision that is relevant in cases of historical pollution is provided in Article 257 TUA, analysed and discussed above.[19]
(b) Traditionally, no felony explicitly intended to preserve environmental resources has been included among the legal tools for environmental protection, since criminal enforcement has long been focused on the aforementioned misdemeanours. More recently something new has emerged: Law of 22 May 2015, n. 68 introduced a new range of environmental crimes into the Italian Criminal Code (Title VI-bis, *Felonies against the environment*). Two articles are particularly relevant for this research, inasmuch as they provide for the criminal enforcement of more severe environmental harms. These crimes, respectively entitled *Environmental pollution* (Art. 452-bis Criminal Code)[20] and *Environmental disaster* (Art. 452-quater Criminal Code),[21] could potentially be of interest for our discussion of historical pollution. Naturally, the attention of this essay is on the interpretation of crimes *of first generation*, since so far

---

[19]See Sect. 4.

[20]Art. 452-bis, *Environmental pollution*:

Shall be punished by imprisonment from two to six years and fine from €10,000 to €100,000 anyone who unlawfully causes significant and measurable impairment or deterioration:

1. of water or air, of extended or significant soil or subsoil portions;
2. of ecosystem, biodiversity, even agrarian, flora or wildlife.

When pollution is produced in a protected natural area or in an area subject to landscape, environmental, historical, artistic, architectural or archaeological constraint, or is produced against protected animal or plant species, the penalty is increased.

[21]Art. 452-quater, *Environmental disaster*:

Apart from the cases designated in the art. 434, whoever unlawfully causes an environmental disaster, shall be punished by imprisonment from five to fifteen years. Alternatively constitute disaster:

1. the irreversible alteration of the ecosystem balance;
2. the alteration of the ecosystem balance which removal is particularly onerous or achievable only through exceptional measures;
3. the offence against public safety assessed on the basis of the extent of the impairment and its negative effects or of the number of people offended or exposed to danger.

When disaster is produced in a protected natural area, or in an area subject to landscape, environmental, historical, artistic, architectural or archaeological constraint, or against protected animal or plant species, the penalty is increased.

no judgement on the new provisions is available for cases of historical pollution. Nevertheless, the last section of the present report will be dedicated to a consideration of how and whether the new regime together with forthcoming judgments on those crimes can lead to a better system of enforcement.

(c) Focusing on what can be considered the *first age* of environmental protection through the criminal justice system, we note that the absence of felonies which concentrate on environmental harm as the key point in the rationale for criminalization means that there are few legal tools available.[22] This may perhaps explain the adaptive attitude of prosecutors and criminal courts, which can be considered as orientated towards a creative approach. Faced with this fork in the road regarding the need for basic justice in cases of harms to the ecosystem and human health, the direction of *enforcement* has been chosen, attempting to apply provisions which were intended to protect neither the environment nor individuals.

Among others,[23] the provision that has been used most frequently to criminalize acts consistent with the empirical model of historical pollution is Article 434 of the Italian Criminal Code, entitled *Collapse of structures and other intended*

---

[22]In fact, Art. 257 TUA talks about excessive risk, but does not refer to the actual causing of environmental harm.

[23]In this perspective, we must consider Art. 437 Criminal Code, *Intentional removal or omission of precautions against industrial accidents*:

> Whoever fails to install appropriate equipment, apparatus or signs for the prevention of industrial accidents or disasters, or removes or damages the same, shall be punished by imprisonment for from six months to five years. If the act results in a disaster or accident, the punishment shall be imprisonment for from three to ten years.

In addition, and particularly in cases were historical pollution overlaps with what can be called a health disaster, Art. 589 Criminal Code, *Negligent homicide* is also relevant:

> Whoever by negligence causes the death of a person shall be punished by imprisonment for from six months to five years.
>
> […]
>
> In the event of the death of more than one person, or of the death of one or more persons and personal injury to one or more persons, the punishment applied shall be that which should be inflicted for the most serious violation committed, increased by up to one-third, but this punishment may not exceed fifteen years.

or even *Negligent personal injury* (Art. 590 Criminal Code):

> Whoever by negligence causes personal injury to another shall be punished by imprisonment for up to three months or by a fine of up to €309.
>
> If the injury is serious the punishment shall be imprisonment for from one to six months or a fine from €123 to €619; if it is very serious, imprisonment for from three months to two years or a fine from €309 to €1239.
>
> […]
>
> In the event of injury to more than one person, the punishment applied shall be that which should be inflicted for the most serious violation committed, increased by up to one-third; but the punishment of imprisonment may not exceed five years.

disasters,[24] along with the corresponding negligent felony provided under Article 449, *Negligent crimes involving damage*[25]; latterly, and only with regard to several cases of groundwater contamination has the charge been shaped consistent with Article 439, *Poisoning water or foodstuffs*,[26] which is also provided as a negligent crime (Art. 452, *Negligent crimes against public health*).[27] However, in both cases there is a clear distance between the basic features of historical pollution—consisting in the damaging of natural resources, although with an appreciable period of latency after the dangerous activity occurred—and the interests that these types of criminal offences aim at protecting, which are undoubtedly different from environmental interests and are rooted primarily in public safety.

Since the last component of the presented strategy of enforcement is predominantly adopted to tackle cases of historical pollution, and at the same time involves many critical aspects, our attention will focus on this case. To briefly anticipate what will be discussed in more detail below, there are two key issues here: first, by interpreting the criminal provisions in question extensively, criminal courts seek an effective remedy in cases of historical pollution, although this goal is not often achieved; second, this extensive—and even creative—interpretation leads to the debasement of principles of criminal justice and the misinterpretation of core aspects of the legal regime.

(Footnote 23 continued)

Of course, the voluntary provisions (respectively, Art. 575, *Homicide* and Art. 582, *Personal injury*) are not considered in this perspective, as it can be assumed that—at least normally—the event is not intentionally caused.

[24]"Whoever, apart from the cases designated in the preceding Articles, commits an act aimed at causing the collapse of a structure or any part thereof, or any other disaster, shall be punished, if his act results in danger to the public safety, by imprisonment for from one to five years. The punishment shall be imprisonment for from three to twelve years if the collapse or disaster occurs."

[25]"Whoever, apart from the cases designated in the second paragraph of Art. 423-bis, through negligence, causes a fire or other disaster designated in the first Chapter of this Title, shall be punished by imprisonment for from one to five years. The punishment shall be doubled in the case of a railway disaster or shipwreck or sinking of a ship used for transporting persons or the crash if an aircraft used for transporting persons."

[26]"Whoever poisons water or any substance which is to be used as food, before it is drawn or distributed for consumption, shall be punished by imprisonment for not less than fifteen years.If the act results in the death of any person, life imprisonment shall be imposed […]."

[27]"Whoever commits, through negligence, any of the acts designated in Articles 438 and 439 shall be punished: […] 2. by imprisonment for from one to five years, in the cases for which these prescribe life imprisonment; 3. by imprisonment for from six months to three years, in the case for which Article 439 prescribes the punishment of imprisonment."

6 Several Critical Points

The courts' extensive interpretations of such criminal provisions have been disputed and contested, primarily as regards the applicability of the offence of *Disaster*—as set out in Article 434 of the Italian Criminal Code—to cases of environmental disaster, which can be considered emblematic in this field. The issue has been the subject of considerable discussion by scholars, who have noted that it is premised on the extension of the crime beyond the limits strictly imposed by the norm (Flick 2015, pp. 12 ff.; Forzati 2015, pp. 1 ff.; Gargani 2008, pp. 468 ff.; Gargani 2010, pp. 418 ff.; Martini 2008, pp. 339 ff.; Piergallini 2004, pp. 278 ff.; Vergine 2013a, pp. 534 ff.; Vergine 2013b, pp. 644 ff.), and the Italian Constitutional Court has also engaged in this discussion, notably through the judgment of 1 August 2008, n. 327 (see Giunta 2008, pp. 359 ff.). The reason is clear: the so-called *Unnominated disaster*[28] (a literal translation of Art. 434 Criminal Code, *Disastro innominato*) may be relevant when a danger to public safety comes about as the consequence of instantaneous behaviour from which serious damage occurs, but it is not pertinent for those cases—very common in the environmental field—in which harm to the ecosystem arises as a cumulative effect of conducts perpetrated over a significant period of time (Gargani 2010, p. 423).

The point is confirmed by the constitutional court judgment already mentioned. In view of the alleged vagueness of its contents, the Court was required to evaluate the consistency of that provision with the constitutional principle of legality. And in fact the norm was confirmed and thus validated, since the judges argued that it was possible to infer, through a process of constitutionally permissible interpretation, a definite meaning to the term *other disaster*: the crime in question has to be placed in relation to the range of further and more defined provisions on disasters located in the same title of the code, and its interpretation has to be bound by homogeneity with the other cases covered in *Crimes endangering the public by means of violence*,[29] as can be deduced by the adjective "other."

Thus, *Unnominated disaster* must share basic features with other types of disaster in order to preserve its constitutional legitimacy. One such feature is *violence*, as a characteristic of the forbidden conduct, since the disaster has to consist in a

---

[28]Such terminology, where the adjective *Unnominated* stands for *nameless*, gives no characterization of the context and is meant to focus on the *generality*—namely the *broadness*—of a disaster as its peculiar feature, so as to distinguish that crime from other provisions on disasters which refer particularly to specific material contexts: *Massacre*, Art. 422 Criminal Code; *Arson*, Art. 423 Criminal Code.; *Flood, landslide or avalanche*, Art. 426 Criminal Code; *Shipwreck, sinking or aviation disaster*, Art. 428 Criminal Code; *Railway disaster*, Art. 430 Criminal Code; and so on.

[29]As Chapter I—Title VI of the Italian Criminal Code (where Art. 434 Criminal Code is located)—is entitled. After all, the necessary consistency of *Unnominated disaster* with common features of the other crimes collected in the same chapter can be deduced by the Ministerial Report which introduces the Italian Criminal Code, according to which that crime was required to answer to all the factors that could endanger public safety, and which could emerge in the future in line with industrial and technical development (see Ardizzone 1989, p. 274).

"destructive event of extraordinary dimensions," and a second is that "the event has to determine [...] danger for the life or the personal safety of an indeterminate number of people."

It is then evident how creative is the attitude of the courts which recur to *Unnominated disaster* to criminalize episodes of environmental disaster and, even more so, of historical pollution. Indeed, although historical pollution can be distinguished in more specific concrete scenarios, consistent with the variegated models thereof, its basic trait is the emergence of long-term consequences of certain activities after a period of latency. Furthermore, the recurring paradigm of historical pollution involves—directly or indirectly (as in the case of waste dumping)—industrial activities characterized by a dynamic of cumulative pollution. Conversely, according to the criteria of the Constitutional Court, a disaster occurs in a precise and delimited spatial/temporal circumstance, since it requires chronological immediacy between the conduct and the event comprising the disaster, which has to endanger public safety (Flick 2015, pp. 15 ff.). And consequently it can be excluded that the criminal response to such episodes could be expressed through Article 434 of the Criminal Code or by means of other crimes against public safety.[30]

Nevertheless, through the analysis of the pertinent case law, it can be surmised that in many cases criminal responsibility has been assigned to subjects who, directly or sometimes even indirectly, were in charge of the polluting activities which caused latent harms, as the analysis performed in Chap. 5 of this volume shows.[31]

Of course such a misinterpretation has significant consequences. First, as already noted, the extension of the perimeter of the crime beyond the limits imposed by the literal wording of the provision is an infringement of the fundamental principle of legality,[32] insofar as such principle forbids interpretation by analogy. Second, there is an issue of *materiality*. In the Italian criminal justice system, no responsibility can be assigned except where the *event*—i.e. the harmful consequences of illegal behaviour—can be considered causally related *beyond any reasonable doubt* to the human conduct which originated it (Stella 2003). It is easily seen how difficult it could be to establish a causal relationship with sufficient certainty in cases of historical pollution, due to the complexity of the processes (physical, chemical, biological, organic) involved in a long-term perspective. Furthermore, according to the second paragraph of Article 41 of the Criminal Code, "supervening causes shall exclude a causal relationship when they were in themselves sufficient to bring about the event." Thus when the causal factor only *contributes* to the event (for instance

[30]In other terms, that interpretation has to be evaluated as correct only when the material case concerns the category of accidents, since under those circumstances a sudden episode normally causes an immediate and acute danger to public health, although combined with long-term consequences to the extent that such an effect consists also in the form of danger to public safety (as in the case of the Seveso disaster).

[31]See Chap. 5 in this volume.

[32]Art. 1 Criminal Code; Art. 25, para. 2, Constitution Cost.

in the case of a non-point-source and diffuse pollution), that relation has to be proved under a standard near to certainty as well, and in order to get that high standard of proof it will be necessary to exclude the possibility that during the long period of latency, a further factor interrupted the causal chain which originated in the polluting activities, since such a factor would be independently sufficient to bring about the harm. Hence, due to the lack of scientific knowledge, and indeed the absence of complete awareness about the material dynamics of the facts, proof of causation is very often difficult.

Another critical aspect concerns the principle of culpability (*principio di colpevolezza*), which specifically stands for an advanced interpretation of the rule that requires the *personality* of criminal responsibility, as stated by the Italian Constitutional Court's interpretation.[33] That principle requires that—to be blameful—the event should be at least foreseeable when the conduct is undertaken. And thus, due to the extended chronological distance between the conduct and the event, it would be extremely difficult to prove negligence even if the link of causation has been demonstrated—not least because of the continuous evolution of scientific knowledge, according to which it is plausible that the subjects in charge of the dangerous activities were not aware of its potentially harmful consequences at the time.

Lastly, and as some scholars have already noted (Forzati 2015, pp. 8–9), we may remark that local courts seem more inclined to assign responsibility for such episodes than higher courts. Courts of first and second instance, which are called to rule over material facts, may be persuaded to give an answer to the quest for justice, sometimes compelled by the gravity of the harmful consequences, though at the cost of a rather forced interpretation of the legal norms. On the other hand, the Italian Supreme Court (*Corte di Cassazione*), which is called upon to evaluate exclusively legal (and not factual) aspects, often underlines the need for proper interpretation of the norms. Particular aspects will be discussed more in detail in the following subsections.

6.1 Causation

In trying to explain the creative attitude adopted by the courts, it is reasonable to opine that it is the complexity of the judicial proof that is crucial here (Gargani 2010, p. 427). And this complexity is surely related to the nature of the causal relationship, and thus the general difficulty of proving environmental harms and personal injuries (meant here as the secondary consequences of pollution, where the phenomenon in question also relates to health disasters).

[33]More precisely, in two relevant judgments the Court inaugurated the more nuanced interpretation of that principle: Italian Constitutional Court Judgment. 364/1988, Italian Constitutional Court Judgment. 1085/1988.

As anticipated above, the main reason for such difficulty consists in the need for specific proof of causation, since to assign criminal responsibility it is not sufficient to prove only general causation. Once this is understood, it is then comprehensible why Italian criminal courts often choose to recur to crimes against public safety, insofar as they carry protection for fundamental interests (i.e. personal safety, life), while at the same time they do not require proof of a causal nexus between the particular harm and the dangerous activity supposed to have determined it.[34]

However, such a scenario can also give rise to misinterpretations, notably in cases where there is only a *presumption* of danger to public safety, and where prosecutors are called upon to prove that an illegal activity has taken place. In fact, the danger always has to be understood as real, and therefore needs to be autonomously verified (Stella 2003, p. 523; D'Alessandro 2012).

To understand this critical remark it can be useful to focus on a material case. The judgment of the *Tribunale di Venezia* in the *Porto Marghera* case[35] is particularly helpful here, as the court analysed the matter deeply (D'Alessandro 2012, pp. 337 ff.): in fact, in the charge formulated by the prosecutors, the contamination of the area was claimed to be causally related to the industrial activity in question on the grounds that the latter had exceeded prescribed limits on emissions.[36] Somehow, an environmental disaster and, in particular, an event which endangered public safety, had been inferred simply from the infringement of emissions limits.

Contradicting the prosecutors, the court stated that such emissions limits only express quality standards which are consistent with a precautionary approach: that is, the purpose of the thresholds is only to avoid risk (or potential risk) to ecosystems and natural resources. As a logical consequence, the infringement of a limit cannot be treated as a basis on which to deduce the actual occurrence of contamination that could endanger public safety (D'Alessandro 2012, p. 339): it is necessary to prove—and not to presume—the *actuality* of the danger.[37]

A critical theoretical point arises here concerning the distinction between general and specific causation. Consistent with a judgment of the Italian Supreme Court in 2002,[38] to prove a causal relationship between behaviour and a subsequent event it is not sufficient to refer to a scientific law which describes in general terms the

[34]Consistent with the interpretation of the Constitutional Court, according to which it is necessary to prove only that there is danger for an indeterminate range of people.

[35]For more details, see Chap. 5, Sect. 1.1 in this volume. For a recent and complete analysis of the case which focuses particularly questions of causation, see Vallini (2015, p. 25 ff).

[36]In particular those contained in the *Protocollo di Intesa per la Laguna di Venezia*.

[37]The point is even more pertinent given that the local court of Venice argued that Arts. 434 and 439 Italian Criminal Code are examples of the most anticipated protection (*reati di pericolo astratto*), so that in order to prove the crime itself it would be sufficient to prove the correspondence of the material fact with the legal provision. That would imply that it is not necessary to verify the material occurrence of the danger (see *Tribunale di Venezia*, 29 May 2002, pp. 497–498). Note also that the Italian Constitutional Court judgment n. 327/2008, which has been already analysed, gives a quite different interpretation of the crime structure.

[38]See *Cassazione penale, Sezioni Unite*, 10 July 2002, n. 30328.

relationship between a casual factor and the relevant effects; rather, it is necessary to demonstrate a particular causal nexus—that is, to verify that the concrete event (evaluated *hic et nunc*) has been caused by the material conduct, understood as the pertinent causal factor. Of course, the implementation of such a model has to take into account that scientific laws frequently cannot be considered universal, since they describe neither certain causal connections, nor relationships with a degree of near certainty. Hence, in order to prove a causal relationship under those circumstances, it is necessary to exclude the incidence of further causal factors: indeed, the method guarantees the achievement of a high level of rational reliability (*alto grado di credibilità razionale*) or, according to another formula, a standard of logical probability (*probabilità logica*).[39]

It should thus be clear how difficult it is to prove a causal link, and particularly so in cases of historical pollution, where the contamination is often the consequence of diffuse pollution, that is, the harmful consequence does not emanate from a unique and pinpointed source (point-source pollution), but is rather the consequence of a combination of several causal factors (non-point-source pollution, diffuse pollution). Moreover, the causation can hardly be proved when the contamination consists of cumulative effects of reiterated pollution over time, in an accretive manner. Finally, faced with a multifactorial dynamic of causation, it would be extremely difficult to rule out a supervening cause which might have blocked the causal relationship, under the regime of the second paragraph of Article 41 Criminal Code (Piergallini 2004, pp. 226 ff.).

However, looking at the case law it appears that difficulties of this sort are often overlooked, and this can lead to a clash between the decision and the legal framework itself. In discussing the local court decision in *Porto Marghera*, a certain inconsistency has been noticed between the premises of the argumentation and its conclusion (Centonze and D'Alessandro 2003, pp. 156 ff.). The judicial decision formally recognized that the existence of a relevant scientific law (which can be named the *explanans*) does not suffice to prove the concrete event *hic et nunc* (the *explanandum*)—that is, the event considered in its uniqueness by considering its particular features and the material context of its occurrence. But the Court also stated that causation in the case of physical injuries to people—and even their deaths—could be considered proved in consideration of epidemiological studies, as they take into account the necessary scientific generalizations (Centonze and D'Alessandro 2003, p. 157). Depending as it does on such an argument, the judgment faces at least two criticisms: first, the proof of causation consisted merely in the description of general causation, with no attention to the need to establish specific causation, and second, the body of scientific knowledge relied upon does not contribute to the causal explication of the event (Stella 2003, pp. 294 ff.;

[39]For a recent application of such criteria to a case of historical pollution (also known as the *Tirreno Power* case), see *Tribunale di Savona*, 11 March 2014, p. 18.

Piergallini 2004, pp. 77 ff.), since epidemiological evidence can only describe the impact of such a toxic substance on the population in general terms.[40]

As the previous example indicates, it is in the field of *health disasters*—as a complementary aspect of historical pollution—that this issue is particularly manifest. In fact the aetiopathogenetic development of a disease that is hypothetically related to exposure to toxins is even more difficult to prove than the dynamics underpinning environmental harms. The Italian case law illustrates a critical issue concerning proof of causation in the case of human—often professional—exposure to asbestos, particularly in cases of pleural mesothelioma, a multifactorial disease which typically reveals itself after a very long period of latency (35–40 years on average) (Fontanella 2012, pp. 1537 ff.; Centonze 2012, pp. 1523 ff.). At least five points are critical in proving causation in cases of such a disease (Fontanella 2012, pp. 1543 ff.): (i) its multifactorial nature, which means that multiple plausible sources can be relevant, and thus factors other than asbestos exposure can activate the pathogenetic process; (ii) the ubiquity of asbestos, such that the related risk can be considered subtle and proof of specific causation is difficult; (iii) the impossibility of ascertaining the precise moment and level of exposure that activates pathogenesis, or of clearly identifying the responsible party; (iv) doubts concerning the causal relevance of subsequent exposure; and finally (v) the question (which is effectively unanswerable) whether *omissions*—namely failures to act on a legal duty ordinarily incumbent on those in charge of production throughout the period of pathological latency—are to be considered causally relevant. Consistent with the legal framework described above, to answer these questions positively it is necessary to prove beyond any reasonable doubt that compliance with a precautionary standard would have impeded the event.

Particularly relevant in this regard is the *Eternit* case, perhaps the most significant of the cases which pertain to this typology (Scaroina 2015, pp. 1 ff.; Zirulia 2013, pp. 471 ff.; Zirulia 2015, pp. 79 ff.).[41] Actually, the inclusion of human exposure—whether professional or not—to toxins as a theme within the panorama of cases of historical pollution requires some further remarks. Although the topics may seem rather distant from each other, in effect they are closer than it appears at first sight: so close, indeed, that scholars have not only compared them but actually associated them with each other. These different paradigms (historical pollution and human exposure to toxins) share several overlapping features, which recall—at their core—the features that are taken to characterize historical pollution within this research project.

First, and most generally, a toxic substance can simultaneously contaminate the environment and injure individuals. Specifically, the consequences for human health can be considered a secondary effect of the contamination, considering, for example, the case of asbestos exposure that can be treated neither as professional

[40]For a different opinion, consistent with a rather different definition of the general regulation of proof of causation, see Masera (2014, p. 343 ff).

[41]See Sect. 1.5 of Chap. 5 in this volume.

nor domestic (regarding the relatives of workers directly exposed) but as properly *environmental*, since it hits the neighbourhoods surrounding the plants where asbestos products are made (Fontanella 2012, p. 1542). Moreover, the consequences can arise significantly later than the behaviour which caused them. Second, the allocation of personal responsibility poses the same critical issues in both scenarios, starting with the proof of a causal link.[42] The point seems to be confirmed by the Eternit case, insofar as the alleged charge contemplated a case of environmental disaster (*ex* Art. 434 Criminal Code), and this was explicitly referred to in the pertinent decisions.[43]

6.2 *Mens Rea: Negligence, the Precautionary Principle, and Oblique Intent*

The long period of time that ordinarily separates the causal behaviour and the subsequent contamination—which typifies cases of historical pollution—also entails difficulty in proving mens rea (even supposing that the causal link between those elements can be ascertained beyond any reasonable doubt). Here we focus on two topics: first, critical aspects pertaining to the alleged charge for negligent crimes —which should be considered the default perspective, since generally speaking it can be excluded that whoever is in charge of an industrial activity actually *wants* on purpose and with full awareness to contaminate the ecosystem; second, the slow process through which case law has developed from responsibility for negligence to alleged charges for wilful behaviour.

Again *Porto Marghera* stands as a milestone, here with regard to negligence (Notaro 2015, pp. 51 ff.). As remarked above, the local court decision excluded a hypothesis of disaster, while at the same time it stated that the causal nexus between the industrial pollution and the physical harm (cases of hepatic angiosarcoma, related to exposure to vinyl chloride monomer (VCM) during the period 1969–1973) had been proved. Nevertheless, the decision was for acquittal, since negligence was excluded. The Court of Appeal judgment overturned this latter point of the first-degree decision, although it confirmed it on further topics: the occurrence of disease depended on negligent omissions by those who had succeeded in management positions over the period in question. In this regard several criticisms emerge (Piergallini 2005, pp. 1684 ff.; Centonze 2004, pp. 125 ff.).

In short, many aspects seem to block a proof of negligence, principally due to the diachronic dimension of the facts. Often the event is not predictable *ex ante*—that is, at the moment when the relevant behaviour was undertaken—because of the

[42]For the similarities between environmental harms and toxic tort suits, see Note 2015, pp. 2256–2257.

[43]*Tribunale di Torino*, 13 February 2012, *Pres*. Casalbore; *Corte d'Appello di Torino*, 3 June 2013; *Cassazione penale* sez. IV, 23 February 2015, n. 7941.

lack of scientific knowledge, which is constantly evolving. As a consequence, the specific risk could neither be perceived, nor was foreseeable, as the legal framework pertaining to the proof of negligence requires (Piergallini 2004, pp. 422 ff.).

According to a scholar who has analysed the judgment in depth, the following aspects are critical: the diligent care required was not intended to avoid the specific risk related to the occurrence of the specific disease (hepatic angiosarcoma, addressed as the event *hic et nunc*), but to avoid different diseases; more radically, at that time it was not even perceivable that such a risk was a particular consequence of exposure to VCM; although some studies were available, they were considered pioneering and were not yet widely recognized, so that there could not be considered to be a duty of care to prevent exposure (Piergallini 2005, pp. 1686 ff.).

Faced with such an obstacle, the Court of Appeal in the *Porto Marghera* case simply *overtook* it, recurring to a rather different criterion in order to replace the proof of negligence: what has been named the *default option* (Pulitanò 2006, pp. 786–787). According to this interpretative scheme, the alleged responsible parties should have behaved in a manner consistent with that peculiar risk, even though it was uncertain and unknown. As has been noted, invoking the *default option* is at its heart intended to—literally—"sterilize" the "diachronic dimension of the facts," due to prolonged extension of the latent causal dynamics (Piergallini 2005, pp. 1694–1695). After all, this argument embodies the adhesion to a presumptive logic in proving negligence, consistent with a precautionary approach (Piergallini 2005, p. 1695).

It is precisely the precautionary principle that plays a significant role in what can be considered a sort of evolution in court decisions regarding mens rea in facts of historical pollution: namely, an increasingly widespread tendency for even more severe charges to be proposed by prosecutors, premised on or assuming the relevance of the precautionary principle for *individuals*, rather than only for public institutions.

Italian scholars have rather recently began to wonder whether the precautionary principle can contribute to defining negligent behaviour, by imposing more severe rules of care under conditions of uncertainty (Castronuovo 2012, ; Forti 2006, pp. 155 ff.; Forti 2007, pp. 573 ff.; Giunta 2006, pp. 227 ff.; Ruga Riva 2006, pp. 1743 ff.). This topic has mainly been related to the environmental field (Ruga Riva 2015a, pp. 10 ff.; Rotolo 2016, pp. 130 ff.), also on the assumption that Article 3-*ter* TUA, which explicitly recalls the precautionary principle, could be applied directly to individuals. On this basis, courts have increasingly used arguments based on that principle to charge polluters under general conditions of uncertainty, sometimes going beyond the commitment contained in emissions permits themselves and requiring more severe duties of care consistent with ALARA ("as low as reasonably achievable") standards and with best available techniques (BAT).[44] Thus, polluters who comply with the emission limits contained in the formal

[44]Which are defined in accordance with the *Best Available Techniques Reference Documents* (BREFs).

authorizations provided by competent public institutions could still be charged, where their conduct was not consistent with the more severe standards imposed by the principle of precaution.[45]

But to the extent that the precautionary principle provides a criterion to handle uncertainty, requiring the avoidance of any activity that could confer even potential risks, it is incorrect to extend it also to individuals with the potential to infer criminal responsibility. To expect private parties to manage situations that are strongly marked by uncertainty eminently entails placing demands on people who do not have the necessary competence or tools; indeed, not even public institutions are fully equipped to face such situations. This smacks of *passing the buck* to citizens (Pardy 2002, pp. 63 ff.; pp. 70 ff.). As Pardy has noted:

> if the precautionary principle is to be applied to prosecutions for violation of this provision, private citizens and commercial operations will be required to do what government cannot or will not do itself, which is to determine and identify the circumstances that fall on either side of the line. The regulator thus avoids applying the precautionary principle to itself, and instead foists it upon parties with fewer investigative resources. (Pardy 2002, pp. 76–77)

After all, the domain addressed by the precautionary principle is characterized precisely by uncertainty, and one of its aspects concerns the chronological distance between the occurrence of the relevant behaviour (understood as the causal factor), and the subsequent harmful event (Pongiluppi 2010, pp. 225 ff.). The point seems to be particularly prominent in cases of historical pollution because of the period which separates the pollution and the emergence of contamination. Indeed, the prolonged latency implies reasonable unpredictability of the consequences, along with the precise nature of the relevant causal process. Moreover, the temporal distance between conduct and event surely will also encompass a period of scientific progress, so that the knowledge available by the time the consequences emerge will necessarily be more extensive than that available when the dangerous activity was carried out.

Looking at the criminal proceedings in historical pollution cases, it seems that the attitude towards defendants has become much more severe as regards the standard of duties of care—through the recourse to the precautionary principle—and that a transformation of the typology of charges has taken place. Slowly, in shifting to the precautionary approach, prosecutors have started to charge intentional crimes, essentially since the proceedings concerning the *Eternit* case (Castronuovo 2015, pp. 107 ff.). According to some scholars, this strategy can be interpreted as a sort of compromise (Masera 2014, pp. 364–365): on the one hand, prosecuting attorneys resort to indicting for crimes which criminalize typologies of facts (crimes of danger) that are different to and less serious than the material episodes actually require,[46] probably to avoid greater difficulties regarding the proof

---

[45]For several examples in this respect, see *Tribunale di Savona, Ufficio del Giudice per le indagini preliminari*, 1 March 2014; *Tribunale di Taranto, Ufficio del Giudice per le indagini preliminari*, 25 July 2012; *Tribunale del Riesame di Taranto*, 7 August 2012, *sez. feriale*.

[46]Consider here cases of life or personal health impairment.

of causation; on the other hand, charges have begun to be more severe on the subjective imputation of responsibility (i.e. mens rea configuration, intentional rather than negligent). As remarked above, from the perspective of the precautionary principle such a configuration of criminal responsibilities implies the foreseeability of harmful events (although they are intrinsically uncertain), and insofar as one assumes the seriousness of the resultant contamination and of the further consequences, the (presumed) foreseeability ends up being transformed into plain awareness. Of course that state of mind then becomes compatible with charges of intentional crimes, mainly applying the figure of the so-called *oblique intent*.

In this respect, it is useful to compare the *Cremona Petrochemical Plant* case[47] and the *Bussi sul Tirino* case, the similarities of which have been outlined above, along with the relevant judgments.[48] In both cases the industrial production and the related activities caused not only contamination of the soil, but also of the whole ecosystem, insofar as groundwater *poisoning* has been verified as a long-term effect.[49]

In the *Cremona* case, oblique intent was deemed as proved (so that a charge of a wilful crime has been confirmed[50]) on the basis of two different arguments: first, although not acting with the purpose of causing an environmental disaster, defendants foresaw such an event as a concrete outcome; second, they did not work to prevent its occurrence and so they exacerbated the situation.[51] Finally, the court stated that the defendants accepted the risk of the occurrence of an environmental disaster, and that—according to that opinion—this consideration satisfied the burden of proof of oblique intent.

In this respect, the decision also emphasizes aspects that to some extent recall criminological concerns about *green-washing*. Indeed, the company reported the relevant circumstances, although such a fulfilment has to be considered insufficient and incomplete, since it tried to cover up the extent of the groundwater contamination; moreover, the purpose of this under-reporting was to avoid the economic impact of the remediation activities that would thus have been activated, and also the reputational damage in the context of the community.[52]

---

[47]See Chap. 5 in this volume, respectively, Sects. 2.4 and 2.3.

[48]*Tribunale di Cremona*, 18 July 2014, Judge Salvini; *Corte d'Assise di Chieti*, 19 December 2014 (ud.).

[49]This point also includes cases at issue as regards historical pollution. In fact, in the decision on the *Cremona Petrochemical Plant* case, several arguments were advanced that it was not to be considered a case of historical pollution, since the contamination had not been inherited from the past, due to the fact that different companies had succeeded each other on the same property (*Tribunale di Cremona*, 18 July 2014, p. 134 ff.). After all, in the light of the definition of historical pollution adopted in the present research, groundwater contamination has to be the result of a diachronic causal process, which as a consequence could emerge only after a long period and after a prolonged state of latency (as the judgment also recognizes: p. 385).

[50]Although not for all the defendant parties.

[51]*Tribunale di Cremona*, 18 July 2014, p. 386.

[52]*Tribunale di Cremona*, 18 July 2014, pp. 376–377.

But the decision in the *Bussi sul Tirino* case is completely different. Given the ontological difference between awareness of the contamination and the actual intention of poisoning the groundwater, no significance can be assigned to the attempts to cover up environmental harm in proving intention.[53] Moreover, oblique intent cannot be inferred from the presumed awareness of the probability that the event will occur, since it requires proof of a state of mind consistent with the intention, and as such cannot be considered simply as the acceptance of risk.[54]

In addition, according to the recent decision on this topic by the Italian Supreme Court, in its more extended composition (*Cassazione a Sezioni Unite*), *Frank's formula*[55]—among other criteria—has to be considered the legal test of oblique intent (Eusebi 2015, pp. 823 ff.). Such a formula upholds that oblique intent is verified only if, in the hypothetical situation in which the defendant realized the occurrence of an event as certain (and not only probable), it can be proved that under those circumstances he or she would have behaved in the same unlawful way.[56] Of course this is an extremely hard test, and in addition it has been considered necessary to verify the particular state of mind that the "intention" requires. And in fact, in *Bussi sul Tirino* the charge of intentional disaster was not confirmed.

Critically evaluating the two decisions, we note that they challenge the rationale of the above-mentioned decision of the *Sezioni Unite*, 24 April 2014, n. 38343, in the *ThyssenKrupp* case, which reconsidered the element of oblique intent and expressed the need for a more accurate (and not presumptive) proof of intention. Unlike in the former case (*Cremona*), in the latter (*Bussi sul Tirino*) the judges more strictly followed the path traced by the decision of the *Sezioni Unite*.

6.3 The Widening of the Range of "Responsible Parties"

Under the circumstances so far described, the crime is generally not the consequence of a personal choice; rather, the offence originates in more complex processes involving many individuals, according to the organizational structure of a corporation (Piergallini 2004, pp. 301 ff.).[57] Responsibilities are thus not centralized in one or even a few persons in managerial or administrative roles, but are diffused among the decision centres involved in each decision-making process. More precisely, the decentralization of this process can follow two different vectors: vertical and horizontal (Piergallini 2004, pp. 313 ff.). While the former dimension evokes the distribution of competences to subordinates by mandate, the latter implies the responsibilities of those who, although not directly involved in the

[53] *Corte d'Assise di Chieti*, 19 December 2014 (ud.). p. 149.

[54] *Corte d'Assise di Chieti*, 19 December 2014 (ud.). pp. 152–153.

[55] The prominence of this formula as legal test of the oblique intent was contested in *Tribunale di Cremona*, 18 July 2014, p. 384 ff.

[56] *Corte d'Assise di Chieti*, 19 December 2014 (ud.), p. 154.

[57] For aspects concerning corporate liability, see Chap. 6 in this volume.

management sphere, still play a significant role in the decision-making, mainly through internal or external control of the process at issue.

The theme has been widely investigated (Centonze 2009) and debated under the provision of the second paragraph of Article 40 of the Criminal Code, according to which "[f]ailing to prevent an event which one has a legal obligation to prevent shall be equivalent to causing it." This raises a number of questions: Under what conditions are those who do not have direct responsibility for decision-making also bound by the obligation of preventing the event? And, accordingly, among the contents of such an obligation could there also be considered the duty to impede a crime committed by others? Finally, are the *warning signs*[58] sufficient to activate the obligation to prevent the offence for those who are so obliged?

The topic has a significant role in the overall criminal framework and is common to all contexts characterized by complex organization—for instance, cases related to crimes of bankruptcy (Caputo 2015, pp. 905 ff.)—as well as environmental crimes (Ruga Riva 2010, pp. 859 ff., 866 ff.). As already discussed above, the misdemeanour provided by Article 257 TUA raises critical issues concerning the extension of criminal responsibilities to the owner of contaminated soil, even if he or she was not directly involved in the polluting activity.[59] And the point is even more critical still, considering that in the administrative field a recent decision of the European Court of Justice has pointed out that under the Italian regulation an *owner who is not responsible* is not obliged to undertake remedial procedures.[60]

Furthermore, the specific features of historical pollution introduce additional issues pertaining to the same core problem: notably, that a further dimension (in addition to the vertical and the horizontal) emerges as regards the extension of responsibility, namely the chronological dimension. This represents the fragmentation of the alleged fact into many segments, each related to the particular person with relevant responsibility over the whole period of time. In order to recap the difference as compared to the issues already analysed, in cases of historical pollution the extension of responsibilities over many individuals imposes not only a synchronic evaluation, but also a diachronic one.

Effectively, the courts resort to the second paragraph of Article 40 of the Criminal Code to criminalize those who successively failed in their duty to avoid the occurrence of the harmful event: according to such an interpretation, those supervising also had a duty to impede the event and to minimize the risk, although they did not directly activate it. Moreover, another general rule has also been used with the same purpose in regard to negligent offences: Article 113 of the Criminal Code, entitled *Collaboration in crimes of negligence*.[61]

---

[58]We translate the term *segnali d'allarme* as "warning signs": more precisely, these are material signals consistent with the likelihood that an offence is going to be committed, ordinarily by others.

[59]For the relevant case law, see Chap. 5, Sect. 2.6 in this volume.

[60]Decision ECJ, 4 March 2015, proceeding C-534/2013.

[61]The full text of the provision is as follows:

In both cases the intended objective is to draw a sort of chain that connects all of the subjective positions over time and links them together, from the period when the polluting activity started until the end of latency of the effects when the contamination emerges, or, sometimes, when proceedings are activated. To grasp this point let us also observe the limit of the attitude in question: to splinter the whole fact into many small fragments is inconsistent with the need to precisely define the causal contribution of each person's conduct (by action or by omission), which has to be individually appreciated and cannot be generalized in the frame of collective responsibility (Piergallini 2004, pp. 336 ff.).

According to some scholars, the logical conclusion of the charge as built up in *Porto Marghera* was to criminalize the whole history of the industrial activity in question, right back to its establishment in 1917 (Perini 2002, p. 398). Referring to Article 113 Italian Criminal Code, the underlying aim of the prosecutors was to avoid having to prove specific causation of the event and of the contributions of each of the twenty-eight people charged in having succeeded in relevant positions over that time (Piergallini 2004, p. 76).

Nevertheless, such an approach conflicts with a rigorous interpretation of the Italian legal framework, as the local court of Venice pointed out in the case described: criminal responsibilities cannot be intended as cumulative since they cannot respond to the logic of spreading the allegations over a wide range of indistinct people whose duties are only blurrily stratified over time. Against the converse opinion are ranged both the culpability principle and, even more so, the fundamental value according to which the criminal responsibility has to be *personal* as stated in the Italian Constitution.[62]

6.4 Statute of Limitation

A further weakness of criminal enforcement in cases of historical pollution concerns the statute of limitation, which is provided in Italian criminal law at Articles 157 ff. of the Italian Criminal Code. Particularly relevant under this regulation is the concept pertaining to the *time of perpetration of the offence*—the *tempus commissi*

(Footnote 61 continued)

When, in a crime of negligence, the event has been brought about through the collaboration of more than one person, each of them shall be subject to the punishment described for that crime.

The punishment shall be increased for anyone who has induced others to collaborate in the crime when the conditions prescribed in Article 111 and in subparagraphs (3) and (4) of Article 112 are present.

[62] *Tribunale di Venezia*, 29 May 2002, pp. 171, 486.

delicti—from which the period of limitation begins according to Article 158 Italian Criminal Code.[63]

For reasons already noted, prosecutors prefer to indict for crimes of danger against the public safety. Accordingly, the perpetration of the crime coincides with the moment when a (real) danger occurs with respect to the community in general, and not to pinpointed individuals. Thus, consistent with the basic rule of the *immediacy* of the relation between the conduct—defined as a huge episode of damage—and the event (the danger), the offence has to be considered *instantaneous*. For this typology of crime the perpetration overlaps with the fact and cannot be shifted to a later moment, as in cases of the so-called *persisting* or *continuing* offences.

Applying this reasoning to cases of historical pollution, one can easily see how statute of limitation regimes could threaten effective enforcement of the indicted crime, insofar as according to this legal regime the relevant period of time starts to run at the moment the dangerous pollution occurs. Of course, under those circumstances the prosecutor faces a trade-off: indictment for offences of danger—rather than of homicide, for instance—could avoid the difficulties in proving the causal nexus; conversely, the statute of limitation could impede sentencing according to the facts in those cases.

The described scenario seems to fit perfectly with the judicial case known as the *Eternit* case.[64] The prosecutors' strategy was to describe the alleged crime as a *continuing offence*, or alternatively an *offence with continuing effects*: although through different definitions, the purpose in both cases was to postpone the moment from which the period of limitation starts to run. According to the former interpretation (*continuing offence*), the crime should be deemed as ongoing so long as the indirect consequences of the fact (the further risks activated through the danger) persist.[65] According to the latter (*offence with continuing effects*), although the fact has to be considered as concluded, its perpetration should be progressively extended in relation to each disease or to each death that occurs in relation to the causal factor.[66]

The Supreme Court decision[67] on this case denied both interpretations and, consequently, deemed the crime as time-barred. In general the offence may be

---

[63]Art. 158 Italian Criminal Code, *Beginning of the period of limitation*:

> The period of limitation shall run, in the case of a completed offense, from the day of consummation; in the case of an attempted offense, from the day on which the offender ended his activities; in the case of a persisting or continuing offense, from the day on which its persistence or continuation ended.
>
> When the law makes punishability for the offense depend on the occurrence of a subsequent condition, the period of limitation shall run from the day on which the condition occurred. However, in cases of offenses punishable on complaint, petition or demand, the period of limitation shall run from the day the offense was committed.

[64]See Chap. 5, Sect. 2.5 in this volume.

[65]Consistent with this interpretation, see Tribunale di Torino, 13 February 2012, p. 509 ff.

[66]As a consequence, the end of the period of perpetration coincides with the end of the statistical excess in disease or death, according to the relevant epidemiological studies. See Corte d'Appello di Torino, 3 June 2013, p. 588.

considered as *persisting* only when the legal provision expressly consents to that interpretation, and case law interpretation has to be bound to the legal provision in this respect. Consequently, specifically in this case the perpetration of the crime coincides with the disaster (understood as a danger to public health) and not with the further episodes of disease or death, which have to be considered merely as aggravating elements (Castronuovo 2015, pp. 127–128).

7 A Brand New Regulation: Will It Be Enough?

In connection with the new legal regulation of environmental crimes, it has been argued that the acquittal sentence in the *Eternit* case, and the subsequent public outcry were factors that could have influenced the introduction of the Law of 22 May 2015, n. 68, which was approved in the immediate aftermath of that decision, although a reform project had been discussing it in the Italian Parliament for a long time (Siracusa 2015, p. 2).

At first sight, the two new environmental crimes seem to be relevant to the perspective adopted in the present research: these are *Environmental pollution* (Art. 452-bis of the Italian Criminal Code) and *Environmental disaster* (Art. 452-quater), where offences are defined as negligent under Article 452-quinquies, *Negligent crimes against environment*.[68]

In fact, no meaningful arguments can yet be offered on this matter, due in general to the lack of case law, and in particular to the lack of decisions referring to cases of historical pollution. Nevertheless we will make several remarks in an attempt to sketch out why the new regime seemingly cannot assure more effective enforcement in cases of historical pollution, at least in the immediate term.

According to the Italian legal framework on the application of the criminal law over time, neither crime will be applicable in the near future, since they are not applicable to pollution that occurred before the promulgation of the law, and since the new discipline is generally less favourable to the defendant than the existing offences.[69] Indeed, assuming a connection between the abstract facts described in Article 434 of the Italian Criminal Code and the new crime of *Environmental*

---

[67] *Cassazione penale*, sez. IV, 23 February 2015, n. 7941.

[68] Clearly, environmental contamination can also cause health disasters. In this perspective, it is worth pointing out the new crime introduced under Art. 452-ter Italian Criminal Code, *Death or injury as a result of the crime of environmental pollution*. Of course, the offence also requires proving causation between the illegal behaviour (independently relevant as *Environmental pollution*, under Art. 452-bis Italian Criminal Code) and the harm, in terms of injury or death.

[69] See the third paragraph of Art. 2, Italian Criminal Code, entitled *Successive penal laws*: "If the law in force at the time an offense was committed and subsequent laws are different, that law shall be applied the provisions of which are more favourable to the accused, unless a final judgement has been pronounced." The new regime could prevail—assuming it is more favourable—only when the extenuating circumstance introduced in Art. 452-decies Italian Criminal Code is applicable.

disaster,[70] it is easy to see that the recently introduced offences entail a more severe regime (Ruga Riva 2015b, p. 27).[71] The same conclusion has to be extended also to the negligent crimes.

There is a further reason why doubt must be cast on the applicability of the new crimes. Paying careful attention to the legal wording, it emerges that it is not every conduct that can be relevant under that norm, but only *unlawful* behaviour.[72] And as it has been interpreted (Ruga Riva 2015b, pp. 5 ff., 29 ff.), the adverb *unlawfully* refers to a category of situations defined through contrast to the legal framework: regulation, authorization contents, and any legal regime which defines the conditions necessary to lawfully perform an activity. Hence, the application of the crime at issue to *legal* behaviours—also intended as segments diachronically unified in a wider *fact*—is not permitted.

Thus we may presume that the new provisions will not be useful for criminalizing historical pollution stemming from the past (conversely, they probably will work for cases of future historical pollution), when the causal conducts are not to be considered unlawful. Such a circumstance could occur in cases of historical pollution for two different reasons: first, because sometimes contamination represents the cumulative effect of *lawful* conduct, inasmuch as it is not in violation of the limits of emissions; second, because it is possible that such behaviour originated from a time when no legal framework for industrial activities or for environmental protection had yet been introduced.[73]

Moreover, some of the aspects that undermined the effectiveness of criminal enforcement in the previous regime are still significant here, and thus deserve deeper scrutiny.[74] Both new provisions require proof of the *event*, understood as the harmful consequences for the environment: specifically the "significant and measurable impairment or deterioration" of environmental resources in the case of *Environmental pollution* (Art. 452-bis Italian Criminal Code), and "the irreversible alteration of the ecosystem balance," "the alteration of the ecosystem balance which removal is particularly onerous or achievable only through exceptional measures," or "the offence against public safety assessed on the basis of the extent of the

---

[70]Conversely, because of the innovative structure of the crime, *Environmental pollution* is incommensurable with pre-existing legal provisions.

[71]Only the provision under Art. 439 Italian Criminal Code may be considered more severe than *Environmental disaster*, due to the comparison of the sanctions: in the former case "imprisonment for no less than fifteen years," while in the latter "imprisonment from five to fifteen years."

[72]The Italian term translated as "unlawfully" is *abusivamente*.

[73]See Sect. 3.

[74]Attention will be focused especially on causation and the statute of limitation, since these represent the most pertinent traits within the new regime, but this shall not preclude interest in aspects concerning mens rea and the possible extension of personal responsibility to several parties over time. The critical aspect regarding both these latter points is the interpretation of a general rule given by the criminal courts, so that only an analysis of the forthcoming case law on the new crimes will enable the formation of any opinion. When it comes to the new legal regime, the introduction of negligent crimes under Art. 452-quinquies Italian Criminal Code will introduce an alternative criminal charge.

impairment and its negative effects or of the number of people offended or exposed to danger," with respect to *Environmental disaster* (Art. 452-quater). Hence, doubts about the improved effectiveness of the new regime again arise concerning the complexity of the causal processes and the associated dynamics that are typical in the production of environmental harms (D'Alessandro 2016, pp. 94 ff., 102). Furthermore, and again consistent with the basic features of historical pollution, it could even be harder to prove a causal link when the event is the consequence of diffuse, non-point-source pollution; or when it is the effect of a cumulative dynamic of production; or in cases where, because of the lack of scientific knowledge, it is not known whether further (undetermined) causes could autonomously have brought about such an event.

Conversely, at the level of the statute of limitation regime, three different aspects have been introduced, apparently to facilitate criminal enforcement. First, the moment from which the limitation period starts to run has been postponed, as it depends on the occurrence of the event, although it remains difficult to precisely determine in concrete terms the perpetration of the offence (D'Alessandro 2016, pp. 102–103). Actually, such an evaluation turns on the new crimes' structure, which requires the occurrence of the event in terms of the different scenarios of environmental harms considered in the crimes' wording. Secondly, the period of limitation has been extended, as a consequence of the more severe sanctions introduced. Indeed, according to Article 157 of the Italian Criminal Code, that period depends on the maximum punishment provided for each crime. And thirdly, the new regime has introduced a rule (Art. 157, para. 6 of the Italian Criminal Code) according to which the period of limitation is doubled when new environmental crimes occur (all those included in the new Title VI-bis of the Italian Criminal Code, *Crimes against the environment*).

However, from a critical perspective it must be noted that these aspects of the new legal regime do not ensure effectiveness of criminal enforcement. Of course, there will be a longer period available to pursue criminal proceedings in these cases, before the time-bar comes into force. But the critical elements concerning the proof of the elements that constitute the crimes could undermine their effectiveness, even despite the extension of the period of limitation.

In conclusion, we may observe a relevant innovation brought in by the recent regulation, namely that *Unnominated disaster* is now expressly mentioned at the beginning of Art. 452-quater of the Italian Criminal Code wording: "Apart from the cases designated in Art. 434 […]." Because of the reference to that pre-existing offence and the (supposed) limited pertinence of the new crimes to cases of historical pollution, it may be presumed that under the new regime the so-called *creative approach* of the courts discussed above, rather than being superseded, will be vindicated and perhaps even incentivized. In other words, since the effectiveness of criminal enforcement in cases of historical pollution seems still undermined, the explicit reference to Article 434 of the Italian Criminal Code provision—which has to be considered as prevailing over the application of the new crime when a single fact is relevant to both felonies—could be understood as an element that both authorizes and legitimates extensive legal interpretation of that provision, consistent with the case law attitude discussed above. After all, faced with the choice between being charged under the new crime of *Environmental disaster* (Art. 452-quater of the Italian Criminal Code), or the pre-existing provisions under Article 434 of the Italian Criminal Code, it is clear that the defendant party would prefer the latter, inasmuch as it is more favourable and, unlike the former charge, does not imply corporate liability.[75]

References

Ardizzone, S. (1989). Crollo di costruzioni e altri disastri dolosi (voce). *Digesto discipline Penalistiche, III* (pp. 274–279). Torino: Utet.

Brisman, A. (2014). The visual acuity of climate change. In P. Davies, P. Francis, & T. Wyatt (Eds.), *Invisible crimes and social harms* (pp. 61–80). New York: Palgrave Macmillan.

Caputo, M. (2015). Dalla teoria dei "segnali di allarme" alla realtà dell'imputazione dolosa nel concorso dell'amministratore non esecutivo ai reati di bancarotta. *Rivista Delle Societa*, 905–940.

Castronuovo, D. (2012). *Principio di precauzione e diritto penale. Paradigmi dell'incertezza nella struttura del reato*. Roma: Aracne.

Castronuovo, D. (2015). Il caso Eternit: Omissione di cautele infortunistiche e disastro ambientale dolosi. In L. Foffani & D. Castronuovo (Eds.), *Casi di diritto penale dell'economia. II. Impresa e sicurezza (Porto Marghera, Eternit, Ilva, ThyssenKrupp)* (pp. 107–147). Bologna: Il Mulino.

Cemeteri, L. (2006). *Ritorno a Seveso. Il danno ambientale, il suo riconoscimento la sua riparazione*. Milano: Bruno Mondadori.

Centonze, F. (2004). *La normalità dei disastri tecnologici. Il problema del congedo dal diritto penale*. Milano: Giuffrè.

Centonze, F. (2009). *Controlli societari e responsabilità penale*. Milano: Giuffrè.

Centonze, F. (2012). Il problema dell'accertamento del nesso di causalità tra esposizione professionale ad amianto e mesotelioma pleurico. *Rivista Italiana Di Medicina Legale*, 1523–1535.

Centonze, F., & D'Alessandro, F. (2003). La sentenza del Tribunale di Venezia sul petrolchimico di Porto Marghera. *Rivista Giuridica Dell'Ambient*, 156–159.

Corona, G. (2015). *Breve storia dell'ambiente in Italia*. Bologna: Il Mulino.

D'Alessandro, F. (2012). *Pericolo astratto e limiti-soglia. Le promesse non mantenute del diritto penale*. Milano: Giuffrè.

D'Alessandro, (2016). La tutela penale dell'ambiente tra passato e futuro. *Jus*, 83–112.

Di Landro, A. (2014). *Bonifiche: il labirinto della legislazione ambientale dove le responsabilità "si perdono"* (pp. 1–48). www.penalecontamporaneo.it, February 28, 2014.

Davies, P., Francis, P., & Wyatt, T. (2014). Taking invisible crimes and social harms seriously. In P. Davies, P. Francis, & T. Wyatt (Eds.), *invisible crimes and social harms* (pp. 1–25). New York: Palgrave Macmillan.

Eusebi, L. (2015). Formula di Frank e dolo eventuale in Cass., S.U., 24 Aprile 2014 (ThyssenKrupp). *Rivista italiana di diritto e procedura penale*, 623–645.

Fontanella, M. F. (2012). L'accertamento del nesso causale tra condotta del datore di lavoro ed evento morte per mesotelioma maligno: linee di un'evoluzione giurisprudenziale. *Rivista Italiana Di Medicina Legale*, 1537–1575.

[75]For more details, see Chap. 6 of this volume.

Flick, G. M. (2015). Parere pro-veritate sulla riconducibilità del c.d. disastro ambientale all'art. 434 c.p. *Cass. pen.*, 12–25.
Forti, G. (2006). "Accesso" alle informazioni sul rischio e responsabilità: una lettura del principio di precauzione. *Criminalia*, 155–225.
Forti, G. (2007). La "chiara luce della verità" e "l'ignoranza del pericolo." Riflessioni penalistiche sul principio di precauzione. In *Scritti per Federico Stella* (pp. 573–674). Napoli: Jovene.
Forzati, F. (2015). *Irrilevanza penale del disastro ambientale, regime derogatorio dei diritti e legislazione emergenziale: i casi Eternit, Ilva ed emergenza rifiuti in Campania. Lo stato d'eccezione oltre lo stato di diritto* (pp. 1–49). www.penalecontamporaneo.it. March 11, 2015.
Gargani, A. (2008). *Trattato di diritto penale. Parte speciale: vol. IX. Reati contro l'incolumità pubblica. Tomo I: Reati di comune pericolo mediante violenza*. Milano: Giuffrè.
Gargani, A. (2010). La protezione immediata dell'ambiente tra obblighi comunitari di incriminazione e tutela giudiziaria. In S. Vinciguerra & F. Dassano (Eds.), *Scritti in memoria di Giuliano Marini* (pp. 403–430). Napoli: ESI.
Giunta, F. (2006). Il diritto penale e le suggestioni del principio di precauzione. *Criminalia*, 227–247.
Giunta, F. (2008). I contorni del "disastro innominato" e l'ombra del "disastro ambientale" alla luce del principio di determinatezza. *Giur. Cost.*, 3539–3544.
Martini, R. (2008). Il disastro ambientale tra diritto giurisprudenziale e principi di garanzia. *Leg. pen.*, 339–355.
Masera, L. (2014). Evidenza epidemiologica di un aumento di mortalità e responsabilità penale. Alla ricerca della qualificazione penalistica di una nuova categoria epistemologica. *Dir. pen. cont.—Riv. Trim.*, 343–365.
Michalowski, R. J. (1985). *Order, law and crime: An introduction to criminology*. New York: Random House.
Micheletti, D. (2007). Sub Art. 257. In F. Giunta (Ed.), *Codice commentato dei reati e degli illeciti ambientali* (pp. 329–380). Padova: Cedam.
Notaro, D. (2015). Il caso del Petrolchimico di Marghera: esposizione a sostanze tossiche e colpa. In L. Foffani & D. Castronuovo (Eds.), *Casi di diritto penale dell'economia. II. Impresa e sicurezza (Porto Marghera, Eternit, Ilva, ThyssenKrupp)* (pp. 51–72). Bologna: Il Mulino.
Note [n.a.]. (2015). Causation in environmental law: Lessons from toxic torts. *Harvard Law Review, 128*, 2256 ff.
Panni, M. (2007). Inquinamento storico e obblighi attuali di bonifica. *Rivista Giuridica Dell'Ambient*, 844–855.
Pardy, B. (2002). Applying the precautionary principle to private persons: Should it affect civil and criminal liability? *Les Cahiers de droit, 43*, 63–78.
Patrono, P. (1980). *Inquinamento industriale e tutela penale dell'ambiente*. Padova: Cedam.
Perini, C. (2002). Rischio tecnologico e responsabilità penale. Una lettura del caso Seveso e del caso Marghera. *Rassegna Italiana di Criminologia*, 389–412.
Perrow, C. (1999). *Normal accidents: Living with high-risk technologies*. Princeton: Princeton University Press.
Pesatori, A. C., Consonni, D., Rugabotti, M., Grillo, P., & Bertazzi, P. A. (2009). Cancer incidence in the population exposed to dioxin after the "Seveso accident": Twenty years of follow-up. *Environmental Health, 8*(39), 1–11.
Piergallini, C. (2004). *Danno da prodotto e responsabilità penale. Profili dommatici e politico-criminali*. Milano: Giuffrè.
Piergallini, C. (2005). Il paradigma della colpa nell'età del rischio: prove di resistenza del tipo. *Rivista Italiana di Diritto e Procedura Penale*, 1684–1703.
Pongiluppi, C. (2010). Principio di precauzione e reati alimentari. Riflessioni sul rapporto "a distanza" tra disvalore d'azione e disvalore di evento. *Rivista Trimestrale Di Diritto Penale Delleconomia*, 225–270.
Pozzo, B. (2008). Il disastro ambientale e le responsabilità civili e penali: percorsi giurisprudenziali. In B. Pozzo (Ed.), *Seveso trent'anni dopo: Percorsi giurisprudenziali, sociologici e di ricerca* (pp. 1–53). Giuffrè: Milano.

Prati, L. (2003). La responsabilità per l'inquinamento pregresso e la 'posizione di garanzia' nella normativa sulla bonifica di siti contaminati. *Rivista Giuridica Dell'Ambient*, 159–163.
Prati, L. (2012). Imputabilità dell'inquinamento e nesso causale tra giurisprudenza nazionale e comunitaria. *Rivista Giuridica Dell'Ambient*, 89–91.
Pulitanò, D. (2006). Gestione del rischio da attività professionali. *Cass. pen.*, 778–796.
Rotolo, G. (2014). Verità, informazione e conoscenza nelle strategie politico-criminali. In G. Forti, G. Varraso, & M. Caputo (Eds.), *«Verità» del precetto e della sanzione penale alla prova del processo* (pp. 387–411). Napoli: Giuffrè.
Rotolo, G. (2016). Modelli 'dinamici' di tutela dell'ambiente e responsabilità penale: problemi e prospettive. *Jus*, 113–150.
Ruga Riva, C. (2006). Principio di precauzione e diritto penale. Genesi e contenuto della colpa in contesti di incertezza scientifica. In E. Dolcini & C. E. Paliero (Eds.), *Scritti in onore di Marinucci, II* (pp. 1743–1777). Milano: Giuffrè.
Ruga Riva, C. (2010). L'obbligo di impedire il reato ambientale altrui. Osservazioni sull'asserita posizione di garanzia del proprietario. In S. Vinciguerra & F. Dassano (Eds.), *Scritti in memoria di Giuliano Marini* (pp. 859–874). Napoli: ESI.
Ruga Riva, C. (2013). *Diritto penale dell'ambiente*. Torino: Giappichelli.
Ruga Riva, C. (2015a). *Dolo e colpa nei reati ambientali* (1 ff). www.penalecontamporaneo.it, January 19, 2015.
Ruga Riva, C. (2015b). *I nuovi ecoreati. Commento alla legge 22 maggio 2015, n. 68*. Torino: Giappichelli.
Ruga Riva, C. (2015c). Il caso Ilva: avvelenamento e disastri dolosi. In L. Foffani & D. Castronuovo (Eds.), *Casi di diritto penale dell'economia. II. Impresa e sicurezza (Porto Marghera, Eternit, Ilva, ThyssenKrupp)* (pp. 149–176). Bologna: Il Mulino.
Ruzzenenti, M. (2001). *Un secolo di cloro ...e PBC: storia delle industrie Caffaro di Brescia*. Milano: Jaca Book.
Salanitro, U. (2011). I criteri di imputazione della responsabilità ambientale. *Amb.&Svil.*, 19–24.
Scaroina, E. (2015). Ancora sul caso Eternit: la "giustizia" e il sacrificio dei diritti. *Archivio Penale*, 1–27.
Siracusa, L. (2015). La legge 22 maggio 2015, n. 68 sugli "ecodelitti"; una svolta "quasi" epocale per il diritto penale dell'ambiente (1 ff). www.penalecontamporaneo.it, July 9, 2015.
Stefanizzi, S., Niessen, B., & Scisci, D. (2008). Seveso 30 anni dopo: costruzione sociale della memoria e rappresentazione del rischio. In B. Pozzo (Ed.), *Seveso trent'anni dopo: percorsi giurisprudenziali, sociologici e di ricerca* (pp. 55–110). Giuffrè: Milano.
Stella, F. (2003). *Giustizia e modernità. La protezione dell'innocente e la tutela delle vittime* (III ed.). Milano: Giuffrè.
Vallini, A. (2015), Il caso del Petrolchimico di Marghera: esposizione a sostanze tossiche e nesso di causalità. In L. Foffani & D. Castronuovo (Eds.), *Casi di diritto penale dell'economia. II. Impresa e sicurezza (Porto Marghera, Eternit, Ilva, ThyssenKrupp)* (pp. 25–49). Bologna: Il Mulino.
Vanetti, F. (2007). Bonifiche: successione tra imprese e inquinamenti storici. *Amb.&Svil.*, 1081–1088.
Vergine, A.L. (2013a). Il c.d. disastro ambientale: l'involuzione interpretative dell'art. 434 cod. pen. (parte prima). *Amb.&Svil.*, 534–541.
Vergine, A.L. (2013b). Il c.d. disastro ambientale: l'involuzione interpretative dell'art. 434 cod. pen. (parte seconda). *Amb.&Svil.*, 644–647.
Wise, E. M. (in collaboration with A. Maitlin) (2005). *The Italian penal code*. Buffalo: William S. Hein & Co.
Zirulia, S. (2013). Caso Eternit: luci ed ombre sulla sentenza di condanna in primo grado. Rivista Italiana di Diritto e Procedura Penale, 471–509.
Zirulia, S. (2015). Il caso Eternit: profili generali in tema di amianto e responsabilità penale. In L. Foffani & D. Castronuovo (Eds.), *Casi di diritto penale dell'economia. II. Impresa e sicurezza (Porto Marghera, Eternit, Ilva, ThyssenKrupp)* (pp. 73–106). Bologna: Il Mulino.

Chapter 5
Historical Pollution Under Scrutiny in the Italian Criminal Case Law

Benedetta Venturato and Eliana Greco

Contents

Sects. 1.1–1.5 were drafted by Benedetta Venturato, while Sects. 1.6–2.5 were drafted by Eliana Greco.

B. Venturato (✉) · E. Greco
Università Cattolica del Sacro Cuore, Milan, Italy
e-mail: benedetta.venturato@unicatt.it

E. Greco
e-mail: eliana.greco@unicatt.it

F. Centonze and S. Manacorda (eds.), *Historical Pollution*,
DOI 10.1007/978-3-319-56937-6_5

1 Analysis of the Case Law

1.1 The Porto Marghera Petrochemical Plant *Case as a Point of Reference (Cassazione penale, sez. IV, 17* May *2006, n. 4675)*

In the *Porto Marghera Petrochemical Plant* case—which may certainly be considered a seminal case of historical pollution in Italy—the Venice Prosecutor placed under scrutiny the industrial activity carried out at the Marghera petrochemical plant over more than thirty years.[1]

In particular, a number of successive board members, executives, and plant managers faced, *inter alia*, two main sets of charges: on the one hand, they were charged with *Involuntary manslaughter, Injury, and Health disaster*, while on the other hand they were charged with *Environmental disaster* as well as *Water poisoning* and *Adulteration*.

The first set of charges related to the effects the petrochemical industrial activity would have produced inside the plant and on the workers' health. In the Prosecutor's reconstruction, the defendants, although aware of scientific studies which since 1975 had reported risks arising from exposure to vinyl chloride monomer (CVM) and polyvinyl chloride (PVC), failed to (i) stop or suspend the plant activity, (ii) put in place adequate measures to prevent workers from contracting extremely serious health diseases, including cancer, (iii) inform workers on the health risks connected with the plant's activity, and (iv) intervene to globally redevelop the plant and the surrounding area and monitor the status of workers' health.

Conversely, the second set of charges related to the effects produced on the environment by allegedly illegal protracted waste management practices. Allegations included both the creation of illegal dumps and the subsequent failure to (i) put in place adequate measures to prevent the deterioration of hygiene and health conditions of the plant and its surroundings, (ii) inform the authorities, and (iii) proceed with the cleanup of the relevant site, which only began in 1995. This set of allegations was based on the assumption that all the defendants had a legal duty to prevent any harm or threat to the environment arising from the activity of the petrochemical plant, eventually—for those who subsequently took up roles in the plant—including the legal duty to remove the consequences of previous illegal waste management. The allegations were also based on the assumption that the defendants were all fully aware of the extreme toxicity of the substances spread both inside and outside the plant, and that they nevertheless took the risks arising from the petrochemical production and the relevant waste. In doing so, according to the Prosecutor, they also contributed to, respectively, starting and aggravating the

[1]The decisions of the Court of Venice and of the Court of Appeal of Venice are published on www.petrolchimico.it, while the Italian Supreme Court's decision is published on *Foro it.*, 2007, II, p. 550 *ff.*

poisoning of the groundwater underlying the Porto Marghera site (which was allegedly destined also for agriculture and other human activities), as well as to contaminating the Venice lagoon. The contamination would, then, have resulted in the adulteration and poisoning of the water and the fauna, threatening public safety and health.

With respect to the first set of charges, the Court of Venice found that there was no causal connection between the contested conduct and most of the diseases, which, in the Prosecution's reconstruction, would have allegedly derived from them. More precisely, apart from liver sarcoma and Raynaud Syndrome, it had been impossible to prove the existence of a causal connection, not only at an individual assessment level, but—in some cases—also in terms of general causation. In relation to liver sarcoma, the Court found that, although the exposure to CVM had effectively caused some of the plant workers to contract liver sarcoma, this could not be attributed to the defendants because, based on the scientific knowledge available at the time of the relevant facts, a causal link between exposure to CVM and liver sarcoma was not foreseeable. Consequently, with respect to these allegations, the Court acquitted the defendants of both *Involuntary manslaughter* and *Health disaster* allegations due to a lack of the relevant mens rea requirement. To the contrary, with respect to the Raynaud Syndrome the Court did find that a causal link between exposure to CVM/PVC and the disease could have been foreseen even at the time, but because the *Injury* charges had become time-barred, the Court's findings could not lead to conviction.

A clarification is needed here: in finding that, from an objective viewpoint, the requirements for the legal test of the *Disaster* offense had been met and that the defendants could not be convicted just for lack of mens rea, the Court laid down a notion of "disaster" which despite being then belied by the Court of Appeal in this case was followed in many of the subsequent judgments on historical pollution cases. According to such a notion, for the offense of *Disaster* to be committed, a violent origin or immediate aftermath is not needed. Rather, the offense could occur even in cases where a disaster that threatens public health and safety is the aggregate effect of a number of acts spread out over a long period of time, provided that this can be shown to have produced serious and diffuse effects on a wide community.[2]

With respect to the second set of charges, the Court of Venice acquitted all the defendants of both the *Environmental disaster* and the *Water poisoning* and *Adulteration* charges. However, before entering into the merits of the Court's

[2]This notion of "disaster," according to which for the crime to be committed it is not necessary that the relevant conduct and its effects be closed in time nor that the disaster has a violent origin, has also been affirmed by the Italian Constitutional Court (see *Corte Costituzionale*. 327/2008, on *Giur. cost.*, 2008, p. 3529 *ff.*) and has been widely followed by subsequent case law (with the exception of *Corte d'Appello Venezia*). See, among others, *Cassazione penale, sez. IV*, 23 February 2015, n. 7941 (the *Eternit* case), on *Dir. pen. contemporaneo*, online, 24 February 2015; *Cassazione penale, sez. IV*, 5 May 2009, n. 18974, on *Ced* Rv 243992; *Cassazione penale,, sez. IV*, 17 May 2006, n. 4675 (the *Porto Marghera* case), on *Foro it.*, 2007, II, pp. 550 *ff.*; *Cassazione penale, sez. V*, 7 December 2006, n. 40330, on *Ced* Rv. 236294.

findings, it is worth briefly presenting the position expressed by the Court with respect to the Prosecution's reconstruction of the defendants' legal duties, on which the charges were based. The Court, in fact, pointing out the limits of a reconstruction of legal duties which is far from infrequent in historical pollution allegations, noted that the defendants' behaviors appeared to have been considered in an out-of-context dimension strongly influenced by hindsight bias and which made it very difficult to identify which conduct and effects were to be attributed to each single defendant. Distancing itself from such an approach, the Court then ruled that no legal duty can arise for executives or managers that obliges them to remove the consequences of the contamination produced by their predecessors within the company's management, as this would result in an infringement of the principle of criminal liability for one's own actions provided for by Article 27 of the Italian Constitution.

On the merits of the *Environmental disaster* allegations, the Court acquitted the defendants on the basis that it had not been proved that the illegal waste management (and the relevant contamination) had engendered a threat to public health and safety, as required under Article 434 of the Italian Criminal Code.

With respect to the *Water poisoning* allegations, the Court first clarified that groundwater contamination may be punished as *Water poisoning* under Article 439 of the Italian Criminal Code, to the extent that the groundwater may be used in human activities.[3] However, it also found that, in the case at issue, the contaminated groundwater contrary to what had been contended by the Prosecutor could neither be destined for agricultural use nor any other human activity. In deciding on the Venice lagoon contamination, the Court set out a general principle according to which, although the offense of *Water poisoning* does not require the Judge to ascertain that an actual threat to public health arose from the poisoning, the offense should nevertheless be construed as aimed at preventing a "real," albeit only potential, threat to public health.[4] Based on this assumption, the Court found that no such "real" threat had arisen from the contamination of the lagoon and acquitted the defendants. Finally, and very importantly, the Court also clarified that the evidence of such a "real" threat cannot be taken from the mere circumstance that the pollution exceeded threshold limit values, because these are precautionary and "ultra-prudent" indicators,[5] which by definition identify levels of contamination that are far removed from any threat to public health and safety.

---

[3]On the possibility of making groundwater contamination fall within the scope of the offense of *Water poisoning* under Art. 439 of the Italian Criminal Code, see *Cassazione penale, sez. I*, 14 October 2005, n. 41983, on *Ced* Rv. 232874; *Cassazione penale, sez. III*, 27 May 1997, n. 7170, on *Cass. pen.*, 1998, pp. 1624 *ff.*; conversely, *Corte d'Assise di Chieti*, 19 December 2014 (ud.), n. 2, p. 80, on *Dir. pen. contemporaneo*, online, 9 February 2015.

[4]See also *Corte d'Assise di Chieti*, 19 December 2014 (ud.), n. 2, pp. 73–75; *Tribunale di Cremona*, 18 July 2014, Judge Salvini, p. 357 (unpublished).

[5]See also *Cassazione penale, sez. V*, 26 April 2005, n. 23465, on *Cass.* pen., 2007, p. 615 *ff.*; *Corte d'Assise di Chieti*, 19 December 2014 (ud.), n. 2, p. 75; dissenting, *Tribunale di Cremona*, 18 July 2014, Judge Salvini, p. 329.

The Court of Appeal overruled the first instance judgment on the *Involuntary manslaughter* allegations. It found fault based on the precautionary principle,[6] stating that if information available at the time suggested that exposure to CVM may have posed a threat to workers' health, albeit uncertain and partially undefined, it was the defendants' legal duty to take adequate measures to prevent that risk.

The Court of Appeal upheld the first instance judgment on all other issues, but it took a different view on the notion of *Disaster*. In particular, it contended that no *Disaster* offense can have been committed if the defendant's conduct did not result in a "harmful macro-event" (such as a shipwreck or a fire) that posed a threat to public health and safety.

The Italian Supreme Court (*Corte di Cassazione*) issued its judgment only on the first set of charges, as the decisions taken on the second set had not been challenged and upheld the judgment of the Court of Appeal, restating the applicability of the precautionary principle.

1.2 Water Poisoning and the Bussi sul Tirino Disaster (Corte d'Assise di Chieti, 19 December *2014 (ud.), n. 2)*

In the *Bussi sul Tirino* case, the Prosecution Service of Chieti placed under investigation and subsequently prosecuted nineteen individuals on allegations of *Intentional water poisoning* and *Disaster*, under Articles 439 and 434, respectively, of the Italian Criminal Code. These individuals, including corporate executives, plant managers, environmental security managers, and corporate advisors who succeeded one another from 1963 to 2003, were all accused of having participated in the commission of the alleged crimes—although at different times and, at least in part, independently of one another.

The *Poisoning* and *Disaster* allegations related to activities they had carried out in their respective capacities within the *Bussi sul Tirino* chemical plant and, once more, to the relevant waste management practices. Specifically, the defendants were accused of building, between 1963 and 1972, a huge dump (called the Tre Monti Dump) close to the Pescara River, where chemical waste was regularly disposed of illegally; they were alleged to have then built a second dump (called the A2 Dump)

---

[6]This legal principle is restated by the Italian Supreme Court Judgment closing the *Porto Marghera Petrochemical* case and is also referred to within the *Porto Tolle* case (*Tribunale di Rovigo*, 22 September 2014, *Pres.* Angeletti, ric. Arrighi, pp. 41–42, on *Dir. pen. contemporaneo*, online, 16 October 2014) and within the *Vado Ligure* case (*Tribunale di Savona*, 11 March 2014, Judge Giorgi, p. 36, on *Dir. pen. contemporaneo*, online, 16 October 2014), with respect to the *Disaster* offense. In particular, in the latter case the judge stated that the managers of a thermo-electrical plant should adjust its emissions to below threshold limit values in accordance with the precautionary principle. For a more detailed analysis of the precautionary principle, see Chap. 4, Sect. 6.2 of the present volume.

upstream of the plant, where until around 1990 further toxic waste was deposited, in breach of the relevant regional authorization. In addition, two more dumps had been built for the illegal disposal of waste from the chemical plant. Furthermore, between 1966 and 1997 approximately, the plant itself was alleged to have leaked significant amounts of lead into the surrounding soil.

Finally, the defendants were accused of having put in place a corporate strategy aimed at circumventing the legal duty to remove the effects of the above-described practices and showing misleading environmental conditions. As a consequence, both the surface water and the groundwater feeding a number of wells were alleged to have been contaminated by hazardous substances, resulting in *Poisoning* and *Environmental disaster* and posing a threat to public safety.

At the end of the first instance trial, the Court (which was a *Corte d'Assise*, and therefore included a jury) acquitted all the defendants of all charges. With respect to the *Water poisoning* allegations, the Court distinguished between surface water and groundwater. In the case of the surface water, the Court found that, although contaminated, it had not been poisoned. According to the Court, the offense of *Water poisoning* requires contamination to result in a real threat to public safety,[7] and in ascertaining this, merely exceeding threshold limit values should be deemed irrelevant.[8] In adopting this view, the Court stuck to the position already taken by the case law in the *Porto Marghera Petrochemical* case. It then concluded that, in the case at issue, no real threat to public safety had been posed. Conversely, the Court distanced itself from previous case law when taking a decision on groundwater. In this case, in fact, the Court construed the offense so as to make groundwater contamination generally fall outside its scope, unless proven that the groundwater was actually drawn for human use.[9] The Court's finding was that groundwater had been poisoned (and not just contaminated), but also that it was not directly drawn and that, consequently, it had been diluted before reaching the wells, losing most of the hazardous substances. Although unnecessary, given this premise, the Court decided to analyze the possible existence of a causal connection between the behavior of each of the defendants and the groundwater poisoning.

The Court ascertained that the contamination of both surface water and groundwater had certainly been caused by the pollution arising from the creation of the Tre Monti Dump, while no actual causal connection could be proven between the pollution arising from the creation of the other dumps and the water contamination. The Tre Monti Dump, in turn, was found to have been built at a time when only one of the defendants was in charge, and the evidence gathered during the trial suggested that none of the other defendants was even aware of its existence. For the purposes of ascertaining this one defendant's liability, it was also established that back in the 1960s and 1970s no legislation prohibited the creation of dumps, and he was found to have acted in good faith.

---

[7]See n. 5 above.

[8]See n. 6 above.

[9]See n. 4 above.

As far as the *Environmental disaster* allegations were concerned, the Court found that the defendants lacked intent, but that the legal test for the offense of *Involuntary environmental disaster* had been met. Due to this latter offense being time-barred, however, it could not lead to conviction.

In this respect, the Court—recalling the Italian Constitutional Court judgment n. 327/2008 and consistently with the *Porto Marghera* judgment—clarified that what matters in order to ascertain whether a disaster offense was committed are not the modalities of the defendant's behavior, but rather its effects: these must consist in a threat to public safety that is particularly serious and diffuse, which may also result from the aggregate effect of several acts spread out over time. However, when the disaster stems from the behavior of several people acting within different contexts and at different times, this may affect the actual perception and knowledge, by each of them, of the contamination that results from their polluting activities. This latter factor may then, in turn, impact on the defendants' relevant state of mind, ruling out intent. Another important principle affirmed by the Court on the mens rea requirement is that intent cannot be proved inherent in the corporate profit objective and in the consequent cost/benefit approach.

Finally, the Court pointed out that the Prosecutor's construction of the offense as continuing until the harmful effects on the environment persist was incorrect. Rather, when the disaster is the aggregate result of several acts, the offense should be deemed complete at the time by which the pollution had reached a sufficiently serious, diffuse, and threatening stage. As a result, the possible persistence over a longer period of time of the effects of the pollution is not to be taken into consideration for the purposes of the statute of limitations.[10] On these grounds, the Court declared even the lesser offense of *Involuntary environmental disaster* to be already time-barred. Against the first instance judgment an appeal was filed directly before the Italian Supreme Court (so called *ricorso per saltum*); however, the Supreme Court converted it into an ordinary appeal and the file was then sent to the Court of Appeal of L'Aquila (*Corte d'Assise d'Appello dell'Aquila*). On 17 February 2017, the Court of Appeal overruled the first instance judgment with a decision which, in its outcome, seems to reveal once more the complexity of historical pollution cases and the struggle of the Italian Judges in trying to manage these cases through criminal law. Neither the decision nor the relevant full opinion have been published yet, which makes it hardly possible to provide here a full picture of the case and of the developments of the above presented arguments in the context of the appeal proceedings. However, from publicly available information, it results that the Court of Appeal found the *Water poisoning* allegations to be grounded, but the relevant offence to be time-barred, while it deemed the *Involuntary disaster* offense to be aggravated and, as a consequence, the relevant limitation period not to have expired. The Court then convicted ten out of the nineteen individuals originally charged to imprisonment up to two and three years

[10]Statute of limitation issues in relation to the disaster offense issues were also dealt with in the *Eternit* case described in Sect. 1.5 of this chapter.

respectively (depending on the individual) and at one time it also declared these convictions all subject to "pardon" (*indulto*). Nevertheless, based on the findings of guilt, the Court also sentenced the defendants to the payment of damages to the civil plaintiffs for an overall amount of approximately 4 million euro.

1.3 Waste Management as a Source of Historical Pollution in the Pitelli Dump *Case (Tribunale della Spezia, 6* June *2011, n. 10007)*

In the *Pitelli Dump* case,[11] eleven individuals were indicted on, among others, *Environmental disaster* allegations under Article 434 of the Italian Criminal Code. Both corporate officers and public officials were prosecuted: the former for having illegally dumped toxic waste in the Pitelli area between 1977 and 1996; the latter for having (in breach of their legal duties) failed to prevent the commission of the crimes by the corporate officers and, in so doing, aided and abetted their conduct.

The corporate officers acted—throughout a twenty-year period—within two companies that specialized in waste management, which either in the absence or in breach of the relevant authorizations had allegedly dumped toxic substances deriving from the industrial activities of various other companies operating in the area. Some of them were also accused of having corrupted public officials to obtain the release of authorizations and permits. In addition, the public officials had been accused of forgery in relation to certain reports and permits granted.

The Court of La Spezia acquitted all the defendants of *Environmental disaster* allegations and declared the *Forgery* offenses time-barred.

For the purposes of the present research, the case appears worth noting in that it shows that irregular—and even illegal—waste management (even when clearly resulting in contamination) does not automatically equal environmental disaster. The Court, in fact, found that the management of both the waste and the dumps had been "characterized by several anomalies," "at least irregular," "to some extent illegal," and "in any case in breach of most of the public authorities' provisions," and that the conduct also showed willful neglect toward the environment and the principles of waste management.[12] Nonetheless, no *Environmental disaster* was found, because in the course of the trial no evidence had suggested that the toxic substances had been conveyed or spread so as to affect sources for agriculture or human nutrition, thereby posing a threat to public safety.

Indeed, the Court of La Spezia, consistently with previous case law, pointed out that the concept of *Environmental disaster* under Article 434, Paragraph 2 of the Italian Criminal Code[13] identifies a large-scale event consisting in the use, the

---

[11]The decision is published on *Dejure*.

[12]See *Tribunale della Spezia*, 6 June 2011, n. 10007, p. 58.

[13]See n. 3 above.

manipulation, or the disposal of significant amounts of hazardous substances on sites intended to support human life or agriculture and that such an event must be of a duration, scale, and intensity as to pose a threat to the life or to the health of an undefined number of people. Conversely, for the purposes of the correspondent offense provided for by Paragraph 1 of Article 434 of the Italian Criminal Code, the threat to public safety must consist in the concrete possibility that a community be exposed to an environmental disaster (as defined above), even if no actual change of circumstances occurs.

1.4 The Search for a Definition of Historical Pollution in the Cremona Petrochemical Plant *Case (Tribunale di Cremona, 18* July *2014, Judge Salvini)*

At a glance, the inclusion of the *Cremona Petrochemical* case[14] among the leading cases of historical pollution in Italy may cause puzzlement, especially if one considers that a number of the company's managers and directors have been indicted for alleged polluting activities which took place only between 2001 and 2007 and that the issue of whether the case should be classified as a case of "historical pollution" was expressly debated at trial.

The circumstance, however, is easy to explain: a petrochemical plant owned by a certain company and located near Cremona had released harmful substances into the soil surrounding the plant and the underlying groundwater since the 1960s. In 1983, the company that owned the plant changed its name and continued conducting its activity as it had done before. With the passing of time, polluting activities (including both the industrial activity itself and the management of the waste produced by the plant) resulted in the contamination of the soil and groundwater in the vicinity of the plant, and also of certain surface water. In 2001, the company self-reported to the Italian authorities. In the report it was stated that the company had identified contamination of the soil and water close to the plant, which they would recover with a cleanup, but that, as no harmful substances had moved from the site, no further intervention was needed. In addition, it was stated that the contamination would have been caused by Second World War bombings and by the polluting activities carried out before 1983. In this respect, the company pointed out that before that year the plant was owned by a different legal entity and, as a consequence, the contamination should be referred back to this other entity.

Charged with *Water poisoning*, in the present case, the defendants themselves tried to contend that "historical pollution" had occurred, with the aim of distancing themselves from the facts. Accordingly, they were initially deemed not responsible for the polluting activities carried out before 2001, but they were still indicted for failing to prevent the contamination from worsening after self-reporting, up until

[14]The decision is currently unpublished.

2007. In the Prosecutor's reconstruction, toxic substances had indeed moved from the site and the company's executives would have then had the legal duty to perform further cleanup interventions, whose need had been denied (and hidden) by them with the 2001 report.

In the course of the trial, the Judge found out that the legal entity which owned the plant until 1983, was in fact the same entity that filed the report in 2001, but under a different name. This led him to clarify that the company could not be deemed a "third party" whose activity was unrelated to the origin of the contamination. Consequently, more stringent remediation obligations should have been placed upon the company's executives. Moreover, according to the Judge, no historical pollution had really taken place, as back in 2001–2007 polluting activities were still ongoing, and in his view only cases in which the polluting activity leading to the contamination had ceased could be classified as proper cases of "historical pollution."

On the merits, the Judge found that both surface water and groundwater drawn for human use had been permanently contaminated as a consequence of the defendants' behavior. Moreover, in disagreement with previous case law, the Judge went on to clarify that the concept of a "threat to public health" in environmental criminal law is not descriptive of an empirical event, but should rather be placed within a legal and social context: exceeding threshold limit values would therefore prove the existence of a threat; in that it would represent overstepping the socially accepted level of risk.[15]

On this basis, the Judge found that a threat to public safety had indeed been caused, but also stated that, for the offense of *Water poisoning* to be deemed to have been committed, the behavior must produce effects which are immediately harmful to the organism, and this, in the case at issue, had not been proved.[16] Consequently, he found that the legal test for *Water poisoning* had not been met, but convicted all the defendants (but one), classifying their conduct as *Environmental disaster*, for which the posing of a more general threat to public safety, in his opinion, would have sufficed.

The first instance judgment was challenged before the Court of Appeal of Brescia (*Corte d'Assise d'Appello*), which issued its decision on 20 June 2016. Neither the decision nor the full opinion have been published yet. However, on the basis of publicly available information, it results that the Court of Appeal of Brescia convicted the former CEO of the company (in charge from 1999 to 2006) to three years of imprisonment on *Involuntary disaster* allegations and acquitted all the other defendants, while also confirming the sentencing to pay damages to the civil plaintiffs issued at the end of the first instance trial. In particular, the appeal judgment would appear to have confirmed the ascertainment of a serious, widespread, and lasting contamination of both the groundwater and the soil in the plant's surroundings, which would have posed a threat to public health and safety.

---

[15]This notion clearly contrasts with the relevant previous case law. See n. 3 above.

[16]See n. 5 above.

In addition, the Court would have grounded the conviction of the former CEO especially on the basis of an alleged reticent behavior, consisting in repeatedly failing to fully disclose to the authorities - even upon their specific requests - the actual status of the pollution and in delaying the necessary measures.

1.5 Exposure to Toxic Substances and Environmental Disaster in the Eternit Case *(Cassazione penale, sez. IV, 23* February *2015, n. 7941)*

Involving a large number of people and concerning one of the most important companies operating in the asbestos industry, the *Eternit* case represents the most significant trial relating to exposure to toxic substances in the Italian case law. A twofold rationale lies behind its inclusion among the leading cases on historical pollution: on the one hand, the charges referred to industrial activities that took place from the first years of the twentieth century up to 1986, and whose effects are still in place; on the other hand, the Prosecutor himself construed the events as falling within the scope of the *Environmental disaster* offense and of the crimes posing a threat to public safety, rather than, as it usually happens in such cases, within the scope of *Manslaughter* and *Injury* offenses aimed at protecting individual life and safety.

The main allegation concerned two corporate managers—who held the office, respectively, from 1952 to 1972 and from 1973 to 1986—who were accused of having intentionally caused an environmental disaster which had then actually taken place across at least four industrial sites. In the Prosecution's reconstruction, the disaster was alleged to have been caused by a runaway dispersion of asbestos over a long period of time. This, in turn, was deemed to have posed a threat to the safety of an undefined number of people, including those who had worked in the plants, their families, and those who lived close to the sites.

As mentioned before, while in previous cases[17] the Prosecutor had—not always successfully—tried to prove a causal connection between the dispersion of asbestos into the environment and the harm to life and safety suffered by specific individuals, in the *Eternit* case the defendants were accused of having caused a "collective event" consisting in a threat to the health of a group of people or a population. In order to prove any such causal relationship, the Judges initially deemed evidence of general causation through epidemiological studies to be sufficient.[18]

---

[17]See, *ex multis*, *Cassazione penale, sez. IV,* 21 November 2014, n. 11128, on *Riml*, 2015, p. 1139 *ff.*; *Cassazione penale, sez. IV,* 27 February 2014, n. 18933, on *Ced* Rv. 262139; *Cassazione penale, sez. IV*, 24 May 2012, n. 33311, on *Riml*, 2013, p. 297 *ff.*; *Cassazione penale, sez. IV*, 22 March 2012, n. 24997, on *Foro it.*, 2012, II, 517 *ff.*

[18]On the use of epidemiology in criminal law, see, also for the relevant references, Chap. 4, Sect. 6.1 of the present volume.

This approach led to conviction both at the end of the first instance trial and on appeal, but the Italian Supreme Court found that the crime had been, in fact, time-barred since the very beginning of the trial and even before the criminal case was brought.

The Supreme Court quashed the appeal decision (and also criticized the first instance judgment as well as the Prosecutor's position) with a judgment which, based on a different interpretation of the structure of the *Disaster* offense, provides insight into the way the rules on the statute of limitations should properly apply to cases of disasters with latent effects.

Everything revolves around the identification of the point at which the crime can be deemed completed, for the purpose of ascertaining when the relevant limitation period should start running. Both the Prosecutor and the Judges held that that the limitation period should start running from the moment at which the disaster occurred.[19] However, within this one approach, the Prosecutor and the Judges took different views. The Prosecutor classified the *Disaster* as a continuing offense which would not cease while the effects of the pollution were ongoing and found that, besides, the defendants' behavior would also have been in place at the time, in the form of a failure to prevent the disaster from continuing (and, in particular, a failure to collaborate in the cleanup). Conversely, the Court of Turin deemed the Prosecutor's interpretation unfounded, denying the existence of a legal duty on the part of the defendants to eliminate the consequences of their own behavior.[20] However, the Court also took the view that the crime was not time-barred (at least in relation to the pollution of two out of the four contaminated sites) because the offense should be deemed continuing until the existence of the "macro-event" of pollution.

The Court of Appeal stuck to yet another thesis, according to which the *Disaster* would not have been time-barred. The Court deemed that the legal test for the *Disaster* offense would include not only environmental pollution, but also an increase in the occurrence of asbestos-related diseases within the population, arising from the pollution. As a consequence, as the Court found that such an increased occurrence of asbestos-related diseases was still ongoing, it concluded that the offenses were far from being time-barred in relation to each of the four relevant sites.

The Italian Supreme Court put an end to the debate, challenging all positions. In the Supreme Court's view, the reasoning of the Judges in the first instance trial was incorrect because it appeared to be based on a substantial confusion between the notions of "continuing offense" and "offense with continuing effects." For its part,

---

[19]The conclusion was based on the theory that the second paragraph of Art. 434 of the Italian Criminal Code (which punishes whoever engages in conduct aimed at causing a disaster, when the disaster subsequently occurs) would provide for an autonomous offense and not just for an aggravating factor of the basic *Disaster* offense provided by Para. 1 of the same provision. On this issue, see also *Tribunale di Avellino*, 15 June 2013, Judge Riccardi, 8, on *Dir. pen. contemporaneo*, online, 15 November 2013. Dissenting, *Tribunale della Spezia*, 6 June 2011, n. 10007, 28. See Chap. 6, Sect. 6.4.

[20]See also, in the same case, *Cassazione penale, sez. IV*, 23 February 2014, n. 7941, pp. 78–79.

the Court of Appeal had erroneously included in the notion of disaster the need for an increased occurrence of diseases as an effect of pollution, which was not required under the law.

Accordingly, the Supreme Court ruled that the offense of *Disaster* should be deemed completed when the defendant's conduct that poses a threat to public safety is interrupted. As in the present case the activity of the plants stopped in 1986, the Supreme Court referred to that moment to ascertain whether the limitation period had expired and concluded that the offense was to be considered time-barred.[21]

1.6 The Responsibility of the "Innocent Owner" in the Comit *Case (Cassazione penale, sez. III, 16* March *2011, n. 18503*[22]*)*

The *Comit* case concerns the conduct of Agent "B" (later owner, not responsible for the contamination) who purchased a site and replaced Agent "A" (previous owner, polluter) as director, with powers of control, supervision, and administration over the contaminated area. The fact originated from the conviction imposed on the defendant by the Court of Milan, for the misdemeanor of Legislative Decree 152/2006 (*Testo Unico Ambiente*), Article 257 (*Failure in remediation proceedings*),[23] Paragraphs 1 and 2, in relation to Article 242 (*Operational and administrative proceedings*).

[21]The legal principle affirmed by the Italian Supreme Court in the *Eternit* case with respect to the statute of limitations found confirmation also in the *Pozzo di Gotto* case (see *Cassazione penale, sez. IV*, 21 July 2014, n. 32170; *G.U.P. Barcellona Pozzo di Gotto*, 11 March 2013, Judge Adamo: both decisions are published on *Dir. pen. contemporaneo*, online, 9 December 2014). This is a minor case of historical pollution, very similar to the *Eternit* case in terms of the key features of the relevant factual background and structure of the indictment: the charges, in fact, related to the diffusion of asbestos while carrying out an industrial activity which, many years later, resulted in contamination and a health disaster. The case is worth mentioning—albeit only briefly—as it affirmed the above-described principles on the statute of limitations, since the preliminary hearing phase of the proceedings affirmed them in the *Eternit* case even before the Italian Supreme Court. Another case where these principles were affirmed is the *Bussi sul Tirino* case described in Sect. 1.2 (see *Corte d'Assise di Chieti*, 19 December 2014 (ud.), n. 2, p. 185). Conversely, in the *Porto Tolle* case (see *Tribunale di Rovigo*, 31 March 2014 (*dep.* 22 September 2014), *Pres.* Angeletti, *ric.* Arrighi et al., p. 72) the Court applied the legal principles on the statute of limitations affirmed by the Court of Appeal in the *Eternit* case.

[22]This decision is published on *Cassazione penale*, 2011, pp. 4405 *ff.*; it is also published on *Rivista Giuridica dell'Ambiente*, 2011, pp. 809 *ff.*

[23]This misdemeanor is described by case law as "a free-form—causal or purely causally oriented —offense with event of damage," where remediation would be an "objective condition of punishment with a negative content." This structure influences the *mens rea* requirements, especially with reference to knowledge of statutory criminal law and to the subjective perception of negative value inherent to polluting conduct. On this topic, among others, see *Cassazione penale, sez. III*, 29 November 2006, n. 9794, on *DeJure*.

The Court sentenced him to a €10,000 fine and ordered him to pay damages to the plaintiff (the province of Milan), because "as a formally delegated person to the Comit pension fund" he had not carried out the performance foreseen by the law (i.e., relevant communications with the authorities in charge): he failed to report the presence of polluting factors (hydrocarbons exceeding 1000 mg/kg) on the site, which was in a state of deterioration due to polluting acts carried out by the previous owners.

The defendant challenged the judgment by reporting "non-compliance or incorrect application of criminal law": the Court had erroneously considered his conduct relevant under Article 257. In particular, he pointed out that "the current wording of Article 257 distinguishes the duty of remediation from the duty of notification and provides for criminal penalties just for the polluter. For the subsequent, innocent owner no crimes are configured."[24]

This framework brings to light a considerable number of questions pertaining to various topics in criminal law: from the culpability principle, to succession in duty of care and the selection criteria for punishable conduct. On these matters, the Italian Supreme Court gives some directions.

According to the defendant's remarks, the Supreme Court actually denies the transfer of liability from the real polluter to the current owner.[25] The reasons supporting this thesis are summarized in the following findings. First of all, the role

---

[24]On this issue see also *Cassazione penale, sez. III*, 2 July 2010, n. 35774, on *DeJure*. In this case "the trial Court of Udine declared M.C. guilty of the offense under Legislative Decree April 3, 2006, No. 152, Art. 257 because, as director of the company *Cromo Friuli* [...], had failed in cleaning up procedures [...]. The defendant proposes further appeal [...] and stressed [...] that Mr. B.B., the previous director of the company, was the person who had caused the contamination [...]. In fact, the defendant was appointed manager only on 14 June 2005, about a year after the alleged offense." Nevertheless, the Court convicted him because "the contamination had not been completed during the years that preceded the appointment of M.C. as director of the company, but it was protracted until February 2008, when the periodic monitoring found dangerous substances exceeding the threshold limit values." The trial Court also found that the polluting effects "would have been related to responsibility of the defendant, even if the thesis of the defense had been correct: according to this thesis, it would not be a new chromium spill over the production process, but a runoff accumulated in the basement during the previous years and operated by rainwater or by rising groundwater. Even in this case, the current release of hexavalent chromium is related to a still active source of contamination located within the plant of *Cromo Friuli* [...]. In other words, the trial judge considered plausible that the contamination coming from inside the plant has not been terminated before Mr. M. took the position of director, but it was protracted until the last exceeding threshold limit values, occurred in February 2008. Therefore, the misdemeanor was attributable also to the wrongful conduct of the applicant, who had negligently failed to delete the active, contaminant source at its plant." On these topics see also: *Cassazione penale, sez. III*, 29 November 2006, cit.; *Cassazione penale, sez. III*, 14 March 2007, n. 26479; *Cassazione penale, sez. III*, 30 May 2007, n. 28525; *Cassazione penale, sez. III*, 11 October 2007, n. 40191; *Cassazione penale, sez. III*, 15 December 2010, n. 11498; *Cassazione penale, sez. III*, 17 January 2012, n. 17817; *Cassazione penale, sez. III*, 22 January 2013, n. 19962 (all decisions can be found on *Dejure*).

[25]Conversely, among others, *Cassazione penale, sez. III*, 9 July 2007, n. 26479, on *Amb. e Svil.*, 2007, pp. 852 *ff.*.

of who causes pollution is autonomous from the role of who ascertains the existence of contaminants on the soil. So, either the obligation to report or the obligation of remediation (both foreseen by Art. 257) could, at most, reside, for the involved but not responsible owner, in Article 245 and not in Article 242 (the latter "invoked by Article 245 just for the regulation of procedural aspects and not invoked by Article 257").[26]

More exactly, Article 245 provides for a set of administrative duties in the case of an innocent owner who discovers contamination, but Article 257, which establishes criminal penalties, links only to Article 242 and not to Article 245.[27] Therefore, "if the legislator had intended to refer, in Article 257, even to those who did not cause the contamination […], not only they would have been mentioned as possible authors of the offense, but Article 257 would also have referred to Article 245 (and not to Article 242) to identify the imposed notification requirements."[28]

So if we consider Article 257 to apply also to the innocent owner (that is, if we consider Art. 245 relevant under Art. 257, which instead roundly refers to Art. 242), there will be a case of *analogia in malam partem*, which—as a part of the principle of legality—forbids the application of rules provided for similar cases, when the matter is specifically regulated by no rule.

In any case, "it might be puzzling to have a system that provided for the same type of sanction against the person guilty of polluting conduct (who is the 'first person' to whom the obligation to remove the consequences of what he caused is addressed) and the party who finds the contamination without having contributed to it in any way."[29] According to case law assessment, that is the wrong way to "give specific effect to the 'polluter-pays principle.'"[30]

2 Hypothetical Scenarios

A short analysis of five hypothetical scenarios is provided in the following paragraphs. Each scenario is related in various ways to the phenomenon of historical contamination and is characterized by an increasing level of harmfulness. For all

[26]*Cassazione penale, sez. III*, 16 March 2011, cit. In particular, Art. 245 (*Intervention and notification obligations by persons not responsible for potential contamination*) states as follows: "The procedures for the interventions of securing, remediation, and environmental restoration covered by this title can still be activated by non-responsible parties […]."

[27]Even though Art. 242 imposes the obligation to report historical contamination existing on the site also upon nonresponsible people who have detected it, case law considers that "in formal terms, the 'obligation to communicate' for the 'involved but not responsible' person actually resides in Article 245 and not in Article 242."

[28]*Cassazione penale, sez. III*, 16 March 2011, cit.

[29]Ibid.

[30]Ibid.

scenarios we will try to give brief solutions, based on the previously examined case law and also on the statutory framework.

It should be noted that every situation typified within each scenario is a mere abstract hypothesis whose evaluation under Italian criminal law could change according to the concrete facts that may occur in reality. Consequently, the principles listed below have to be considered as general guidelines and not as specific directions.

2.1 Hypothetical Scenario 1

Agent A owns a piece of land and contaminates it over a long period of time. No damages or threats to human health occur as a consequence of the pollution. Scenario 1 envisages the case of polluting conduct without any negative effects on human health. According to Italian law, the lack of a "detrimental attitude" does not imply that Agent A might not be criminally liable.

His conduct should be evaluated under Article 257, which does not require health damage to have been caused. In order to be thus appraised, the conduct must have produced at least two other conditions. First, an *excessive concentration* of polluting factors: namely, a quantity of dangerous substances that exceeds the threshold limit values foreseen by the law. Second, Agent A must have failed: (i) in remediation and cleanup measures; and (ii) in reporting about the existence of potential polluting substances on the soil. Moreover, after the *Environmental Crimes Reform* (Law of 22 May 2015, No. 68), Agent A might be found responsible also under Article 452-bis (*Environmental pollution*) or—depending on circumstances—under Article 452-terdecies (*Failure in environmental remediation*) of the Italian Criminal Code. In particular, the latter constitutes a further hypothesis of failure in remediation proceedings, which cannot concur with the offense foreseen by Article 257 *Testo Unico Ambiente*. Although there are no relevant judgments on this subject yet, it should be noted that the provision which was introduced at the starting of Article 257 by the Reform ("unless the offense does not constitute a more serious one") would attribute to Article 257 a limited space of application: since the latter is a misdemeanor while the offense provided for by Article 452-terdecies is a crime.

2.2 Hypothetical Scenario 2

Agent A owns a piece of land and contaminates it. The land is subsequently purchased by Agent B. No damages or threats to human health occur as a consequence of the pollution. Scenario 2 concerns the criminal liability of an innocent owner for contamination produced by others, without causation of health damages.[31]

In order to identify the relevant offenses under Italian criminal law, the considerations related to the previous scenario can also be taken into account in this one. Therefore, Agent A might be criminally liable for the misdemeanor under Article 257 of the *Testo Unico Ambiente*.[32] He is the only person to whom the fact should be imputed.

No responsibilities are generally attributed to Agent B: in most cases the substantive content of Article 257 is interpreted as being addressed to the polluter and not to the involved but not responsible person. A transfer of criminal liability from the previous owner to the current one is actually denied, if the latter did not contribute to the contamination.[33] Conversely, a part of case law is apt to consider that also the conduct of the "second owner" could be relevant under Article 257, but only with reference to the duty of reporting and not to the duty of remediation.[34]

Further analysis of the main considerations on this topic is provided in Sect. 1.6 above. In addition, the offenses under Articles 452-bis and 452-terdecies of the Italian Criminal Code could also be relevant in this case, with only reference to the conduct of Agent A; indeed, on the one hand, the crime under Article 452-bis is explicitly addressed to the person who caused the contamination; on the other hand, Article 452-terdecies—which is applicable if Article 452-bis is not—does not refer to the subsequent owner as the possible author of the crime. Nevertheless, according to the content of the provision, the conduct of the "second owner" would potentially fall within the scope of Article 452-terdecies if Agent B was considered as a person obliged to remediation by the law, by the order of a Court or of a public authority. In any case, it should be stressed that the case law has not given any specific guideline on this topic yet.[35]

2.3 *Hypothetical Scenario 3*

Agent A owns a piece of land and contaminates it. The polluted land is subsequently purchased by Agent B. In the meantime, the land contamination results in contamination of the groundwater, which is regularly used in agriculture. With reference to this scenario, the conduct of Agent A should be relevant under Article

---

[31]Some examples can be found in: *Cassazione penale, sez. III*, 16 March 2011, cit.; *Cassazione penale, sez. III*, 2 July 2010, cit.; *Cassazione penale, sez. III*, 29 November 2006, cit.; *Cassazione penale, sez. III*, 14 March 2007, cit.; *Cassazione penale, sez. III*, 30 May 2007, cit.; *Cassazione penale, sez. III*, 11 October 2007, cit.; *Cassazione penale, sez. III*, 15 December 2010, cit.; *Cassazione penale, sez. III*, 17 January 2012, cit.; *Cassazione penale, sez. III*, 22 January 2013, cit.

[32]See Sect. 1.6 above for further analysis of this topic.

[33]See Sect. 1.6 above.

[34]Consistently, among others, *Cassazione penale, sez. III*, 9 July 2007, n. 26479, cit.

[35]See Sect. 2.1 above.

439 of the Italian Criminal Code—namely, *Poisoning of water or substances for supply*, which will be taken into special account within the following section[36]—only if a real threat to public safety occurs (more exactly: if the groundwater is used in human activities and in human nutrition).[37]

Conversely, if this condition does not exist, the conduct of Agent A could be considered consistent with the scope of the offense foreseen by Article 434 of the Italian Criminal Code: *Collapse of buildings or other fraudulent disasters* ("*Unnominated disaster*").[38]

As has been shown previously (Sects. 1.1–1.5), case law has tried to clarify the exact sense and implications of the notion of "disaster," which—as the Italian Supreme Court affirmed—is "poorly defined" and "results in a brief expression capable of assuming, in common parlance, a broadly diversified range of meanings."

Generally the concept is identified as "a destructive phenomenon [...] which [...] affects the ecosystem [...] by causing a massive, long-lasting deterioration process of the human habitat" (so as "a macro damage event with potential detriment to public safety"). It must consist of "a destructive event of extraordinary proportions—not necessarily huge—capable of producing serious, complex, and extensive damaging effects." Moreover, "in terms of offensive projection, the event must have caused [...] a danger to life or physical integrity for an unspecified number of people."[39]

The conduct of Agent A might also be evaluated under Articles 452-bis and—only if an environmental disaster as described by law derives from the conduct—452-quater of the Italian Criminal Code; these offenses can potentially concur with the crime of Article 439 of the Italian Criminal Code.[40]

No responsibilities can be attributed to Agent B without the existence of a connection between his activities and the event: "each defendant can be held liable for prior, simultaneous or successive facts caused by others, only if there is [...] a relation with the duty of care, such as a link between its own conduct and the occurrence."[41]

---

[36]See Sect. 2.4 below.

[37]On this topic, *Cassazione penale, sez. I*, 14 October 2005, n. 41983, on *Ced* Rv. 232874, cit.; *Cassazione penale, sez. III*, 27 May 1997, n. 7170, on *Cass. pen.*, 1998, cit.; *Corte d'Assise di Chieti*, 19 December 2014, cit. This issue is analyzed in Sects. 1.1 and 1.2 above.

[38]Italian *Corte Costituzionale*, decision No. 327/2008, cit. On this topic see also: *Cassazione penale, sez. IV*, 7 December 2006, cit.; *Cassazione penale, sez. IV*, 5 May 2009, cit.; *Tribunale di Savona*, 11 March 2014 (*Tirreno Power* case), cit., p. 18; *Cassazione penale, sez. IV*, 17 May 2006 (the *Porto Marghera* case), cit. On this subject, see Sects. 1.1–1.5 above.

[39]On this topic, see *Cassazione penale, sez. IV*, 23 February 2015, n. 7941, cit.

[40]See Sect. 1.6 above.

[41]*Corte d'Appello di Venezia* 15 December 2004; on this topic see also: *Corte d'Assise di Chieti*, 19 December 2014, cit.; *Cassazione penale, sez.* IV, 23 February 2015, cit.

2.4 Hypothetical Scenario 4

Agent A owns a piece of land and contaminates it. The polluted land is subsequently purchased by Agent B. In the meantime, the land contamination results in the contamination of groundwater which is regularly used in human nutrition. The conduct of Agent A should be evaluated under Article 439, Paragraph 1 of the Italian Criminal Code (*Poisoning of water or substances for supply*)[42] when the behavior is intentional, and under Article 452 of the Italian Criminal Code (*Unintentional crimes against public health*) when the conduct is negligent. The first offense punishes "whoever poisons waters or substances for supply before they are drawn or distributed for consumption"; the second one punishes the same acts, foreseen by Articles 438 and 439 of the Italian Criminal Code, when they are committed as a result of negligence.

Case law has identified a number of conditions for the application of Article 439, Paragraph 1. Some of these are listed below.

(i) "The poisoning of water […] must not necessarily have lethal potential, but it is sufficient that it has the potential to harm health."[43]
(ii) "The water table must be currently polluted and the level of contamination encountered at the point of drawing must constitute a real danger for the consumers."[44]
(iii) "The notion of 'poisoning' takes into account not the mere introduction of toxic substances into the water, but refers to the effect of this conduct, consisting in a situation of concrete, real danger for human health."[45]
(iv) "General knowledge of the presence of organohalogen is not proof of willful misconduct; at least it shows a negligent underestimation of the situation."[46]

If these conditions exist, the legal test for the offense of *Poisoning of water or substances for supply* will be met and the fact will be imputed to Agent A; otherwise, it will not.

With regard to the conduct of Agent B, the observations related to the previous scenario can be taken into account also in this one.

[42]Para. 2 punishes the crime of *Death or injury as a consequence of poisoning*.

[43]*Corte d'Assise di Chieti*, 19 December 2014, cit., p. 70.

[44]Ibid. In particular: "The law does not incriminate the conduct of those who use toxic substances and introduce them into the water for human consumption, but the conduct of the person who produces a poisoning, which is the final effect of the conduct" (this "final effect" consists in making the water dangerous for consumers).

[45]Ibid.

[46]Ibid.

2.5 Hypothetical Scenario 5

Agent A owns a piece of land and contaminates it. The polluted land is subsequently purchased by Agent B. There is evidence of a causal link between the pollution and the death or injury of a number of people. According to Italian criminal law, the conduct of Agent A should be evaluated under the following rules.

(i) Article 439 of the Italian Criminal Code, *Poisoning of water and substances for supply*,[47] if the contamination causes the poisoning of surface and groundwater (when the conduct is negligent, Art. 452 of the Italian Criminal Code, *Unintentional crimes against public health*).[48]
(ii) If the conduct causes an "uncontrolled, harmful and still ongoing"[49] exposure of population to dangerous substances, Article 434 of the Italian Criminal Code, *Collapse of buildings or other fraudulent disasters* ("*Unnominated disaster*") can be applied.[50] If the conduct is negligent, Article 449 of the Italian Criminal Code—*Crimes of negligent damage*—is applicable.
(iii) Article 589 of the Italian Criminal Code—*Involuntary murder*.
(iv) Article 590 of the Italian Criminal Code—*Involuntary personal injuries*.[51] Nevertheless, the *Environmental Crimes Reform* introduced Article 452-ter into the Italian Criminal Code for the case of *Death or injury as a result of the crime of environmental pollution*: it may be relevant in this case.
(v) Article 452-quater of the Italian Criminal Code (*Environmental disaster*)—which is only applicable if Article 434 is not—and Article 452-bis (*Environmental pollution*) may possibly concur with the other crimes. Both these offenses were also introduced by the *Environmental Crimes Reform*.

Agent B will be liable for the same offenses—eventually, by cooperating with A—only if:

(i) He "has a position related to a duty of care within the company that managed the business risk during the etiologically relevant period."[52]
(ii) He carries out a causally efficient conduct, that increases the prior contamination (that is, if his conduct is a "condition" of the event).[53]

---

[47]See Sect. 2.4 above.

[48]See Sect. 2.4 above.

[49]*Cassazione penale, sez. IV*, 23 February 2015, n. 7941, cit.

[50]See Sects. 1.1–1.5 and 2.3 above (on this subject: *Cassazione penale sez. IV*, 7 December 2006, cit.; *Cassazione penale sez. IV*, 5 May 2009, cit.; *Tribunale di Savona*, 11 March 2014 [*Tirreno Power* case], cit., p. 18; *Cassazione penale, sez. IV*, 17 May 2006 [*Porto Marghera* case], cit.).

[51]The most helpful case on this subject is *Porto Marghera case*, cit. (see Sect. 1.1 above).

[52]*Tribunale di Venezia*, 29 May 2002, p. 171.

[53]*Tribunale di Venezia*, 29 May 2002, cit.

(iii) There is "a causal connection between his activities and the fact".

Agent B cannot be held responsible just because of his role, if there is no etiological link between his behavior and the event. In the same way, the existence of the *mens rea* requirements (and, in particular, the element of the intent) cannot be deduced by the "sole reason that the [defendants] are 'aware' of contamination coming from the past."[54] There must be a purposeful direction of acts toward the event figured as possible.

With regard to offenses of negligence (especially murder and personal injuries), Agent B would not be liable if he had not violated any precautionary rule that:

(i) was aimed at preventing the risk realized by a *hic et nunc* event;
(ii) was knowable by the Agent at the time of the facts;
(iii) arose from strong and common scientific knowledge in the relevant field;
(iv) whose violation is causally linked to the event (according to assessment criteria on negligence-causality).

For further analysis of these topics see Chap. 4. For a summary of the relevant case law, see Sects. 1.1–1.5 above.

References

Barcellona Pozzo di Gotto, G. U. P. (2013). 11 March 2013, Judge Adamo, on *Dir. pen. contemporaneo*, online, 9 December 2014.
Cassazione penale, sez. III, 27 May 1997, n. 7170, on *Cass. pen.*, 1998, pp. 1624 *ff.*
Cassazione penale, sez. V, 26 April 2005, n. 23465, on *Cass. pen.*, 2007, pp. 615 *ff.*
Cassazione penale, sez. I, 14 October 2005, n. 41983, on *Ced* Rv. 232874.
Cassazione penale, sez. IV, 17 May 2006, n. 4675 (*Porto Marghera*), on *Foro it.*, 2007, II, pp. 550 *ff.*
Cassazione penale, sez. III, 29 November 2006, n. 9794, on *DeJure.*
Cassazione penale, Sez. V, 7 December 2006, n. 40330, on *Ced* Rv. 236294.
Cassazione penale, sez. III, 14 March 2007, n. 26479, on *DeJure.*
Cassazione penale, sez. III, 30 May 2007, n. 28525, on *DeJure.*
Cassazione penale, sez. III, 9 luglio 2007, n. 26479, in *Amb. e Svil.*, 2007, pp. 852 *ff.*
Cassazione penale, sez. III, 11 October 2007, n. 40191, on *DeJure.*
Corte Costituzionale, 30 July 2008, n. 327, on *DeJure.*
Cassazione penale, sez. IV, 5 May 2009, n. 18974, on *Ced* Rv 243992.
Cassazione penale, sez. III, 2 July 2010, n. 35774, on *DeJure.*
Cassazione penale, sez. III, 15 December 2010, n. 11498, on *DeJure.*
Cassazione penale, sez. III, 16 March 2011, n. 18503, on *Cass. pen.*, 2011, pp. 4405 *ff.*, on *Rivista Giuridica dell'Ambiente*, 2011, pp. 809 *ff.*
Cassazione penale, sez. III, 17 January 2012, n. 17817, on *DeJure.*
Cassazione penale, sez. IV, 22 March 2012, n. 24997, on *Foro it.*, 2012, II, pp. 517 *ff.*

[54]*Tribunale di Venezia*, 29 May 2002, cit.

Cassazione penale, sez. IV, 24 May 2012, n. 33311, on *Riml*, 2013, pp. 297 *ff.*
Cassazione penale, sez. III, 22 January 2013, n. 19962, on *DeJure.*
Cassazione penale, sez. IV, 27 February 2014, n. 18933, on *Ced* Rv. 262139.
Cassazione penale, sez. IV, 21 July 2014, n. 32170, on *Dir. pen. contemporaneo*, online, 9 December 2014.
Cassazione penale, sez. IV, 21 November 2014, n. 11128, on *Riml*, 2015, pp. 1139 *ff.*
Cassazione penale sez. IV, 23 February 2015, n. 7941 *(Eternit)*, on *Dir. pen. contemporaneo*, online, 24 February 2015.
Corte d'Assise di Chieti, 19 December 2014 (ud.), on *Dir. pen. contemporaneo*, online, 9 February 2015.
Corte d'appello di Venezia, 15 Decemebr 2004, on www.petrolchimico.it
Tribunale della Spezia, 6 June 2011, n. 10007, on *DeJure.*
Tribunale di Avellino, 15 June 2013, Judge Riccardi, 8, on *Dir. pen. contemporaneo*, online, 15 November 2013.
Tribunale di Savona, 11 March 2014, Judge Giorgi *(Vado Ligure)*, on *Dir. pen. contemporaneo*, on-line, 16 October 2014.
Tribunale di Savona, 18 July 2014, Judge Salvini (unpublished).
Tribunale di Rovigo, 22 September 2014, *Pres.* Angeletti, ric. Arrighi, on *Dir. pen. contemporaneo*, online, 16 October 2014.
Tribunale di Venezia, 29 May 2002, on www.petrolchimico.it

Chapter 6
Historical Pollution and Corporate Liability in the Italian Criminal Law

Rossella Sabia

Contents

1 Introduction to Corporate Liability in Italy: Focus on Environmental Crimes Under Legislative Decree 231/2001

With the Legislative Decree of 8 June 2001, n. 231, corporate liability for administrative infringements depending on crimes was introduced into the Italian legal system.[1] Corporate liability is related to the commission of a crime—among those specifically laid down by the Legislative Decree itself—by specific parties (top managers or employees), committed in the interest of or for the benefit of the corporation.

[1]On Legislative Decree 231/2001, see among others De Maglie (2002), Bassi and Epidendio (2006), De Vero (2008), Presutti et al. (2008), Cadoppi et al. (2010) and Lattanzi (2010).

R. Sabia (✉)
LUISS Guido Carli University, Rome, Italy
e-mail: rsabia@luiss.it

F. Centonze and S. Manacorda (eds.), *Historical Pollution*,
DOI 10.1007/978-3-319-56937-6_6

Law 300/2000, with which the Italian Parliament had delegated to the government power to introduce this type of liability, did include environmental crimes among the predicate offences of corporate liability.[2] The government instead preferred a "minimalist" option, exercising its delegation only in part and excluding environmental crimes from the scope of the new law.[3]

Thus, when applying it, in accordance with the principles of "legal certainty" (*tassatività*) and "typicality" (*tipicità*),[4] the national judiciary denied that under Legislative Decree 231/2001 corporations could be liable for environmental crimes.

In particular, even though Article 192, Paragraph 4 of Legislative Decree of 3 April 2006, n. 152 (Testo Unico Ambiente—TUA, also known as *Codice ambientale*)—concerning the abandonment and/or uncontrolled disposal of waste[5]—contains express reference to the provisions of Legislative Decree 231/2001, the Italian Supreme Court (Corte di Cassazione) denied that the law on corporate liability could apply in relation to environmental crimes, because these crimes were not included in the catalogue of the predicate offences and because there were no indications regarding the related sanctions.[6]

---

[2]Law 300/2000 ratifies and implements a series of international acts prepared on the basis of Article K.3 of the Treaty of the European Union. It also contains the delegation to the government on regulations regarding the administrative liability of legal persons and entities without legal personality. Art. 11, Para. 1, Lett. (d) of Law 300/2000 provided for corporate liability related to the commission of crimes in the field of environmental and land protection listed therein.

[3]See §12 of the Report accompanying Legislative Decree 231/2001. The Italian Government chose to limit the exercise of the delegation to the list of offences included in the Conventions ratified by Law 300/2000 (and therefore to the crimes of bribery and fraud), and it considered that a period of time was to be provided for the companies to assimilate the new type of liability before extending it to other types of offences of high impact on the business world; see Bricchetti and Pistorelli (2011, p. 48).

[4]According to the prevalent interpretation among criminal law scholars, the principle of "legal certainty" (*tassatività*) concerns the technique of formulating the offences and expresses (a) the duty of the legislator to provide as accurate description as possible of the offences and (b) the duty of the judges to provide an interpretation that reflects the facts as they are described by the law (Mantovani 2013, p. 62; Fiandaca and Musco 2014, p. 85 ff.; Marinucci and Dolcini 2015, p. 71 ff.). The principle of "typicality" (*tipicità*) expresses the need under criminal law to identify each offence through a legislative description of a human fact that should be offensive (Palazzo 2016, p. 81). On these principles, with reference to corporate liability, see Lottini (2007a, p. 2294), Brunelli (2008a, p. 85 ff.), De Vero (2008, p. 137 ff.), Di Giovine (2010, p. 47 ff.) and Scoletta (2014, p. 885 ff.).

[5]Pursuant to Art. 192, Para. 4 of Legislative Decree 152/2006, if directors and representatives of a legal person are responsible for the illegal act under and for the purposes of Para. 3, the legal person and the persons surrogated to the rights of the legal person itself will be jointly and severally liable, pursuant to the provisions of Legislative Decree of 8 June 2001, n. 231, concerning the liability of legal persons, companies, and associations. According to some scholars, Testo Unico Ambiente (TUA) turned out to be a missed opportunity to extend corporate liability for breaches in environmental matters: see Arena and Cassano (2007, p. 34) and Romolotti (2007, p. 38).

[6]Cass. Pen., Sec. III, 7 October 2008, No. 41329 (filed on 6 November 2008), *Galipò*, rv. 241528.

The national legislative scenario on environmental matters changed when the Legislative Decree of 7 July 2011, n. 121, entered into force. With this decree, Italy belatedly implemented Directive 2008/99/EC on the protection of the environment through criminal law and Directive 2009/123/EC on ship-source pollution and on the introduction of penalties for infringements. As regards changes to the relevant rules, Legislative Decree 121/2011 was limited to a few amendments of the Italian penal code norms, while much more significant innovations concerning corporate liability were made by means of the introduction of Article 25-undecies, entitled *Environmental crimes*,[7] into the corpus of Legislative Decree 231/2001 (Madeo 2011, p. 1062 ff.; Ruga Riva 2011, p. 7 ff.; Scarcella 2011, p. 854 ff.; De Santis 2012, p. 687 ff.; Ramacci 2012b, p. 639 ff.; Scoletta 2012, p. 17 ff.; Pelissero 2013, p. 44 ff.).

This new provision extends corporate liability with reference to a wide and varied range of environmental crimes. Among the offences that are currently included in the catalogue of the so-called 231 predicate offences are numerous misdemeanours in the field of pollution of water, soil, and air, as well as the failed remediation of sites (offences already provided for by Testo Unico Ambiente and currently referred to in the aforementioned Article 25-undecies of Legislative Decree 231/2001).

The analysis here will focus on some of these offences (see Sects. 3 and 4) selected, given the perspective of this research, as relevant for the analysis of the phenomenon of *historical pollution* in relation to corporate liability.[8] In particular, the relevant offences will be those envisaged in Article 137 TUA on *Water pollution*, Articles 256 and 260 TUA on *Waste*, Article 279 TUA on *Air pollution*, and Article 257 TUA on *Remediation of contaminated sites*.[9]

---

[7]Art. 25-undecies of Legislative Decree 231/2001, in its original wording, recalls among the predicate offences the two new misdemeanours provided for in Art. 727-bis of the Italian penal code (killing, destroying, capturing, removing, or holding protected wild animal or plant species) and Art. 733-bis of the Italian penal code (destruction or deterioration of habitat within a protected site); some crimes on international trade in protected fauna and plant species (Law 150/1992); crimes on production and use of ozone-depleting substances (Law 549/1993); crimes related to ship-source pollution (Law 202/2007); and finally, and most significantly, misdemeanours on pollution of water, soil, and air are identified and included among the predicate offences of corporate liability provided for by Testo Unico Ambiente.

[8]In accordance with the definition adopted by the present research, the offences listed in Testo Unico Ambiente, identified in order to analyse the phenomenon of historical pollution in relation to corporate liability, could all cause pollution of sites and natural resources with effects emerging a long time after the moment at which the conduct was carried out.

[9]In more detail, the offences to be examined provided for in Art. 25-undecies of Legislative Decree 231/2001 are the following: Art. 137, Paras. 2 and 3 TUA: *Discharge of industrial waste water containing hazardous substances without the required authorization and discharge of the same substances in violation of the requirements or authorization of the competent Authority*; Art. 137, Para. 5 TUA: *Discharge of industrial waste water over the limits*; Art. 137, Para. 11 TUA: *Failure to observe prohibitions on discharges into the soil, subsoil and underground water*; Art. 256, Paras. 1 and 3 TUA: *Activities of management of waste and unauthorized construction or operation of unauthorized landfill*; Art. 260 TUA: *Organized activities for illegal trafficking*; Art. 279, Para. 5 TUA: *Exceeding the limit values of air quality provided for by the legislation in force*; Art. 257, Paras. 1 and 2 TUA: *Sites remediation.*

Thus, the legislative action of 2011 represented a first attempt to provide a *criminal* response to the problem of environmental crimes committed by corporations, as a typical expression of corporate crime. Scholars, however, while recognizing the "good intentions" of the Italian legislator, have widely criticized the new provisions, highlighting the ineffectiveness of the protection granted[10] (in accordance both with European political and criminal indications and with the established judicial domestic practice) and the overall irrationality of the system (Plantamura 2011, p. 496; Scoletta 2012, p. 37).

The criminal law protection against pollution has remained focused on misdemeanours (*contravvenzioni*) and on *abstract endangerment* offences (as stated above, they were already provided for in Testo Unico Ambiente) that are not sanctioned with strong penalties. The exclusion of certain *crimes*—whose introduction was expected and hoped for—from the list of predicate offences of corporate liability in environmental matters caused much debate at the time. Reference is here made to the most serious crimes (*delitti*) provided for in the Italian penal code to protect public safety, traditionally used by the judicial authorities to prosecute conduct damaging the environment carried out by natural persons (such as, for example, the "unnominated disaster"—a literal translation of *disastro innominato*—under Article 434, and the poisoning of drinking water under Article 439, both also in their negligent forms under Articles 449 and 452 of the Italian penal code).[11]

From these considerations, the objection arises that this regulatory approach has created an asymmetric and not very rational set of legal provisions. As for the type and the number of offences individuals and corporations can be liable for, while individuals could (also) be held responsible for the above-mentioned *crimes* under the penal code, the collective entity could only be liable for less significant offences (*misdemeanours*) provided for in Testo Unico Ambiente and referred to in Article 25-undecies of Legislative Decree 231/2001. However, at the same time, with regard specifically to the misdemeanours provided for in Testo Unico Ambiente, the consequences for individuals were minor (being misdemeanours subject to a short period of limitation, sanctioned with not severe penalties and extinguishable by the payment of a sum of money—*oblazione*); meanwhile, Legislative Decree 231/2001 provided for a strict system of rules for entities, which receive severe sanctions, not extinguishable through the payment of the just-mentioned *oblazione*, and subject to a statute of limitation similar to that provided in civil law, as laid down in the Decree itself (see Sect. 6).

[10]For example, with reference to the Testo Unico Ambiente, the following offences were not introduced among the predicate offences of Legislative Decree 231/2001: offence of abandoning of waste sanctioned by Art. 256, Para. 2, TUA; almost all the infringements related to air pollution; offences established by special regulations, expressly mentioned in Directive 2008/99/EC and related to activities with high environmental impacts (such as the rules on environmental integrated authorization—*AIA*).

[11]For further details on this issue, see Chap. 4, Sect. I.6; see also Ramacci (2012a), Forzati (2015), Scaroina (2015) and Ruga Riva (2016, p. 283 ff.).

Given these reported problems, the legislator intervened again with the recent reform of environmental crimes (Law 68/2015).[12] As a result of the amendment introduced by Article 1, Paragraph 8, corporate liability pursuant to Article 25-undecies of Legislative Decree 231/2001 has been extended to the new hypotheses of intentional crimes of environmental pollution (Art. 452-bis of the Italian penal code) and environmental disaster (Art. 452-quater of the Italian penal code), and to the corresponding unintentional cases (Art. 452-quinquies of the Italian penal code, with a reduced fine). Association for purposes of delinquency (the common one and the Mafia-related one) with the aggravating environmental circumstance (Art. 452-octies of the Italian penal code) and the crime of trafficking and abandonment of highly radioactive materials (Art. 452-sexies of the Italian penal code) are also introduced in the catalogue of the new criminal offences concerning environmental protection.

Moreover, there is no perfect coincidence between the provisions concerning individuals and those concerning corporations: the Law of 2015 also introduced the crime of failed remediation of sites (Art. 452-terdecies of the Italian penal code), which is not drawn up among the predicated offences of corporate liability.

With regard to the previous non-coincidence of levels between the natural person's liability and the entity's, this intervention has therefore largely remedied the aforementioned asymmetries, without prejudice to the other matters set forth above.

2 General Principles and Their Interaction with the Phenomenon of Historical Pollution

Corporate liability, introduced in Italy by Legislative Decree 231/2001, beyond the formal *administrative* label, has an essentially criminal foundation.[13] In particular, concerning the implications for the theme of historical pollution, it is appropriate to draw attention to Articles 2 and 3 of the Decree. Article 2 of Legislative Decree

---

[12]The Law of 22 May 2015, n. 68 (published in the Official Gazette No. 122 of 28 May 2015) introduced into the Italian system the new Title VI-bis of the penal code, entitled "Crimes against the environment" (*Dei delitti contro l'ambiente*). For an overview of the main innovations and for a comment on the reform, see Chap. 4, Sect. I.7; and see among others Masera (2015), Ramacci (2015), Ruga Riva (2015) and Siracusa (2015); in particular, with reference to corporate liability, see Amarelli (2016).

[13]Legal doctrine and jurisprudence have discussed the nature of this liability at length. It may be held to be administrative liability, according to the wording of Legislative Decree 231/2001 (Marinucci 2002, p. 1203 ff.; Romano 2002, p. 398 ff.; Palazzo 2016, p. 56 ff.), or criminal liability, in light of the fact that it is established by a criminal court and due to the afflictive nature of the sanctions (ex multis, Paliero 2001, p. 845; Carmona 2002, p. 208; Fiorella 2003, p. 85 ff.; Manna 2003, p. 1103; De Vero 2008, p. 323; De Simone 2012, p. 21). The thesis of the "tertium genus" of liability is supported by Pelissero (2007, p. 867) and Di Giovine (2010, p. 14). The Italian Supreme Court in its more extended composition (Cassazione a Sezioni Unite) adhered to the latter approach in its judgement of 24 April 2014 (filed on 18 September 2014), No. 38343, *Thyssenkrupp*, rv. 261112.

231/2001 (the principle of legality—*principio di legalità*) states that the corporation cannot be held responsible for a fact constituting a crime if its administrative liability in relation to that crime and the related sanctions are not expressly foreseen by a law that came into force prior to the commission of the fact.

In this Article, two fundamental principles of criminal law are expressly set out: the principle of legality—in relation to the precept and to the sanctions and with the particular meaning of "reservation for law" (*riserva di legge*)—as the corporation can be held liable under Legislative Decree 231/2001 only if its administrative liability in relation to the alleged offence and the related sanctions are expressly established by a law; and the principle of "non-retroactivity" (*irretroattività*), as that law should have entered into force before the offence was committed.[14]

In the provision outlined, any form of automatic liability is excluded: corporate liability is entailed only in the case of a fact qualified as a criminal offence by the law, and only in the case of offences previously "authorized" by the legislator and laid down by the Decree (Brunelli 2008a, p. 86; Scoletta 2009, p. 371 ff.).

Moreover, because of the principle of "non-retroactivity", any legislative modification with incriminating function can concern only future facts. On corporate liability, the introduction of a new type of crime into the catalogue of predicate offences must always qualify as a "new incrimination": even where a fact has been already foreseen as a crime in relation to natural persons, it can be considered as a predicate offence of corporate liability only *after* its introduction into the catalogue of "corporate offences" and only if the other conditions provided by the Decree do occur.

This observation implies that, also in this case, the sequence *law–fact* does not tolerate any modifications, in the sense that the offence shall be committed *entirely after* the entry into force of the law.[15] Consider for example the introduction *ex novo* of a new case of "habitual crime" (*reato abituale*; a crime which comprises a series of many similar conducts, where crime commission occurs with the last one of the series):[16] under such circumstances, the previous acts, that have been carried out in the absence of a legislative provision, cannot be placed in connection with those which occurred after the entry into force of the new law, and cannot be

---

[14]Here there is a clear reference to the content of Art. 1 of the Italian penal code concerning natural persons, which states that "no one may be punished for an act which is not expressly designated as an offense by law, nor with punishments which are not thereby prescribed". This norm, as well as the further norms of the Italian penal code reported, comes from the translation provided for by Wise (1978).

[15]Scholars have criticized courts' decisions that have held corporations punishable in the case of fraud against the State (Art. 640-bis of the Italian penal code) where the offence was committed *before* and the profit achieved *after* the entry into force of Decree 231; similarly, they have disputed that interdictory sanctions can be applied to the corporation in consideration also of profits achieved *before* the entry into force of the Decree for the purposes of calculating the entity of the profit: see Pistorelli (2006, p. 132 ff.) and Brunelli (2008a, p. 87).

[16]The so-called *reato abituale* is a crime in which the law requires the *reiteration* of a series of homogeneous actions or omissions, even after a long interval of time (Mantovani 2013, p. 511 ff.; Marinucci and Dolcini 2015, p. 245 f.).

considered among those necessary to complete the so-called minimum set (*serie minima*) of conducts (Fornasari 1991, p. 7; Petrone 1996, p. 203; Romano 2004, p. 348; Mantovani 2013, p. 97).[17] In fact, for the application of the new incriminating law it is not sufficient that the last act should have occurred under it; it is already required that this act can assume criminal relevance (Mantovani 2013, p. 97), thus completing the "minimum set" required for the occurrence of damage to the legal asset protected (Fornasari 1991, p. 7).

As a consequence, a new "habitual crime" cannot be considered to have occurred when just one further episode happens after the entry into force of the new law. Otherwise, criminal relevance would be placed also on the previous conducts, which must not have any criminal relevance in order to avoid the violation of the principle of "non-retroactivity" *in malam partem*, and in light of the fact that, in that event, the criminal law could not have any function in guiding the choices of the individual (Petrone 1996, p. 203; Mantovani 2013, p. 97; Valsecchi 2009, p. 1409).

In the matter of corporate liability, special attention for complying with the principle of legality is also confirmed by the latest jurisprudence of the Italian Supreme Court. The judges of the Supreme Court annulled the preventive seizure aimed at the "confiscation for equivalent" (in accordance with Articles 19 and 53 of Legislative Decree 231/2001) of money and other property at disposal of the companies *Riva F.i.r.e S.p.A.* and *Riva Forni Elettrici S.p.A.* in the well-known *Ilva* affair.[18] On that occasion, it was strongly reaffirmed that crimes not listed in the catalogue of predicate offences referred to in the Decree[19] cannot be considered for the purposes of corporate administrative liability. Moreover, the Court did exclude that those offences that are unfit to found liability can be "recovered" through Article 416 of the Italian penal code (*Association for purposes of delinquency*, which is *included* among the predicate offences ex Article 24-ter of the Decree), in *violation of the principle of legal certainty* enshrined in Legislative Decree 231/2001. Otherwise, the provision of Article 416 of the penal code would be transformed into an "open" provision, with an elastic content that is potentially adequate to include any type of crime in the catalogue of the predicate offences.[20]

---

[17]In this respect, it should be also clarified that, in case individual conducts do already constitute separate criminal offences and the new law simply introduces the new "habitual crime" (the so-called *reato abituale improprio*), scholars come to a different conclusion. Without prejudging the necessary occurrence of a "minimum set", in such an event the previous conducts assume relevance and the occurrence of *a single act* under the new law could also be sufficient for the application of that law, even if it is more severe: Marinucci and Dolcini (2015, p. 246); see also Romano (2004, p. 348) and Valsecchi (2009, p. 1407 f.); conversely, Petrone (1996, p. 204) states that it should be determined which is the *most favourable* law. On these issues, see Sect. 3.

[18]Cass. Pen., Sec. VI, 20 December 2013 (filed on 21 January 2014), No. 3635, *Riva F.i.r.e. S.p.A.*, rv. 257789. For a comment on this judgement, see Piergallini (2014) and Trinchera (2014).

[19]In this case, the crimes (provided for by the penal code, and not included among the predicate offences of Decree 231) were the so-called unnominated disaster, the intentional removal or omission of precautions against industrial accidents, and the poisoning of water or foodstuffs.

[20]The Supreme Court seems to have "softened" its position with a later decision (Cass. Pen., Sec. III, 14 October 2015, filed on 23 November 2015, No. 46162, published on www.iusexplorer.it).

Moreover, the Supreme Court pointed out that in accordance with the principles of legality and non-retroactivity pursuant to Article 2 of the Decree, the application of sanctions to corporations is subject to the existence of an express legal provision that came into force *before* the offence to be committed. In the present case, this is relevant for the purposes of the application of the confiscation for equivalent (Art. 19 of the Decree), and therefore the Court observed that the application of the *in rem* precautionary measure and of the following confiscation cannot be retroactive or applicable with reference to conducts carried out before the entry into force of corporate administrative liability; in fact, only conducts carried out from the moment at which organized and environmental crimes were included in the catalogue of the predicate offences are considered relevant in this issue.

From this perspective, account should be taken of *the date when conducts constituting a crime* that should be made punishable *have been committed*, and not of the moment in which the corporation achieved the profit. Therefore, on this point the Corte di Cassazione ruled that for a correct determination of the profit liable to confiscation, the decree that authorized the preventive seizure should not have considered either conducts related to organized crimes committed before the entry into force of the law that extended the catalogue of the predicate offences to the crime referred to in Article 416 of the Italian penal code (Law of 15 July 2009, n. 94), nor any violations of the environmental legislation committed before the entry into force of Legislative Decree 121/2011.

Legislative Decree 231/2001 reveals its criminal law nature also with reference to the regulation of *succession of criminal laws* (Art. 3 of the Decree) which, in line with the provision enshrined in Article 2 of the Italian penal code for natural persons, establishes in Paragraph 1 that the corporation cannot be held liable for a fact that according to a later law no longer constitutes a crime, or in relation to which the corporation's administrative liability is no longer contemplated, and, if

(Footnote 20 continued)

The case concerns a company charged—with reference to Art. 416 of the penal code and Arts. 3–4 of Law 146/2006—with organizing, promoting, and being part of a transnational criminal conspiracy aimed at committing an undetermined number of crimes involving tax evasion and fraud. It is to be noted that the crime of association (also the transnational one) is provided as a predicate offence for corporate liability, while tax crimes are not included in the catalogue. The Supreme Court, in its ruling mentioned above, states that the preventive seizure aimed at the confiscation for the equivalent was properly applied in relation to the crime of association and that the profit gained was to be identified as the profit resulting from the offences of tax fraud included in the programme of the transnational criminal organization. This interpretation does not violate Art. 2 of Legislative Decree 231/2001 (*principle of legality*), in the sense that a profit gained through fiscal offences—not listed in the closed catalogue provided for by the Decree—would be subject to seizure for equivalent. As a matter of fact, the profit, meant as the set of benefits derived from the crime, may also consist of the complex of the benefits resulting directly from the set of the organisation-related crimes (tax crimes in this case), from which the offence of criminal conspiracy is completely independent. Ultimately, the Supreme Court does not consider the preventive seizure in relation to offences excluded from the catalogue of Legislative Decree 231/2001 admissible as such, but it considers the measure applicable in relation to tax crimes only because the profit resulting from tax offences is the benefit for which the crime of association was conceived.

there has been a conviction, the enforcement and the juridical effects thereof shall cease.

Corporate liability (together with the enforcement of the conviction and of the juridical effects) shall cease both in case of *abolitio criminis* and in case the later law, without prejudice to the criminal relevance of the fact, no longer contemplates the corporation's liability pursuant to Legislative Decree 231/2001, relating to the commission of that crime. In case of a new formulation of the predicate offence, the abolishing effects will mirror those of *partial abolitio criminis* identified in the structural comparison among the criminal laws in succession (Scoletta 2014, p. 890).

Paragraph 2 of the above-mentioned provision regulates the hypothesis of a just amending succession (*successione modificativa*), ruling the retroactivity of the most favourable law if the law in force at the time in which the infringement has been committed and the following ones are different (unless a final decision has been issued). In case of many laws in succession, the primacy of the most favourable intermediate law is stated, even if it was no longer in force at the time of the process, considering the wording of the provision (Para. 2 says "successive *laws*" using the plural) and that this is the rule applied for natural persons (Lottini 2007b, p. 2297).[21]

For the purposes of the application of the rules concerning succession of criminal laws, it would seem preliminary to determine the time when the administrative infringement depending on a crime can be considered committed. The law generically refers to the time in which the *infringement* has been committed, but most scholars, to provide major guarantees, retain that the moment in which the predicate offence was committed by natural persons should be considered. This conclusion was reached taking into account that it is easier to determine the *tempus commissi delicti* of the crime the natural person can be held liable for, than to determine the moment when the facilitation or the non-adoption of the compliance model by the corporation took place. Therefore, active or omissive conducts of the corporation have to be evaluated with reference to the specific predicate offence committed by the individual (De Simone 2002, p. 94; Brunelli 2008b, p. 88; Di Giovine 2010, p. 54). As a consequence, also for the succession of laws concerning the administrative infringements depending on a crime, it is necessary to consider criteria used to determine the *tempus commissi delicti* in relation to individuals (Lottini 2007b, p. 2298).

---

[21]The rule refers, even in this case, both to the succession of criminal laws related to the predicate offence and to the succession of laws concerning the administrative infringement depending on crimes: see De Vero (2008, p. 140).

3 Historical Pollution and Offences Which Entail Corporate Liability

Given the necessary theoretical framework in relation to the origin of corporate liability and to the criminal law principles ruling this matter, it is appropriate to ask what are the relevant offences in relation to historical pollution in the perspective of corporate liability: as anticipated, Article 25-undecies of Legislative Decree 231/2001 includes a wide range of environmental crimes, but only some of those have characteristics of interest for the purposes of this research.

It has already been noted that the most significant cases referred to in Article 25-undecies of the Decree are essentially misdemeanours provided for in Testo Unico Ambiente. Among them, the most significant provision relating to historical pollution is that laid down in Article 257, Paragraphs 1 and 2 of TUA on remediation (see Sect. 4).[22] These are hypotheses that can be extended to some or all of the three mentioned fields (water, soil, airspace) on which the protection of the environment through criminal law is based.

As previously stated, among the provisions of Testo Unico Ambiente added to the catalogue of the predicate offences of Decree 231/2001, Article 137 concerning discharges, Articles 256 and 260 on waste, and Article 279 concerning air pollution are also relevant.

Finally, with respect to the issue of historical pollution, the impact of the new criminal types of environmental pollution and environmental disaster (Arts. 452-bis and 452-quater of the Italian penal code) must be assessed. Corporate liability is now provided with reference to these offences, by reason of the reform introduced by Law 68/2015.

The question to be answered is whether within the outlined legislative framework there is a possibility to apply Legislative Decree 231/2001 in relation to phenomena of past pollution.

First, in accordance with the principles of legality and retroactivity examined above, corporate liability must be excluded with reference to environmental crimes committed *before* the entry into force of Article 25-undecies of the Decree.

This provision introduced the above-mentioned environmental crimes among the predicate offences of corporate liability, since 16 August 2011: in compliance with the principle of legality provided for in Article 2 of Legislative Decree 231/2001, a judge must necessarily apply Legislative Decree 231/2001 when deciding on the liability of corporation lacking an appropriate compliance programme for environmental crimes (committed by top managers and employees, in the interest of or for the benefit of the corporation) committed *after* that date.

[22]On this, see also Chap. 4, Sect. I.4.

Conversely, as regards offences committed and ended—thus, completed offences —*before* the entry into force of the law which introduces environmental crimes in Decree 231, as already said there is no corporate liability under Legislative Decree 231/2001.

Moreover, as far as historical pollution is concerned, it is difficult to draw a clear dividing line, as it is a dynamic phenomenon with diachronic dimensions in progressive development, often with deferred exteriorization. To give an example, consider multiple conducts, repeated over time, contaminating the environmental media (discharges of hazardous substances, transfer of waste to an illegal dump). Some conducts were carried out in a time prior to the introduction of environmental crimes among the predicate offences of corporate liability, while others were carried out later. Think of a site now polluted due (also) to conducts that occurred in a more-or-less distant past. Quid juris?

In trying to address this issue, it should be first pointed out that the relevant cases that engage corporate liability in environmental matters do already raise a number of complex issues about the natural persons who commit the criminal offences. For the purposes of this research, the main issue here is related to the identification of the nature of the offences (whether they are "instantaneous", "persisting", or "habitual" crimes)[23] with consequent effects on the commission of the crime and on the running of the period of limitation.

It should further be noted that, in order to prosecute particularly serious pollution cases attributable to conducts carried out over a significant period of time (and/or with delayed effects), in Italy the public prosecutors usually define charges based on criminal offences (*delitti*)—in order to circumvent the difficulties[24] considered above—and that these criminal offences do not fall—even today—within those engaging corporate liability (here reference is made to cases laid down in Arts. 434 and 439 of the Italian penal code: see Sect. 1). Thus in Italy, as regards criminal liability, the most serious environmental offences have traditionally been prosecuted through the incrimination of the individual author of the crime (even where the offence was part of a business context) and through the above-mentioned crimes laid down in the Italian penal code (rather than through misdemeanours provided for in the TUA); and this trend has not stopped with the introduction of environmental crimes among the predicate offences of corporate liability in 2011.[25]

---

[23]A crime is *instantaneous* when the offence is instantaneous, as it arises and ends at the same time, and may not extend over time; a *persisting* crime is a crime for which the law requires that the offence does last over time due to the wilful conduct of the agent (Mantovani 2013, p. 438). As regards *habitual* crime, see n. 16 above.

[24]For a more detailed examination of these issues, see Chap. 4, Sects. I.5–I.6.

[25]Once again, the recent *Ilva* affair seems exemplary: the public prosecutor used the crime of association (Art. 24-ter of Legislative Decree 231/2001) to prosecute violations in environmental matters not provided among the predicate offences of corporate liability: see Sect. 2.

Against this background, the central issue concerns the identification of the *tempus commissi delicti*: namely, the challenge is to determine under which law the fact was committed. The issue therefore arises with respect to events that took place not entirely under one of the laws, but partly under one, partly under the other; this is a relevant aspect in relation to certain types of crimes, such as "habitual" and "persisting" crimes.

From this particular point of view, it is necessary to examine the individual offences that can trigger corporate liability. As regards the specific cases of TUA that are relevant for the purposes of corporate liability, it would seem necessary to draw some distinctions on the basis of the nature of the criminal offences and of the *tempus commissi delicti*.

For the purposes of this analysis, the distinction between "instantaneous" crimes (*reati istantanei*) on the one hand, and "persisting" (*reati permanenti*) and "habitual" crimes on the other becomes important for determining the commission of the offence, and therefore for the identification of the law applicable in cases of the succession of criminal laws, as well as of the moment when the statute of limitation begins to run.

Misdemeanours on water pollution laid down in Article 137 TUA, namely the hypotheses referred to in Paragraphs 2 and 3 (disposal of industrial waste waters containing dangerous substances in the absence of an authorization, and disposal contrary to the requirements of the authorization) are considered *persisting* offences, as it can be assumed that the unlawful situation has extended over time, as a result of the persisting conduct of the author, with the consequence that the commission of the crime—together with the *dies a quo* for the period of limitation—has to be individuated at the end of this situation (for example, after the closing of the plumbing or after gaining the authorization) (Bernasconi 2012a, p. 454).

The hypothesis referred to in Article 137, Paragraph 5 TUA does occur when table limits are exceeded and it has an *instantaneous* nature, with the consequence that as many offences may be individuated as the number of conducts exceeding table limits. The sole exception is the continuous discharge, which occurs when the alteration of the ecological acceptability of the receiving body of water lasts seamlessly over a long period and as a result of persistent voluntary conducts (Bernasconi 2012a, p. 475).

The misdemeanour provided for in Article 137, Paragraph 11 (failure to observe prohibitions on discharges into the soil, subsoil, and underground water) may take the form of a *persisting* or *habitual* criminal offence depending on the implementing modalities (seamless or not) of the discharge (Costalunghi 2007, p. 104).

Provisions on waste provided for in Article 25-undecies of Legislative Decree 231/2001 are the most numerous. Article 256, Paragraph 1 TUA, concerning non-authorized waste management activities, does sanction several conducts: thus, the time of commission cannot be identified in a uniform way. Criminal offences which occur during collection and transportation activities without the required registration, as well as during illegal trade and intermediation activities, may fall within the "potentially habitual" offences (*reati eventualmente abituali*), meaning that the typical offence may be considered completed just by the execution of the single conduct, and its possible reiteration would, however, give rise just to one criminal offence (Bernasconi 2012b, p. 917 f.). The situation is different in case of activities focusing on abusive recovery and disposal activities, which have a long-lasting nature and should be considered as *persisting* offences (at least when these activities are long-lasting or are articulated in tightly connected phases which form an indissoluble whole), with the period of limitation starting from the end of the situation of persistence (Bernasconi 2012b, p. 918).

The construction of an unauthorized landfill provided for in Article 256, Paragraph 3 TUA has been considered by the Supreme Court in its more extended composition (Sezioni Unite) as a *persisting* criminal offence until the completion of the works; after that moment the crime should be considered with *persisting effects* (*a effetti permanenti*).[26] The period of limitation should start from the moment in which the landfill is built. However, some scholars noted that the construction of an unauthorized landfill could be completed also with only one act, therefore having an instantaneous nature, when the landfill is the result of a single act of abandonment of waste that spoils the environment (Bernasconi 2012b, p. 930). Instead, the criminal offence of management of an unauthorized landfill (also provided for in Art. 256, Para. 3 TUA) should be considered as *persisting*, as it lasts for as long as the organization has been operating. In such a situation, the criminal offence would be no longer persisting from the moment in which authorization is granted or the landfill is closed or the judgment at first instance is given.[27] As a further consequence, it is likely that the continuation of non-compliance with the law could extend liability to whoever succeeded a previous person in the management of the landfill and continued his activities although he knew (or should have known) that the landfill was unauthorized.

In accordance with the prevailing case law and most scholars, the specific case of crime provided for in Article 260 TUA (organized illicit waste trafficking) falls within the *habitual* criminal offences, as the criminal relevance of the fact is subordinate to the required recurrence of illicit management activities. The criminal offence will no longer be habitual when illicit activities stop or the organization is dismantled or the activities are no longer illicit as the required authorizations are given (Bernasconi 2012c, p. 974).

Article 279, Paragraph 5 TUA in the field of air pollution punishes those who, when operating a plant, exceed the emission limit values, thus also exceeding the air quality limits provided for by the legislation in force. Jurisprudence considers such conducts as *persisting* criminal offences (Monaldi 2012, p. 1035).

This analytical framework is necessary, as relevant distinctions shall be made based on the nature of the offences provided for in the Testo Unico Ambiente, which can engage corporate liability.

---

[26]Cass. Pen., Sezioni Unite, 5 October 1994 (filed on 28 December 1994) No. 12753, *Zaccarelli*, rv. 199385. For a comment, see Paone (1995).

[27]Cass. Pen., Sec. III, 11 April 1997 (filed on 29 April 1997) No. 4013, *Vasco*, rv. 207613.

As already noted, jurisprudence considers these infringements mainly as persisting or habitual ones: as a consequence, according to most scholars and to the jurisprudence itself, the moment of commission of the crime is identified as the moment when the last unlawful conduct that completes the fact constituting the offence has occurred (Mantovani 2013, p. 96; Fiandaca and Musco 2014, p. 213; Marinucci and Dolcini 2015, p. 246).

Considering these crimes in relation to Legislative Decree 231/2001, according to the general principles of criminal matters, corporate liability could be assumed even if the crime did start in a moment prior to the introduction of environmental crimes into the predicate offences list, provided that it has been completed under the new provisions.

It would be worthwhile reporting a Corte di Cassazione judgement[28] referring to an order in which the Tribunale del Riesame annulled the preventive seizure for equivalent decree on properties of a corporation, in relation to the commission of a number of crimes concerning waste, carried out *before* the entry into force of Legislative Decree 121/2011. According to the judges of Tribunale del Riesame, since there was no evidence of the continuation of illicit activities up to July 2011, date of entry into force of the new law, based on the principle of legality (Art. 2 of Legislative Decree 231/2001) the corporation could not be liable in relation to environmental crimes.

The Prosecutor appealed before the Italian Supreme Court on the ground of an infringement of Articles 256 and 260 TUA, asserting that the Court should have considered the persisting nature of these environmental crimes, thus ruling out that the facts alleged could be prior to the 2011 legislative reform (as degraded sites have not been restored).

The Italian Supreme Court drew some distinctions based on the specific alleged offences. As regards waste abandonment, unauthorized landfill, and illegal trafficking provided for in Articles 256 and 260 TUA, the Court claimed that the alleged order is correct, considering, however, on the basis of an appraisal of the facts, that there was a lack of evidence regarding the continuation of infringements up to July 2011 or even beyond and, consequently, that the legislation provided for in Legislative Decree 121/2011 on corporate liability also concerning environmental crimes cannot be applied. Indeed, crimes of waste abandonment and unauthorized landfill are potentially persisting crimes, whose illegality ceases with the last illicit waste disposal or with the property seizure or with the judgement at first instance. The offence of illegal waste trafficking is a habitual crime, as mentioned above, since necessarily it has to be completed by many similar conducts, and crime commission occurs with the last one of these events.

[28]Cass. Pen., Sec. III, 14 April 2015 (filed on 30 September 2015) No. 39373, published on www.penalecontemporaneo.it. For a comment, see Perini (2015).

The Court noted that the order is wrong in law as regards storage of waste, considered by the jurisprudence as a crime with persisting nature, whose commission occurs with the removal of the factual illegal situation. The order is wrong in law also in relation to the failed remediation of the polluted sites (Art. 257 TUA), as it established that the commission of the crime had occurred before July 2011. Also in this case it can be considered as a persisting crime, as the seizure of the polluting site (prior to the elimination of the damage) is not sufficient to interrupt the conduct, and safety operations, remediation, and environmental restoration activities are needed—repair measures that, if put into effect prior to the judicial decision, make the crime not punishable.

As regards such crimes, the judges of Tribunale del Riesame should not have verified the lack of evidence regarding the continuation of infringements up to July 2011 or even after, but, as far as the first crime is concerned, it should have ensured that the factual illegal situation had ceased, and as far as the failed remediation is concerned, that the cessation of the illegal conduct had occurred before July 2011, the sole date relevant to determining crime commission.

Thus, if in theory the rules provided for by Legislative Decree 231/2001 on corporate liability may be considered applicable also to past pollution phenomena, within the stated limits, it should be noted that on this issue there is no established case law, and that this judgment is certainly a significant statement by the Italian judges, but an isolated one.

From another angle, criminal offences recently introduced with Law 68/2015 do have some implications for the present issue, especially as regards issues of transitional law and its possible impact on pending lawsuits.[29]

In particular, it is important to point out that the newly introduced norm relating to environmental disaster provided for in Articles 452-quater and quinquies of the Italian penal code, in both intentional and negligent cases, lacks sufficient references to resolve doubts about the nature of the crime, specifically whether it is persisting or instantaneous with persisting effects (Ruga Riva 2016, p. 266).

Future developments depend on the interpretation of the new provision, in the sense that the interpretation of the crime as persisting or not is relevant to the individuation of *tempus commissi delicti* and to calculating the period of limitation. If the crime is considered to be persisting, each case of environmental disaster—occurred after the crime commission or the significant omissions—could be punishable; if the crime is considered to be instantaneous with persisting effects, it should be considered time-barred (Milita 2015a).

[29]On the scope of the new rules and in particular on the crime of environmental disaster, see Chap. 4, Sect. I.7.

In fact, scholars (Masera 2015, p. 15 f.) have noted that with reference to the specific profile now under consideration, the new rules cause problems very similar to those which arose in the *Eternit* trial[30] as regards "unnominated disaster". All the cases considered here (on the one hand, disaster as provided for by Art. 434 of the penal code, on the other hand the new crimes of environmental pollution and environmental disaster) are in fact "result crimes" (*reati di evento*), where the typical event consists of an unclear phenomenon extended over time. Can the persisting nature of the typical event of the crime affect the determination of the moment of commission, even when the conduct which caused the event has stopped? The Italian Supreme Court, with respect to the offence of "unnominated disaster", has provided a negative response, stating that the phase of commission ends with the cessation of the conduct.

The Court stated that the moment of commission of the crime coincides with the event, and therefore with the occurrence of the disaster. The disaster is a destructive fact of extraordinary proportions which causes a danger to public safety, but the persistence of the danger and its materialization as a concrete damage are unrelated to it and can be considered further effects (and so any subsequent deaths or injuries, although related to the disaster, cannot be taken into account in order to identify the start point for the running of the period of limitation).

There is a structural similarity between the new environmental crimes and the crime on which the Supreme Court has ruled, in two respects: they are all "result crimes", and the typical events (that is the impairment or deterioration, in the case

[30]Cass. Pen., sec. I, 19 November 2014 (filed on 23 February 2015) No. 7941, *Schmidheiny*, rv. 262789, with the comment of Zirulia (2015). The *Eternit* trial concerns the occurrence of an environmental disaster caused by the dispersion of asbestos over time, which took place in Italy, involving an undefined number of people (plants' workers and population living close to the four sites of Cavagnolo, Casale Monferrato, Bagnoli, and Rubiera). The defendants (two corporate managers who led the company until 1986, when it went bankrupt), accused of having caused a threat to the safety and health of the population, were convicted both by the first instance Court and on appeal. The key legal point is the identification of the moment in which the crime of "unnominated disaster" can be considered as having been committed, and so the moment in which the period of limitation starts to run. In this context, different interpretations of the structure of the crime arose. The Prosecutors qualified the disaster as a *persisting offence* represented by the ongoing situation of pollution; the conducts of the defendants were considered still in progress too, in the form of failure to cooperate in sites remediation. The Court of first instance, with a different approach, stated that the crime was not time-barred because the offence continues as long as the *event* of pollution persists, despite the absence of active or omissive conducts of the defendants. The Court of Appeal, with another description, stated that the legal test for the disaster offence should include not only the environmental pollution, but also the *increase of asbestos-related diseases* within the population. On the basis of the incidence of such diseases in the areas surrounding the plants, the Court found that the disaster was far from being time-barred. These decisions were all overturned by the Supreme Court, which ruled that the crime had been time-barred since the beginning of the trial: the offence of disaster should be considered as committed when the defendants' *conduct posing a threat to public safety is interrupted* (in the *Eternit* case, the activity of the plants stopped in 1986 and so the Supreme Court referred to that moment to determine that the period of limitation had expired). For further details, see also Chap. 5, Sect. I.5.

of environmental pollution; the irreversible alteration or the alteration particularly difficult to eliminate, in the case of environmental disaster) cannot be considered as resulting from an instantaneous event, or be easily constrained within specified time limits. Therefore, in all likelihood, the judges who will rule on the newly introduced crimes, in order to determine the moment of their commission, will necessarily have to take into account the above-mentioned judgement of the Supreme Court in the *Eternit* trial. So, with particular regard to the hypothesis of environmental disaster, where the polluting conduct did stop *before* the entry into force of Law 68/2015 but its *effects* do last *afterwards*, the moment of commission of the crime should be identified in the moment in which the conduct ceased (to this end, neither the fading away of the disastrous effects nor the situation of non-removal of the effects are relevant) and the new law should not be deemed to apply, neither in relation to the individual, nor, *a fortiori*, in relation to the corporation (Ruga Riva 2016, p. 267).[31]

A different solution should be proposed in case of conducts of serious environmental contamination begun before the entry into force of the law, which, however, *do last afterwards*, thereby causing an exacerbation of their own effects. As was anticipated, regardless of whether or not the new hypothesis of environmental crime provided in Article 452-quater of the Italian penal code qualifies as a persisting crime, in light of the jurisprudence's orientation that emerged in relation to the crime under Article 434 of the Italian penal code, the relevant point in order to determine the moment of commission is that the persisting of the damage is accompanied by the *persisting of the conduct*. As a consequence, if all the necessary conditions are met, where the polluting conducts do last during the period of validity of the law introduced in 2015, it is believed that this new and stricter law can apply (Ruga Riva 2016, p. 267).

In this type of hypothesis, in which the moment of commission of the predicate crime takes place during the period of validity of the new law, the possibility that the corporation can also be held liable under Legislative Decree 231/2001 should also be considered. Bearing in mind the jurisprudential positions described above, it is, however, necessary to specify that, in compliance with the principles of legality and non-retroactivity, as far as the position of the corporation is concerned, only conducts carried out after the introduction of the new environmental crimes into the catalogue of predicate offences should be considered, in particular for the purposes of the determination of the profit liable to confiscation.

[31] A similar conclusion should be reached in case both the conduct that caused the disaster and its effects ceased *before* the entry into force of Law 68/2015: as far as the natural person is concerned, the new law will not apply, both by virtue of the safeguard clause at the beginning of Art. 452-quater of the Italian penal code ("except cases provided for in Article 434") and because the new law provides for stricter sanctions. Since in similar hypotheses the natural person could not be held responsible for the new crime of environmental disaster and this crime would, however, constitute a new incrimination for the corporation, not even liability under Legislative Decree 231/2001 should apply.

4 The Offences Provided for by Article 257 TUA (*Sites Remediation*)

The practical implementation of the rules on corporate liability provided for by Legislative Decree 231/2001 has revealed some concerns that could usefully be tackled in connection with the analysis of historical pollutions theme. The most interesting issues, as regards criminal offences punishable under the Testo Unico Ambiente and on the specific point regarding corporate liability, concern the practical implementation of Article 257 TUA.[32] This provision, entitled *Sites remediation* (*bonifica dei siti*) foresees in Paragraph 1 two specific crimes:

Failed remediation, in conformity with the project approved by the competent authority (in the context of specific administrative procedures as prescribed by Arts. 242 ff. TUA), in the event of soil, subsoil, surface, and groundwater pollution with concentrations exceeding the risk thresholds (CSR) (since a detrimental situation has been detected and it has to be removed: Art. 257, Para. 1, first part, TUA);

Failure to communicate that an event which has the potential to contaminate the environment occurred (where a danger is possible, and it is still not possible to know whether the remediation is compulsory: Art. 257, Para. 1, last part, TUA).

As regards corporate liability, the question firstly was raised whether this latter case (failed communication of the polluting event) should fall within predicate offences under Legislative Decree 231/2001. Article 25-undecies, Paragraph 2, Letter (c) of Legislative Decree 231/2001 provides for the application of pecuniary sanctions for crimes provided for in Article 257 TUA in relation to "the *infringement* of Paragraph 1" of that provision.

Thus, Article 25-undecies refers to "infringement", *in the singular form*, of Paragraph 1: this was considered by many as meaning that *only* the crime of *failed remediation* is among the predicate offences of corporate liability, and not the crime of failure to communicate the potentially polluting event (Ruga Riva 2012, p. 412).[33]

Instead, other scholars are inclined towards a more extensive interpretation, considering that corporate liability is to be referred *both* to failed remediation and to failed communication (Fimiani 2012a, b; Di Landro 2014, p. 9). In fact, Article 25-undecies of the Decree refers to *crimes* under Article 257 TUA, and Article 257 TUA prescribes autonomous specific crime cases (failed remediation and failed

---

[32]For the administrative provisions concerning remediation, see Chap. 7.

[33]In support of this thesis, Ruga Riva (2012) notes that punishment limits established by law are unique (fine up to two hundred and fifty quotas): should this provision also cover the offence of failure to communicate the potentially polluting event, it would end up unreasonably merging conducts with a different disvalue under the same sanction (as shown, indeed, by the differentiated penalties provided for individuals). On the whole, the interpretation most consistent with the principle of legality, in the case of possible divergent interpretations, should impose the strict reading which provides all possible guarantees: therefore, according to the author, the crime of failed communication is not among those included in the catalogue of predicate crimes.

communication) just in Paragraph 1. Most consider infringement under Paragraph 2 (pollution caused by dangerous substances) as a mere aggravating circumstance.[34]

So, where it refers to "crimes", Article 25-undecies of Legislative Decree 231/2001 should necessarily include both the cases laid down in Article 257 TUA in the list of predicate offences, thus both failed remediation and failed communication.

Those scholars who consider the crime of failed communication as a predicate offence of corporate liability, also specify that if the corporation "inherits" a *previous contamination* (or if the offence is committed by third parties unrelated to the corporation who do not fall within the categories of top managers nor employees mentioned under Arts. 6 and 7 of Legislative Decree 231/2001), it cannot be held liable for failed communication, as this was already stated in relation to the owner/natural person not being responsible for the potentially dangerous event. In fact, on this point the Corte di Cassazione excluded in general terms that the owner/non-polluter is obliged to notify the territorially competent offices about pollution on his property: only the person who *caused* the pollution is obliged to give notification of the polluting event.[35]

In relation to corporate liability, as far as crime of failed remediation is concerned, another question that arises is whether the implementation of the approved remediation project of the contaminated site, by the natural person liable for the offence, will benefit only the latter or also the corporation. In order to solve this problem, it seems appropriate to briefly examine the stages of the debate on the nature of the crime of failed remediation and on the nature of the remediation itself.

---

[34]See Cass. Pen., Sec. III, 13 April 2010 (filed on 9 June 2010) No. 22006, *Mazzocco*, published on www.lexambiente.it, and also most scholars: for example, according to Di Landro (2014, p. 26), this view is supported by the formalistic criteria of the *nomen iuris* and of the so-called topographic location of the provisions (i.e. both provisions are included in the same Article); it is also supported by the structural *principium individuationis* of circumstances, with reference to the provision of Para. 2, which is described through a referral to the offence described in Para. 1 (and not autonomously). If Paras. 1 and 2 of Art. 257 TUA were reconstructed not as autonomous offences, but, respectively, as a simple (Para. 1) and an aggravated (Para. 2) type of the same offence, it could be concluded that the remediation could extinguish this offence regardless of the nature (dangerous or not) of the pollutants. However, there is also a contrary opinion according to which the best solution is instead the one that configures the case of pollution by hazardous substances as an autonomous offence: see Cass. Pen., Sec. III, 14 March 2007 (filed on 9 July 2007) No. 26479, *Magni*, rv. 237134.

[35]Cass. Pen., Sec. III, 16 March 2011 (filed on 11 May 2011) No. 18503, *Comit* case, rv. 250143. For comments on this decision, see Ariolli and Cappai (2011, p. 4405 ff.) and Fasani (2011, p. 809 ff.). Paone (2012, p. 523) agrees with this judgment: the obligation to communicate on interested subjects who are not responsible lies in Art. 245 TUA and not in Art. 242 TUA. Therefore, if in Art. 257 TUA the legislator had intended to make a reference also to those who caused the pollution, he would have included those who have that feature among the active subjects of the offence; or he would have referred to Art. 245 TUA to detect the different punishable obligation to communicate. This judgement is analysed in Chap. 5, Sect. I.6.

There was already debate on this during the period in which Article 51-bis of Legislative Decree 22/1997 (*Decreto Ronchi*) was applied, this being the normative antecedent of the specific case that today is provided for in Article 257 TUA.[36] According to a first line of approach (Severino 2000, p. 417; Giampietro 1998, p. 69) the crime under Article 51-bis of Legislative Decree 22/1997 had to be considered as an "unrestricted result crime" (*reato di evento a forma libera*); the fact of having caused "the pollution or the real and actual danger of pollution" was to be considered as an integral part of the conduct and a relevant element of the typical fact. The implementation of remediation was only considered as an objective condition of non-punishability with a negative structure (*condizione obiettiva di non punibilità negativa*), or, according to others, as a supervening cause of non-punishability (*causa sopravvenuta di non punibilità*).

Other scholars (Anile 1999, p. 125) considered the specific case laid down in Article 51-bis of Decreto Ronchi as an "omissive crime", which lacks an event in a naturalistic sense, since it would have occurred because of the failure to comply with the duty of remediation (under Art. 17 of Legislative Decree 22/1997). The previous (potentially also non-culpable) pollution did not fall within the criminal conduct, since it was a premise of the criminal conduct.

According to the latter approach, Article 51-bis was considered applicable to those situations in which the pollution (or risk of pollution) had occurred before the entry into force of the provision itself (or after the entry into force, but in any case before the adoption of Ministerial Decree 471/1999, namely of the implementing legislation which provided the maximum acceptable contamination thresholds). On the contrary, the other approach which considers the crime a "result crime" did exclude its application to previous pollution cases, occurring before the entry into force of these norms, in order to avoid infringing the principle of non-retroactivity.

At first, the Corte di Cassazione[37] had agreed with those who considered the provision under Article 51-bis of Legislative Decree 22/1997 as an omissive crime, whose criminal disvalue is based on the danger alleged *ex lege*, due to the failure to remediate in line with deadlines provided for in Article 17 of the same Decree.

---

[36]As regards the relationship between Art. 51-bis of Legislative Decree 22/1997 and Art. 257 TUA, the main case-law orientation considers that Art. 257 (regarding remediation of sites) is less severe than the previous provision under Art. 51-bis of Decreto Ronchi, because the significant criminal offence area is reduced and the sanctions are mitigated. As a consequence, in the case of historical contamination, Art. 257 TUA can have a possible retroactive effect *in mitius* and can also apply to conducts realized before the law entered into force. With reference to the time of the commission of the offence and, consequently, to the period of limitation, the failed remediation ex Art. 257 TUA—in continuity with Art. 51-bis of Decreto Ronchi previously in force—is a hypothesis of a persisting offence (and not instantaneous with persisting effects), as the unlawful conduct takes place according to the procedure provided for by Art. 242 TUA; it continues as long as the omission of the offender lasts, and in any case as long as the asset is available for the offender.

[37]Cass. Pen., Sec. III, 28 April 2000 (filed on 7 June 2000) No. 1783, *Pizzuti*, rv. 216585. For comments, see Prati (2001) and also Brunelli (2002, p. 241).

Consequently, pollution (or danger of pollution) fell within the factual premises and not within the essential elements of the crime.

Thus, after the entry into force of the Testo Unico Ambiente, as regards Article 257 TUA the Supreme Court[38] rejected the previous interpretation by jurisprudence in relation to Article 51-bis of Decreto Ronchi, considering the specific case as an "unrestricted result crime", which is subject to the objective condition of non-punishability with a negative structure. The incriminating event is pollution caused by intentional or negligent conduct whose punishability is subject to failed remediation.

As mentioned above, the question was raised once again after the introduction of Article 25-undecies into Legislative Decree 231/2001, and thus of the crime of failed remediation in the list of predicate offences: it has been questioned whether the implementation of a remediation project is also useful to the corporation or not, in relation to *its* liability.

Thus, on the one hand some scholars (Ruga Riva 2012, p. 412 f.) consider remediation as a *supervening cause of non-punishability*, namely as an extinguishing condition in a broad sense that would exclusively benefit the natural person and not (also) the corporation (in compliance with the principle of autonomy of corporate's liability ex Art. 8 of Legislative Decree 231/2001), while on the other hand some scholars (Fimiani 2012b, p. 62) consider remediation as an intrinsic *objective condition of non-punishability*, which affects the level of damage to the legal asset protected. In that event, the corporation would not be liable in case of implementation of the restoration project of the contaminated site, as the predicate offence (under Art. 257 TUA) would not have occurred or at least the pollution would have not achieved further levels of damage.[39]

Given the many uncertainties in reconstructing the crime, it has to be noted that corporate liability is linked to an organizational fault or to a criminal enterprise policy, which do not cease thanks to the restoration. Legislative Decree 231/2001 seems not to allow reward considerations for any conduct that can lead to non-punishability, especially in the absence of compliance programmes. Such conducts have only mitigating effects, with a decrease of pecuniary sanctions (Art. 12, Para. 2, Lett. (a) of Legislative Decree 231/2001), and/or the exclusion of any

---

[38]This approach began with the following judgement: Cass. Pen., Sec. III, 29 November 2006 (filed on 8 March 2007) No. 9794, *Montigiani*, published on www.lexambiente.it; for comments, see Vergine (2009). This approach was later confirmed: see Cass. Pen., Sec. III, 14 March 2007 (filed on 9 July 2007) No. 26479, *Magni*, rv. 237134 and Cass. Pen., Sec. III, 13 April 2010 (filed on 9 June 2010) No. 22006, *Mazzocco*, published on www.lexambiente.it.

[39]A third interpretation is offered by Di Landro (2014, p. 25), who deems that the exemption of the remediation *may apply* to entities, enhancing Art. 257, Para. 4 TUA: according to this provision, the compliance with the remediation project is a condition for exemption from punishment for environmental crimes provided by *other laws* for the same event. Para. 2, Lett. (c) of Art. 25-undecies of Legislative Decree 231/2001 (which is surely a law that is *different* from the TUA) punishes the same event provided for by Art. 257, Paras. 1 and 2 TUA, extending to a different subject (the corporation) the liability for some environmental crimes that do already exist (Art. 257 TUA).

interdictory sanctions (Art. 17), requiring for such purposes not only the *restitutio in integrum* but also the implementation of compliance programmes and the achieved profit being made available for its confiscation (Di Landro 2014, p. 28).

The legal doctrine noted that, in relation to certain crimes, it might prove appropriate (through explicit reward rules) to provide impunity for corporations whose conduct, even if belated, is in opposition to the crime which has already occurred: this is especially true in the present case, since remediation costs (often serious) are mainly borne by the corporation and not by the natural person (who, under these conditions, benefits from impunity under Art. 257 TUA) and since restoration is the main aim of the criminal law (Ruga Riva 2012, p. 413).

5 Sanctions Applicable to the Corporation: Confiscation and the Notion of "Profit"

Efforts have been made to explain how, in relation to the research topic and in the context of Legislative Decree 231/2001, the heart of the matter lies in the possibility of identifying corporate liability in cases of polluting activities that have begun in the past and are still ongoing, thus causing potentially harmful effects. In view of an affirmative answer on a theoretical level—to the extent clarified thus far—it is consequently necessary to address the question of the practical applicability of precautionary measures and sanctions provided for by Legislative Decree 231/2001 in such cases. The recent position taken by the Corte di Cassazione in relation to the *Ilva* case in Taranto is particularly interesting.[40]

As has already been pointed out,[41] the Court underlined, as regards corporate liability, the need to respect both the principle of legality and the principle of non-retroactivity (referred to in Art. 2 of Legislative Decree 231/2001) which "cover" the precepts and the relative sanctions.

This has significant implications: in the specific case, the Court analysed the implementation under Articles 53 and 19 of Legislative Decree 231/2001 of preventive seizure intended for the purpose of confiscation for equivalent of goods for an amount equal to EUR 8.1 million, considered equivalent to the profit resulting from a number of offences, as well as the notion of profit liable to confiscation.

On a preliminary basis, it can be noted that, as far as the conceptual elaboration of the notion of profit liable to confiscation under Legislative Decree 231/2001 is concerned, by highlighting its structure the jurisprudence took into account both the sanctioning nature of confiscation provided for in Article 19 of the Decree and the fact that the profit is linked to an entrepreneurial economic activity, that is lawful (Silvestri 2014, p. 9 ff.).

---

[40]Reference is made to the above-mentioned judgement of the Corte di Cassazione, No. 3635 of 2014. For a comment about confiscation in this proceeding, see Maiello (2014).

[41]See Sect. 2.

As regards the rules laid down in Legislative Decree 231/2001, it has been stated that the profit liable to confiscation is the so-called net profit, less the "lawful expenses" falling under the corporate budget: such a concept would be in conformity with the re-balancing ratio on which the confiscation in question is based in the Decree—the only concept capable of respecting both the principle of "proportionality" and the principle of "legal certainty".

In the *Ilva* case, according to the prosecutors, the legal and responsible representatives of the corporation had committed environmental crimes in the interest and for the benefit of the corporation, through the failed execution of activities necessary to implement environmental security and protection measures, thereby allowing the corporation to yield a profit consisting of sums of money equal to savings on costs that were necessary for the adaptation of the steel plants.

The Corte di Cassazione claimed that, since under Article 19, Paragraph 2 of Legislative Decree 231/2001 confiscation for equivalent is intended for properties *not* associated with the crime, it could potentially also concern intangible economic benefits, which may include benefits resulting from cost-cutting measures or non-payments; that is, benefits resulting from conducts which do not improve the corporation's financial situation, but do assure its non-decrease, may be included.

However, in the *Ilva* case, the Court referred to judgement no. 26654/2008 of *Sezioni Unite*, which assessed that the notion of *cost-cutting* does presuppose the collection of an income, not reduced by the costs which should have been paid; namely, it presupposes a *positive* economic out-turn, resulting from the alleged criminal conduct (in the present case, a fraud).

The Court therefore assumed that the notion of profit liable to confiscation cannot be understood in absolute terms, as profit which does not include goods materially owned by the person liable for the offence (non-increase of debts), but that it is to be considered as a collected income, not reduced by the costs that should have been paid (non-decrease in assets).

The profit has to be materially yielded, and it has to be higher than the profit that could have been yielded without incurring due expenses, as it is not possible to draw several times from the same sum of money, considering it both in positive terms (as patrimonial growth) and in negative terms (as cost-cutting), thus unfairly doubling the burden imposed on the corporation. In short, only the *positive difference* between the yielded patrimonial growth and the one that should have been yielded without the undue cost-cutting is liable to expropriation.

As far as the identification of the profit liable to confiscation is concerned, in the *Ilva* case the Supreme Court established that the core issue should be *the moment in which the crimes occurred* and not the moment in which the corporation yielded the profit.

Thus, in relation to the issue in question, the decree which provided for preventive seizure aimed at the confiscation for equivalent should have *not* considered—for the purposes of the exact individuation of the profit liable to confiscation—environmental crimes that occurred *before* the entry into force of Legislative Decree 121/2011. This is in recognition of the fact that the confiscation for equivalent has a sanctioning nature, since in this regard only conducts present in the catalogue of predicate offences are relevant.

The notion of profit is also discussed in another judgement concerning corporate liability under Legislative Decree 231/2001, namely the order of 26 September 2013 of the Court of Chieti made upon a request by a corporation for a review of a seizure aimed at confiscation provided for by a GIP (preliminary investigations magistrate) of Lanciano (with the decree of 12 August 2013 adopted under Arts. 19 and 53 of Legislative Decree 231/2001), in a proceeding for unauthorized landfill in which the corporation had been involved under Article 25-undecies of Legislative Decree 231/2001.[42]

The seizure concerned the cost of disposal of waste unlawfully deposited into land by the accused person under Article 256, Paragraph 1, Letter (b) and Paragraph 2 TUA, in relation to Article 192 TUA. According to the Court, the corporation did not incur this expense, and this represents the profit of the illicit conduct in question.

Therefore, in its decision the Court pointed out that, as regards the alleged period of limitation for these offences and the non-retroactivity of Article 25-undecies (the contested conducts were considered as having occurred before the entry into force of Legislative Decree 231/2001), the uncontrolled disposal of waste has a persisting nature and is to be considered as an activity of waste management, which lasts until the disposal or recovery, thus as an existing conduct, that can be contested against the corporation.

6 The Peculiar Statute of Limitation

Article 22 of Legislative Decree 231/2001 provides for a peculiar statute of limitation, inspired by civil law, for the administrative sanctions under that Decree. Paragraph 1 establishes a period of limitation of five years from the date of crime commission. Paragraph 2 provides for cases interrupting the period of limitation, namely requests for the application of precautionary measures and the charging of administrative infringement. By reason of the interruption, a new period of limitation will be provided.

As far as this second hypothesis is concerned, if the interruption has taken place through the charging of administrative infringement depending on a crime, Paragraph 4 provides that the period of limitation does not start until the moment when the sentence defining the trial becomes final. After the prosecutor has correctly notified the charging of administrative infringement, the period of limitation

[42]Trib. Chieti, ord. 26 September 2013, case No. 2022/2012.

will remain interrupted until the judgement defining the trial occurs, with the possibility for the corporation to be involved in the trial for an indefinite period.[43]

This provision should be coordinated with Article 60 of Legislative Decree 231/2001, which states that charging under Article 59 is not possible if the offence on which the administrative infringement of the corporation depends has been extinguished, as it is time-barred. As regards the most serious crimes, the minimum time required under Article 157 of the Italian penal code is six years, so a crime cannot be time-barred before the end of the period of limitation of the corporation's administrative infringement. As regards misdemeanours—the majority of hypotheses provided for in Article 25-undecies of Legislative Decree 231/2001 are misdemeanours—the minimum time required for the period of limitation is four years. This means that in these hypotheses, regardless of the provision of Article 22 of the Decree, the charging shall take place within that period, which is shorter.

The time of the commission of the offence is also essential for Legislative Decree 231/2001, and here the penal code's provisions must be considered—namely, provisions under Article 158 of the Italian penal code.[44] As regards the rules concerning the continuing offences (*reato continuato*), these are provided for in Article 21 of Legislative Decree 231/2001 (plurality of infringements—*pluralità di illeciti*): it is considered that different periods of limitation shall be applied, which begin to run from the time of the commission of each single offence (Lottini 2002, p. 178 f.; Cerqua 2005, p. 242; De Vero 2008, p. 255).

Under Article 6, Paragraph 1 of the newly introduced reform n. 68/2015 on environmental crimes, regulations on the period of limitation of new offences have been tightened, and time-barring limits are doubled in comparison with those provided for in Article 157, Paragraph 6 of the Italian penal code. The issue of *tempus commissi delicti*, and thus also the identification of the moment when the crimes occurred, are also crucial for the recently introduced crimes of environmental disaster and environmental pollution (and now included in the catalogue of the predicate offences under Legislative Decree 231/2001).

In fact, in these types of crimes the typical event could occur at a time after the latest discharge of substances or physical alteration or tampering with the pre-existing environment.[45]

---

[43]On this topic, see Lottini (2002, p. 176 ff.), Rossi (2005, p. 573 ff.), De Vero (2008, p. 253 ff.), Di Giovine (2010, p. 146 ff.), Furlotti (2010, p. 290 ff.), Piergallini (2010, p. 248 ff.), Scaroina (2013, p. 2120 ff.) and Severino (2016, p. 80 ff.).

[44]On the provisions of the Italian penal code concerning the statute of limitation, see Chap. 4.

[45]Molino (2015, p. 34) notes that the investigation and the prosecution of the most serious environmental offences currently have a very long period of limitation (forty years in the case of intentional environmental disaster; up to fifty years in the presence of interrupting acts) and that this is in contrast with the very short period of limitation provided for misdemeanours in environmental matters.

Scholars consider it inappropriate to double the periods of limitation. It is in fact recognized that the reasons behind this choice can now be considered outdated in the light of the high minimum/maximum range of sanctions envisaged for the new crimes, as well as for their nature as "result crimes" (*reati di evento*).

Although the provision is intended to solve problems concerning the short period of limitation for the crime of "unnominated disaster" and the identification of the moment when the period of limitation for the same crime begins to run, the legislators did not consider that the new rules had already removed such problems. In fact, they provide sanctions with high maximum limits for environmental crimes (to which long periods of limitation are linked, consistent with the ordinary rules of the penal code) and the moment when the period of limitation begins to run can be now identified with certainty as the moment in which the harmful event occurs, for both environmental pollution and environmental disaster (Siracusa 2015, p. 220).

The new provisions reduce the risk of conduct occurring after their entry into force being time-barred.[46] The same cannot be said for conduct resulting from activities that occurred before May 2015, whose fate will depend on future interpretations of the identification of the moment when the period of limitation begins to run (Milita 2015b). As said before, this development, with a high probability, and in light of the proximity between old and new provisions, will be affected by the recent positions taken by the Corte di Cassazione on the issue, in relation to the crime of "unnominated disaster" (see Sect. 3).

References

Amarelli, G. (2016). I nuovi reati ambientali e la responsabilità degli enti collettivi: Una grande aspettativa parzialmente delusa. *Cassazione Penale, 1,* 405–424.

Anile, F. (1999). Bonifica dei siti contaminati: Obblighi di ripristino e tutela penale. *Ambiente & Sviluppo, 2,* 121–127.

Arena, M., & Cassano, G. (2007). *La responsabilità da reato degli enti collettivi*. Milano: Giuffrè.

Ariolli, G., & Cappai, M. (2011). Il proprietario o il gestore dell'area non risponde del reato di omessa segnalazione di un evento inquinante cagionato da terzi. *Cassazione Penale, 12,* 4405–4414.

Bassi, A., & Epidendio, T. E. (2006). *Enti e responsabilità da reato*. Milano: Giuffrè.

Bernasconi, C. (2012a). Sub art. 137. In L. Costato & F. Pellizzer (Eds.), *Commentario breve al Codice dell'ambiente* (*D. legisl. 3 aprile 2006, n. 152*) (2nd ed., pp. 450–643). Padova: Cedam.

Bernasconi, C. (2012b). Sub art. 256. In L. Costato & F. Pellizzer (Eds.), *Commentario breve al Codice dell'ambiente* (*D. legisl. 3 aprile 2006, n. 152*) (2nd ed., pp. 908–944). Padova: Cedam.

Bernasconi, C. (2012c). Sub art. 260. In L. Costato & F. Pellizzer (Eds.), *Commentario breve al Codice dell'ambiente* (*D. legisl. 3 aprile 2006, n. 152*) (2nd ed., pp. 972–981). Padova: Cedam.

[46]As noted by Masera (2015, p. 16) with reference to the problems of "denial of justice" in evidence in the *Eternit* case, the legislator, with this event in mind, doubled the ordinary statute of limitation for the new crimes; new cases of environmental disaster shall expire in thirty years, and if such provision had been applicable in the *Eternit* case, the period of limitation would not have been expired even with reference to the moment of the cessation of activities.

Bricchetti, R., & Pistorelli, L. (2011). Danni all'ambiente e inquinamento marino: L'Italia si uniforma agli standard europei. Imprese responsabili: Scattano sanzioni pecuniarie. *Guida al Diritto, 38,* 48–51.

Brunelli, D. (2002). Profili penali nella bonifica dei siti inquinati. In P. M. Vipiana (Ed.), *La bonifica dei siti inquinati: Aspetti problematici* (pp. 221–260). Padova: Cedam.

Brunelli, D. (2008a). Art. 2. In A. Presutti, A. Bernasconi, & C. Fiorio, *La responsabilità degli enti. Commento articolo per articolo al D. legisl. 8 giugno 2001, n. 231* (pp. 85–87). Padova: Cedam.

Brunelli, D. (2008b). Art. 3. In A. Presutti, A. Bernasconi, & C. Fiorio (Eds.), *La responsabilità degli enti. Commento articolo per articolo al D. legisl. 8 giugno 2001, n. 231* (pp. 87–91). Padova: Cedam.

Cadoppi, A., Garuti, G., & Veneziani, P. (Eds.). (2010). *Enti e responsabilità da reato.* Torino: Utet.

Carmona, A. (2002). *Premesse a un corso di diritto penale dell'economia.* Padova: Cedam.

Cerqua, L. (2005). Il trattamento sanzionatorio. In C. Monesi (Ed.), *I modelli organizzativi ex d. lgs. 231/2001. Etica d'impresa e responsabilità degli enti* (pp. 209–250). Milano: Giuffrè.

Costalunghi, D. (2007). Voce Acque. 1. D. legisl. 3 aprile 2006, n. 152 (sub art. 137). In F. Palazzo & C. E. Paliero (Eds.), *Commentario breve alle leggi penali complementari* (2nd ed., pp. 75–107). Padova: Cedam.

De Maglie, C. (2002). *L'etica e il mercato. La responsabilità penale delle società.* Milano: Giuffrè.

De Santis, G. (2012). La tutela penale dell'ambiente dopo il d.lgs. n. 121/2011 di attuazione della direttiva 2008/99/CE. *Responsabilità Civile e Previdenza, 2,* 668–696.

De Simone, G. (2002). I profili sostanziali della responsabilità c.d. amministrativa degli enti: La "parte generale" e la "parte speciale" del d.lgs. 8 giugno 2001, n. 231. In G. Garuti (Ed.), *Responsabilità degli enti per illeciti amministrativi dipendenti da reato* (pp. 56–126). Padova: Cedam.

De Simone, G. (2012). La responsabilità da reato degli enti: Natura giuridica e criteri (oggettivi) d'imputazione. *Diritto Penale Contemporaneo* (pp. 1–48). October 28, 2012. http://www.penalecontemporaneo.it/upload/1351253564De%20Simone%20definitivo.pdf. Accessed April 23, 2017.

De Vero, G. (2008). La responsabilità penale delle persone giuridiche. In C. F. Grosso, T. Padovani, & A. Pagliaro (Eds.), *Trattato di diritto penale. Parte generale* (Vol. IV). Milano: Giuffrè.

Di Giovine, O. (2010). Lineamenti sostanziali del nuovo illecito punitivo. In G. Lattanzi (Ed.), *Reati e responsabilità degli enti: Guida al d.lgs. 8 giugno 2001, n. 231* (2nd ed., pp. 3–152). Milano: Giuffrè.

Di Landro, A. (2014). Bonifiche: Il labirinto della legislazione ambientale dove le responsabilità penali "si perdono". *Diritto Penale Contemporaneo* (pp. 1–48). February 28, 2014. http://www.penalecontemporaneo.it/upload/1393494624DI%20LANDRO_2014.pdf. Accessed April 23, 2017.

Fasani, F. (2011). Sui soggetti attivi del reato di omessa comunicazione delle contaminazioni ambientali. *Rivista Giuridica dell'Ambiente, 6,* 809–810.

Fiandaca, G., & Musco, E. (2014). *Diritto penale. Parte generale* (7th ed.). Bologna: Zanichelli.

Fimiani, P. (2012a). Bonifica e sistema 231: I rapporti nel caso di omessa comunicazione dell'evento di contaminazione. Parte I. *Rifiuti—Bollettino di Informazione Normativa, 198*(8–9), 44–46. Milano: Edizioni Ambiente.

Fimiani, P. (2012b). Bonifica e sistema 231: I rapporti nel caso di omessa comunicazione dell'evento di contaminazione. Parte II. *Rifiuti—Bollettino di Informazione Normativa, 199*(10), 60–62. Milano: Edizioni Ambiente.

Fiorella, A. (2003). Principi generali e criteri di imputazione all'ente della responsabilità amministrativa. In G. Lancellotti (Ed.), *La responsabilità della società per il reato dell'amministratore* (pp. 85–102). Torino: Giappichelli.

Fornasari, G. (1991). Voce Reato abituale. In *Enciclopedia giuridica* (Vol. XXVI, pp. 1–7). Roma: Treccani.

Forzati, F. (2015). Irrilevanza penale del disastro ambientale, regime derogatorio dei diritti e legislazione emergenziale: I casi Eternit, Ilva ed emergenza rifiuti in Campania. Lo stato d'eccezione oltre lo stato di diritto. *Diritto Penale Contemporaneo* (pp. 1–49). March 11, 2015. http://www.penalecontemporaneo.it/upload/1426012622FORZATI_2015a.pdf. Accessed April 23, 2017.

Furlotti, P. (2010). Art. 22—Prescrizione. In A. Cadoppi, G. Garuti, & P. Veneziani (Eds.), *Enti e responsabilità da reato* (pp. 290–292). Torino: Utet.

Giampietro, F. (1998). Bonifica dei siti inquinati: Dal D. Lgs. "Ronchi" al "Ronchi bis". *Ambiente & Sviluppo, 1,* 67–74.

Lattanzi, G. (Ed.). (2010). *Reati e responsabilità degli enti: Guida al d.lgs. 8 giugno 2001, n. 231* (2nd ed.). Milano: Giuffrè.

Lottini, R. (2002). Il sistema sanzionatorio. In G. Garuti (Ed.), *Responsabilità degli enti per illeciti amministrativi dipendenti da reato* (pp. 127–182). Padova: Cedam.

Lottini, R. (2007a). Voce Responsabilità delle persone giuridiche. D. legisl. 8 giugno 2001, n. 231 (sub art. 2). In F. Palazzo & C. E. Paliero (Eds.), *Commentario breve alle leggi penali complementari* (2nd ed., pp. 2294–2295). Padova: Cedam.

Lottini, R. (2007b). Voce Responsabilità delle persone giuridiche. D. legisl. 8 giugno 2001, n. 231 (sub art. 3). In F. Palazzo & C. E. Paliero (Eds.), *Commentario breve alle leggi penali complementari* (2nd ed., pp. 2295–2299). Padova: Cedam.

Madeo, A. (2011). Un recepimento solo parziale della Direttiva 2008/99/CE sulla tutela penale dell'ambiente. *Diritto Penale e Processo, 9,* 1055–1065.

Maiello, V. (2014). La confisca ex D.Lgs. n. 231/2001 nella sentenza Ilva. *Giurisprudenza Italiana, 4,* 969–978.

Manna, A. (2003). La c.d. responsabilità amministrativa delle persone giuridiche: Il punto di vista del penalista. *Cassazione Penale, 3,* 1101–1121.

Mantovani, F. (2013). *Diritto penale. Parte generale* (8th ed.). Padova: Cedam.

Marinucci, G. (2002). «Societas puniri potest»: Uno sguardo sui fenomeni e sulle discipline contemporanee. *Rivista Italiana di Diritto e Procedura Penale, 4,* 1193–1214.

Marinucci, G., & Dolcini, E. (2015). *Manuale di diritto penale. Parte generale* (5th ed.). Milano: Giuffrè.

Masera, L. (2015). I nuovi delitti contro l'ambiente. *Diritto Penale Contemporaneo* (pp. 1–16). December 17, 2015. http://www.penalecontemporaneo.it/upload/1450028287MASERA_2015a.pdf. Accessed April 23, 2017.

Milita, A. (2015a). Il disastro ambientale, da delitto innominato a tipico: La qualificazione del delitto come eventualmente permanente e le responsabilità patrimoniali. *Lexambiente.it*. June 10, 2015. http://www.lexambiente.com/materie/ambiente-in-genere/188-dottrina188/11556-ambiente-in-genere-il-disastro-ambientale,-da-delitto-innominato-a-tipico-la-qualificazione-del-delitto-come-eventualmente-permanente-e-le-%E2%80%9Cresponsabilit%C3%A0-patrimoniali%E2%80%9D.html. Accessed April 23, 2017.

Milita, A. (2015b). Il disastro ambientale nominato. *Lexambiente.it*. September 11, 2015. http://www.lexambiente.com/materie/ambiente-in-genere/188-dottrina188/11715-ambiente-in-genere-il-disastro-ambientale-nominato%E2%80%9D.html. Accessed April 23, 2017.

Molino, P. (2015). Relazione n. III/4/2015. Novità legislative: Legge n. 68 del 22 maggio 2015, recante "Disposizioni in materia di delitti contro l'ambiente". Website of Corte di Cassazione (pp. 1–36). May 29, 2015. http://www.cortedicassazione.it/cassazione-resources/resources/cms/documents/RelIII_4_15.pdf. Accessed April 23, 2017.

Monaldi, F. (2012). Sub art. 279. In L. Costato & F. Pellizzer (Eds.), *Commentario breve al Codice dell'ambiente* (*D. legisl. 3 aprile 2006, n. 152*) (2nd ed., pp. 1034–1036). Padova: Cedam.

Palazzo, F. (2016). *Corso di diritto penale* (6th ed.). Torino: Giappichelli.

Paliero, C. E. (2001). Il d. lg. 8 giugno 2001: Da ora in poi societas delinquere (et puniri) potest. *Corriere Giuridico, 7,* 845–848.

Paone, V. (1995). Il reato di discarica abusiva: Un importante punto fermo della Corte di Cassazione. *Foro Italiano, 6,* 345–352.

Paone, V. (2012). Responsabilità penali e contaminazione dei siti. *Ambiente & Sviluppo, 6,* 520–525.

Pelissero, M. (2007). Responsabilità degli enti. In F. Antolisei (Ed.), *Manuale di diritto penale. Leggi complementari* (13th ed., Vol. 1, pp. 845–944). Milano: Giuffrè.

Pelissero, M. (2013). Reati contro l'ambiente e il territorio. In F. Palazzo & C. E. Paliero (Eds.), *Trattato teorico pratico di diritto penale* (Vol. XI). Torino: Giappichelli.

Perini, C. (2015). Responsabilità dell'ente per reati ambientali e principio di legalità. *Diritto Penale Contemporaneo—Rivista Trimestrale, 2,* 190–196. http://www.penalecontemporaneo.it/foto/3951DPC_Riv_Trim_2_2015.pdf#page=196&view=Fit. Accessed April 23, 2017.

Petrone, M. (1996). Voce Reato abituale. *Digesto delle discipline penalistiche* (Vol. XI, pp. 188–212). Torino: Utet.

Piergallini, C. (2010). I reati presupposto della responsabilità dell'ente e l'apparato sanzionatorio. In G. Lattanzi (Ed.), *Reati e responsabilità degli enti: Guida al d.lgs. 8 giugno 2001, n. 231* (2nd ed., pp. 211–256). Milano: Giuffrè.

Piergallini, C. (2014). Responsabilità dell'ente e pena patrimoniale: La Cassazione fa opera nomofilattica. *Rivista italiana di Diritto e Procedura Penale, 2,* 998–1011.

Pistorelli, L. (2006). Natura della responsabilità degli enti e criteri di imputazione oggettiva al vaglio dei giudici di legittimità. *La Responsabilità Amministrativa delle Società e degli Enti, 4,* 121–138.

Plantamura, V. (2011). Responsabilità individuale e degli enti nel D. lg. 7 luglio 2011, n. 121 di attuazione delle direttive europee sulla tutela penale dell'ambiente. *Rivista Trimestrale di Diritto Penale dell'Economia, 3,* 477–504.

Prati, L. (2001). Bonifica dei siti inquinati in base al Decreto Ronchi: L'art. 51-bis del d.lgs. 22/1997 al vaglio della Cassazione. *Rivista Giuridica dell'Ambiente, 1,* 65–82.

Presutti, A., Bernasconi, A., & Fiorio, C. (2008). *La responsabilità degli enti. Commento articolo per articolo al D. legisl. 8 giugno 2001, n. 231.* Padova: Cedam.

Ramacci, L. (2012a). Il "disastro ambientale" nella giurisprudenza di legittimità. *Ambiente & Sviluppo, 8–9,* 722–725.

Ramacci, L. (2012b). Responsabilità amministrativa degli enti collettivi e reati ambientali. *Ambiente & Sviluppo, 7,* 639–648.

Ramacci, L. (2015). Prime osservazioni sull'introduzione dei delitti contro l'ambiente nel codice penale e le altre disposizioni della legge 22 maggio 2015 n. 68. *Lexambiente.it.* June 8, 2015. http://www.lexambiente.com/materie/ambiente-in-genere/188-dottrina188/11562-ambiente-in-genere-prime-osservazioni-sull-introduzione-dei-delitti-contro-l-ambiente-nel-codice-penale-e-le-altre-disposizioni-della-legge-22-maggio-2015-n-68.html. Accessed April 23, 2017.

Romano, M. (2002). La responsabilità amministrativa degli enti, società o associazioni: Profili generali. *Rivista delle Società, 2–3,* 393–411.

Romano, M. (2004). Pre Art. 39 c.p. In M. Romano, *Commentario sistematico del codice penale* (3rd ed., Vol. I, pp. 299–353). Milano: Giuffrè.

Romolotti, T. E. (2007). Il nuovo testo unico dell'Ambiente e il d. lgs. 231/2001: Un'occasione perduta? *La Responsabilità Amministrativa delle Società e degli Enti, 1,* 37–42.

Rossi, A. (2005). Responsabilità "penale-amministrativa" delle persone giuridiche (profili sostanziali). In A. Rossi (Ed.), *Reati societari* (pp. 511–582). Torino: Utet.

Ruga Riva, C. (2011). Il decreto legislativo di recepimento delle direttive comunitarie sulla tutela penale dell'ambiente: Nuovi reati, nuova responsabilità degli enti da reato ambientale. *Diritto Penale Contemporaneo* (pp. 1–18). August 8, 2011. http://www.penalecontemporaneo.it/upload/RUGA%20RIVA%202011%20AMBIENTE.pdf. Accessed April 23, 2017.

Ruga Riva, C. (2012). Reato di omessa bonifica e D. Lgs. n. 231/2001: La bonifica giova (anche) all'ente? *Ambiente & Sviluppo, 5,* 412–414.

Ruga Riva, C. (2015). *I nuovi ecoreati. Commento alla legge n. 22 maggio 2015, n. 68.* Torino: Giappichelli.

Ruga Riva, C. (2016). *Diritto penale dell'ambiente* (3rd ed.). Torino: Giappichelli.

Scarcella, A. (2011). Nuovi ecoreati ed estensione ai reati ambientali del D. Lv. n. 231/2001 sulla responsabilità degli enti. *Ambiente&Sviluppo, 10,* 854–859.

Scaroina, E. (2013). La societas al cospetto del tempo: Il regime della prescrizione dell'illecito amministrativo dipendente da reato nel d. lgs. n. 231 del 2001. *Cassazione Penale, 5,* 2108–2133.

Scaroina, E. (2015). Ancora sul caso Eternit: La giustizia e il sacrificio dei diritti. *Archivio Penale Online, 3,* 1–27.

Scoletta, M. (2009). Nuove ipotesi di responsabilità amministrativa degli enti. In S. Corbetta, A. Della Bella, & G. Gatta (Eds.), *Sistema penale e "sicurezza pubblica": Le riforme del 2009* (pp. 373–380). Milano: IPSOA.

Scoletta, M. (2012). Obblighi europei di criminalizzazione e responsabilità degli Enti per reati ambientali (note a margine del D.Lgs. 121/2011 attuativo delle Direttive comunitarie sulla tutela dell'ambiente). *Rivista Giuridica dell'Ambiente, 1,* 17–48.

Scoletta, M. (2014). La responsabilità da reato delle società: Principi generali e criteri imputativi nel d. lgs. n. 231/2001. In G. Canzio, L. Cerqua, & L. Luparia (Eds.), *Diritto penale delle società* (pp. 861–928). Padova: Cedam.

Severino, P. (2000). I profili penali connessi alla bonifica dei siti inquinati. *Ambiente & Sviluppo, 5,* 417–424.

Severino, P. (2016). Il sistema di responsabilità degli enti ex D. Lgs. 231/2001: Alcuni problemi aperti. In F. Centonze & M. Mantovani (Eds.), *La responsabilità «penale» degli enti: Dieci proposte di riforma* (pp. 73–86). Bologna: Il Mulino.

Silvestri, P. (2014). Relazione n. 41/2014. La nozione di profitto confiscabile nella giurisprudenza della Sezioni Unite della Corte di Cassazione. Website of Corte di Cassazione (pp. 1–27). June 17, 2014. http://www.cortedicassazione.it/cassazione-resources/resources/cms/documents/Relazione_pen_41_14.pdf. Accessed April 23, 2017.

Siracusa, L. (2015). La legge 22 maggio 2015, n. 68 sugli "eco delitti": Una svolta "quasi" epocale per il diritto penale dell'ambiente. *Diritto Penale Contemporaneo—Rivista Trimestrale, 2,* 197–224. http://www.penalecontemporaneo.it/foto/3951DPC_Riv_Trim_2_2015.pdf#page=203&view=Fit. Accessed April 23, 2017.

Trinchera, T. (2014). Caso ILVA: La Cassazione esclude la confisca per equivalente del profitto dei reati ambientali. *Diritto Penale Contemporaneo.* February 11, 2014. http://www.penalecontemporaneo.it/materia/-/-/-/2800-caso_ilva__la_cassazione_esclude_la_confisca_per_equivalente_del_profitto_dei_reati_ambientali/. Accessed April 23, 2017.

Valsecchi, A. (2009). Il delitto di atti persecutori (il cd. stalking). *Rivista Italiana di Diritto e Procedura Penale, 3,* 1377–1414.

Vergine, A. L. (2009). Il reato di omessa bonifica: Due decisioni interrompono un prolungato silenzio. *Ambiente & Sviluppo, 11,* 981–990.

Wise, E. M. (trans.) (1978). *Italian penal code*. Littleton, Colorado: F.B. Rothman.

Zirulia, S. (2015). Eternit, il disastro è prescritto. Le motivazioni della Cassazione. *Diritto Penale Contemporaneo.* February 24, 2015. http://www.penalecontemporaneo.it/d/3709-eternit-il-disastro-e-prescritto-le-motivazioni-della-cassazione. Accessed April 23, 2017.

Chapter 7
Italian Administrative Regulation and Responsibility for Historical Pollution

Calogero Miccichè

Contents

1 Environmental Regulation and the Complexity of the Administrative Functions Involved in the Governance of Historical Pollution

In recent years, the Italian legal system has been supplemented by many laws whose aim, direct or indirect, is to protect the environment and to (re)establish a balance between, on the one hand, the protection of ecosystems and human health, and on the other, the demands of work and production, all with a view to greater sustainable development (Renna 2012, p. 73).

The adoption of these laws, unfortunately, has seldom anticipated or guided the development of the many productive sectors and otherwise polluting activities undertaken in the country, and rather has represented a reaction to the negative externalities which those activities produced at the expense of the environment.

C. Miccichè (✉)
Università Cattolica del Sacro Cuore, Milan, Italy
e-mail: calogero.micciche@unicatt.it

F. Centonze and S. Manacorda (eds.), *Historical Pollution*,
DOI 10.1007/978-3-319-56937-6_7

Indeed, as was already evident by the middle of the twentieth century, the introduction of new environmental legislation has often been opposed by economically interested parties, and such resistance has only been overcome once public opinion about the unacceptability of the situation had matured (Giannini 1973, p. 38).

This has had serious consequences, since it has favored pollution and the degradation of land and of other environmental matrices, handing on to current and future generations the costs of managing historical pollution and of remedying the damage to environmental systems.

Responding to the problems posed by historical pollution, and more generally by environmental pollution itself, the law has assigned new ecological functions to the public administrative authorities, and these functions have come to form a complex and pervasive system of environmental governance.

In their land planning activities[1] (known in Italian as *piani regolatori, piani per il governo del territorio* or *piani urbanistici*), municipalities must now not only organize their territory and the activities undertaken therein, but contribute to defining the minimum levels of environmental quality that soil must meet, and consequently handling their environmental obligations (for more detail, see Sect. 2).

Moreover, through municipal planning, public authorities may encourage private investment in the cleaning up of contaminated or brownfield sites, if necessary through negotiating planning solutions with the operators, in the knowledge that without the intervention of private capital it can be impossible to meet the costs of restoring the most polluted areas (see most recently the Tribunale Amministrativo Regionale [TAR], Campania, Napoli, Sec. I, 22 March 2016, no. 1471, regarding environmental remediation and brownfield management in the former industrial area of Bagnoli).

Another important tool for environmental governance is represented by mandatory preventive assessments of the environmental impact of major interventions, of which the most important are the environmental impact assessment (EIA) and the strategic environmental assessment (SEA). Such assessments are useful instruments for managing historical pollution, even though they have a mainly preventive value, being based on the assumption that "the best environmental policy consists in preventing the creation of pollution or nuisances at source, rather than subsequently trying to counteract their effects" (Directive 85/337/CEE). EIAs and SEAs, indeed, are often issued conditional on the prior execution of environmental actions to remediate degraded areas which may be different from the specific object of intervention and where such degradation may derive from historical contamination.

[1]Numerous planning instruments have immediate environmental effects: for instance, regional landscape plans, spatial plans of provincial coordination, hydrogeological plans, programs for localization of strategic infrastructure, plans for water protection, for waste management, for noise recovery; and also plans for the rehabilitation of areas with a high risk of environmental damage or for the prevention of industrial risks.

Particularly relevant are the authorizations named Integrated Pollution Prevention Control (IPPC) and imposed on operators of the most polluting industrial facilities (such as landfills, incinerators, chemical and mining industries, and others as identified in Annex 7 to Part 2 of Legislative Decree 152/2006, the so-called Environmental Code or TUA). These measures serve not only as a comprehensive form of authorization, but also allow the administration to periodically[2] assess the technological efficiency of the relevant installations, outlawing those which do not ensure a sufficient level of environmental protection and thereby eliminating some of the traditional causes of pollution.

The administrative regime governing productive activities and service delivery has also taken on an environmental value, since the issuing of permits or *nulla osta*[3] assumes the environmental sustainability of the activities in question (Dell'Anno 2014, p. 313). These conditions must persist over time, and the administration can revoke permissions already granted if it finds that the requirements for their lawful exercise have vanished. The relevant audits must be effective, periodic, widespread, impartial, and comprise the legal prerequisites for the exercise of further sanctioning power usually afforded to the administrative authorities.

This sanctioning power pertains to environmental violations regardless of whether they have led to contaminations or alterations that are particularly serious or back-dated, and takes the form of sanctions that may have a real effect on the exercise of activity (such as an order to cease the irregularities within a certain period, suspension and withdrawal of authorization), or a pecuniary nature (in the case of fines).

The exercise of the sanctioning power in the European Court of Human Rights (ECHR) case law is brought back to the more general punitive power of the State and is thus subject to the general principles of legality, predictability, proportionality, and reasonableness, with regard to the definition of the regulatory offenses and the corresponding sanctions. This affects the character of the sanctioning procedure, which must ensure an acceptable level of transparency and publicity, as well as respect for the principle of *ne bis in idem*, so as to avoid the choice of sanction determined by the legislature (being administrative rather than criminal) infringing the fundamental rights of the person (ECHR Second Sec., judgment dated 4 March 2014 in the case of *Grande Stevens and others vs. Italy*, applications no. 18640/10).

The above are just some of the instruments available to the administrative authorities for the management of human-caused pollution, and such examples are sufficient to make plain the complexity of the current system. There is no doubt,

[2]The requirement is for verification every five years, and this is to be undertaken using a single technical standard within the European Union: the Best Available Techniques (BAT) published by the European IPPC Bureau, with a BAT Reference Report (BREF).

[3]These authorizations are regulated by different laws, have different natures and release processes, and, because they usually require complex pre-assessments, must be concluded by the release of an express measure (see CJEC, 28 February 1991, C-360/87 and CJEC, 6 November 2008, C-381/07; Portaluri 2014).

however, that the most important administrative functions performed as regards the management of historical pollution are those designed to contain and remove contamination that poses the most serious health and ecological risks, possibly placing the costs of operations on the persons liable, in compliance with the principle of "polluter pays."

2 The Administrative Management of Injury Caused to Environmental Matrices in Cases of Historical Pollution

2.1 Remediation Obligations Associated with the Exceeding of Threshold Values

2.1.1 The CSC and CSR System with Reference to Remediation Obligations

To understand the administrative management of historical pollution, one needs to consider that all human activities release into the environment substances that are not present *ab initio*, which, in turn, cause the various phenomena of pollution, and which may produce more permanent effects in time. Not all pollution, however, causes contamination that is legally relevant, because for this it is necessary that the environmental matrices contain concentrations of pollutants which exceed certain predefined thresholds.

The foregoing is true with regard to contaminations subsequent to the entry into force of the environmental regulations, but also with regard to those that occurred in the past and which can thus be described as historical contaminations. On this issue, it may be observed that before the assessment of the Environmental Code, the Legislative Decree of 5 February 1997, no. 22 (*Decreto Ronchi*), like the Ministerial Decree of 25 October 1999, no. 471, defined an area as contaminated where pollutants in excess of a predetermined concentration threshold were detected (Cingano 2013, p. 40). The Environmental Code, however, changed this definition, introducing a new one consistent with the new objective of sustainable development.

As regards locally confined environmental matrices (namely soil and water),[4] only pollution which possesses an element of real hazard (considering the specificities of each site) are now considered relevant. This result was achieved by introducing two autonomous significance thresholds: the *threshold concentrations of contamination* (*concentrazioni soglia di contaminazione* or CSC), and the *risk threshold concentrations* (*concentrazioni soglia di rischio* or CSR).

[4]The double threshold system does not apply to the air matrix, which continues to be protected notably by limiting potentially harmful emissions (Art. 268, Para. 1, Lett. (a) of the Environment Code) or the commercialization and use of polluting products (automobiles, boilers, etc.).

The CSC identifies levels of contamination in environmental matrices, just as in the previous legislation. Thresholds are expressed abstractly, as numerical values which refer to individual contaminants and respective land uses, and they are usually determined by the tables contained in Annex 5 of Title V of Part 4 of the Environment Code (but on the problems of the integration and updating of tables, see Sect. 2.1.2).

The exceeding of these thresholds, to be ascertained through specific laboratory tests on samples taken at the site, does not make a site (soil, subsoil, or water) contaminated, but only *potentially contaminated* (Art. 240 TUA). Where a situation of potential contamination occurs, the code requires the polluter and also the stakeholders to undertake a study (*caratterizzazione*) of the site and to conduct a *site-specific risk analysis*. In other words, if an agent exceeds the CSC, the code provides that the agent should: (a) carry out further geological surveys to determine the extent and type of potential contamination in soil and groundwater (*characterization plan*); and (b) calculate—according to the predetermined principles of the code and the process of analysis of health and environmental risks—the CSR, namely the thresholds of contamination acceptable with regard to the specific characteristics of the site examined.

All investigations should be conducted jointly with the administration, following guidelines which are updated constantly by the administrative bodies with specific technical and scientific skills relevant to the environment and to health, especially the Istituto Superiore per la Protezione e la Ricerca Ambientale (ISPRA).

The site can be considered *contaminated* only if the CSRs are exceeded. In this case, making-safe, remediation, and in some cases also environmental restoration operations are necessary. On the other hand, no actions are described as mandatory for private actors, because the site does not present risks for goods (especially human health) that are subjected to protection (Grassi 2014, p. 692).

2.1.2 The Setting of the Thresholds and the Relationship Between the Principles of Legality and of Precaution

Using the tables annexed to the Environmental Code, it is possible to identify the pollutants (i.e., dangerous substances) most often dispersed in environmental matrices and to predetermine their significance thresholds, ensuring the simultaneous defense of the precautionary, legality, and legal certainty principles. The tables identify different thresholds depending on the environmental matrices involved and the specific needs of protection. The tables concerning the CSCs for soil and subsoil show different thresholds for each use of the site (residential or productive) and of the consequent exposure of people to risk. This dual nature of the thresholds is not provided for the water matrix.

Although effective, the tabular system raises some delicate questions. Firstly, the table concerning the CSCs for soil and subsoil does not take into account all the potential uses of a site, nor the frequent mix of functions, which creates tensions regarding the applicable thresholds which need to be solved case by case, through administrative or judicial channels.

Secondly, the division of the tabular thresholds into two parts does not consider soils used for agricultural production and animal husbandry, which prompted the administration and the administrative courts to apply in these cases—in accordance with the precautionary principle—the more restrictive CSCs provided for soils with residential use (TAR Lombardia, Brescia, Sec. I, 21 January 2013, no. 50).

But the biggest problems come from the fact that the tables are neither exhaustive nor flexible, since (i) they do not identify the CSCs of all existing substances, and (ii) they do not define a procedure for their future integration through administrative action. This regulatory gap has been tackled by allowing the administration to work analogically, extending the CSCs of toxicologically similar substances to those not (yet) mentioned in the tables in accordance with the precautionary principle (this occurred, for example, for the CSC applicable to Methyl *Tert*-Butyl Ether or MTBE). Jurisprudential and administrative practice has clarified that the extension must proceed on the basis of technical assessments, taking into account the prevailing views in the international scientific community.

Although the solution identified confers on the administration the power to impose obligations on private parties to undertake expensive interventions of making-safe and remediation based on assessments carried out by the administration itself, this is considered legitimate because administrative law requires neither the prohibition of analogy (as provided for in criminal law), nor the most stringent interpretations of legal principle. The duty to ensure the legality of the administrative action does not in fact bind such action to provisions which have the character of precision and definiteness (formal legality), because this would thereby exclude the possibility of recognizing the discretionary powers of public bodies (Consiglio di Stato [Cons. St.], Sec. VI, 20 May 2014, no. 2526). The key to the lawfulness is, however, that the integrative power is exercised on a strictly technical basis and through consideration of the prevailing views of the international scientific community as expressed in the regulations of other countries and the opinions issuing from national and international technical organizations (TAR Lombardia, Brescia, Sec. I, 6 May 2015, no. 630; but see also Cons. St., Sec. III, 14 July 2014, no. 3664).

With regard to the second problem noted above, namely the absence of legal processes for updating the CSC table, the possibility of attributing to the public authorities the power to override the legal force of CSCs seems excluded. The predetermination of a substance's CSC does indeed fix a legislative choice that cannot be overcome by the administration, not even by applying the precautionary principle, because the principles of separation of powers, rule of law, and legal certainty must also be respected. Up to now, therefore, it seems reasonable to suppose that the updating of the tables would involve a new exercise of regulatory power (though compare TAR Lombardia, Brescia, Sec. I, 6 May 2015, no. 630).

In this regard, however, one should note that the inability to alter the tables does not seem to fully exclude the administrations' and the courts' power of intervention when new scientific evidence signals a danger to human health or other assets, subject to constitutional protection for concentrations of pollutants that do not exceed the relevant CSC. In this case, in fact, the constitutional relevance of the goods put at risk could justify the administration in permitting specific (i.e., case by case) non-applications of tabular thresholds, not in order to impose the usual actions of remediation and environmental restoration, but to adopt measures necessary to provide for reasonable and adequate protection of such goods.

A confirmation of this solution lies in the Environmental Code itself, which while imposing the obligation to adopt emergency safety measures in response to events that immediately endanger health or the environment, does not assign relevance either to the exceeding of tabular thresholds or at the time of contamination (Art. 242 TUA).

2.1.3 Remediation Management: Administrative Competence and the Duties of Responsible Parties

The administrative procedures for the management and removal of historical contamination are regulated by Part 4 of the Environmental Code. The Environmental Code primarily regulates the measures that must be implemented in emergencies, that is when a situation of pollution is discovered which, although dating from the past, presents actual risks of spreading or exacerbation (Art. 242, Para. 1). A situation of this kind places on private agents (the owners or site managers) the obligation both to communicate the discovery to the public authority, and to take measures that may have a fact-finding (environmental characterization) or an operating (preventive measures) character, depending on whether it is necessary to take immediate action on the possible contamination in order to limit its harmful effects.

As regards the preventive measures that can be implemented in the event of historical contamination which is subject to risk exacerbation, a central role is played by emergency safety measures. These consist of actions that would be technically capable of containing the spread of contamination (mitigation measures), for example, the construction of physical or hydraulic barriers.

For the implementation of preventive measures, the overcoming of the CSCs is irrelevant. The obligation to monitor the overcoming of the CSCs, and subsequently of the CSRs, arises only at the end of the emergency phase, with the aims of determining the need to proceed to remediation or to safe-making actions and the targets of those interventions.

In this regard, it may be noted that even (and especially) in the case of historical contamination, Italian legislation allows a certain flexibility in determining the results to be obtained from an environmental intervention. In fact, every remediation project and/or environmental restoration must take into account the specific nature of the site in question and its actual intended use, present and future, as well as the type of pollution, the technical feasibility of removing it, the risks associated with materials handling, and, not least, the economic sustainability of the interventions.

It follows that on sites where productive activities are still being undertaken it will be possible to handle contamination only through safety interventions which limit pollution without entirely removing it (Art. 242, Para. 9 TUA). In the case of contaminated sites which do not have a productive use, or where the production activities have permanently ceased, what is instead required is the removal of the contamination through remediation or, in case the first action is not practicable for technical or economic reasons, permanent safety measures on the site. All these operations must preferably be made at the expense of the persons responsible for the historical pollution, its omitted removal, or its aggravation.

The search for the responsible persons falls to the administration. This search must be carried out in debate with the individuals involved and is particularly important when the contamination occurs on a site on which various potentially polluting activities have been carried out, or where different agents have operated. In fact, if the contamination is the result of actions or omissions by several operators (concurrently or over years), each must take action proportional to their contribution, therefore *pro quota* and without joint and several liability (CJEU order of 9 March 2010 C-478/08 and C-479/08, and CJEU 24 June 2008, due C-188/07).

This search for the responsible party is a duty of the administration, which can be sanctioned in the event of inaction because it serves the protection both of the public interest, as expressed in the "polluter-pays" principle, and of the assets of innocent owners or of those who will have to bear the economic burden of restoration, directly or indirectly (TAR Lombardia, Milano, Sec. IV, 15 April 2015, no. 940).

The identification of the responsible party, however, is rarely easy, since it requires a thorough investigation into the history of the site, the type of pollutants uncovered, their reaction processes, and the degree to which they can be attributed to the activities of the suspected party. To attribute the contamination to a particular party, the administration must demonstrate, according to the evidence rule of "more likely than not" (Peres 2012, p. 100), the existence of a causal link between acts or omissions of that party and the contamination or its aggravation, in line with the EU polluter-pays principle (Cons. St., Sec. III, 14 May 2015, no. 2462; CJEU, 9 March 2010, C-378/08, §§ 57, 58).[5] The proof can be given directly or indirectly, since the administration can also use a simple presumption, on the basis of factual information from which serious, precise, and consistent clues can be drawn, concerning the production of pollution and its attributability to certain parties on the basis of a causal link (Cons. St., Sec. V, 22 May 2015, no. 2569; Sec. V, 16 June 2009, no. 3885).

---

[5]According to the case, "the competent authority must have plausible evidence capable of justifying its presumption, such as the fact that the operator's installation is located close to the pollution found and that there is a correlation between the pollutants identified and the substances used by the operator in connection with his activities. Where the competent authority has such evidence, it is thus in a position to establish a causal link between the operators' activities and the diffuse pollution found. In accordance with Article 4(5) of Directive 2004/35, such a situation therefore falls within the scope of the directive, unless those operators are able to rebut that presumption" (CJEU, 9 March 2010, C-378/08).

Indeed, some authors opine that to assign the cost of environmental interventions to a party, it would be necessary to prove that that party acted intentionally or at least negligently (Grassi 2014, p. 707; but see too Cons. St., Sec. V, 22 May 2015, no. 2569). This is a minority position, however, because there is a prevalent thesis that the obligation to carry out remediations pertains even in case of accidental action if there is a causal link between the action (or omission) and the contamination, as required by the rules on environmental damage caused in the exercise of dangerous activities (see Sect. 2.2).

Moreover, the majority jurisprudence also believes that the polluter-pays principle is applicable to historical contamination, irrespective of whether the polluting activities were contrary to law when they were realized (TAR Veneto, Sec. III, 25 February 2014, no. 255). This is because the contamination would give rise to a situation which endures until its causes are removed and the environmental parameters are reported as having returned to within the legal limits fixed first with the Ronchi Decree and then with the Environmental Code (*ex plurimis* Cons. St., Sec. IV, 9 October 2007, no. 5283),[6] with the consequence that the relevant legislation is also applicable to situations arising from actions that were not originally prohibited (Cons. St., Sec. V, 16 March 2016, no. 1054; Sec. V, 23 September 2015, no. 4466).

This reconstruction has been criticized on the assumption that to qualify as permanent an unlawful act must involve not only the permanence of the effect (pollution) but also of the action itself (Cons. St., Sec. VI, 5 March 2015, no. 1109; but see too Corte di Cassazione Civile [Cass. Civ.] Sec. II, 1 February 1995, no. 1156, concerning an infiltration of oil products into the soil and then into the ground). Contrary to this, an agent can only be held responsible for an instantaneous act if, although producing permanent effects, it was already unlawful when it was committed.

Although it may seem reasonable, it does not seem appropriate to exempt from liability parties who have committed instantaneous polluting actions in the past. This is primarily because the Italian legal system, in implementing the polluter-pays principle, and contrary to what has happened in other European legal systems, has deliberately considered irrelevant the moment of commission of the polluting act with regard to liability for remediation.[7] Furthermore, even when it was caused without violating specific rules, pollution which is not the result of specifically authorized activities should be considered illegal because it is contrary to the constitutional provisions in force since 1948 to protect the environment and human health in preference to economic activities (see, among others Arts. 2, 9, 32, 41 of

---

[6]Judgment concerning the legitimacy of intervention on a site in Trieste contaminated by the operation of a refinery closed in 1965 and having maintained a deposit of products until the early 1980s.

[7]On the possibility of application of the remediation process also to historical pollution that occurred before the entry into force of environmental regulations, see Balloriani 2016, p. 320, according to which there is continuity between the general rules governing civil liability and those relating to remediation.

the Constitution). Finally, a party who in the past has carried out highly polluting activities cannot be considered always in good faith, because the legality of his actions can be attributed, in some cases, to lobbying activities, delaying the adoption of environmental disciplines (see Sect. 1).

To assign the costs of remediation to the person responsible for historical contamination, however, is not always possible, since that party may no longer exist, may be bankrupt, or, again, may have transferred ownership of the company to third parties before the entry into force of the latest environmental regulations.

The disappearance of the responsible person, however, does not always extinguish the liability. There are cases in which the person responsible only disappears formally, for example, because the company merged into a new legal entity, or because the company that actually did the pollution no longer exists, but not the company that controlled the activities and benefited from the activities undertaken by the former. In these cases, the courts have sometimes applied the polluter-pays principle anyway, stating that the liability for the pollution caused by the subsidiaries may fall on the company or on the industrial group, applying the substantialist definition of "company" given by the CJEU relating to competition and applicable when there is evidence of dependence between the subsidiaries and the group (see TAR Abruzzo, Pescara, Sec. I, 30 April 2014, no. 204, Cons. St., sect. V, 5 December 2008 no. 6055).

Moreover the liability for historical pollution is not attenuated by the fact that the person responsible has transferred the contaminated properties and the liability for future reclamations to third parties. The remediation obligations are in fact mandatory, because they implement the European polluter-pays principle and cannot be affected by contracts that put the costs of environmental interventions on the community (Cons. St., Sec. VI, 10 September 2015 no. 4225). For this reason, the contractual clauses concerning the transfer or assumption of liability for remediation are judged not enforceable against the administration, but only valid between the parties (Cons. St., Sec. V, 23 September 2015, no. 4466; TAR Friuli Venezia Giulia, Trieste, Sec. I, 28 January 2008, no. 89; TAR Lombardia, Milano, Sec. I, 13 February 2001, no. 987). The responsibility for remediation, therefore, is never generated simply by the properties of the contaminated site.

It must be noted that, in accordance with the jurisprudence of the European Union (CJEU, 5 March 2015, C-534/13), the law does not place any remediation obligation on owners or site managers which did not contribute to the pollution or to its aggravation. Nevertheless, it is their responsibility alone to take emergency measures (if necessary) and to communicate the discovery of pollutants. The lack of obligation, however, does not forbid the owner or other private stakeholders from voluntarily executing environmental interventions (TAR Lombardia, Milano, Sec. II, 9 April 2013, no. 883).

There are various reasons why a non-responsible party could decide to take on the costs of environmental intervention. They could, for example, be concerned to safeguard their rights over an area, since when the individuals responsible for the contamination are not identified or do not provide, the remediation actions will be carried out by the administration (Art. 250 TUA), which can then recoup their costs either by acting against the responsible person (if identified and solvent) or by exercising a legal right on the value of the reclaimed area, regardless of its ownership. To this it can add, though, that when a non-responsible party decides to anticipate the cost of environmental intervention, it retains the right to recoup those costs from the responsible party as identified by the administration.

In thus clarifying the criteria that legitimize the assignment of remediation and safety operation costs to a particular party, it is observed that the responsible party (as well as other stakeholders) must always do these works jointly with the administration and its technical agencies, where the latter are tasked with verifying and preliminarily approving the projects and also supervising their implementation and verifying and certifying the results achieved.

In addition to the above, it is noted that Legislative Decree no. 152/2006 also requires partially different procedures (with respect to expertise, time, and manner of execution) in respect of the size of the contaminated site and the type of contamination. Among these, special procedures mention should be made of those related to *sites of national interest* (*siti di interesse nazionale or SIN*), since these sites are often affected by particularly serious historical contamination as regards quantity and harmfulness of pollutants, the importance of the impact on the surrounding environment in terms of health and ecological risk, as well as damage to cultural and environmental heritage. The classification of a site as an SIN is the responsibility of the Ministry of Environment and allows these areas to be treated unitarily with the help of the most authoritative national technical scientific agencies (Istituto Superiore di Sanità and ISPRA).[8] Among the recognized sites of national interest, of particular note are those described as possessing *overriding public interest for industrial conversion* (for example, the SIN of the steel plant of Taranto), because their environmental rehabilitation is structured into their production processes, based on program agreements signed between the private operators and the public authorities (Art. 252-bis TUA).

Moreover, with regard to SINs, the law promotes the identification of agreed solutions, providing the possibility of concluding global transactions with those responsible for contamination, in order to meet the costs of clean-up and restoration as well as of environmental or other damage for which the State or other regional or local authorities can claim compensation. This is an important tool, because it permits the identification of financial settlements case by case, ensuring not only the feasibility of interventions, but also the formal attribution of part of the costs to the community and that seems—under certain conditions—generally appropriate in consideration of the social, fiscal, and economic benefits that the community has derived from the polluting activities over time.

Note, though, that the possibility of derogating to the polluter-pays principle, combined with the definitive nature of agreements, imposes on the competent authorities a duty to make thorough investigations into the situation of site and on

---

[8]Sites affected by asbestos mining and production activities are always treated as sites of national interest (Art. 252, Para. 2-bis TUA).

the real cost of all necessary interventions, in order to prevent the transfer of unforeseen or excessive liabilities to the public finances.[9] The code pursues this result by providing that the conclusion of the agreements shall be preceded by a complex procedure (Art. 306-bis TUA), yet this is still not very transparent. In this regard, it would be desirable that in the negotiation of the transaction agreement it is regulated, providing a degree of civic participation—allowing, for example, for the submission of comments on the proposals—in order to promote a wider degree of control over the agreements, as well as the identification of solutions tailored to the communities involved.

2.2 *Compensation for Damage Caused to the Environment as a Common Good*

The Environmental Code of 2006, in accordance with Directive 2004/35/EC and in partial continuity with the first environmental guidelines (Law 349/1986, Art. 18), provides an obligation for compensation on the responsible party in addition to those reviewed in the preceding paragraphs. Indeed, the regulations on environmental damage have a wider normative scope because they consider the environment as a unitary good that, on one hand, includes not only environmental matrices but also the habitats, flora, and fauna (Art. 300 TUA),[10] and on the other is integral to the protection of individual rights. This second regulation does not duplicate the obligations on the responsible party, because it has a subsidiary nature and relates to damages outstanding at the close of the preliminary safety and remediation operations.

In truth, with regard to historical contaminations, it must be noted that this subsidiary liability is limited in scope, because Directive 2004/35/EC and

[9]In this regard, see the special report No. 23/2012 of the European Court of Auditors, *Have EU Structural Measures Successfully Supported the Regeneration of Industrial and Military Brownfield Sites?*, www.europarl.europa.eu.

[10]Pursuant to Art. 300 TUA, environmental damage is: (1) any significant and measurable deterioration, direct or indirect, of a natural resource or the utility provided by the latter (§ 1); (2) deterioration, in comparison with the original conditions, caused: (a) to protected species and natural habitats, as well as flora and fauna, as well as protected natural areas; (b) to inland waters, by actions that significantly adversely affect the ecological, chemical, and/or quantitative status or ecological potential of the waters concerned, as defined in Directive 2000/60/EC with the exception of adverse effects where applicable as in Art. 4, § 7 of that directive; (c) to coastal waters and those included in the territorial sea by the above acts, even if carried out in international waters; (d) to the ground, by any contamination that creates a significant risk of adverse effects, even indirectly, to human health due to the introduction into the ground, on or under the land, of substances, preparations, organisms, or microorganisms harmful to the environment (§ 2).

As clarified by Castronovo 2013, p. 1027, the two figures of environmental damage have a special relationship between them, such that the second figure specifies the first without exhausting it, but identifying its essential nucleus.

Legislative Decree no. 152/2006 expressly exclude its applicability to damage occurring before their entry into force. For this reason, the only historical contaminations to which the rules on environmental damage are applicable are those produced by acts which, although begun before the entry into force of the guidelines, continued afterward (CJEU, 4 March 2015, C-534/13). In the presence of contaminations of this kind, Article 311 of the Code provides (i) strict liability in case of damage from activities dangerous to human health and the environment (activities identified in Annex 5 to Part 6 of the Environment Code), and (ii) liability for fraud or fault in any other case.

The strict liability of operators engaged in hazardous activities is not a liability position by the fact that they have worked or currently work in that area. Once again, the liability exists only if the administration proves a causal link between the activity and the damage. Strict liability, moreover, is excluded *ex ante* when the environmental damage or the imminent threat of such damage (a) was caused by a third party and occurred despite the existence of security measures appropriate in principle to prevent it; (b) resulted from compliance with a compulsory order or instruction emanating from a public authority; (c) was caused by an emission or event expressly permitted by a license legally awarded at the date of issue or event and in full accordance with the conditions specified therein; and (d) was caused by an emission or activity or any other usage of a product in the course of an activity which the operator can demonstrate was not considered likely to cause environmental damage according to the state of scientific and technical knowledge at the time of release or the activity (Art. 308, Para. 4 and 5 TUA).

In the event that the environmental damage is a result of competing activities by more than one subject, as is often the case in connection with old or diffuse contaminations, it is provided that each responds to the extent of their own personal liability, and the related debt is transmitted, according to law, to their heirs to the extent that they actually profited from it (Art. 311, Para. 3 TUA).

The action for damages is subject to a limitation period of thirty years, starting from the event or incident that caused the damage (Art. 303, Lett. (g) TUA). It is a *sui generis* compensation, which has the simultaneously private and public nature of sanction and compensation.

Because the compensation should preferably be done through performing specific actions, it is up to the administration to determine the concrete measures to be adopted, in compliance with the procedures of the Environmental Code (see Salanitro 2009, p. 148). These environmental measures may also include interventions on areas other than the one that is actually contaminated.

In case the adoption of such remedial measures results is omitted, incomplete, or dissimilar to the terms and conditions prescribed, in whole or in part, the current legislation allows monetary compensation, the value of which must be equivalent not to that of the damage but to the costs of activities necessary to achieve full and correct implementation of the remedial measures (Art. 311 TUA). Equivalent compensation represents the *extrema ratio* of environmental disputes and is no longer an alternative to compensation in the specified form, by monetizing the nonfulfillment of the repairs (Bonelli 2014, p. 15; Comporti 2011, p. 22).

As anticipated, the aforementioned legislation does not apply to historical contamination caused by polluting events concluded before the entry into force of the Environmental Code. With regard to such contamination, however, the inapplicability of the current legislation on environmental damage does not exclude all responsibility conferring on the relevant party. Indeed, the prevailing opinion is that, in addition to the obligation of making-safe and remediation, the polluter is liable for further damage to the environment according to previous legislation, that is (i) Article 18 of Law 349/1986[11] for damage caused between 1986 and 1997, and (ii) the ordinary liability regime provided for in Articles 2043 and 2050 of the *Codice Civile* (Cod. Civ.) for previous damage (see recently Cass. Civ., Sec. I, 7 March 2013 no. 5705).[12]

The basis of this reconstruction is the view that Law no. 349/1986 has introduced a special regime of the civil liability already in place. In particular, jurisprudence states that, since the natural environment is a public good with constitutional status, injury to it confers on the public authorities responsible for its protection the right to seek compensation for non-pecuniary damage caused. This right derives from Article 9 of the Constitution (Para. 2) and from Article 2059 Cod. Civ., and it existed before the introduction of the Law of 8 July 1986, no. 349, because Article 18 did not introduce into the Italian legal system a notion of environmental damage, but is limited to sharing between the state, local authorities, and environmental groups the legitimacy to intervene in those proceedings as regards compensation (Cass. Civ., Sec. III, 10 October 2008, no. 25010).

In relation to hypotheses of historical responsibility, with regard to the prescription of the right to credit, it was reported that the long latency of the damage in question and the potential for its harmfulness to worsen, reasonably induces a postponement of the *dies ad quem* of the maturing of the prescription (in virtue of the provisions of Arts. 2935 and 2947 Cod. Civ.). The beginning of this term is therefore postponed to the moment when the contamination is discovered, or could have been discovered (Cons. St., Sec. V, 5 December 2008, no. 6055). In this regard, the civil courts have also held that, since in matters of environmental damage the unlawful conduct consists in keeping the environment in a damaged

[11]This is a more restrictive rule that concerns any environmental damage, regardless of whether it has exceeded the threshold value, when it derives from acts that are intentional or negligent and in violation of laws or administrative measures.

The rule provides that: (1) any malicious or negligent act in violation of the law or action taken based on the law that would jeopardize the environment, causing damage to it, by altering, spoiling, or destroying in whole or in part, obliges the perpetrator to pay compensation to the State; (2) for matters referred to in § 1 the jurisdiction belongs to the ordinary courts, without excluding the Court of Auditors, in Art. 22 of the Decree of the President of the Republic of 10 January 1957, no. 3; (3) action to compensate for environmental damage, although practiced in the criminal courts, is to be promoted by the state as well as by local authorities as regards the assets covered by the harmful act.

[12]According to Prati 2008, p. 172, in the absence of transitional arrangements, the ordinary rules of tort apply also to environmental damage which materialized after the repeal of Art. 18 of Law 349/1986, if it is the result of earlier emissions, events, or incidents.

condition, the term of action for compensation shall begin to run only from the time when such conditions have been voluntarily removed by the injurer or the conduct has been made impossible by the innocent loss of the good, for example, in case of seizure (Cass. Civ., Sec. III, 6 May 2015, no. 9012 and 19 February 2016, no. 3259).

Irrespective of the regulatory regime applicable to single cases of historical contamination, the only subject entitled to claim compensation for environmental damage in Italy is the State, through the Ministry of the Environment (Giracca 2014, p. 600; Cass. Civ., Sec. III, 6 May 2015, no. 9012). Local authorities, citizens, and associations are only permitted to solicit the intervention of the Ministry by presenting complaints, observations, documents, and information, in case of environmental damage or any imminent threat of environmental damage. This intervention request, while forcing the Ministry to proceed, does not restrict its autonomy in decision-making. The reasons for this choice differ and must be sought in the need to promote the identification of solutions that enable environmental restoration without compromising economic competitiveness.

If the Ministry of Environment decides to take action to obtain compensation for environmental damage, it will do so through the courts (through civil action in criminal process or addressing the question to the administrative judge), or by administrative action (issuing the environmental injunction for restoration expected by Art. 313 TUA). The two modes of ministerial intervention are alternatives, and the choice is subject to judicial review.

3 Possible Extension of Liability Resulting from Historical Pollution

3.1 Possible Liability for Third Parties

Economic or moral loss caused to third parties is also the object of liability for the polluters, and here it is not relevant that the polluter has already performed clean-up or has paid compensation for the environmental damage. This is pursuant to the Environment Code (Art. 313, Para. 7), but more generally is an application of the rule that every injury caused to others *contra ius* engenders responsibility.

Differing from the assumptions discussed above, the responsibility to third parties damaged by contamination is not subject to special rules, but is governed by the common rules of tort (Art. 2043 Cod. Civ.). Those who have suffered damages can take action in criminal trials, or through a specific civil action, within the period of limitation of five years from the day when the fact occurred, was discovered, or could have been discovered (a hypothesis that is especially important in the case of historical pollution).

It is generally accepted that action for compensation for economic and moral loss requires trying both the alleged damage as well as the environmental violation that produced it, as well as an immediate and direct causal link between the event and the harm and thus its attributability to intent or negligence of the author. These assumptions of liability are heterogeneous, and private parties, as well as public entities (usually local authorities) and some environmental associations, can claim compensation. In particular—and to simplify—contamination can legitimate private parties in seeking compensation for damage caused by depreciation of their property, deprivation of its use, and for the cost of health care. Damage to health, both in the case of proven medically injury (biological damage) and in the presence of psychological disorders due to justified fear of getting sick (Cass. Civ., Un. Sec., 21 February 2002, no. 2515), is also compensable.[13] Likewise, local authorities can take action for lower land values caused by land degradation (Corte di Cassazione Penale, Sec. III, 23 May 2012, no. 19437) or reputation damage to local communities (Cass. Civ., Sec. III, 22 March 2012, no. 4542), but also when the environmental emergency has forced the institution to divert human and material resources to deal with it, diverting these from ordinary activities.

The cases mentioned are not exhaustive, however, because the legal system provides more instruments for the protection of damaged third parties. Those affected by a polluting activity may apply to the courts for the adoption of legal measures that restrict or arrest the relevant pollution sources, such as ordering structural and technological changes to installations or the adoption of appropriate measures to resolve the case.

Because these measures seriously affect the property rights and economic freedom of the operators, however, the possibility of obtaining them is subject to stringent conditions, and those who claim to have been harmed must prove both the damages suffered as well as the causal link to the polluting activities, and their unlawfulness due to the absence of authorization or violation of prescriptions on acceptable levels established by law or by regulation.

3.2 *Possible Liability of Public Administration and Officials*

In closing this review of the governance system for historical pollution and the concomitant responsibilities, it is necessary to consider the position of the administration. The Italian legal system does not reserve a position of privilege for the public administration. Just like any other private operator, the State and other public bodies may be called to account for contamination caused in the exercise of a

[13]On the occasion of the environmental disaster caused by ICMESA in Seveso, for example, the psychological damage (stress, worry) has been compensated even in the absence of physical injury or other financial loss, on the condition of proof that a psychological disturbance had taken place. In this regard, it has been noted that the discretion in the quantifying of moral damage and the connection to the crime have given the compensation a punitive nature (Patti 2013, p. 135).

directly polluting activity, both with regard to the responsibility for clean-up and for damage to third parties, and with regard to administrative sanctions. It is of no import that the contamination occurred in the exercise of strictly public functions or of normal economic activity, but what matters is that the exercise of the activity caused injury to the environment and to the rights of third parties.

The responsible subjects for the relevant damages will be the administration and its officials who are identified as the perpetrators of the offense, for having caused or having contributed to provoking it through their actions or omissions, whether intentional or grossly negligent (not in case of ordinary negligence). The liability, moreover, is not diminished even when the pollution was due to the voluntary action of officials, provided they were acting within their institutional roles (Renna 2016, p. 804). The administration and its agents may also be called to answer for damages caused by the bad use (or by the omission) of functions assigned to them by law, whether unlawfully and intentionally or negligently.

The jurisprudence pertaining to such cases is extensive and can be traced to the so-called responsibility by administrative measure, an expression that unites otherwise distinct cases, namely damages caused by the contents of an unlawful measure, damages resulting from breach of procedural rules that leave the public administration discretion in their implementation, and damages resulting from a breach of rules which impose procedural obligations (Renna 2016, p. 809).

Consider the case of an administration that, in exercising the powers of tabular integration referred to above, requires the execution of expensive environmental interventions on the false assumption that there is scientific evidence providing a rationale for the application of a very restrictive CSC. Or consider a case in which the execution of environmental actions has been imposed on subject who is not responsible for the pollution. And consider again the case of damage caused by delays or failure in execution of administrative responsibilities, such as environmental policing (see Corte dei Conti [Court of Auditors, C. Conti], Sec. Lombardia, 31 July 2015, no. 137), or those required to track down the party responsible for the contamination. With regard to the latter case, indeed, a delay in the identification of the responsible party is sufficient to create serious prejudice to an owner who is not guilty, regardless of whether or not he has executed voluntary environmental interventions, because in the first case he cannot act to recover the costs supported, and in the second case he cannot freely make use of the polluted area.

Moreover, the administration's responsibility is not only relevant with regard to its external relations, that is to say relations between individuals and the administration itself, but also to its internal relations, that is relations between the institution and its functionaries. The fact that the administration is in most cases the subject that pays the immediate costs of environmental actions and compensation, does not exempt the officials themselves from responsibility. On the contrary, all their conduct, when they cause damage to the institution, provides legitimate grounds for a judgment of liability against them before the Court of Auditors. Indeed, all persons exercising public functions are subject to a judgment of responsibility by this Court, whether they are employees, elected officials, or external staff assigned to perform specific tasks (Schlitzer and Imposimato 2012, p. 80). The facts that generate liability may be several, and on the whole they confer on to the Court of Auditors a very considerable influence over the administrative activity, here as well as in the field of environmental governance. The liability actions against public agents, indeed, are promoted also for financial losses caused by lost revenue or non-pecuniary damage. For example, the Court of Auditors pronounced judgements of liability due to the non-exercise of organizational functions, both as regards surveillance of environmental violations (C. Conti, Sec. Abruzzo, 23 April 2007, no. 432) and of waste collection services (C. Conti, Sec. Campania, 9 December 2009, no. 1492; C. Conti, Sec. Liguria, 23 May 2013, no. 23). Although in these cases the environmental harm is always in the form of a specific financial loss (sanctions that would be imposed on transgressors of environmental legislation, in the first case; higher costs of service management of waste, in the second), it is clear that the responsibility affirmed by the Court of Auditors concerns refunding/compensatory acts, as well as sanctions (for the nullification of the efforts of the police in the first case; for having caused the failure of sectorial planning in the second). This sanctioning function is particularly evident in the case of injuries to noneconomic goods, such as the administration's image. In these cases, the abstractness of the asset damaged provides the Court a power of settlement that allows quantification of the (noneconomic) damage in proportion to the seriousness of the offense.

We have just noted that by means of these functions, the Court of Auditors exerts a vast influence on the overall public administration system, sufficient to affect—significantly—the decisions of officials, and extending even to the conceptualization of the public interest assumed by the administration. With regard to the problem of historical pollution, this power has been exercised carefully, because the Court of Auditors can claim the merit of having recognized—in advance of the legislator—the environment as a public good that should be the object of specific constitutional protection, thus paving the way for the new environmental policies that led in 1986 to the institution of the Ministry for the Environment and to the first regulations on environmental damage. The work carried out by the accounting Courts in punishing environmental violations by public officials appears to close the circle of discourse on administrative environmental governance, restoring to those working for the administration the duty to daily commit to the management of historical pollution, and to identify the best measures for the restoration of violated ecosystems.

References

Balloriani, M. (2016). L'obbligo di bonifica dei siti inquinati e il risarcimento in forma specifica del danno ambientale. Alcune questioni aperte sugli inquinamenti storici. In F.G. Scoca & F. Di Sciascio (Eds.), *Le proprietà pubbliche. Tutela, valorizzazione e gestione.* Napoli: Editoriale Scientifica.

Bonelli, F. (2014). Il risarcimento del danno all'ambiente dopo le modifiche del 2009 e del 2013 al T.U. 152/2006. *Diritto del commercio internazionale*, 1, 3.

Castronovo, C. (2013). Il danno all'ambiente rivisitato. *Vita Notarile, 3,* 1029.
Cingano, V. (2013). *Bonifica e responsabilità per danno all'ambiente nel diritto amministrativo.* Padova: Cedam.
Comporti, G. D. (2011). La responsabilità per danno ambientale. *Rivista quadrimestrale di Diritto dell'Ambiente, 1,* 2.
Dell'Anno, P. (2014). Ambiente (Diritto Amministrativo). In P. Dell'Anno & E. Picozza (Eds.), *Trattato di diritto dell'ambiente*. Cedam: Padova.
Giannini, M. S. (1973). «Ambiente»: saggio sui diversi suoi aspetti giuridici. *Rivista trimestrale di diritto pubblico*, 1973, 15.
Giracca M.P. (2014). Danno ambientale. In R. Ferrara & M.A.Sandulli (Eds.), *Trattato di diritto dell'ambiente*, vol. I, Milano: Giuffrè.
Grassi, S. (2014). La bonifica dei siti contaminati. In R. Ferrara & M.A.Sandulli (Eds.), *Trattato di diritto dell'ambiente*, vol. II, Milano: Giuffrè.
Patti, S. (2013). Danno ambientale e risarcimento del patema d'animo. In F. Busnelli & S. Patti (Eds.), *Danno e responsabilità civile*. Torino: Giappichelli.
Peres, F. (2012). Il risarcimento del danno all'ambiente nella matrice terreno e la disciplina per la bonifica dei siti contaminati. In G. Perulli (Ed.), *Il danno ambientale*. Torino: Giappichelli.
Portaluri, P.L. (2014). Autorizzazioni ambientali: tipologie e principi. In R. Ferrara & M.A. Sandulli (Eds.), *Trattato di diritto dell'ambiente*, vol. II, Milano: Giuffrè.
Prati, L. (2008). *Il danno ambientale e la bonifica dei siti inquinati. La nuova disciplina dopo il d. lgs. 152/2006 e la sua riforma*. Assago: Ipsoa.
Renna, M. (2016). Responsabilità della Pubblica Amministrazione. *Enciclopedia del Diritto. Annali* (Vol. IX). Milano: Giuffrè.
Renna, M. (2012). I principi in materia di tutela dell'ambiente. *Rivista quadrimestrale di Diritto dell'Ambiente, 1–2,* 62.
Salanitro, U. (2009). *Il danno ambientale*. Roma: Aracne.
Schlitzer, E. F., & Imposimato, C. (2012). L'ambiente e la sua tutela risarcitoria. In G. Perulli (Ed.), *Il danno ambientale*. Torino: Giappichelli.

Chapter 8
Historical Pollution and Criminal Liability in the United States

Joseph F.C. DiMento and Ava Badiee

Contents

1 Definition of Historical Pollution

"Historical pollution" is the behavior of polluting sites, which either started in the past and is still ongoing, or which took place discretely in the past. The behavior often relates to industrial or other productive activities. Historical pollution usually

J.F.C. DiMento (✉) · A. Badiee
University of California, Irvine, USA
e-mail: jdimento@law.uci.edu

A. Badiee
e-mail: abadiee@lawnet.uci.edu

F. Centonze and S. Manacorda (eds.), *Historical Pollution*,
DOI 10.1007/978-3-319-56937-6_8

manifests its contaminating impacts or effects a while after the polluting behavior. "Pollution" is the conduct rather than the conduct's effects, while "contamination" is the consequence of such behavior and refers to the contaminated, compromised, or damaged conditions of the sites and natural resources which have been polluted.

The basic feature of historical pollution is the period of time between the beginning of the relevant behavior (pollution) and its observation. It is also usually reflective of a period of time between the behavior (pollution) and the related consequences (contamination of sites and natural resources). Historical pollution can include harms and threats to environmental resources and human health.

For the purposes of our research, the concept includes "legacy pollution." The concept of "ongoing pollution" will not be used; rather, it will be treated as a type of legacy pollution, identifying past conduct whose effects are ongoing. The concepts of cumulative pollution and diffuse pollution will not be used as separate categories.

The "historical" feature reflects the fact that time is a very important factor within the analysis. At this stage, it does not appear appropriate to identify a specific amount of time between the relevant conduct and its effects as an element of the definition of historical pollution. However, possible changes in legislation, culture, public perception, or scientific knowledge may be considered important factors when identifying cases of historical pollution.

Since the concept could implicate a wide range of situations, it is necessary to limit the focus of the research. Attention will be paid only to pollution caused by industrial and other production-related activities.

2 Sociological and Economic Context

In this section we describe the evolving realization of the challenge of historical pollution in the United States. This description includes public awareness, media attention, and government recognition. "Until the mid-1970s the issue of toxic waste sites was not on the systemic or institutional agendas of most local and state governments, not on the federal government's agenda. Government leaders and the public became aware of the issue through incidents such as Love Canal" (O'Brien et al. 1984, p. 335).

2.1 Love Canal

No treatment of historical pollution in the United States could ignore the notorious case of the Love Canal. Other cases that became infamous in the early years of the discovery of the general phenomenon were Valley of the Drums, Hopewell, Stringfellow Acid Pits, Memphis, and Times Beach. Love Canal is the best-known American event, and the discovery highlighted the need for a new legal framework to address historical pollution.

In the 1940s, Hooker Chemicals and Plastic Corporation (later known as Hooker Chemical Company, and more recently as Occidental Chemical Corporation or OxyChem) bought a site in Niagara Falls, New York, from one William Love after determining it was a good location for a chemical waste landfill.

In 1892, Love had began the work of the development of an ambitious housing project by connecting the upper and lower Niagara Rivers via a seven-mile canal (Logan 2012). In 1910, construction of the Love Canal was abandoned because of an economic downturn, and by 1920, the canal had been transformed into a municipal and industrial chemical dumpsite. At the time, landfills were "an environmentally acceptable method of hazardous waste disposal" (Beck 1979). From between 1920 and 1978, the area fostered a number of facilities and companies that generated hazardous waste, including facilities of the local and federal government and Hooker Chemical Company. After taking over the land, Hooker Chemical Company took protective measures to prevent the release of chemicals in the dumpsite, including digging deeper holes for ultra-hazardous materials and lining the canal with thick clay. The company used the canal until 1953, by which time it had dumped 21,000 tons of chemicals there. It eventually sold the property to the Niagara Falls City School District for $1 after the school district hinted at the possibility of eminent domain proceedings despite being warned that the site was a chemical waste dump. The school district was apparently strapped for cash and needed the space in the face of the increasing population of the city. In 1978, the Love Canal ruptured, resulting in one of the biggest environmental disasters in U.S. history. Unfortunately, the damage affected many families living in the area, not only causing many to relocate, but also causing serious physical ailments (Warren 1980). Subsequently,

> [a] controversy [arose] over who—Hooker or the Niagara Falls Board of Education—was legally responsible for the exposure of the public to the chemical wastes and the resulting illnesses experienced by families who resided in the area. Despite the restrictions and risk stipulations in the deed conveyed to the Board of Education by Hooker, the Board decided to develop the land above the fill site and its surrounding area. The Board authorized construction of an elementary school on the site in 1955, the city constructed a sewer line through the canal in 1960, and developers constructed homes and streets next to the site. The toxic chemicals stored there eventually seeped from their ruptured and deteriorated containers into the soil, basements, and storm sewers. [...] By August 1978, the New York health commissioner declared a state of emergency in the area, and 239 families were evacuated. [...] State and federal investigations into the conditions at the landfill identified 248 different chemicals and 82 chemical compounds, 11 of which were known carcinogens. These toxins included benzene, toluene, chloroform, carbon tetrachloride, lindane, and trichlorophenol that was contaminated with the carcinogen dioxin. Residents exposed to these and other chemicals reported miscarriages, birth defects, cancer, and asthmatic, urinary, and convulsive disorders.[1]

[1] *Encyclopedia.com*, s.v. "Love Canal," http://www.encyclopedia.com/places/united-states-and-canada/us-political-geography/love-canal#1G23045301389.

Most residents decided to leave, while some stayed with the hope the area would eventually redevelop (Usiak 1982). Newspapers at the time began referring to the Love Canal event as a disaster—"Military Dumped Radioactive Wastes, Gas at the Love Canal, Investigators Say"[2]—and federal and state officials became immensely concerned. President Carter issued a declaration of emergency and the federal and state government scrambled to blame Hooker Chemical Company for the damage (Usiak 1982). A state task force tried to place blame for some of the pollution on the U.S. military (Usiak 1982). Meanwhile, many interested parties investigated the safety of the area; the Environmental Protection Agency (EPA) suggested that only the core was dangerous, a contention rejected by the Congressional Office of Technology Assessment (Kevles 1983). A *Los Angeles Times* article stated that it would be safer for humans to live in space than near the Love Canal; a series of tests had been run on rats which revealed large amounts of pesticides in the fatty tissue, among other things. Love Canal was considered by some to be the greatest environmental disaster caused by humans.[3]

The residents living on the canal became citizen-activists, pushing for solutions by lobbying on Capitol Hill and broadcasting their anger (Warren 1980). On one occasion, residents from the Love Canal detained two EPA representatives for six hours while demanding that the federal government relocate the 900 families living in the area (The Gilder Lehrman Institute of American History 2014). President Carter's declaration of Love Canal as a national emergency was the first time a president declared a non-natural federal emergency (Finding Dulcinea 2011). He proposed the creation of a $1.6-billion fund, financed mostly by fees on oil and chemicals, to clean up hazardous waste dump sites in the United States such as the Love Canal. Some literature suggests that this citizen-activism is what transformed the hazardous waste issue into a nationwide public environmental problem (Brook 1994). The EPA ultimately cleaned up twenty-one tons of toxic chemicals on the sixteen-acre site (DeCarvalho 2013).

The effects of the Love Canal prompted investigations into other sources of toxic waste. Among the most notorious was the Valley of the Drums.

2.2 *Valley of the Drums*

The Valley of the Drums drew national attention in 1979 as one of the country's worst abandoned hazardous waste sites. Drums accumulated during a ten-year period were strewn in pits and trenches over a twenty-three-acre site in Bullitt County, Kentucky. The drums were deteriorating and leaked into Wilson Creek,

---

[2]"Military Dumped Radioactive Wastes, Gas at Love Canal Investigators Say," *L.A. Times*, May 3, 1980.

[3]"Gli 8 disastri ambientali più gravi al mondo causati dall'uomo," http://www.ecoblog.it/post/9169/gli-8-disastri-ambientali-piu-gravi-al-mondo-causati-dalluomo.

a tributary of the Ohio River. The waste chemicals included benzene, toluene and methyl methacrylate.

Around 832 drums were found floating down the Ohio River after being swept away from an industrial site. In addition, thousands of others were scattered from four sites where the drums were held. The fourth site, called the Valley of the Drums, held between 20,000 and 100,000 additional drums on its surface. One of the alleged dumpers, A.I. Taylor, had been paid a fee to haul the drums to a dump about five miles from his valley, plus a fee he was supposed to give to the operator of the disposal site. Taylor pocketed all the money and dumped the drums in his valley. Taylor also spilled the contents of some of the drums onto the ground in order to resell the drums. He died the night before his trial (Bukro 1980). John White administrator of EPA's Region IV office in Atlanta said at the time that he was certain the groundwater was contaminated.

The costs of the cleanup were a major concern, and the Region IV office spent $100,000 for scattered drum cleanup and transportation. A lack of funds prevented total cleanup (EPA 1981). Officials wanted to collect some of the cleanup costs from those who caused the contamination under existing statutes, but no one could be held liable.

Cases like these raised concern, and often outrage, in the United States. It is difficult to communicate now the fear, distrust, and desperation that were linked to cases like Love Canal and Valley of the Drums. Also, these disasters were discovered at a time when the chemical industry as a whole was being criticized—including through the widely read *Silent Spring* by Rachel Carson, but also in many more sensational attacks. Families were facing clusters of cancer that affected their children and the prospect of having their homes and entire neighborhoods razed. Someone, it was widely felt, had to be held responsible for these terrible results of industrial society (DiMento 1986).

Of course, this was not the official response of all policymakers (although it was of many) but public opinion provided the background for the creation of new law.

> "Industry had shown laxity, not infrequently to the point of criminal negligence, in soiling the land adulterating the waters with its toxins". So says a recent report from a House investigative subcommittee.[4] The report adds that as deposits of dangerous industrial wastes proliferate, the authority charged with eliminating the hazards, the Environmental Protection Agency, has done little to search out such sites and compel offending companies to clean them up. Meanwhile, as much as 35 million tons of toxic waste continues to be improperly disposed of every year, and, charges one environmental watchdog group, another year and a half may pass before the EPA puts proper regulations into force. There may be as many as 34,000 seriously hazardous waste dumps spotted about the country (Brown 1979).

[4]*How to Solve the Nation's Infrastructure Problem: Hearings Before the Comm. on Pub. Works and Transp.*, 99th Cong. 217 (1986).

3 Responses to Disasters

What some consider the third wave of environmentalism began in the United States as a result of tragedies such as Love Canal (Brook 1994). The government's role was first seen as being to study hazardous waste rather than regulate solid waste.[5]

Historical pollution strains the law and legal systems. It often engenders strong and passionate views of what is *fair*, what should be done, and who should be

[5]A report was called for in the 1970 Resource Recovery Act, though this later became the basis for the Resource and Conservation Recovery Act (RCRA) of 1976. In this note, we briefly summarize the statute. Among other lessons it points out that the American approach employs several different strategies—not only based on criminal law. The RCRA governs the treatment, storage, and disposal of solid and hazardous waste. The purpose of the RCRA is to address and minimize the present, recent, and future threats to human health and the environment. Unlike CERCLA, the goal of the RCRA is not to promote the cleanup of toxic waste sites or allocate those costs. EPA is responsible for ensuring that institutions are in compliance with the regulations promulgated under RCRA. Pursuant to this role, the EPA can bring administrative actions, civil suits, and criminal enforcement suits against persons in noncompliance with these regulations. Of the three routes for recovery, the EPA most often brings administrative actions against persons in noncompliance.

Administrative actions can range from informal written notifications of a violation to formal unilateral and consent orders under various sections of the RCRA, including sections 3008, 7003, 9006, and 3013. Usually a formal action is taken only where there is a significant violation or if an entity does not comply with an informal notification. Of the various provisions of the RCRA, Section 3008 is invoked most frequently in enforcement actions. The Section 3008 administrative order is used to: (1) impose a civil penalty for current RCRA violations; (2) impose penalties for past violations; or (3) mandate compliance with RCRA requirements.

The EPA can also bring civil enforcement actions, including actions under Section 7003, which authorizes the EPA to stop "an imminent and substantial endangerment to health or the environment" by restraining certain activities or taking necessary actions when one's actions "may present an imminent and substantial endangerment to health or the environment."

The EPA can pursue criminal penalties for persons responsible for knowing and serious violations. The following are situations where a criminal penalty might be pursued, if at all, where a person:

> (1) knowingly transport[s] … RCRA waste to a facility that has no active RCRA permit; (2) knowingly "treats, stores or disposes" of … RCRA waste either with no permit, or in knowing violation of a "material condition or requirement" of a permit or of an interim status rule; (3) knowingly makes a fraudulent statement or omits material information in a permit, waste manifest, report, or other document regarding RCRA compliance; (4) knowingly "generates, stores, treats, transports, disposes of, exports, or otherwise handles" hazardous waste; (5) knowingly transports without a manifest any hazardous waste; (6) knowingly violates export requirements; [and] (7) knowingly handles used oil that is a hazardous waste without full compliance.

Finally, the RCRA authorizes citizen suits forcing parties responsible for RCRA violations to "take action" by cleaning up and disposing of toxic waste. One can also seek a prohibitory injunction under RCRA to restrain a responsible party from further violating the RCRA. Broun and O'Reilly (n.d.), *RCRA and Superfund: A Practice Guide*, 3d § 5:19 (2015 and October 2016 update).

responsible for mitigation, cleanup, damages, and compensation for the injured. Our focus here, however, is on major *legal* questions.

Which characteristics of a legal system allow for the production of law that effectively addresses historical pollution? The challenge is to find solutions and resolutions that, while fair (as that term is used in the vernacular), are consistent with the controlling law. Historical pollution often invites responses that may counter the fundamental principles of a legal system: ***nullum crimen sine lege*** (no crime without law), non-retroactivity, no bill of attainder, no *ex post facto* laws, advance notice of obligations, *mens rea*, causation, presumption of innocence.

3.1 Preliminary Broad Summary of the United States Legal Strategy

Because the United States is a common law country that is also a federal system, describing its understanding of legal responses to historical pollution is a massive undertaking. Historical pollution is addressed by statutory federal law, by federal state cooperative law, and by the individual law of the fifty states when not pre-empted (that is, when the federal government has not closed the door either explicitly or implicitly to state controls). In the area of historical pollution, the federal government leaves considerable room for states to undertake more stringent controls than those of the federal government (Environmental Law Institute 2015). The common law also provides a strong arsenal of strategies for addressing some types of historical pollution.

3.2 How the Problem Was Seen by Americans: Opinion Polls

To understand U.S. willingness to address historical pollution with a broad arsenal of law it is helpful to describe the extent of concern—and indeed outrage—over pollution like that of the Love Canal and the Valley of the Drums. Below we summarize several findings from public opinion polls from the period that tapped into American citizens' concerns over specific historical pollution cases and the general attitudes on responsibility.

Have you heard of the Love Canal incident, or not?[6]

63% Yes
29% No
8% Not sure.

[6]Methodology: Conducted by Cambridge Reports/Research International during July 1981 and based on 1500 personal interviews. Sample: National adult. Sample size is approximate [USCAMREP.81JUL.R216].

As you know, residents near the Love Canal, in the Niagara Falls, New York, area, were reported to have stillbirths, cancer, deformed children, and chromosome damage as a result of the dumping of hazardous chemical wastes. The people who live near Love Canal want to move out and are suing the chemical company there and the federal government for $3 billion for damages done to them. How serious a problem do you think the dumping of toxic chemicals is in the country today—very serious, only somewhat serious, or hardly serious at all?[7]

76% Very serious
20% Only somewhat serious
2% Hardly serious at all
2% Not sure.

If you had to pick one, who do you think is most to blame for conditions such as those found at Love Canal—companies which dump their toxic wastes near where people live, agencies of the federal government which have also dumped hazardous wastes in such places, the federal government for not enforcing safety standards on toxic chemicals, real estate people who have built housing near places where such dumping has taken place, or local and state government for not being alert to hazardous waste dangers?[8]

29% Federal government for not enforcing safety standards
23% Companies which dump toxic wastes
22% Local and state government
5% Real estate people
4% Agencies of the federal government which have dumped hazardous wastes
13% All (vol.)
* None (vol.)
4% Not sure
* Less than 0.5%.

Would you favor or oppose […]. The federal government investigating places like Love Canal, where chemical waste has been dumped, even though this might lead to well over $10 billion in suits against the federal government and the chemical companies?[9]

83% Favor
12% Oppose
5% Not sure.

[7]Methodology: Conducted by ABC News/Louis Harris and Associates, 5–9 June 1980 and based on 1493 telephone interviews. Sample: Likely voters [USABCHS.070780.R3D].

[8]Methodology: Conducted by ABC News/Louis Harris and Associates, 5–9 June 1980 and based on 1493 telephone interviews. Sample: Likely voters [USABCHS.070780.R2].

[9]Methodology: Conducted by ABC News/Louis Harris and Associates, 5–9 June 1980 and based on 1493 telephone interviews. Sample: Likely voters [USABCHS.070780.R2].

In cases like Love Canal, who should be responsible for any damages to the community and to human health caused by the disposal of poisonous chemical wastes—the community itself, the company responsible for safely disposing of these wastes, the company that produces the wastes, or the government?[10]

3% The community
29% The company responsible for safely disposing of wastes
23% The company that produces these wastes
7% The government
28% Combination (vol.)
1% Other (vol.)
9% Don't know

(Source: Cambridge Reports National Omnibus Survey, July 1980).

4 The Federal Response

4.1 The Political Background of Federal Legislation

The major federal legal response was the passage of the Comprehensive Environmental Response Compensation and Liability Act (CERCLA), commonly referred to as Superfund. Establishing the Superfund program was a calculated political decision by President Jimmy Carter's administration and the EPA, which took advantage of the election year opportunity to exploit volatile public concern at the expense of large industries. Legislators were motivated by the existence of superfund sites in nearly every congressional district. Also, the orphan sites issue was so significant to voters that any opposition to new aggressive federal action could be politically very risky.

4.2 Legal Challenges to the Innovative New Law

Nonetheless, lawmakers confronted doctrines of U.S. law that could be used to vitiate innovative law (Hird 1993). Most fundamentally, the U.S. Constitution, under Article 1, Section 9, states: "No Bill of Attainder or ex post facto law shall be passed." Furthermore, constitutionally required due process includes fair advance notice of laws before sanctions can be imposed for disobedience of the legal

[10]Methodology: Conducted by Cambridge Reports/Research International during July 1980 and based on 1500 personal interviews. Sample: National adult. Sample size is approximate [USCAMREP.80JUL.R182].

command. There were also general concerns with fairness, and the need to hold the right people accountable for harms occurring in the past.

Challenges facing the law included how to impose retroactive liability on a broad group of actors deemed responsible, using a system that secures prompt environmental cleanups of releases of hazardous substances into the environment; how to impose a "polluter-pays principle"—re-internalizing the cost of environmental harm by holding to account those whose actions caused the harm; and how to establish remedies for past releases of harmful substances with effects that are ongoing.

President Carter had stated during his 1981 State of the Union address that improper waste disposal was "one of the most pressing problems" to come to light during his presidency, a view included in his forward-thinking environmental agenda. He described three of his initiatives aimed at preventing another Love Canal and stressed the importance of maintaining the U.S. commitment to preserving and enhancing the quality of the environment (Carter 1981). Later, his successor, Ronald Reagan, in signing H.R. 2005 (Superfund Amendments and Reauthorization Act of 1986) said, "[m]y overriding concern has been the continuation of our progress to cleanup hazardous waste sites that endanger the health and safety of our citizens. All Americans can expect no less from their government."

4.3 The Comprehensive Environmental Response, Compensation, and Liability Act (Superfund)

In this section we describe the several strategies available for addressing historical pollution. Standing in importance above them all is Superfund. This most innovative U.S. initiative has also been very controversial. This section also covers more briefly general criminal law and U.S. tort law, tools still available to clean up and compensate for historical pollution. We have chosen to focus on CERCLA, along with the Superfund Amendments and Reauthorization Act (SARA), because they cover a large percentage of the historical pollution concerns, and the United States has a long and deep experience with the subject and the response to it.

4.3.1 CERCLA's Main Provisions

The statute creates a multibillion dollar Superfund to pay for cleanup of releases (42 U.S.C.A. § 9611). It requires the parties who generated and handled hazardous waste to pay for cleanup costs (42 U.S.C.A. §§ 9607, 9613). CERCLA allows for judicial and administrative enforcement actions against responsible parties for remediation at sites (42 U.S.C.A. §§ 9607, 9606).

The statute provides both criminal and civil penalties (42 U.S.C.A. § 9607 (b), (c) and (d)(2)) and also provides Natural Resource Damages (NRDs): CERCLA defines "natural resources" as "land, fish, wildlife, biota, air, water, ground water, drinking water supplies, and other such resources belonging to, managed by, held in trust by, appertaining to, or otherwise controlled by the United States, any State or local government, any foreign government, any Indian tribe, or, if such resources are subject to a trust restriction on alienation, any member of an Indian tribe" (42 U.S.C. § 9601).[11] Citizens are also authorized to commence actions.

Potentially responsible parties (PRPs) are subject to liability: these are (1) the current owner or operator of a site; (2) the owner or operator of a site at the time of hazardous material disposal; (3) the arranger for the disposal or treatment of a hazardous material; or (4) the entity that chose a disposal site and transported hazardous material there. Both individuals and corporations can be liable. Scenario 2 (Sect. 6 below) provides examples of the potential liability of PRPs (42 U.S.C.A. § 9606).

The law creates a list of national priorities in an attempt to ensure that sites of the greatest risk to human health are remediated first (Plater et al. 2010).

4.3.2 Liability and Defenses Under CERCLA

PRPs are held liable without regard to fault for damages, which include government cleanup costs, removal, and restoration of resources. When polluters are not found or are unwilling to pay, damages can be filed against the fund.

Payment of claims by the fund subrogates the fund to the rights of the claimant.

The standard of strict liability is not explicitly found in the text of the statute, but Section 101 of CERCLA provides that the standard of liability under CERCLA shall be construed as the standard of liability under Section 311 of the Federal Water Pollution Act. Courts have read Section 311 to impose strict liability on responsible parties.

Courts have concluded that CERCLA permits but does not require joint and several liability. Some courts have held that joint and several liability should always be imposed unless defendants can show that a reasonable basis exists for apportioning the harm against them: see *United States v. N.E. Pharm & Chem Co.*, 579 F. Supp. 823, 844 (W.D. Mo. 1984). Also regarding divisibility of harm and apportionment in *Burlington N. & Santa Fe Railway v. United States*, 556 U.S. 599 (2009), the Supreme Court held that in spite of the fact that a contamination event may cause a single harm, such harm may still be divisible if the defendant demonstrates a rational basis for apportionment. CERCLA defendants bear the burden of proving that a rational basis for apportionment exists.

CERCLA liability is premised on governmental response pursuant to Section 104 and the National Contingency Plan, which limits liability.

[11]Provisions denoting Type A and Type B NRD regulations are found at 43 C.F.R. § 11.13(d)–(e) (2006).

"Act of a third party defense": When a hazardous release or threatened release is caused solely by the act of a third party with whom a defendant has little or no relation, this can serve as a defense. The related defense of innocent landowner is described in Scenario 2. Derivative liability and piercing of the corporate veil have also been addressed by the U.S. Supreme Court, in *United States v. Bestfoods*, 524 U.S. 51 (1998). The issue was whether a parent corporation that actively participated in, and exercised control over, the operations of a subsidiary may, without more, be held liable as an operator of a polluting facility owned or operated by the subsidiary. The Court held that a parent corporation can be charged with derivative CERCLA liability for its subsidiary's actions only when the corporate veil is pierced. But a corporate parent that actively participated in, and exercised control over, the operations of the facility itself may be held directly liable in its own right as an operator of the facility.

The Court reasoned that the question is not whether the parent operates the subsidiary, but rather whether it operates the facility, and that operation is evidenced by participation in the activities of the facility, not the subsidiary. "Control of the subsidiary, if extensive enough, gives rise to indirect liability under piercing doctrine, not direct liability under the statutory language" (*United States v. Bestfoods*, 524 U.S. 51, 68 (1998)).

4.3.3 The Criminal Provisions of CERCLA

The government can seek criminal penalties for violations of CERCLA, including substantial fines and imprisonment. Criminal enforcement is sought in cases involving the most serious environmental violations or those with egregious negligence or conduct involving intentional, willful, or knowing disregard for the law. Corporate or other high-level officials are often targets of this kind of criminal enforcement.

Congress initially authorized EPA's criminal enforcement program in 1982 and expanded the program with the 1990 Pollution Prosecution Act (Pub. L. No. 101–593, 104 Stat. 2954 (1990); 42 U.S.C.A. § 4321 (1990)).

One who knowingly divulges information designated as confidential may be fined, imprisoned, or both (42 U.S.C. § 9604(e)).

42 U.S.C. § 9603 provides criminal liabilities where these elements are proved: (1) knowledge of waste disposal; (2) a release into the environment of an amount of hazardous substance in excess of the reportable quantity limit of § 102 without a federal permit; (3) the person charged is actively "in charge" of the facility at which the release occurred; and (4) this person knowingly failed to report the release after learning of it. Section 9603(a) provides that a person in charge of a facility or vessel who knowingly fails to notify the appropriate agency of the U.S. Government, immediately upon becoming aware of such release shall be fined or imprisoned (for specified periods) upon conviction. CERCLA requires the submission of reports to the EPA about the existence of hazardous waste sites and the occurrence of a hazardous substance spill. Both the individual who submitted the report and the

company can be subject to criminal or civil penalties for failure to report, pursuant to 9603(b).

Targets can include low-level employees of a corporation who could have reported an incident. For example, in *United States v. Carr*, 880 F.2d 1550 (2d Cir. 1989), the Court found that a maintenance foreman was "in charge" of a facility where he had *any* supervisory authority over the vehicle from which or the area in which the dumping occurred, regardless of his control of the facility. Because a person liable under 9603(c) is required to maintain records setting forth identities, characteristics, and quantities of wastes deposited at the site for fifty years, criminal penalties can be imposed where one knowingly destroys or erases these records, pursuant to 42 U.S.C. § 9603(d). Failure to inform and false statement liability are noted in Scenario 2.

In CERCLA criminal prosecutions, defendants have a right to a jury trial. Punishments for failing to report releases under CERCLA are based on the U.S. Sentencing Guidelines.

Criminal fines are available against a corporation and the parent company if the corporation cannot pay. Corporations may be held liable under the vicarious liability doctrine. They can sometimes avoid liability where they have strong ethics self-reporting and training programs.

Forty-two U.S.C. § 9607 provides for criminal penalties for failing to notify EPA of the existence of hazardous waste treatment, storage, and disposal facilities. Further, any person who knowingly fails to provide the required notice shall not be entitled to any limitation of liability or to any defenses to liability set out in Section 9607.

Forty-two U.S.C. § 9612 provides for criminal enforcement where parties knowingly give false information as part of a claim for reimbursement from the Superfund. Upon conviction, the party will be fined in accordance with the terms of Title 18, or imprisonment for not more than three years (or five years if a subsequent conviction), or both.

4.3.4 EPA Administrative Order Process

CERCLA Section 106 allows the EPA to bring administrative or judicial enforcement actions against responsible parties to force them to perform remediation when it determines that a site may present an "imminent and substantial endangerment to public health or … the environment." EPA can issue "such orders as may be necessary to protect public health and welfare and the environment." The EPA provides a summary of civil cases and settlements under CERCLA. For the period 2007 through 2015, a large number of cases were settled by consent decree or administrative order on consent.[12] These options are described in Scenario 3.

[12]See Web site of the United States Environmental Protection Agency, https://cfpub.epa.gov/enforcement/cases/.

4.3.5 The Cleanup Process: Removal, Remediation

U.S. law authorizes short-term removals, where actions may be taken to address releases or expected releases in a prompt fashion. It also authorizes long-term removals that address the dangers associated with the release or threat of release of hazardous substances that are serious, but not immediately life-threatening.

Several factors are used in determining first steps (removal, remediation) and follow-up steps in the cleanup process. After EPA independently discovers a site, or a site contaminated by possible releases of hazardous substances has been reported, the site is entered into the Comprehensive Environmental Response, Compensation, and Liability Information System (CERCLIS), EPA's computerized inventory of potential hazardous substance release sites. EPA then follows these steps to determine the potential for a release of hazardous substances from the site: (1) preliminary assessment/site inspection; (2) National Priorities List (NPL) site listing process; (3) remedial investigation/feasibility study; (4) record of decision; (5) remedial design/remedial action; (6) construction completion; (7) post-construction completion; (8) NPL deletion; (9) site reuse/redevelopment.

During the preliminary assessment, an investigation determines whether a site poses a threat that requires investigation. The preliminary assessment also assesses possible response actions.[13]

If there is a recommendation for further action, a site inspection is performed to identify which sites enter the NPL; this provides data needed for the Hazard Ranking System (HRS). Section 105 of CERCLA required EPA to establish criteria for determining priorities among releases or threatened release of hazardous substances for the purpose of taking remedial action. To meet this requirement, EPA developed the Hazard Ranking System (47 FR 31180, July 16, 1982) to evaluate sites for possible inclusion on the NPL. The NPL includes those sites that appear to pose the most serious threats to public health or the environment. Sites on the NPL are eligible for Superfund-financed remedial action.

SARA required EPA to revise the HRS to more accurately "assess the relative degree of risk to human health and the environment posed by sites." SARA also required the HRS to take into account recreational use of surface waters, contamination of the human food chain and drinking water supplies, and potential contamination of ambient air.

During a site inspection, investigators collect waste and environmental samples to determine the substances present at the site and ascertain whether they are being released into the environment. This is done to identify whether an emergency response is necessary and which sites qualify for the NPL. At the end of the preliminary assessment and site inspection, the EPA creates a site score to determine whether further investigation is necessary or whether the site should receive a "no further action planned" recommendation (NFRAP). A site that receives an

[13]In general, see the Web site of the United States Environmental Protection Agency, https://nepis.epa.gov.

NFRAP recommendation can be re-examined later. After a two-stage site inspection is performed, the HRS score is compiled.

Remedial action financed by Superfund can only occur after a site is placed on the NPL (notice the distinction between remedial and removal actions—EPA has the authority to use the Superfund for cleanup pursuant to a removal action, as long as it meets certain requirements). EPA has listed the worst sites on the NPL; sites not on the NPL tend to pose fewer public health hazards and show problems of a more localized nature than those for example of Love Canal and Times Beach; they do not awaken the potential for public emotional reaction (Broun and O'Reilly 2015).

Sites are placed on the NPL using three mechanisms pursuant to Section 300.425(c) of the NCP. The first mechanism is EPA's HRS. An HRS score of 28.50 makes a site eligible for proposal to the NPL. The second mechanism allows States or Territories to designate one top-priority site, regardless of score. The third mechanism allows a site to be listed if it meets all three of these requirements: (1) the Agency for Toxic Substances and Disease Registry (ATSDR) of the U.S. Public Health Service has issued a health advisory that recommends removing people from the site; (2) the EPA determines the site poses a significant threat to public health; and (3) the EPA anticipates it will be more cost-effective to use its remedial authority (available only at NPL sites) than to use its emergency removal authority to respond to the site.

After a site is placed on the NPL, a remedial investigation/feasibility study is performed at the site. This process includes these phases: Scoping, Site Characterization, Development and Screening of Alternatives, Treatability Investigations, and Detailed Analysis.[14]

The Record of Decision is a public document that explains which cleanup alternatives will be used to clean up a Superfund site. Remedial Design (RD) is the phase in Superfund site cleanup where the technical specifications for cleanup remedies and technologies are designed (EPA).[15]

The Act and its judicial interpretations have addressed how PRPs may recover cleanup costs. CERCLA does provide potentially responsible parties with a cause of action to recover from other PRPs costs incurred in voluntarily cleaning up a contaminated site (*United States v. Atlantic Research Corp.*, 127 S. Ct. 2231 (2007)). However, a private party who has not been sued in a CERCLA administrative or

[14]"Superfund Remedial Investigation/Feasibility Study (Site Characterization)," *U.S. Environmental Protection Agency*, https://www.epa.gov/superfund/superfund-remedial-investigationfeasibility-study-site-characterization.

[15]"Superfund: Remedial Design/Remedial Action," *U.S. Environmental Protection Agency*, https://www.epa.gov/superfund/superfund-remedial-design-remedial-action.

The final assurance rules have not been everywhere forthcoming, slowing down the incentives for operators to clean up sites. See *In re Idaho Conservation League*, 811 F.3d 502 (2016). In general incentives to comply with CERCLA have also been influenced by complex insurance coverage rules and opinions. See, for example, *Ash Grove Cement Co. v. Liberty Mut. Inc. Co*, 2016 WL 2731656, 9th Cir. May 11, 2016.

cost-recovery action could not obtain contributions from other liable parties pursuant to §113. The statute specifies that a party may obtain contribution "during or following any civil action under CERCLA §106 or §107 (*Cooper Industries, Inc. v. Aviall Services, Inc.*, 125 S. Ct. (2004); Aviall was contractually responsible for cleanup costs)."

In *United States v. Atlantic Research Corp.*, 551 U.S. 128 (2007) the United States Supreme Court resolved a disagreement among the circuit courts of appeal regarding cost recovery under CERCLA. Christoper Rycewicz provides a detailed summary of this decision:

> The statute provides a number of routes by which parties that have incurred cleanup costs may seek to share the burden of remediation with other liable parties. The ability to spread costs among other PRPs encourages parties to actively and voluntarily address contaminated sites, and ensures that other polluters pay their fair share of investigative and cleanup costs.
>
> In 2004, however, the Supreme Court cast doubt on this cost-sharing system in cases in which PRPs voluntarily remediated contaminated sites without having first been sued by the federal government. In *Cooper Industries, Inc. v. Aviall Services, Inc.*, 543 U.S. 157 (2004), [...], the Supreme Court held that because of the plain language in Section 113 of CERCLA, in order to avail itself of the "contribution" under Secton 113 and force other liable parties to share cleanup costs, a PRP must first have been sued by the federal government. This decision left many PRPs that had voluntarily incurred substantial costs cleaning up sites without access to statutory contribution, [...]. The decision [...] meant that many parties that had already voluntarily agreed to remediate sites would be stuck bearing the entire cost of cleanup. In addition, it meant that going forward no PRP would ever remediate a contaminated site unless first sued by the federal government. Such a result would be inequitable and undermine any incentive for voluntary cleanup.
>
> The *Atlantic Research* opinion restored the polluter-pays principle to CERCLA. The Court did not reverse its opinion in *Cooper*; rather, it clarified that PRPs voluntarily cleaning up contaminated sites have an alternative route to forcing other liable parties to bear their fair share of costs. Rather than seeking contribution under Section 113, a party that has voluntarily incurred response costs can seek "cost recovery" under Section 107 from other liable parties [...] as the Court explained, "a PRP that pays money to satisfy a settlement agreement or a court judgment [resulting from a Section 107 claim] may pursue § 113(f) contribution". The Court noted that "[r]esolution of a § 113(f) counter-claim would necessitate the equitable apportionment of costs among the liable parties, including the PRP that filed the § 107(a) action". As a result, the courts are now able to distribute liability among all the PRPs.[16]
>
> The Court's opinion has also removed the monopoly that the federal government sought to impose on CERCLA cases.

4.3.6 Status of the Superfund

As of November 2010, there were 1280 sites listed on the NPL; an additional 347 have been delisted, and 62 new sites have been proposed. Approximately 70% of

[16]Christoper Rycewicz, "United States Supreme Court Restores 'Polluter Pays' Principles to Superfund Law," http://www.millernash.com/united-states-supreme-court-restores-polluter-pays-principles-to-superfund-law-06-26-2007/, accessed July 1, 2015.

Superfund cleanup activities have been paid for by parties responsible (PRPs) for the cleanup of contamination. The only time cleanup costs are not borne by the responsible party is when that party either cannot be found or is unable to pay. For sites where the party was not found, the Superfund program used the trust fund to pay for cleanup. Taxes on crude oil and certain chemicals, together with an environmental tax on corporations based on taxable income, were initially used for cleanup until the authority for these taxes expired in 1995, and this income source was eventually exhausted by 2003. Since then, the trust fund continues to receive revenues in the form of cost recoveries, interest on the fund balance, fines and penalties, and general revenue fund appropriations that supplement the trust fund balance (GAO 2008). CERCLA pre-empts any special state tax fund intended to pay for the same type of expenses that would be paid by Superfund, but use of state funds to compensate, for example, equipment costs or to conduct research is beyond CERCLA's scope and is not pre-empted (*Exxon Corp. v. Hunt*, 475 U.S. 355 (1986)).

4.3.7 An Historical Assessment

A thirty-year retrospective on Superfund, one of the major parts of U.S. historical pollution law, provides a good overall orientation. As EPA summarized:

> It's easy to forget that there was a time in the United States when EPA lacked the legal authority to clean up hazardous waste sites like Love Canal, New York. […] Even though the EPA had been established for ten years, it was not until December 11, 1980, that President Jimmy Carter signed into law the Comprehensive Environmental Response, Compensation and Liability Act (CERCLA or Superfund). This historic new statute gave EPA the authority to clean up uncontrolled hazardous waste sites and spills […]
>
> The Superfund law authorizes the Agency and its partners to address abandoned, accidentally spilled, and illegally dumped hazardous wastes that pose current or future threats to human health or the environment […] sites where EPA has […] authority to remediate sites where the hazardous waste release did not occur through a sudden tragedy like the Columbia shuttle disaster or through natural causes like hurricanes, but, rather, through years of poor and sometimes illegal waste management practices. Some of these sites can involve hundreds of chemicals with tons of contaminated waste spanning hundreds of acres; often the contamination affects groundwater in addition to soil.[17]

5 Other Law Relevant to Historical Pollution

In addition to the variety of criminal, civil, and administrative actions under federal and state statutory law, in the U.S. legal system other options are available to address the problems associated with historical pollution.

[17]"Superfund 35th Anniversary," Web site of the United States Environmental Protection Agency, https://www.epa.gov/superfund/superfund-35th-anniversary.

5.1 *Tort Law and Liability*

Independent of the federal regime and state criminal law, civil law continues to apply to historical pollution. There is a vast body of toxic tort law (Craig 2010). Some see tort law as a public regulatory tool and view public regulation and tort law as complements, not substitutes (Abelkop 2013). Here we highlight some tort strategies from the Restatement (Second) of Torts.

Pollution of Waters. The pollution of water refers to the alteration of the physical, chemical, or biological qualities of water so as to make it harmful to domestic, commercial, industrial, agriculture, recreational, or other beneficial uses of water or land, or detrimental to the public health, safety, and welfare, or to livestock, wild animals, birds, fish, or other aquatic life. Pollution may result from contamination by the addition of industrial, chemical, human, or animal waste, refuse, or other noxious substances.

A case note in the Restatement refers to *Anderson v. W.R. Grace & Co.*, 628 F. Supp. 1219, the basis for the phenomenally successful book and movie that has become a teaching tool in many courses in U.S. law schools, *A Civil Action*. In that action, plaintiffs "who contracted leukemia and other illnesses, as well as family members of children who died of leukemia and other illnesses, sued the company that allegedly contaminated plaintiffs' well water by chemical dumping."[18]

Invasions of Interests in Land Other than by Trespass. "An interference with the use of water caused by an act or conduct that is not itself a use of water but that affects the quality or quantity of water may subject the actor to liability if the act or conduct (a) constitutes a nuisance, (b) constitutes a trespass, or (c) is negligent, reckless or abnormally dangerous in respect to the use."[19]

More traditional tort law actions such as for battery are also at least theoretically available (i.e. O'Reilly 2015).

However, it is also the case that "traditional norms of tort liability are inadequate to obtain remedial cost remediation," and they have other limitations (Plater et al. 2010). Scenario 5 elaborates the availability of and challenges to successful use of tort law for historical pollution.

5.2 *Other Criminal Law and Liability*

It remains the case in the United States, as it did before modern environmental law, that the conventional criminal law can be used in pursuit of historical pollution. And it is being used. Crimes implicated include conspiracy, mail fraud, the giving of false statements, and others. For example, Title 18 of the U.S. code "makes it a

---

[18]Restatement (Second) Of Torts § 832, Case Citations—by Jusridiction at D.Mass (AM. LAW. INST. 1979).

[19]Restatement (Second) of Torts § 849 (AM. LAW. INST. 1979).

crime for any person to make false statements to the government, to conspire to defraud the government, to cheat the government out of money, or to conduct mail or wire fraud" (Brickey 2008). Prosecutions face the normal challenges in criminal law including burden of persuasion, statutes of limitations, and *mens rea*, including strict notions of scienter. Again, the law is federal and state, with some states creating specific environmental crimes statutes (Periconi 2008).

The Supreme Court has held in *CTS Corp. v. Waldburger*, 134 S.Ct. 2175 (2014) that although CERCLA, in §9658, preempted statutes of limitations, it did not preempt statutes of repose because statutes of limitations and statutes of repose serve different purposes, and there is other evidence that the statutory text supports the exclusion of statutes of repose from §9658. The case involved the application of a North Carolina statute of repose that prevents subjecting a defendant to a tort suit brought more than ten years after the last culpable act of a defendant. Property owners who owned land that a corporation had used to operate an electronics plant filed a state-law nuisance action against the corporation in 2011, after they learned in 2009 that their well water was contaminated, allegedly due to the presence of toxic chemicals the corporation had stored on the land. The district court found that the lawsuit was barred by a state statute. The Fourth Circuit found that the state statute was preempted by §9658 of CERCLA.

6 Comparative Research Based on Hypothetical Scenarios[20]

Scenario 1

Agent A owns land and contaminates it over a long period of time. No damages or threats for human health occur as consequences of the pollution.

Agent A is strictly liable for the contamination on its land even if no threats to human health or damages occur as a consequence of the pollution.[21] Thus, A is liable for cleanup. CERCLA §107 provides: "[any responsible party (PRP) associated with a 'facility'] from which there is a release, or a threatened release which causes the incurrence of response costs, of a hazardous substance, shall be liable for … [any] necessary costs of response incurred by any other person consistent with the national contingency plan." Release has been construed broadly to include the mere presence of hazardous substances in soil and groundwater. Therefore, A is liable for the contamination even if the pollution is minimal.

[20]In what follows, our analysis in places involves the assumption of various additional facts to elaborate the U.S. law. The factual patterns are simple and we have focused on assumptions of circumstances which best illuminate CERCLA and other sources of law.

[21]Section 101 of CERCLA says that the standard of liability under CERCLA shall be construed as the standard of liability under Section 311 of the Federal Water Pollution Act. Courts have inferred Section 311 to impose a strict liability on responsible parties.

Scenario 2

Agent A owns land and contaminates it. The land is subsequently purchased by Agent B. No damages or threats to human health occur as consequences of the pollution.

Assuming the government has assessed the site and placed it on the National Priorities List, both Agent A and Agent B might be responsible under Section 9607 of CERCLA. Absent an exception, Section 9607 of CERCLA holds liable: "(1) the owner and operator of a vessel or a facility, and (2) any person who at the time of disposal of any hazardous substance owned or operated any facility at which such hazardous substances were disposed of."[22] Thus, current owners are strictly liable for all cleanup costs associated with their property, while past owners are liable for cleanup only if disposal occurred during a period where they owned and operated the site, *see Bancamerica Commercial Corp. v. Trinity Industries, Inc.*, 900 F. Supp. 1427, 1462–63, 26 Envtl. L. Rep. 20757 (D. Kan. 1995), aff'd in part, rev'd in part and remanded, 100 F.3d 792, 43 Env't. Rep. Cas. (BNA) 2021, 27 Envtl. L. Rep. 20397 (10th Cir. 1996), opinion amended, 103 F.3d 80, 36 Fed. R. Serv. 3d 1065 (10th Cir. 1996). Furthermore, courts have held that the same parties designated as liable under §107 are liable under §106, which allows the federal government to obtain injunctive relief and to issue administrative cleanup orders, *see U.S. v. Hardage*, 750 F. Supp. 1444, 1457, 21 Envtl. L. Rep. 20714 (W.D. Okla. 1990).

Since there is some type of chemical contamination perpetrated by A, and B is held responsible for the cleanup costs, the burden is on B to prove that hazardous substances were disposed of at the site during the time A owned the facility. The Supreme Court in *Atlantic Research Corporation* held that a PRP has a cause of action to recover, from other PRPs, costs incurred in voluntarily cleaning up a contaminated site. Therefore, assuming B can prove A is liable for the contamination, he may be able to recover a portion of the costs from A. Furthermore even if B is not subject to a civil action, if B remediated a contaminated site pursuant to a consent order, B can bring a CERCLA cause of action to recover costs; see *W.R. Grace & Co.-Conn. v. Zotos Intern., Inc.*, 559 F.3d 85 (2d Cir. 2009).

B might be able to avoid liability using the defense of an innocent landowner. Pursuant to the SARA amendments to CERCLA, innocent landowners have a defense where they have "[undertaken] all appropriate inquiries" before purchasing the land, and "[have] no actual or constructive knowledge" of the hazardous substance. Therefore, if B exhausted all appropriate inquiries before buying the land from A, B might not be held liable.

Finally, A might be criminally liable for his or her actions. First, 18 U.S.C.A. §§ 1001 et seq. makes it a crime for a person to make false statements to the government, to conspire to defraud the government, to cheat the government out of money, or to conduct mail or wire fraud. Thus, if a person submits an intentionally false statement to the EPA regarding the existence of any hazardous waste sites or

---

[22]This is not an exhaustive list. There are other covered persons under Sect. 9607 of CERCLA.

spillage of a hazardous substance, he or she can be criminally liable. In this scenario, both the person who made the claim and the corporation involved can be responsible. Thus, if A or any of his or her workers filed a misstatement with the EPA, he or she may be responsible under the federal criminal code. Under this scenario, it is not stated whether A intentionally deceived the government. Since this was not indicated under the hypothetical facts, we assume that no false statements were made to the government, and A is not responsible.

Second, CERCLA imposes criminal sanctions on any person in charge of a facility from which a reportable quantity of a hazardous substance is released who fails to immediately notify the appropriate federal agency (42 USC § 9603(b)(3)). If A knew about the contamination and still failed to report the contamination to the proper officials, A might be criminally liable under 42 U.S.C. § 9603(b)(3) for failing to report the release of a hazardous substance into the environment. Third, CERCLA §103(c) provides penalties where an owner and operator of non-RCRA-permitted facilities at which hazardous materials were disposed, as well as transporters who hauled hazardous materials to disposal sites, fails to notify the EPA of the contamination in a timely manner (42 U.S.C. § 9603(c)). Finally, § 104(e) authorizes the EPA to require any person to furnish information pertaining to actual or threatened releases (42 U.S.C.A. § 9604(e)(7)(B)). Thus, A might be liable under all four of these criminal provisions assuming he or she was aware of the contamination or made misstatements about contamination on the site. If any of the contamination continued into B's ownership and B knew, then B might also be liable under these provisions.

Ultimately, both A and B are likely liable under the civil provisions of CERCLA, whereas A might also be liable under the criminal provisions.

Scenario 3

Agent A owns land and contaminates it. The polluted land is subsequently purchased by Agent B. In the meantime, land contamination results in contamination of ground waters, which are regularly used in agriculture.

Both Agent A and Agent B are likely liable under CERCLA. First, A and B may be liable under Section 106 of CERCLA. Under Sect. 106(a), the EPA can seek cleanups in a number of different ways, including through an administrative order, a lawsuit in federal district court, or other relief. The EPA can issue a UAO (Unilateral Administrative Order) pursuant to CERCLA §106 where someone might be exposed to groundwater contaminated by hazardous substances from a site that can cause imminent and substantial endangerment to his or her health; *U.S. v. E.I. du Pont de Nemours & Co., Inc.*, 341 F. Supp. 2d 215 (W.D. N.Y. 2004). Thus, A and B might be subject to a cleanup order. A or B may be subject to treble damages for failing to comply with an administrative order. If only B is served with the UAO, B may pursue reimbursement for the costs of cleanup from A, a PRP. A and B can also be liable in district court.

Second, A and B may be liable under Section 107, provided the government undertakes cleanup of a site and then later seeks recovery from A and B as PRPs. Section 107 of CERCLA imposes liability on owners or operators of a facility where a hazardous substance has been released. This includes previous owners and current owners. The government can initiate cleanup proceedings of sites against owners and operators under § 107. These parties can be responsible for "(A) all costs of removal or remedial action incurred by the United States Government or a State or an Indian tribe not inconsistent with the national contingency plan; (B) any other necessary costs of response incurred by any other person consistent with the national contingency plan; (C) damages for injury to, destruction of, or loss of natural resources, including the reasonable costs of assessing such injury, destruction, or loss resulting from such a release; and (D) the costs of any health assessment or health effects study carried out" under Section 42 U.S.C.A. § 9604(i).

For there to be prima facie liability under Section 107, the plaintiff must prove (1) that defendant is a "covered" or statutorily liable person who falls within one of the four classes of PRPs; (2) that a "release" or "threat of release" of a hazardous substance occurred; (3) that the release caused the plaintiff to incur "necessary costs of response" consistent with the NCP; and (4) that the hazardous substance emanated from a facility or vessel. Here, A and B are covered as owners of the facility. Even though A was a previous owner, A is liable because it owned the facility when the hazardous contamination occurred. The contamination of the groundwater is evidence of a release from A and B's facility. Therefore, both A and B can be liable under Section 107. Section 107 cost-recovery actions also may be initiated by private parties if such parties are "innocent." Therefore, if B is not responsible for the contamination, he can pursue recovery costs from A.

Ultimately, A and B are both likely liable under Sections 106 and 107.

Scenario 4

Agent A owns land and contaminates it. The polluted land is subsequently purchased by Agent B. In the meantime, land contamination results in contamination of ground waters, which are regularly used in human nutrition.

Scenario 4 will likely have the same outcome as Scenario 3.

Scenario 5

Agent A owns land and contaminates it. The polluted land is subsequently purchased by Agent B. There is evidence of a causal link between the pollution and the death or injury of a number of people.

Depending on the nature of both A's and B's behavior, other than liability under Superfund and tort law, it is possible that the agents could be prosecuted under criminal law, including that of negligent homicide or even more serious crimes (Humphreys 1990).

CERCLA does not provide a cause of action for persons without liability for the cleanup costs of the site containing the hazardous substances; *New Castle Cnty. v. Halliburton NOS Corp.*, 111 F.3d 1116 (3rd Cir. 1997). The facts here do not show that any of the victims has liability for the cleanup costs of hazardous materials. Therefore, no action can be brought under CERCLA. However, as for earlier scenarios there are several liability possibilities if assumptions are made, for example, about failures to report.

The victims could bring toxic tort suits against A for personal injuries caused or contributed to by exposure to any hazardous substance released into the environment.

To prevail in a toxic tort action, the plaintiff must show (1) that he or she was exposed to chemicals released by the defendants; (2) that those chemicals can cause the types of harm they suffered; and (3) that the chemicals did in fact cause the harm; *In re Paoli R.R. Yard PCB Litigation*, 35 717, 752 (3d Cir. 1994). Each of these elements requires expert testimony and expensive tests. If the plaintiffs can bring evidence proving these elements, they may be able to prevail in a toxic tort claim. However, causation is often difficult to prove in these cases.

The victims can bring an action for assault if an actor intended harmful or offensive contact upon another or put another in apprehension of such contact. In other words, A must have known that it made a release of chemicals to which the victims were exposed.[23]

To prove assault, the plaintiff must show intent. Here, if A knew or was substantially certain that he or she was contaminating the land and that others could be exposed to or particularly harmed by this toxic contamination, A would have the requisite intent. B, on the other hand, likely will not be liable because he or she presumably lacks the requisite intent, from the facts provided; see *Barth v. Firestone Tire and Rubber Co.*, 673 F.Supp 1466 (N.D. Cal. 1987).

A may also be liable to the victims for battery if harmful contact is shown. The plaintiffs must establish harmful or offensive touching. A toxic tort action for battery requires that the actor knows with substantial certainty that his actions will cause another, directly or indirectly, to come into contact with a foreign substance in a manner that a reasonable person would regard as offensive. Thus, the victims can argue that the contact caused by the emission of the toxic waste is offensive. However, again, causation in toxic tort cases can be difficult to prove (Reilly 2015).

A landowner is strictly liable to others for harm caused by toxic wastes stored on his property that flow onto the property of others; *Rylands v. Fletcher*, 3 Law Reports—English and Irish Appeals 330, 338–39 (House of Lords 1868). A and B might be strictly liable for engaging in an activity that causes harm to a plaintiff due to an abnormally dangerous activity. An abnormally dangerous activity is one that creates a risk of serious harm that cannot be eliminated even through the utmost care. However, plaintiffs need to prove that the activity is within the scope of the abnormal risk that is the basis for the liability, and that a causal relationship exists between any actions by the defendant and plaintiffs' claimed damages; *Pinares v. United Technologies Corp.*, 2011 WL 240522 (S.D. Fla. 2011). Assuming that the toxic contamination was the result of an abnormal risk and the victims can show causation, A and B might be strictly liable for the harm caused.

[23]Restatement (Second) of Torts § 29 (1965).

A and B might also be liable, under the theory of negligence per se, to compensate the public for damages and injury resulting from the disposal of hazardous wastes. If there is a statute imposing a duty on persons, negligence per se can be invoked where a plaintiff alleges that the defendant violated the law; the violation proximately caused harm to the plaintiff; and the law was intended to protect persons such as the plaintiff from the harm alleged. If the victims can prove these elements, then A and B may be liable for negligence per se, especially because they might be liable under environmental laws such as the RCRA and CERCLA; *Bagley v. Controlled Environment Corp.*, 127 N.H. 556, 503 A.2d 823, 827-28, 23 Env't. Rep. Cas. (BNA) 1955, 16 Envtl. L. Rep. 20606 (1986).

Ultimately, A is likely liable for assault, battery, negligence per se, and engaging in an abnormally dangerous conduct.

7 Conclusion

Has the approach of the United States adequately addressed the historical pollution challenge? While the record is mixed, in part because of the daunting nature of the phenomena, there have been successes. On historical pollution, contributions have been made in several areas.

There has been an evolution of broad liability notions (which may not everywhere meet the test of layperson fairness), which seemed essential to address the magnitude of the challenges of orphan dumps and related historical pollution. Addressing the injuries and damages of hundreds of communities, often poor and minority, called for the pursuit of environmental justice even when traditional legal principles might be compromised. The United States recognized that practical answers, even if messy or legally without precedent, needed to be tried.

Application has been made of a reasonably fair version of the polluter-pays principle, where the polluter is not always and everywhere an organization or individual defendant. Also, recognition was made that "clean" is not everywhere pristine, and that economic development and other societal goals can benefit from brownfield development. Also, there has been recognition that regardless of fault and liability, removal and remediation must be undertaken.

Evaluation of Progress of Historical Pollution Legislation

The EPA's own assessment on addressing historical pollution is as follows:

> For the past 30 years, the Superfund program has been making substantial progress protecting thousands of communities by cleaning up the Nation's most serious hazardous waste sites and by responding to thousands of oil and chemical spills. EPA has completed construction of cleanup remedies at 67.5% of final and deleted sites on the National Priorities List. The agency has readied nearly 1.3 million acres of land for return to productive use, and more than 455,800 acres are ready for anticipated use. With the passage of the American Recovery and Reinvestment Act (ARRA), the Superfund program has applied $600 million in ARRA funds to accelerate clean-ups at 31 ongoing construction projects and to start new construction projects at 26 sites [...] as the program enters its

> fourth decade, we believe we have the opportunity to improve Superfund's efficiency and management, as well as that of EPA's other cleanup programs. To that end, EPA has begun implementation of an initiative to better use the Agency's land cleanup authorities to accelerate clean-ups where possible, address a greater number of contaminated sites, and put these sites back into productive use while protecting human health and environment.[24]

This is only one assessment, and that of the federal agency primarily responsible for cleanup of this form of historical pollution. But the positive view is more realistic today than it was twenty years ago when evaluations of U.S. law, especially of the Superfund, were highly critical: cleanup took too long, many of the funds were going to lawyers rather than sites, and citizens were not fully enough involved. The law has evolved; it has become more fair from the layperson's perspective; sites have been cleaned up, although still at a very expensive rate; and cooperation with the private sector and with the states has increased.

A Word on Brownfields

The United States in its pursuit of historical pollution came to realize that perfection may be the enemy of action. The pervasive question in environmental law has been: *how clean is clean?* It was answered through ambient air-quality standards in air law and through the goal to "restore and maintain the chemical, physical and biological integrity of the Nation's waters" in the Federal Water Pollution Act (33 U.S.C. 1251). It became clear with the use of CERCLA that it would not be feasible to restore all contaminated sites to an environmental quality that would allow for residential, school, and other human use purposes. It also became clear that seeking such standards was an obstacle to the redevelopment of many sites. Plater et al. summarize:

> Perhaps the best example of CERCLA's unintended consequences concerns sites known as 'brownfields' [...] Fearing the [...] liability provisions of CERCLA, many prospective purchasers, developers, and lenders historically chose to avoid brownfields redevelopment altogether, with the result that a number of brownfields sites became major community and taxpayer burdens. Recognizing that CERCLA's remedial implementation strategy unfortunately has contributed to the perpetuation of brownfields rather than their remediation, EPA and the states have developed strategies to work more closely with regulated communities to encourage the redevelopment of brownfields [... a ...] shift of focus, toward a more cooperative approach in partnership with the regulated community. (Plater et al. 2010)

The EPA began its Brownfields program in 1995. Brownfields are real property, the expansion, redevelopment, or reuse of which may be complicated by the presence or potential presence of a hazardous substance, pollutant, or contaminant (EPA).[25] The EPA Brownfield goal is:

> to empower states, communities, and other stakeholders in economic redevelopment to work together in a timely manner to prevent, assess, safely clean up, and sustainably reuse brownfields. EPA's Brownfields Program provides financial and technical assistance for

[24]"How Superfund Works," mhsenviro1.wikispaces.com/file/view/Superfund.docx.

[25]"Brownfield Overview and Definition," U.S. Environmental Protection Agency, https://www.epa.gov/brownfields/brownfield-overview-and-definition.

brownfields activities [...] through an approach based on [...] goals [of public health and economic development].[26]

In addition to the federal program, state programs exist, although the nomenclature may differ from brownfield (Tang 2003).[27] California is an example.[28] Its law authorizes the state to clean up sites that do not qualify for cleanup under CERCLA. The state also provides funds to the state to pay its required share of CERCLA costs and provides compensation for persons injured by exposure to hazardous substances.

References

Abelkop, A. D. K. (2013). Tort law as an environmental policy instrument. *Oregon Law Review, 92,* 381.

Beck, E. C. (1979). The love canal tragedy. EPA website. http://www2.epa.gov/aboutepa/love-canal-tragedy

Brickey, K. F. (2008). *Conventional criminal statutes.* Environmental Crime: Law, Policy, Prosecution.

Brook, D. (1994). Review of Andrew Szasz's article EcoPoulism: Toxic waste and the movement for environmental justice. *Journal of Political Ecology: Case Studies in History and Society, 1* (1). http://jpe.library.arizona.edu/volume_1/brookvol1.htm

Broun, C. N. & O'Reilly, J. T. (n.d.). *RCRA and superfund: A practice guide* (2015 and October 2016 update).

Brown, M. H. (1979, December). Love Canal and the poisoning of America. *Atlantic Magazine.* http://www.theatlantic.com/magazine/archive/1979/12/love-canal-and-the-poisoning-of-america/376297/

Bukro, C. (1980, July 8). Valley of the Drums plays deadly chorus. *Chicago Tribune.* http://archives.chicagotribune.com/1980/07/08/page/1/article/valley-of-the-drums-plays-deadly-chorus#text

Carter, J. (1981). *State of the union address.* http://www.jimmycarterlibrary.gov/documents/speeches/su81jec.phtml

Craig, R. et al. (1979, 2010). *Toxic and environmental torts: Cases and materials.*

DeCarvalho, J. P. (2013). Love Canal disaster. *Toxipedia.* http://www.toxipedia.org/display/toxipedia/Love+Canal+Disaster

DiMento, J. F. (1986). *Environmental law and American business: Dilemmas of compliance.* New York: Plenum.

Environmental Law Institute (2015). *Law of environmental protection,* Spring, Sect. 7.10.

EPA (1981, September 18). *EPA schedules emergency cleanup at Valley of the Drums.* https://www.epa.gov/aboutepa/epa-schedules-emergency-cleanup-valley-drums

Finding Dulcinea. (2011). *On this day: President Carter approves emergency financial aid for love canal.* http://www.findingdulcinea.com/news/on-this-day/July-August-08/On-This-Day–Love-Canal-Declared-Federal-Emergency.html

GAO, Funding and Reported Costs of Enforcement and Administration Activities. (2008). http://www.gao.gov/products/GAO-08-841R

---

[26]Ibid.

[27]Tang counts ten states that have Brownfield programs.

[28]California Environmental Protection Agency (Cal/EPA) Department of Toxic Substances Control (DTSC) Site Mitigation and Brownfields Reuse Program.

Hird, J. A. (1993). Environmental policy and equity: The case of Superfund. *Journal of Policy Analysis and Management, 12*, 324. http://www.jstor.org/stable/pdf/3325238.pdf?acceptTC=true

Humphreys, S. L. (1990). An enemy of the people: Prosecuting The corporate polluter as a common law criminal. *American University Law Review*, *39*, 311.

Kevles, A. (1983, September 28). Of mice and men in Love Canal. *L.A. Times*.

Logan, M. (2012). Dumped on: The messy truth about Love Canal; My America odyssey: Discovering one America at a time. http://myamericanodyssey.com/dumped-on-the-messy-truth-about-love-canal-ny/

O'Brien, R. M., et al. (1984). Open and closed systems of decision making: The case of toxic waste management. *Public Administration Review, 44,* 334.

O'Reilly, J. T. (2015). How is a cause of action established for toxic assault and battery? What common law causes of actions exist. *Toxic Torts Practice Guide*, *1*, § 6.12.

Periconi, J. J. (2008). The state of environmental crimes prosecutions in New York. *Natural Resources & Environment, 23*. http://www.periconi.com/CM/Articles/Environmental_Crimes.pdf. Accessed July 2, 2015.

Plater, Z. et al. (2010). *Environmental law and policy: Nature, law and society* (4th ed., Vols. 680–681).

Stephenson, J. B. (2004). Superfund program: Updated appropriation and expenditure data. http://www.gao.gov/new.items/d04475r.pdf

Tang, J. J. (2003). Public participation in brownfield development: A framework for community empowerment in zoning practices. *Seattle Journal of Environmental Law, 3,* 241.

The Gilder Lehrman Institute of American History (2014). *Timeline: Love Canal hazardous waste disaster*. https://www.gilderlehman.org/sites/default/files/inline-pdfs/LoveCanal_Timeline.pdf

U.S. Environmental Protection Agency (2016). *Brownfields*. EPA.GOV, http://www.epa.gov/brownfields/. Last updated March 29.

Usiak, R. (1982, June 13). Residents, buyers wait for Love Canal rebirth. *L.A. Times*.

Warren, E. (1980, June 20). Crusader for the Love Canal fights on. *L.A. Times*.

Chapter 9
Historical Pollution in the UK (England and Wales): The Residual Role Played by Criminal Law

Valsamis Mitsilegas and Elena Fasoli

Contents

This chapter is a co-authored publication. Prof. Mitsilegas is primarily responsible for Sects. 2–4 and Dr. Fasoli for Sects. 5 and 6.

V. Mitsilegas (✉) · E. Fasoli
Queen Mary University of London, London, UK
e-mail: v.mitsilegas@qmul.ac.uk

E. Fasoli
e-mail: e.fasoli@qmul.ac.uk

F. Centonze and S. Manacorda (eds.), *Historical Pollution*,
DOI 10.1007/978-3-319-56937-6_9

1 Introduction

The long and rich industrial history of the UK is characterized not only by socio-economic development but also by the environmental and health impacts of industrial activities. The regulations to control the use of, for example, raw materials or industrial waste in the UK have been very permissive if not altogether absent in the past, so that industrial practices have allowed the spillage of contaminants into sites of industrial activity and the disposal of waste without consideration of the potential hazardous implications.

Historical pollution is both a health and environmental problem that presents significant challenges from the legislative, political, and social points of view, especially considering that historic industrial sites are often located in city centres and waterfront locations. Contamination can exist at large industrial sites (e.g. large chemical plants) but can also occur on small sites (e.g. the premises of a small dry-cleaning agency).

The Department for Environment, Food and Rural Affairs (DEFRA) has pictured the phenomenon under consideration in the following terms:

> England has a considerable legacy of historical land contamination involving a very wide range of substances: on all land there are background levels of substances, including substances that are naturally present as a result of our varied and complex geology and substances resulting from diffuse human pollution; on some land there are greater concentrations of contaminants, often associated with industrial use and waste disposal; in a minority of cases there may be sufficient risk to health or the environment for such land to be considered contaminated land.[1]

Historical pollution in the UK has been described as a "patchwork system of liability rules spread across a wide range of areas that address specific problems of contamination and clean up" (Bell et al. 2013, p. 586). The matter is regulated under the law relating to land contamination, which is a system of clean-up procedures under Part 2A of the Environmental Protection Act 1990 ("Pt2A").[2] However, experts highlight that the application of Pt2A is limited, as the legal system favours the planning system or other forms of voluntary action. In this context very little space is left to criminal law, as regulators very rarely resort to imposing criminal penalties. As will be described in the following sections, the pragmatic, mainly administrative, approach, combined with the limited role played by criminal law, is the distinctive features of the UK (England and Wales) legal system when dealing with phenomena of historical pollution.

---

[1]DEFRA, *Environmental Protection Act* (1990)*: Part 2A Contaminated Land Statutory Guidance* (2012), p. 5.

[2]http://www.legislation.gov.uk/ukpga/1990/43/contents. A full set of documents providing technical guidance on the management of contaminated land, including how to investigate, assess, and manage the risks, is available at https://www.gov.uk/government/collections/land-contamination-technical-guidance.

The chapter is structured as follows. Firstly, the sociological and economic context and the general legal background of the UK legal system will be described. Then the various definitions of historical pollution contained in the relevant legal texts and other official sources will be addressed. The analysis of the relevant provisions of land contamination law and those related to the planning system will follow, including relevant cases. The extent and the way in which UK (E&W) law criminalizes conduct detrimental to the environment , and the (residual) role played by criminal law will be then addressed. Against the background of the above analysis we will present some tentative concluding evaluations of the phenomenon of historical pollution in the UK (E&W). The results of the legal research will be enriched by the views provided by leading experts on land contamination.[3]

2 Economic and Social Context

In the UK, environmental protection and control has a long history dating back to the industrial revolution. The shift from a society mainly based on agriculture to an industrialized one marked by the use of mechanically driven equipment is generally accepted to have started in Britain in the late eighteenth century.[4]

The growth of the industrial sector was accompanied by a lack of government regulation of industrial development at both national and local levels (McLaren 1983, p. 163). As highlighted,

> the consequences for the population […] and the natural environment were traumatic. Those who toiled in the factories, mills and smelters bore the brunt of the social and environmental abuse. The sites chosen for the dwellings of the labourers were typically low lying, badly drained and unsanitary. The particular misery of the workers was matched by the more general and pervasive environmental abuse caused by the factories. (McLaren 1983, pp. 163–164)

[3]Interview with Mr. Andrew Wiseman on 6 May 2015, Solicitor and Convenor of the UKELA Working Party on Land Contamination (http://www.ukela.org/rte.asp?id=64). The views expressed constitute the personal view of the interviewee and do not necessarily reflect the position of the UKELA Working Party. Interview with Mr. Nick Marks on 5 June 2015, Regulator and Contaminated Land Officer commenting on behalf of Environmental Protection UK (EPUK, http://www.environmental-protection.org.uk/). The views expressed constitute the personal view of the interviewee and do not necessarily reflect the position of EPUK. Comments have also kindly been provided by Mr. Gordon Wignall, Barrister at Six Pump Court Chambers, expert in planning and environmental law, and by Ms. Rachel Mulheron, Professor of Law at Queen Mary University of London. The authors are also grateful to Ms. Anne Brosnan, Chief Prosecutor at Environment Agency, for the cases provided on corporate environmental crime. The authors remain solely responsible for any error or omissions contained in this chapter.

[4]For an historical overview see, among others, Bell et al. 2013, pp. 18 ff.; Johnstone and Horan 1996, p. 212; Thorsheim 2002, pp. 381–401; Mosley 2014, pp. 143–169.

By the middle of the nineteenth century, the state of the environment and contaminated drinking water had already become a serious danger to public health. Two major cholera epidemics occurred in London in 1866 and 1872, although conditions in other parts of the country were just as bad, especially in the industrial Midlands and in the north (Johnstone and Horan 1996, p. 212).

The first significant government regulation was adopted in that period. These laws were primarily concerned with the human consequences of pollution or poor water quality. The objective was to improve human health rather than ameliorating the quality of the environment.

The legislation dealing with public health dates back to 1875 (Public Health Act), and at the beginning of the nineteenth century, legislation on town planning was also introduced in order to meet health needs (Bell et al. 2013, p. 19).

For many years, industrial activity in general was conceived as purely technical, so that causing environmental crimes was seen as without moral blame. Pollution was seen as a natural consequence of industrial activity and was associated with economic development; therefore, there was little political and public support for action.

By way of example, it was not until 1937 that appropriate legislation was passed to tackle the problem of industrial effluent control. The Public Health Act 1937 gave industry the right to discharge trade effluents into a public sewer subject to conditions as to the nature and composition of the effluent, its volume, and its rate of flow. Permission to discharge was provided under a "consent to discharge" arrangement (Johnstone and Horan 1996, p. 215).

Similarly, in order to tackle the problem of concentrations of sulphur dioxide in the air, regulators decided to raise the height of industrial chimneys to better disperse and dilute the harmful pollutant (Mosley 2014, p. 157). As has been noted,

> by 1960, Britain had built more than 60 new power stations and greatly extended the generating capacity of many older installations, with their chimneys reaching heights in excess of 135 m. These tall chimneystacks, intended to reduce local air pollution, transported sulphur emissions over hundreds and even thousands of kilometres. (Mosley 2014, p. 157)

Throughout the years, there has been a shift in the emphasis of the laws to enhance environmental protection. This shift was also mirrored in judicial practice.

Until the 1970s, for example, water discharge activity "was not criminal in any real sense," even though "in the public interest [it was] prohibited under a penalty" (*Alphacell Ltd v Woodward* [1972] AC 824).

From the 1970s onwards, the State started to play a more active role in the control of pollution and protection of the environment (Johnstone and Horan 1996, p. 23). The mid-1970s also coincided with the progressive recognition that contaminated land was a problem for redevelopment of lands previously used for industrial-related purposes (Zhao 2013, p. 118).

Later, in the 1980s, the House of Commons Select Committee on the Environment started an inquiry on the topic. The Committee concluded that there was a general lack of reliable information on the scale, nature, and distribution of the problem of land affected by contamination in the UK, and encouraged the Government to take action (Harris and Denner 1997, p. 25).

Subsequently, the attitude of judicial practice changed and the breach of provisions protecting the environment was considered criminal in nature (*National Rivers Authority v Alfred McAlpine Homes East Ltd* [1994] 4 All ER 286). More recently, this interpretation in favour of an enhanced protection of the environment has been further reinforced. For example, it has been stated that

> the environment in which we live is a precious heritage and it is incumbent on the present generation to preserve it for the future. Rivers and watercourses are important parts of the environment and there is an increasing awareness of the necessity to preserve them from pollution. (*R v Anglian Water* [2004] Env LR 10)

However, the level of change in attitude at different levels of society towards environmental crimes is still not enough to assure full protection of the environment and public health. As has been observed, the more the public becomes aware of—and demands action on—environmentally harmful activities, the more such activities are likely to be subject to criminalization (Potter 2014, p. 9).

This shift in the emphasis of the protection of the environment is also reflected in the three domestic cases that will be analysed more in detail in Sect. 5.4. It is in fact worth anticipating the social and economic contexts in which those cases arose.

The first two cases (*Bawtry Gas Works* and *Redland Minerals and Crest Nicholson Residentials*) describe the phenomenon of the succession of companies during the nationalization, privatization, and then reorganization of the gas industry in the UK. Bawtry and District Gas Company, for example, was initially a private gas company that started operating gasworks in 1915 and was then nationalized and acquired by another company, the East Midlands Gas Board. Later, in 1986, with the privatization of the gas industry, the responsibilities of the latter company passed on to British Gas and finally to National Grid Gas. The case shows the difficulties in finding a company liable for groundwater contamination, which happened at a time when the statutory provisions on contaminated land (i.e. "Pt2A," see Sect. 5.1) were not yet in force.

The third case (*Corby Steelworks*) refers to soil contamination caused by some major steel companies in the UK between the 1970s and 1980s. More precisely, the case deals with the decontamination activity that was later carried out by the local authority in the 1980s and 1990s when full knowledge about the possible dangerous effects of the release of toxic dust into the air was still lacking.

3 Definition of Historical Pollution

The present study adopts a working definition of historical pollution, which refers to phenomena of contamination of sites and/or natural resources deriving from past polluting activities. In UK (E&W), this phenomenon falls under the "contaminated land regime," which will be described in detail in the following sections.

It is useful to anticipate, though, that the UK (E&W) system gives much more relevance to the *effects* of the conduct (i.e. the fact that the land is currently contaminated) rather than to the *location in time* of the source of the polluting activity. This emphasis on the *effects* emerges from the definition itself of "contaminated land" that can be found in Sect. 78A9(2) of the 1990 Environmental Protection Act (see Sect. 5.1). "Contaminated land" is in fact "any land which *appears* to the local authority in whose area it is situated *to be in such a condition*, by reason of substances in, on or under the land," that

(a) Significant harm is being caused or there is a significant possibility of such harm being caused; or
(b) Significant pollution of controlled water is being or is likely to be caused; in determining whether any land appears to be such land, a local authority shall act in accordance with guidance issued by the Secretary of State with respect to the manner in which that determination is to be made.

In addition, Sect. 78A(4) of Pt2A of the 1990 Environmental Protection Act defines "harm" as that caused to "the health of living organisms or other interference with the ecological systems of which they form part and, in the case of man, includes harm to his property."

Contaminated land is also defined in the dedicated governmental website in the following terms: "land where substances could cause significant harm to people or protected species or significant pollution of surface waters or groundwater."[5] It is noteworthy that this definition refers especially to contamination caused in the past by factories, mines, steelworks, refineries, and landfills. Quite in the same vein, in 1996 the then Department of the Environment published forty-seven industry profiles describing the main sources of contamination in industry: storage and dispensing of fuels, especially from aviation, asbestos manufacturing works, and fertilizer manufacturing works.[6]

Furthermore, Planning Policy Statement 23 (although now superseded by the National Planning Policy Framework),[7] giving guidance to the planning authorities on development activities undertaken on contaminated lands, refers to "land affected by contamination" to cover "all cases where the actual or suspected presence of substances in, on or under the land may cause risks to people, human activities or the environment, regardless of whether the land meets the statutory definition in Part 2A" (Marion 2010, pp. 14–15).[8] The last part of the definition seems to introduce a degree of flexibility in the identification of land as contaminated.

---

[5]https://www.gov.uk/contaminated-land/overview.

[6]All industry profiles are available at https://www.gov.uk/government/publications/department-of-environment-industry-profiles.

[7]https://www.gov.uk/government/uploads/system/uploads/attachment_data/file/6077/2116950.pdf.

[8]See also *Planning Practice Guidance on Contaminated Land* at http://planningguidance.planningportal.gov.uk/blog/guidance/land-affected-by-contamination/land-affected-by-contamination-guidance/.

Further definition of the phenomenon under consideration is to be found in the Model Procedures for the Management of Land Contamination, which define lands affected by contamination as "areas with elevated levels of naturally occurring substances; specific sites that have been occupied by former industrial uses, which may have left a legacy of contamination from operational activities or from waste disposal; areas of land in which substances are present as a result of direct or indirect events, such as accidents, spillages, aerial deposition or migration."[9] The Contaminated Land Rehabilitation Network for Environmental Technologies (CLARINET) considers instead "*contaminated land* a general term to describe sites, and wider areas of *land*, which have elevated concentrations of chemicals or other substances (*contamination*), usually resulting from man's use of the land."[10]

From the above analysis, it appears that the approach followed is not only that of giving precedence to the *effects* of the conduct (rather than to the location in time of the origins of these conducts) but also that of requiring the *conduct* in question to pose threats to humans, to property, or to the wider environment.

The threshold of the contamination is also relevant. Normal levels of contamination do not amount to land contamination, as only unacceptable ones are considered. In this regard, the 2012 *Contaminated Land Statutory Guidance* by DEFRA, which is part and parcel of the statutory provisions Pt2A, specifies that "terms, such as 'land affected by contamination' or 'land contamination' are used to describe the much broader categories of land where contaminants are present but usually not at a sufficient level of risk to be contaminated land."[11] In fact, normal levels of contaminants in soil should not necessarily make land contaminated unless there is a particular reason to consider it as such. In practice, normal levels of contaminants in soil exist when there is a natural presence of contaminants that do not pose a unacceptable risk to health or to the environment, or the presence of contaminants caused by a low level of diffuse pollution (e.g. historic use of leaded petrol and the presence of benzopyrene from vehicle exhausts) and common human activity other than specific industrial processes (e.g. spreading of domestic ash throughout gardens).[12]

Some indications on the time factor as an essential feature of historical pollution phenomena can be found in the legal literature. While the relevant national legislation does not fix a precise time limit for an activity to be considered "past" activity, legal doctrine tries to draw a distinction between "historical pollution" on the one hand and "current contamination" on the other. According to Bell, McGillivray, and Pedersen (2013), for example, the latter refers to existing activities that are considered unlawful and regulated by the statutory provisions setting

---

[9]https://www.gov.uk/government/uploads/system/uploads/attachment_data/file/297401/scho0804bibr-e-e.pdf.

[10]CLARINET, *Sustainable Management of Contaminated Land: An Overview* (2002), p. 1. On CLARINET, see also below.

[11]DEFRA, *Environmental Protection Act* 1990*: Part 2A Contaminated Land Statutory Guidance* (2012), p. 3.

[12]Ibid., p. 12.

out pollution control, whereas the former refers to soil contamination as a result of past activities that fall outside the present-day control but whose effects still persist nowadays.

More precisely, the existing pollution control regimes (such as waste law, water law) deal mainly with the *ex ante* control (or prevention) of polluting activities. In other words, contamination results from a failure to comply with modern standards (i.e. *unlawful* conducts). By contrast, land contamination law deals with contamination that happened in the past when the activity was (presumably) considered *lawful* (Bell et al. 2013, p. 588).

As it has been highlighted, waste disposal facilities are nowadays regulated by strict standards but, in the past, open voids, typically those left by mineral extraction, have been filled with waste, with limited control or documentation of the materials deposited. These can leach toxic materials or decay products, which contaminate surface or groundwater (Lewis 2012, p. 420).

These two regimes (the past and the current one) coexist and can potentially interact (or overlap). An example of interaction is in the area of water pollution. Contamination of controlled waters (territorial waters, coastal waters, inland waters, and groundwater) is primarily regulated under the Water Resources Act 1991, which deals with cases of water pollution that *have not* been historically contaminated (see also Sect. 6).[13] Sections 161–161D of the Water Resources Act 1991 provide that the Environment Agency, which is the most important regulatory and enforcement agency in England,[14] can serve a works notice on the responsible person requiring that person to clean up pollution or remedy the harm caused. Failing to comply with the works notice constitutes a criminal offence, and the person could be liable (a) on summary conviction, to imprisonment for a term not exceeding three months or to a fine not exceeding £20,000 or to both; (b) on conviction on indictment to imprisonment for a term not exceeding two years or to a fine or to both. The Agency could also decide to do what that person was required to do and recover from them any costs or expenses reasonably incurred.

At the same time, the above-described powers of the Environment Agency to issue a work notice can overlap with those of the local authorities to issue a remediation notice under the Pt2A regime. Usually, a memorandum of understanding between the local authorities and the Environment Agency is adopted in order to regulate the way these powers need to be used so as to avoid duplication of regulatory efforts.

In practice, when historic pollution has occurred and the pollutants have completely entered the relevant body of the controlled waters (in other words, when the controlled waters are not subject to contamination coming from historic pollution) or where the source site cannot be identified anymore so that "significant pollution

[13]http://www.legislation.gov.uk/ukpga/1991/57/contents.

[14]The Environment Agency is an independent corporate body (Environment Act 1995, s. 1(1)) whose main aim is to minimize harm to the environment. It has a range of enforcement and other powers available. It is responsible, *inter alia*, for waste management, water pollution, the integrated pollution prevention and control regime, and radioactive substances.

linkage" no longer exists, the regime of the pollution of controlled waters under the Water Resources Act takes precedence over the Pt2A and a works notice (instead of a remediation notice by the local authorities) is issued by the Environment Agency (Bell et al. 2013, p. 589).

Difficulties could arise when neither the Agency nor the local authority has yet taken action. Under these circumstances, the local authority seems to be under a "duty to act," whereas the Agency seems to have only the "power to act."

Finally, this enquiry about the definition of historical pollution as applied to the UK (E&W) context must be enriched by the views of the experts. In everyday practice, the regulators in fact consider as current events (for which the relevant pollution control regime applies) those having occurred within six months, whereas for all others the contaminated land regime applies. Therefore, without putting much attention on whether the conduct was lawful or unlawful at the time of the occurrence, the regulator looks pragmatically at the factual situation as it stands at the time of assessment (even where, for example, the waste was buried fifty years ago) and takes action accordingly, applying the regime which is considered most appropriate in that specific case. Usually, in cases of ongoing pollution the regulator deals with the situation without applying the Pt2A.

4 General Legal Background: The Influence of International and European Law

As described in the previous section, the protection of the environment originally derived from statutory public-health-based controls from the Victorian and Edwardian eras, but with time it developed towards a less anthropocentric and more environment-oriented approach. The law applicable to specific cases, including cases of historical pollution, derives from the stratification of statutory provisions over time, but also from case law that has been developed by judges (common law).

This law has been also influenced by European Union legislation and the international agreements setting environmental standards. In this section, we analyse the most relevant ones from the perspective of land contamination in the UK (E&W).

As far as the international level is concerned, the most important instrument is the 2001 Stockholm Convention on Persistent Organic Pollutants (POPs), which deals with chemical substances that persist in the environment, bioaccumulate through the food web, and pose a risk of causing adverse effects to human health and the environment.[15] The Stockholm Convention was ratified by the UK on 17 January 2005. The obligations of the Stockholm Convention are implemented in UK by the 2007 Persistent Organic Pollutants Regulations[16] which contain

[15]http://chm.pops.int/default.aspx.

[16]http://www.legislation.gov.uk/uksi/2007/3106/made.

provisions regarding production, placing on the market, and use of chemicals, management of stockpiles and wastes, and measures to reduce unintentional releases of POPs.[17] In this regard, it has to be noted that the 2007 Persistent Organic Pollutants Regulations constitute, at the same time, the transposition of the EU Regulation 850/2004 on persistent organic pollutants.[18]

As to the comparison between the contaminated land regime under Pt2A and the Stockholm Convention, the former, which also identifies POP-contaminated sites, relies on the existence of a pollutant linkage (i.e. source, pathway, and receptor, as will be explained in more detail below), but most importantly on the presence of the likelihood to cause significant pollution or harm to the following receptors: people, controlled waters, specific ecological systems, and specific types of property. By contrast, sites with POPs affecting other types of receptors, or simply storing POPs, are not caught by the Pt2A regime.

In addition, the contaminated land regime under Pt2A refers to broader groupings of substances compared to those contained in the Stockholm Convention, such as metals and metalloids, inorganic compounds, organic compounds, and gases.[19]

As far as the European level is concerned, Pt2A mirrors the provisions contained in Directive 2004/35 on environmental liability with regard to the prevention and remedying of environmental damage (ELD), at the same time differing in some relevant aspects. It seems preferable to highlight these aspects only after describing Pt2A in detail (see Sect. 5.1.6).

A final consideration has to be made with regard to regional initiatives concerning historical pollution that have an impact on UK. By way of example, the UK actively participated in CLARINET, an interdisciplinary network on the sustainable management of contaminated land in Europe that ran from 1998 to 2001. Among the various activities, national UK experts, including scientific experts, actively participated in the CLARINET working groups.[20]

With regard to more recent activities, the UK is currently a partner of EUGRIS, which is a web portal offering information and services on topics related to soil and water, initially focusing on the problems of contaminated land and water.[21] In addition, a representative from Contaminated Land: Applications in Real

---

[17]See the National Implementation Plan for the Stockholm Convention on Persistent Organic Pollutants (2012) at https://www.gov.uk/government/uploads/system/uploads/attachment_data/file/82680/consult-nip-pop-doc-20121121.pdf.

[18]The Regulation also contains provisions regarding the production, placement on the market, and use of chemicals, management of stockpiles and wastes, and measures to reduce unintentional releases of POPs. Furthermore, Member States must set up emission inventories for unintentionally produced POPs, national implementation plans (NIPs), and monitoring and information exchange mechanisms. For further details, see http://ec.europa.eu/environment/pops/.

[19]https://www.gov.uk/government/uploads/system/uploads/attachment_data/file/313964/geho0109bpha-e-e.pdf.

[20]http://www.eugris.info/newsdownloads/final%20report%20clarinet.pdf.

[21]http://www.eugris.info/.

Environments (CL:AIRE)[22] is also a member of the Network for Industrially Contaminated Land in Europe (NICOLE).[23] CL:AIRE is a UK not-for-profit organization established in 1999 to stimulate the regeneration of contaminated land in the UK by raising awareness of practical and sustainable remediation technologies, whereas NICOLE is a leading forum on contaminated land management in Europe promoting cooperation between industry, academia, and service providers on the development and application of sustainable technologies.

5 Administrative and Private Law

As anticipated, in the UK (E&W) cases of historical pollution are dealt with by the contaminated land regime (Pt2A), which will be described in the following sections. The main administrative clean-up procedures will be addressed in Sect. 5.1, whereas the application of the planning system and the relevance of brownfield redevelopments will be the topics of Sect. 5.2. In addition, given that individuals may suffer injuries as a consequence of the polluting activity, tort law will be also taken into account (Sect. 5.3). The description of the relevant provisions and regulatory practices will be enriched by reference to case law (Sect. 5.4).

5.1 The "Contaminated Land Regime"

The UK (E&W) legislation on land contamination is quite recent. It was adopted in the 1990s after the recognition of the problem of redeveloping sites that had been previously used for industrial activities.

After several legislative modifications,[24] the system currently in force consists in a regulatory process composed of different phases: identification of the site, investigation, assessment, and remediation. The contaminated land regime applies to all land—whether residential, commercial, industrial, or agricultural—and to "controlled waters," i.e. surface, ground, and coastal waters. It can affect owners, occupiers, developers, and lenders.

The bulk of legislation is composed of the Environmental Protection Act (1990) Pt2A, as amended in 1995,[25] and of the Contaminated Land Regulations 2006, as amended in 2009 and subsequently in 2012.[26]

---

[22]http://www.claire.co.uk/.

[23]http://www.nicole.org/.

[24]For the history of the contaminated land liability regime in the UK, see in particularly Zhao (2013, pp. 117–150). See also https://www.gov.uk/contaminated-land/overview.

[25]http://www.legislation.gov.uk/ukpga/1990/43/contents.

[26]http://www.legislation.gov.uk/uksi/2006/1380/contents/made.

The statutory provisions are complemented by Statutory Guidance by DEFRA (the last one adopted in 2012) and other secondary legislation (e.g. DEFRA's *Guidance on the Legal Definition of Contaminated Land* 2008). These documents are legally binding and form an essential part of the regime along with the other guidance and regulations.

In addition, the public consultation documents "Paying for Our Past"[27] and "Framework for Contaminated Land,"[28] both adopted in 1994, are also part of the regulatory background. Key outcomes of these consultations have been, for example: commitment to the polluter-pays principle; focus on past pollution only (as legislation to control new pollution already existed); action to be taken only in cases where actual or potential risks to health or the environment existed and remediation was financially viable.

5.1.1 The Powers of (and Cooperation Between) Enforcing Authorities

Pt2A requires primarily local authorities in England and Wales to investigate lands in order to identify those that are significantly contaminated. Usually, local authorities ask developers to conduct investigations (and also to clean up in case the land is found to be at risk) as a condition to being granted planning permission. Once the investigation (and clean-up) is concluded, the developer reports back to the local authorities regarding the activities undertaken. Each local authority prepares a contaminated land strategy[29] that is reviewed and updated periodically.

In practice, local authorities have to: prepare a strategy to identify contaminated land in their area and implement it; inspect land (investigations) to identify contaminated land (and special sites for enforcement by the Environment Agency or Natural England); identify responsibility for the remediation; bring about the remediation voluntarily or through enforcement action; maintain a public register of the remediation; and provide the Environment Agency with local land contamination and remediation information.

Once contaminated land has been identified, local authorities have also to pay attention as to whether the site falls into the category of "special site" because, for example, of the seriousness of the harm under the Contaminated Land Regulations 2006. For special sites, the competence goes to the Environment Agency in England or Natural Resources in Wales. In practice, the competence stays within the local authorities unless the site is considered a "special site."

---

[27]Department of the Environment, "Paying for Our Past" (London, Department of the Environment, 1994).

[28]Department of the Environment, "Framework for Contaminated Land: Outcome of the Government's Policies Review and Conclusions from the Consultation Paper Paying for Our Past" (London, Department of the Environment, 1994).

[29]See, for example, the Contaminated Land Strategy of London Borough of Hounslow (2008) at http://www.hounslow.gov.uk/contaminated_land_strategy.pdf.

5.1.2 The Concept of "Pollutant Linkage"

The presence of a "pollutant linkage" characterizes a piece of land as contaminated. There must be a link between "contaminant, pathway, and receptor," whereby the contaminants are the potential source of the damage; the receptor can be, for example, a nature conservation site, the water, human beings suffering birth defects or serious injury, buildings that can potentially be polluted; and the pathway is the means by which the receptor can be damaged. A pollutant linkage is formed when there is a link between a contaminant and a receptor by means of a route that drives that contaminant (that otherwise would stay underground) into the wider environment. The linkage has to be "significant" because of the types of receptor or the seriousness of the damage (Bell et al. 2013, p. 596).

5.1.3 Liability for Remediation

Sites might have complex histories regarding site operation and ownership. If land presents a "significant risk of harm to human health" the local authority must take action. Practice shows that there are difficulties in identifying the presence of unacceptable risk that would require remediation (Environment Agency, Reporting the Evidence 2009, pp. 22–24). Local authorities usually follow guidance documents, such as the Model Procedures for the Management of Land Contamination. Under this guidance the local authorities are asked to verify (a) whether the contamination is a problem or (b) whether it could become one (ibid., p. 31).

Once a site has been classified as "contaminated land" the local authorities have to notify the relevant parties, and a three-month consultation period starts in order to allow them to reach a negotiated solution, which testifies to the preference of the system for voluntary remediation. The regulator must in fact use reasonable endeavours to consult every owner or occupier of the land and must also make reasonable endeavours to consult the person on whom the notice is to be served before issuing it (Lees 2015, p. 66).

After the expiration of the three-month period, a remediation notice can be served by the regulator on the liable parties ("appropriate persons") that thus have to undertake the clean-up according to the terms contained in the remediation notice. The standard to which land must be remediated is that the land must no longer be considered as contaminated under its current use, applying best practicable techniques.

The recipient of a remediation notice has the right to appeal to the Secretary of State within 21 days of the notice (Sect. 78L(1) of Pt2A). The appeal suspends the enforcement of the notice until the determination or withdrawal of the appeal.

5.1.4 The "Appropriate Person"

Section 78F(2) of Pt2A defines the "appropriate person" as the "person, or any of the persons, who *caused* or *knowingly permitted* the substances, or any of the substances, which have been the cause of the contamination to be in or under the land." Therefore, whoever caused or knowingly permitted the contamination is considered a "Class A" person and liable.

Liability rests on the principle that actions of companies and individuals undertaken many decades ago, long before Pt2A was in force, can still give rise to liability for remediation today. Liability is therefore *retrospective* or *retroactive*. In addition, there is no limitation period with regard to regulatory action and enforcement under Pt2A.

Courts do not ask whether a person had knowledge, is at fault, or has been negligent, but requires that they have been involved in the active operation (or that they failed to act). This is referred as to "having caused" the event. "Knowingly permitted" requires instead that a person "[has] knowledge of the existence of contaminating substances in or under land as well as ha[s] permitted their existence to continue in, on or under that land" (Bell et al. 2013, p. 56).

When assessing "causation" the local authorities have to answer questions such as: What did the alleged appropriate person do in order to cause the presence of the substance? Was this action a cause of the presence of the polluting substance?

In practice there is a presumption that the operator/occupier of the site at the time of the pollution caused the presence of the contaminant. However, in order to mitigate this strict and retrospective liability, the regime provides for exclusions of liability for some categories of Class A persons, such as: landlords, when tenants were the cause of the pollution; persons who have transferred the responsibility for the contamination by reducing the price of the land in order to reflect the real conditions; and persons who have sold the land in question "with information" about the real conditions so that the buyer should have been aware of it. In addition, another basis for exclusion from the category of a Class A person is, for example, when the original polluter who dumped waste in the past later on sold the land, and the buyer autonomously introduced a new "receptor" on the site, such as a new housing development (*Circular Facilities (London) Ltd. v. Sevenoaks District Council* [2005] EWHC 865).

If, after a reasonable inquiry, the Class A persons cannot be found, the regulator seeks the owners or occupiers of the land (so-called Class B persons) and obliges them to pay for remediation. In this case, there is no need to establish a causal link as liability directly follows from the person's "status." This system has been criticized for its strictness towards innocent owners or occupiers. It has been highlighted in fact that in case Class A persons are found after the identification of Class B persons, even if the latter in the meantime has already started to pay the costs of remediation, no mechanism has been established for them to recover the expenses already incurred from the "appropriate" Class A persons (Fogleman 2005, p. 1221). However, the regime provides for some exclusions in favour of certain categories of Class B persons, such as tenants paying rent, given the fact that they do not have interest in the value of the land in the long term and do not benefit from any increase in the value of land should it be cleaned up.

In addition, the system applies proportionate liability. Liability is imposed on those who have separately contributed to the problem according to a proportion principle so that every person has to bear the costs of remediation according to the share of their own responsibility for the contamination (Fogleman 2014, p. 133) (see the *Redland Minerals Ltd.* case in Sect. 5.4.2).

Finally, if neither Class A nor B persons can be found liable for a "linkage," that linkage becomes "orphan" and the enforcing authority "*should* itself bear the costs of any remediation which is carried out" (*Statutory Guidance* 2012, p. 61). The use of the term *should* instead of *shall* has to be noted. In fact, like in the ELD, the public authority is not under obligation to intervene to take remedial measures ("the competent authority *may* take these measures itself, as a means of last resort": Art. 6.3 ELD. See also Sect. 5.1.6).

Experts have highlighted that buying a property in the UK (E&W) entails the need to investigate whether it is located on former industrial land (about two-thirds of all new houses are built on land that has already been developed in the past). It is usually the task of private solicitors involved in the property transaction to check the history of the land. In this regard, the Law Society, which is an independent professional body for solicitors in England and Wales, issues guidance for practitioners when dealing with land contamination.[30]

However, usually the authorities try not to let the costs of remediation fall on innocent owners or occupiers.[31] In the *Bawtry Gas Works* case (see Sect. 5.4.1), for example, the costs of the clean-up were borne by the public.

Interestingly, until recently a government fund was operating in order to finance the local authorities to inspect the lands and to pay for clean-up costs when it was not possible to find the person who originally contaminated it. In financial year 2009/2010, DEFRA made a £17.5 million contribution via the Contaminated Land fund. In 2013/2014, the funding dropped significantly to £2 million per annum. Unfortunately, the fund has been recently cut down so much that the local authorities have now to use their own resources, and experts have highlighted that the remediation of contaminated lands is not always considered a priority for local administrations.[32] The level of public knowledge and concern is not often enough to induce public authorities to put it at the top of their political agenda.[33]

---

[30]See, for example, http://www.lawsociety.org.uk/support-services/advice/practice-notes/contaminated-land/; http://www.bbc.co.uk/news/uk-england-bristol-10746865; http://www.bathchronicle.co.uk/Council-bid-government-cash-clean-contaminated/story-12982057-detail/story.html; http://www.dailypost.co.uk/news/north-wales-news/wrexham-residents-wont-pay-contaminated-8888726; http://www.bbc.co.uk/news/uk-scotland-tayside-central-26249311.

[31]http://www.bbc.co.uk/news/uk-scotland-tayside-central-26249311.

[32]The funding programme for contaminated land closed on 1 April 2014. For a three-year period, a small amount of money will be still available to local authorities for ongoing remediation projects and "absolute emergency cases." Funding will then cease completely from 1 April 2017. See http://www.environmental-protection.org.uk/committees/land-quality/policy/defra-contaminated-land-capital-grants-to-be-axed/.

[33]http://www.bathchronicle.co.uk/Undefined-Headline/story-11331037-detail/story.html.

5.1.5 Insurance

During property transactions, it is not mandatory to set up environmental insurance. Where there is the risk of contamination, solicitors may advise buyers to obtain an environmental insurance policy should any polluting activity have occurred in the past and be subsequently discovered. Such policies are widely available to cover the costs of remediation of both undetected and disclosed contamination and related liabilities. However, they may also contain limitations on "what and whom" to cover. In the 1990s, the Association of British Insurance, for example, recommended to its members to exclude coverage for gradual pollution (i.e. pollution that happened slowly over time) from liability policies. The current insurance system in fact covers only sudden events (e.g. a tank that suddenly bursts).[34]

5.1.6 Comparison with the Environmental Liability Directive

The UK (E&W) regime on contaminated land is similar in several aspects to the provisions contained in the Environmental Liability Directive (ELD),[35] although it differs in some others. First of all, provided that the Pt2A regime has a specific focus on contamination from historical pollution incidents, it follows that it is concentrated on remediation measures rather than preventive aspects. The ELD addresses both these aspects equally.

In addition, the regime of liability under Pt2A is not limited to the responsibility of the "operator" as in the ELD, but includes that of the owners and occupiers. The Pt2A regime does not exclusively address contamination from specified and listed activities, as it happens under the ELD system.

Most importantly, under the Pt2A there is no statute of limitations, which means that there is no limitation period with regard to regulatory action and enforcement for historical pollution. By contrast, the ELD provides that actions carried out before 30 June 2007 are excluded from the scope of liability.[36]

5.1.7 Evaluation

The contaminated land regime constitutes a useful tool that has been given to enforcing authorities to deal with phenomena of historical pollution. Nevertheless, it has been criticized in several respects. First of all, there are different degrees of implementation of Pt2A throughout the country, depending on the level of

---

[34]See, for example, http://www.cli.co.uk/pages/contaminated-land.aspx.

[35]Directive 2004/35/EC on environmental liability with regard to the prevention and remedying of environmental damage [2004] OJ L143, 30.4.2004, p. 56.

[36]European Commission, DG Environment, *Annex—Part A: Legal Analysis of the National Transposing Legislation, Implementation Challenges and Obstacles of the Environmental Liability Directive* (2013, p. 360).

resources, expertise, training, and political influence of each enforcing authority. Councils with more financial resources employ experienced contaminated land officers that co-ordinate the application of the planning or Pt2A systems. Other councils lack sufficient resources even for investigation. It appears that the lack of resources constitutes a major problem for local authorities. Therefore, restoring the contaminated land fund could represent an important step forward.

Secondly, the system has been criticized for its strictness towards "innocent" owners or occupiers when the original polluter cannot be found. The authorities usually try to find solutions to avoid the costs of remediation falling on them. The difficulty is in fact finding the right balance between having the State (tax-payers) bear the costs of remediation, on the one hand, and having innocent occupiers/owners pay for clean-up costs, on the other. An option could be putting a charge over the land so that when the owner sells the land (or passes it on to relatives) the State can recover the costs incurred. Conscious buyers need, therefore, to conduct environmental checks before buying land if they do not want to incur the risk of being considered responsible for an activity that they did not contribute to causing.

In addition, it has been highlighted that Pt2A would not accord enough protection to executors or administrators of a probate estate. In fact "if the former owner of the contaminated land has died and the current owner is the probate estate, the estate may be an appropriate person. Similarly, if a company is in liquidation, the insolvency estate may be an appropriate person" (Zhao 2013, p. 132).

Finally, the system has attracted criticism as it leaves legal uncertainty as to the meaning of "reasonable inquiry" and "found"—for example, when a person dies or a company ceases to exist after the issuance of the remediation notice, can they still be considered "found"? (Zhao 2013, pp. 131–132).

5.2 *Brownfield Measures and Voluntary Remediation*

The UK is one of the most populated countries in Europe; land is therefore a precious resource. In 1996, the Government set a national target to ensure 60% of all new developments be built on brownfield land. This target was withdrawn in 2011 without being replaced with "a clear enough statement of the importance of using previously-developed sites first."[37] However, experts highlight that even without such official indication, the re-use of lands still remains highly encouraged.

As already stated, the UK legal system favours voluntary remediation. Pt2A, Sect. 78H(5)(b) provides that the enforcing authority shall not serve a remediation notice if it is satisfied that "appropriate things are being, or will be, done by way of remediation without the service of a remediation notice on that person." This is precisely the reason why the contaminated land regime has not been extensively

[37]See http://www.publications.parliament.uk/pa/cm201012/cmselect/cmcomloc/1526/152611.htm.

used. The Environment Agency reports that only 144 out of 746 sites "identified as contaminated" had been remediated by March 2007, while 90% of contamination was dealt with through the planning system.[38]

Voluntary remediation is usually associated with the concept of brownfield land. Brownfield land means previously developed land, whether or not it is currently used, and it occurs in rural and urban areas, but does not include agricultural or forestry land or buildings. Brownfield measures (i.e. measures adopted in relation to a brownfield site) provide that site owners or operators are given incentives to undertake (voluntarily) the necessary remediation actions. This is usually a condition for being granted approval for development by the public authority.

In practice local authorities are required to develop local strategies, such as tax incentives, to favour companies that are willing to undertake remediation measures.[39] The National Planning Policy Framework encourages local authorities to plan to reuse brownfield land before using greenfield sites, as long as the brownfield site is no longer environmentally valuable. This approach informs the UK planning policy.[40]

As to the comparison between the application of the planning system and Pt2A, both of them assess the risk arising from contamination and set up remedial measures. However, Pt2A focuses on existing uses of land that are affected by past activities, whereas the planning system looks not only at current, but also at future or proposed uses of the land. In addition, Pt2A requires demonstration that an unacceptable risk exists, whereas the planning regime requires developers to ensure that the development is safe. In other words, the local authorities have to demonstrate that land is posing a high risk before it can be determined as contaminated land, whereas the developer or owner has to show that the proposed development will be safe in order to avoid the need to remediate.[41] Finally, once remediated under the planning system, land can no longer be identified as contaminated land under Pt2A.[42]

Experts have reported that nearly 75% of contamination cases are cleaned up voluntarily through the planning process. If someone applies for planning permission to redevelop land, the planning authority requires the clean-up of the land

---

[38]Environment Agency, *Reporting the Evidence: Dealing with Contaminated Land*, p. 3.

[39]E.g., the *Contaminated Land Inspection Strategy* in the London borough of Hillingdon (2013–2018), p. 22. http://www.hillingdon.gov.uk/media.jsp?mediaid=29431&filetype=pdf).

[40]Department of Communities and Local Government, 2011, https://www.gov.uk/government/news/department-for-communities-and-local-government-publishes-2010-to-2011-annual-report-and-accounts. See in particular Catney 2006, p. 333.

[41]Land Quality Management Ltd, *SP1004 International Processes for Identification and Remediation of Contaminated Land, Report no. 1023-0* (2013) 13-1, available online via the RANDD/DEFRA website, http://randd.defra.gov.uk.

[42]https://www.gov.uk/government/uploads/system/uploads/attachment_data/file/297014/geho0512bupg-e-e.pdf. See also *Planning policy Statement 23—Annex 2—Planning and Pollution Control: Development of Land Affected by Contamination*, although replaced in 2012 by the National Planning Policy Framework.

before allowing the development to take place.[43] Under the planning system, land contamination is a "material consideration." Planning authorities consider the potential or actual presence of contamination before authorizing any particular changes to land use (Luo et al. 2009, p. 1128).

From the analysis conducted above it can be inferred that the system follows a pragmatic approach. The land is expected to be "safe" (not creating risks for future users) rather than completely "clean." There is a common practice of avoiding moving contamination from one place to another and of favouring the underground isolation of the contamination, rather than complete elimination. As to the level of clean-up, experts report that it depends on the prospective purposes of the land. Usually, a higher level of clean-up is required when building houses as compared to building a car park. As has been highlighted, "land contamination problems arise because land previously used for industrial uses is now being used for residential uses, but most of those who live on residential land have an expectation that residential land is 'safe'" (Fisher et al. 2013, p. 974). However, experts have highlighted that the general public is usually more concerned about property value rather than actual possible health effects of past contamination. The consequences of site contamination for human health and the environment tend in fact to be underestimated. In other words, until someone is injured by contamination, residents usually do not ask the local authorities to start an investigation.

These last considerations speak to **Hypothetical Scenarios 1 to 4**. In fact, if there is no significant risk of harm to human health, for example to future users of the land, and if there is neither damage nor threat to human health as a consequence of the pollution (scenarios 1 and 2), the relevant authorities would not be required to take action under the procedures analysed above. By contrast, should a risk to human health exist following, for example, a contamination of groundwater used in agriculture or in human nutrition (scenarios 3 and 4), the relevant authorities would be required to take action accordingly.

5.3 Tort Law

Private law remedies, such as negligence, private or public nuisance, and breach of statutory duties, are also relevant. In fact, historical pollution cases can at the same time have endangered the health of the public and caused personal injuries to individuals who are therefore entitled to be awarded damages. It has been reported, though, that it is very difficult to prove in court that contamination has caused personal injuries.

[43]Updated figures prepared by DEFRA as to the contaminated land sector activity in England and Wales are available at http://randd.defra.gov.uk/Default.aspx?Menu=Menu&Module=More&Location=None&Completed=0&ProjectID=136.

These considerations speak to **Hypothetical Scenario 5**, whereby in the presence of a causal link between the pollution and the death or injury of a number of people, the latter would be allowed to initiate judicial action in order to be awarded damages for personal injury caused by the negligent conduct of the operator (see the *Corby* case, Sect. 5.4.3).

Civil procedures, such as nuisance actions, are usually tried in the County Court by a circuit judge, whereas cases that are more complicated or of high value are tried in the High Court by a High Court judge. The Technology and Construction Court is a division of the High Court in England and Wales that specializes in technically complex issues.

It has to be noted that, compared to the other remedies, public nuisance (besides being a tort) is always a crime in common law; therefore, it is classified as a criminal offence and is usually remedied by criminal penalties. By contrast, for example, private nuisance is exclusively a civil wrong; therefore, damages are the appropriate remedy to compensate those who have been harmed (Fisher et al. 2013, pp. 231–235).[44] According to Lee, "public nuisance is potentially a high revealing tort in the environmental arena. In particular, public nuisance conceptualizes harm as a public event that has an impact on communities, avoiding the misinterpretation of environmental harm as a series of individual harms, but recognizes that individuals may also suffer, as individuals" (2009, p. 136).

5.4 *Relevant Case Law*

On the basis of the above analysis, the following subsections describe relevant examples. The first and second cases refer to the difficulties related to the succession of company liabilities for clean-up over time. The third analyses the topic from the eyes of the injuries caused to individuals and tort law.

5.4.1 Bawtry Gas Works[45]

The *Bawtry Gas Works* case arose from oil and groundwater contamination by coal tar (a by-product of the gas-making process) and other pollutants, including nickel, by gas companies. It particularly concerns the difficulties related to the statutory succession of company liabilities in the framework of subsequent nationalization, privatization, and then reorganization of the gas industry in the UK.

---

[44]"Private nuisance" deals with protecting the occupier of a land in respect of unreasonable interference with the enjoyment or use of his or her land. Public nuisance is a crime that at the same time is actionable in tort law if the claimant suffers "particular damage" on top of the damage suffered by the public in general. See also (Bishop and Jenkins 2011, p. 285; Lee 2011, p. 986; Hamer 2009, p. 8).

[45]http://en.wikipedia.org/wiki/Bawtry_gasworks_contamination.

In 1915, a private gas company, the Bawtry and District Gas Company, started operating gasworks. It then merged with the South Yorkshire and Derbyshire Gas Company in 1931. Later in 1948 these companies were nationalized by the Gas Act and acquired by the East Midlands Gas Board (EMGB) which inherited its predecessors' assets and liabilities. During the 1960s the EMGB sold the site to developers, who in their turn sold the land to further developers (Kenton Homes Ltd. and Jackson Ltd.), which then started constructing housing in the late 1960s. (During the sale the EMGB made reference to the existence of underground tanks containing coal tar.) Subsequently, in 1972 the EMGB's liabilities passed on to British Gas Corporation and then, in 1986, with the privatization of the gas industry, to British Gas plc. Further transactions devolved such liabilities to Transco and finally to National Grid Gas plc (NGG).

In 2001, a resident living in the area discovered a tar pit, while he was digging in his garden. This led to the determination that the house was situated on a contaminated land. After the discovery the Doncaster Metropolitan Borough Council started an investigation of the site and found that the contamination also affected other adjacent parts and houses. The site was then considered a "special site" under the Environment Protection Act in 2003, and the enforcement powers passed onto the Environment Agency. In September 2005, the Agency identified Kenton Homes Ltd. and Kenneth Jackson Ltd. as Class A "appropriate persons" (Wolf and Stanley 2014, p. 337). However, as both developers had been dissolved in the meantime, the only one left to pursue was NGG, which had applied for a judicial review at the High Court. The owners/occupiers of the properties were instead to be considered "Class B" persons. In relation to them, the Agency had already decided that the costs of the remediation would have caused these individuals considerable hardship so they could not be pursued.

Before the High Court, NGG's main argument was that it could not be considered the "appropriate person" because it had not itself "caused or knowingly permitted" the contamination. In addition, it submitted that

> even if there had been liability at those times, the statutory provisions transferring property, rights, liabilities and obligations under the various reorganisations had referred to such liabilities as those existing immediately before the date of transfer. So, the transfers could not have operated to transfer liabilities under Part IIA, because this had not been in force at the time of the transfers.[46]

In 2006 Mr. Justice Forbes agreed with the Agency's position and stated first of all that according to ministerial statements, the Government had expressly considered the issue of statutory successors to bodies such as British Coal and British Gas, and had intended that Pt2A liability be borne by successors (R (*National Grid Gas plc (previously Transco plc)) v Environment Agency* 2006 EWHC 1083 (Admin)). In addition, considering the nature of Pt2A liability, Mr. Justice Forbes stated that it was self-evident that liability under Pt2A was of its very nature capable

[46]http://en.wikipedia.org/wiki/Bawtry_gasworks_contamination.

of relating to acts, omissions, or activities undertaken before that regime came into force.

Subsequently, the decision of the High Court was appealed at the House of Lords (House of Lords: Judgments—R (on the application of National Grid Gas plc (formerly Transco plc)) (Appellants) v. Environment Agency (Respondents) (Civil Appeal from Her Majesty's High Court of Justice)) where the decision of the lower Court was overturned and the Agency's decision of September 2005 quashed. The Court stated that NGG could not have been an "appropriate person" as it neither caused nor knowingly permitted the contamination in the Bawtry site. The House of Lords clarified that a Class A person is "whoever materially caused or knowingly permitted the contamination on the land and that any attempt to expand this concept would mean walking away from the actual text of the legislation." In a nutshell, the Lords confirmed that liability should remain with the original polluter, provided that it could be found (any amendment to this approach would require a new legislative intervention by the Parliament). The Lords therefore stated that the statutory successor to the gas company was insufficiently connected to the original polluter to constitute an appropriate person.

Eventually the clean-up was funded at public expense; DEFRA provided a grant (Carriage and Ennis 1990, p. 1563). The cost was around £63,253 per property involved, amounting to £695,782 overall. All structures were removed together with any contaminated materials, and in order to prevent residents from coming into contact with residual contamination, separators were installed (0.6 m below the ground). The soil dug out was replaced with new fill of a type suitable for domestic gardens, and there was full restoration of levels of the land, drainage, boundaries, and garden/domestic infrastructure.

5.4.2 Redland Minerals and Crest Nicholson Residentials

This case refers to the contamination of groundwater by bromate and bromide, affecting drinking water supplies. From the 1950s to the 1980s, companies currently known as Redland Minerals Ltd. were carrying out chemical production in Sandridge, Hertfordshire. This production involved the generation and deposition in the soil of bromate and bromide. Over time the contaminants started to work their way down into the aquifer strata near St. Albans (Hertfordshire) and into the watercourse that was used for public supply.

In 1980 Redland ceased its activities and Crest, a developer of residential properties, in the meantime decided to carry out some testing on the site. The tests showed bromide contamination in the upper soil without, however, proving further that the substance had already penetrated into the aquifer. Crest bought the site from Redland in September 1983, and a few months later it started to demolish some buildings and break up the hardstanding.

It has to be noted that during the property transaction Redland informed Crest about the fact that the works already undertaken had contaminated the site with bromide, although not with bromate.

Construction works by Crest began only a few years later with the consequence that for a period of time, even though the presence of the contaminants was well known, the site remained exposed to rain with the consequent worsening of the contamination. In 1986 Crest started work after disposing of a shallow layer of soil across the site. The subsequent excavation works failed to remove all the substances given the fact that in the meantime the pollutants already had reached the lower strata. Eventually, in 1987, Crest sold the land.

The site was identified as contaminated land in 2002, and the matter was referred to the Environment Agency. Redland and Crest were both identified by the Agency as "appropriate persons" (Class A). Redland's defence that "it had sold with information" was not upheld by the Agency as this information was not sufficient to make Crest fully aware of the contamination in the aquifer. Therefore, the Agency issued a remedial notice. The companies appealed to the Secretary of State who then requested an inspector to hold an enquiry. It was decided that Crest should bear 15% of the liability for bromate contamination and 55% of the liability for bromide, with the remaining 85% (bromate) and 45% (bromide) being allocated to Redland.

The Secretary of State subsequently confirmed the notice, and Redland and Crest challenged the decision before the High Court (Redland Minerals Ltd, R (on the application of) v Secretary of State for Environment, Food and Rural Affairs and Ors [2010] EWHC 913 (Admin) (4 February 2010)).

While assessing the causation test "the Court did not engage with issues such as standard of proof for causation, or multiple sufficient causes, etc. Rather, the court simply assessed whether, on the facts, the individuals before the court could be said to have caused the pollution to be in, on, or under the land. No broad guidance was given, despite the dearth of case law in this area" (Lees 2015, p. 86).

The High Court eventually rejected the appeal brought by the companies by stating that

> Redland had caused all the bromide and bromate to be on the land in the first place and allowed them to filter down to the lower strata during its long period in control of the site. Crest had brought no contaminants onto the site but had accelerated the way in which the contaminants already in the land were flushed down to the lower levels.

5.4.3 Corby Steelworks Case[47]

The *Corby Steelworks* case is a class action (Group Litigation Order)[48] brought by the parents of eighteen children who were born with physical defects as a consequence of the flawed remediation of contaminated land.

In the early 1930s the area of Corby (in Northamptonshire) started to become a steelmaking centre. By the 1960s Corby was already one of the biggest

[47]http://en.wikipedia.org/wiki/Corby_toxic_waste_case; http://www.environmentlaw.org.uk/rte.asp?id=266.

[48]https://www.justice.gov.uk/courts/rcj-rolls-building/queens-bench/group-litigation-orders.

industrialized areas in the Midlands, and one of the largest operating companies was Stewarts & Lloyds Ltd.

In the 1980s Stewarts & Lloyds Ltd. became profitless and was shut down. Subsequently, Corby Borough Council decided to acquire the relevant area from the British Steel Corporation which had previously acquired it after Stewarts & Lloyds Ltd. ceased trading. Because of the high contamination with toxic waste from the former steelworks complex, the Council decided to decontaminate and redevelop the land, but the works undertaken were carried out by carrying the toxic waste in open lorries, leading to spillage of sludge all over roads and the consequent release of toxic dust into the air.

Between the 1980s and 1990s several children in the area were reported to have been born with upper-limb malformations (shortened or missing arms, legs, and fingers), although though no such cases were previously detected in the respective family history.

In 2005, after evidence was presented in order to show the higher rate of birth defects caused by the exposure to toxic waste,[49] the High Court allowed the parents to pursue the claim in the form of a class action against Corby Borough Council (the so-called Corby Group Litigation).

The group sued the Council for damages for personal injury allegedly caused by its negligent handling of the contaminated soils. The claims were in negligence, breach of statutory duty, and public nuisance (tort law).

The allegations of public nuisance were set aside by the House of Lords (*Corby Group Litigation v Corby Borough Council* [2008] EWCA 463), and in 2009, the case was brought before the Technology and Construction Court, which is part of the High Court (Corby Group Litigation, Re [2009] EWHC 1944 (Technology and Construction Court)).

Mr. Justice Akenhead stated that the Council had been "extensively negligent" in its clean-up activities. The Judge found that the appropriate standard of care had not been met by the Council between 1985 and 1997 and that it was reasonably foreseeable that the local population could be exposed to hazardous or contaminated substances (McCullough 2010, p. 269). Judge Akenhead held that a considerable quantity of contaminated material present in mud and dust was spread over the whole of Corby and up to four kilometres from the main site, including a further area (McCullough 2010, p. 270). Therefore, the Judge established that the complainants' injuries could have been caused by the Council's negligence. As highlighted:

> it was agreed that CBC owed a duty to the claimants and their mothers to take reasonable care to avoid injury to them in the execution of the reclamation works. The standard of care was that of an ordinarily careful local authority embarked on reclamation works of the type involved. The duty was to be judged by the standards known or reasonably ascertainable, and the knowledge available, at the time of the alleged breach. […] Akenhead also held that it was well established that a defendant could not evade liability for negligence where, such

[49] http://www.independent.co.uk/news/uk/crime/parents-of-30-children-sue-over-birth-defects-they-blame-on-cleanup-of-toxic-waste-dumps-517200.html.

> as in this case, it engaged independent contractors to perform work which, by its very nature, involved special danger to another. (McCullough 2010, p. 270)

Eventually, the Judge did not determine the amount of compensation in favour of the children as this needed to be decided in a subsequent hearing. However, later a non-judicial settlement of the case was reached. In 2010, the Council agreed to pay a confidential amount, while continuing to deny liability.[50] The Council undertook the clean-up works through the planning process.

5.4.4 Evaluation of Cases

The three cases analysed above tackle the phenomenon of historical pollution from different points of view. The first two cases address companies' liability in application of the contaminated lands regime, although with major differences. In the first case, the "appropriate person," who actually caused or knowingly permitted the historical pollution, could not be "found." The Court held that companies could not be held liable for contamination that occurred at the site prior to the privatization and subsequent take over. As a consequence, the Court decided that the clean-up costs should be borne by the general public.

By contrast, in the second case the "appropriate persons" were "found," although with the difficulties inherent in apportioning liability between the companies. It is important to note that, on the one hand, the defence of the selling company based on "selling with information" was not upheld as this information was not sufficient to make the buyer fully aware of the presence of the contamination. On the other hand, the case demonstrates that the conduct of accelerating the contamination can also give rise to liability. In this regard, it could be said that contaminated land is one of a number of environment-related compliances that have to be investigated in the context of corporate transactions (Tromans and Turrall-Clarke 2009, p. 560).

Legal doctrine has pointed out that these two cases testify to the existence of different and to some extent unpredictable and difficult-to-reconcile approaches followed by courts regarding the test of causation. In fact, while in the second case the developers were held responsible for causing contamination simply as a result of failing to "cover" the contaminated land, in the first case the statutory successor to a gas company was considered not sufficiently connected to the original polluter so as to constitute an appropriate person (Lees 2015, p. 88).

In the third case, the phenomenon of historical pollution is seen from the perspective of private injuries. The local authority has been considered negligent for not taking reasonable care in removing the contamination so as to avoid injury according to the standards known, or reasonably ascertainable, at the time of the

[50]http://www.telegraph.co.uk/news/health/news/7598590/Council-to-pay-out-millions-to-birth-defect-children.html; http://www.theguardian.com/society/2010/apr/16/corby-steelwords-compensation-fight.

conduct, although a non-judicial settlement of the case was finally preferred. The case also shows the application of the planning system for clean-up.

5.5 Access to Information and Public Participation

The role of the public with regard to access to justice has already been addressed in the context of the application of private law remedies. An illustrative example of this has been described in the context of the *Corby Steelworks* case where residents living in the vicinity of a polluting industry decided to take legal action. By contrast, in this section the role of the public will be addressed in relation to its demand to have access to environmental information and to participate in the relevant decision-making processes regarding contaminated land.

Given the seriousness of the phenomenon, affected members of the public should be willing to participate in decisions which can have an impact on their wellbeing. However, a general lack of public awareness has been detected in practice.

Against this premise, in the context of the contaminated land regime, public participation materializes in the establishment of site registers at a local level and on the provisions on public consultation.

5.5.1 Site Registries

Under the contaminated land regime, the enforcing authorities are required to keep public accessible registers of data related to Pt2A (Sect. 78R). These registers include data on sites being screened or investigated for possible contamination, background site information, details of the relevant parties, remediation notices, appeals, convictions, notices in relation to special sites and information about remediation work notified to the authority, and voluntary remediation agreements (Sect. 78R3).

5.5.2 Public Consultation

Public consultation aims at presenting the current law and policy on the matter and asking a series of questions regarding suggested approaches to dealing with contaminated land and as to the level of satisfaction with present arrangements. Community-based consultation can help in developing recommendations on the acceptable level of human health risk for the site or identifying the appropriate remediation methods, even though the level of technicality required in such a complex topic may be seen to discourage sensible public participation.

The Royal Commission on Environmental Pollution (21st Report, chapter 7) of 1998 found that public consultation could provide a method both to establish peoples' values regarding environmental issues such as historical pollution and to educate the public so that those values may reflect deeper understanding of the issues involved.[51] The use of surveys, consultation exercises, and public inquiries was therefore recommended by the Royal Commission.

One of the major public consultations in the UK (Paying for Our Past)[52] was carried out in 1994 (see Sect. 5.1). In the aftermath of this consultation process, the Government published a framework for Contaminated Land in 1994.

The most recent public consultation was instead launched in 2010 by DEFRA and the Welsh Assembly Government, and it resulted in the consultation of more than 150 stakeholders, including the UK Environmental Law Association (UKELA) .[53] The participatory process provided advice on how the regime should be updated, in particular addressing a number of changes regarding brownfields potentially associated with historical pollution.[54] The revised 2012 Statutory Guidance took into account the outcomes of the 2010 public consultation (Statutory Guidance 2012, p. 3).

Finally, it has to be mentioned that the 2004 Environmental Information Regulations grant members of the public the right to access environmental information held by public authorities. The request can be made by letter, email, telephone, or in person. Local authorities endeavour to answer questions that may be raised by members of the public, developers, and landowners in relation to Pt2A and planning by providing relevant contact details.[55]

[51] *Setting Environmental Standards, 21st Report* (1998), esp. Chap. 7: Articulating Values. Available at http://webarchive.nationalarchives.gov.uk/20110322143804/http:/www.rcep.org.uk/reports/index.htm.

[52] *A Consultation Paper from the Department of the Environment and the Welsh Office*, March 1994.

[53] http://www.ukela.org/content/doclib/204.pdf.

[54] DEFRA, *Public Consultation on Changes to the Contaminated Land Regime under Part 2A of the Environmental Protection Act* 1990 (2010).

[55] By way of example, the Contaminated Land Strategy of London Borough of Hounslow reports that "[t]he Council receives a growing number of such requests each year, and has developed a standardised report, which forms the basis of the response in most cases. A report will include comments relating to the current priority group allocated to a given site under the Part 2A inspection process. The report will also confirm any formal decisions made, or action taken by the Council in accordance with Part 2A if appropriate" (Contaminated Land Strategy of London Borough of Hounslow (2008) at http://www.hounslow.gov.uk/contaminated_land_strategy.pdf).

6 Environmental Criminal Law

In this section, we analyse the phenomenon of historical pollution through the eyes of criminal law. In order to do so we will first describe the way in which the UK (E&W) legal system criminalizes conducts detrimental to the environment (Sects. 6.1–6.3), including corporate environmental crimes (Sect. 6.4), and then, we will put forward some tentative evaluations of the applicability of the criminal law system to cases of historical pollution.

6.1 Key Features of the Criminal System

In UK (E&W), there are no specific criminal courts that prosecute conducts detrimental to the environment. Environmental criminal cases begin in the Magistrates' Court, whereas more serious cases are tried in the Crown Court. General criminal law and procedure rules apply to breaches of environmental legislation that constitute criminal offences, including historical contamination cases. The main source of criminal law has been the common law, as developed through decisions of the courts and by legal doctrine (Ashworth and Horder 2013).

English criminal law is characterized by the absence of a criminal code. In the majority of the cases the public report to the police a certain offence (Faure and Heine 2000, p. 339). Only in a minority of cases do the police start the investigations autonomously. In this context, the Environment Agency has a proactive role. The Agency can visit premises or building sites to check on compliance with the environmental law. Regulatory agencies usually prosecute as a last resort. They prefer to rely on informal and formal warnings as a means of putting pressure on wrongdoers. In this regard, the Environment Agency complies with a published "enforcement and prosecution policy."[56] The Agency can prosecute incidents or breaches with significant environmental consequences: operating without the required licence, consent, or authorization; persistent or excessive breaches of statutory requirements; failure to comply with formal requirements to remedy environmental harm; reckless disregard of environmental management or quality standards; or obstructing Agency staff in the gathering of information. Influential factors can be whether an incident was foreseeable, the intent of the offender, previous offences, and the deterrent effect of prosecuting (Bell et al. 2013, p. 298).

When the Agency considers that a criminal sanction is appropriate, it assesses the situation in accordance with the requirements of the Code for Crown Prosecutors before starting a prosecution. The Agency must then meet the test set out in the Code in order to determine whether there is sufficient evidence and that

[56]The policy is available online at http://www.environment-agency.gov.uk/static/documents/Business/1430_10_External.pdf).

the prosecution is in the public interest.[57] However, the decision as to whether or not to charge a wrongdoer is taken under the statutory charging scheme introduced by the Criminal Justice Act 2003. Eventually it is the Crown Prosecution Service that takes the decision as to whether to charge the suspect.

The trial is adversarial. The prosecution attempts to prove its case beyond reasonable doubt, and the defence seeks to undermine the prosecution's case and to create reasonable doubt. It should be noted that the task of the defence is not to prove the innocence of the defendant but to prevent the prosecution from proving guilt. The evidence of witnesses in major criminal cases is evaluated by a jury. While the judge may comment on the evidence of fact, the jury is the sole arbiter of the facts in the case. In all other trials, the magistrate, judge, or inspector (reporter) evaluates the evidence given. Where a person is found guilty after a trial in the Magistrates' Court, they may appeal to the Crown Court against the conviction.

As to the limitation period, the prosecution is prevented if the time limit laid down by the law has expired before the commencement of proceedings. Statutory offences may have a specific time limit, and they indicate when the prescriptive period begins to run. By contrast, if no specific time limit is set, a general prescriptive period of six months (from the date of the alleged offence) applies.

It is not compulsory to have a lawyer in any environmental judicial review. Lawyers who specialize in environmental law can be found through Chambers and Partners or the Legal 500. The Environmental Law Foundation (ELF) specializes in environmental law, and the UK Environmental Law Association (UKELA) is a body of environmental lawyers, including experts in land contamination. The Planning and Environmental Bar Association (PEBA) is a body of barristers in England and Wales that deals with environmental and planning cases.

Against this background, in the following section we describe some examples of environmental offences with special regard to soil, water, and air contamination, even though it has to be anticipated that these have never been applied to cases of historical pollution.

6.2 *Environmental Criminal Offences*

In the UK, there is no national code dedicated to environmental crimes. The legislation in the field is highly fragmented and spread across many statutes and regulations. The most relevant instrument is the 1990 Environmental Protection Act, which defines the fundamental structure and authority for waste management and control of emissions into the environment.[58] The complexity of the UK environmental legislation is mainly due to the considerable reforms carried out in the

[57]http://www.environment-agency.gov.uk/static/documents/Business/Enforcement_and_Sanctions_Guidance.pdf.

[58]http://www.legislation.gov.uk/ukpga/1990/43/contents.

last twenty years. However, a certain level of standardization has been reached at the legislative level as a result of the transposition of EU law on the matter, particularly the Environmental Crime Directive.[59]

Of particular relevance first of all is the offence of supplying water unfit for human consumption under the 1991 Water Industry Act, Sect. 70. The conduct is described as follows: “where a water undertaker’s supply system is used for the purposes of supplying water to any premises and that water is unfit for human consumption,” this constitutes a criminal offence.[60]

The more general water pollution offence arises when a person causes or knowingly permits a water discharge activity or groundwater activity except to the extent that this is authorized by an environmental permit. This offence is provided by the Environmental Permitting Regulations 2010, regulation 12.[61] The leading case on the matter is the *Alphacell* case (*Alphacell Ltd v Woodward* [1972] AC 824), where the House of Lords gave guidance on the interpretation of the requirement of “causing” the discharge. That case concerned a paper factory owner that had been convicted for entering a polluting substance into a river above the level of consent. The offence related to an underground pipe, which had become disconnected due to a blockage. The appellant was unaware of the pollution. The House of Lords held that there was no need to prove negligence or fault. The factory was guilty of the general offence of causing pollution simply by carrying on the activity that gave rise to the pollution.

In the waste management sector the 1990 Environmental Protection Act (Sect. 33) makes it a criminal offence to knowingly causing or permitting the deposit of controlled waste in or on land, submit controlled waste to any listed operation (i.e. disposal and recovery operations) unless it is permitted, and treat, keep, or dispose of controlled waste in a manner that is likely to cause pollution of the environment or harm to human health.[62]

---

[59]Directive 2008/99/EC on the protection of the environment through criminal law [2008] OJ L328, 6.12.2008, p. 28.

[60]The offence is prosecuted on summary conviction with a fine not exceeding £20,000 or on conviction on indictment, to a fine to be established and/or imprisonment for a term not exceeding two years. It constitutes a defence for the responsible person the fact that they had no reasonable grounds for suspecting that the water would be used for human consumption; or they took all reasonable steps and exercised all due diligence to secure that the water was fit for human consumption on leaving the primary water undertaker’s pipes or was not used for human consumption.

[61]Acting under statutory authority amounts to a defence. In addition, where the entry of discharge was made in an emergency in order to avoid danger to human health this constitutes a defence as well and the polluter has to take reasonable steps to minimize the pollution as soon as possible.

[62]The maximum penalty for these offences is, on conviction in the Magistrates’ Court, 12 months’ imprisonment and/or a fine of £50,000 or, on conviction in the Crown Court, five years’ imprisonment and/or an unlimited fine. It constitutes a defence the carrying out of the relevant activities in an emergency in order to avoid danger to the public (but danger to the environment is not mentioned). In addition, there is a defence if the operator took all reasonable precautions and exercised all due diligence to avoid the commission of the offences.

As to air pollution and emissions, the 1993 Clean Air Act (Part I) provides that dark smoke shall not be emitted from a chimney of any building, or from a chimney that serves the furnace of any fixed boiler or industrial plant.[63]

It has to be noted, though, that even if potentially applicable in the sense that no legal obstacles exist, none of these criminal offences has ever been used in cases of historical pollution. In fact, as explained in Sect. 5, the system favours a more "pragmatic" approach, using the contaminated land regime or the planning system.

However, criminal law plays a residual role in the context of the contaminated land regime. In fact, if the operator fails to comply with the clean-up instructions contained in the remediation notice, this constitutes a criminal offence under the Environmental Protection Act 1990 (Sect. 78 M).

In practice, if after six months from the issuing of the remediation notice the site has not yet been cleaned up, the regulator asks the Magistrates' Courts to fine the responsible person and oblige them to undertake the appropriate remedial measures. The responsible persons are first given the opportunity to repair the damages caused. Therefore, the system clearly aims at inducing them to clean up the land rather than to face prosecution.

Be that as it may, this offence can be tried in the Magistrates' Courts (Sect. 78 M of Pt2A) and the penalty is a fine. In cases in which the contaminated land is an industrial, trade, or business premises, the maximum penalty is a fine of £20,000, whereas outside these cases the maximum fine is £5,000 (Sect. 78 M.4). As has been highlighted, "the criminal offence is therefore constituted by the failure to comply with a remediation notice, but the pre-requisite for such liability is a valid remediation notice. This demands a) the existence of contaminated land (significant harm or risk of significant harm); b) that the notice be served on an appropriate person; c) that the actions contained within that notice are considered reasonable" (Lees 2015, p. 66).

However, no cases of criminal prosecution have been reported so far. This testifies to a precise choice by the UK legal system to leave only a residual and subsidiary role to criminal law responses in the context of historical pollution phenomena. The rationale behind Pt2A was precisely that the unfairness of the retrospective liability when addressing the long-term consequences of past polluting activities had to be counterbalanced by the absence of criminal law responses.

[63]The offence is prosecuted on summary conviction to a fine not exceeding level 3 on the standard scale if the chimney is of a private dwelling or to a fine not exceeding level 5 on the standard scale in any other case. It constitutes a defence, for example, proving that the alleged emission was solely due to the lighting up of a furnace which was cold and that all practicable steps had been taken to prevent or minimize the emission of dark smoke; that the alleged emission was solely due to some failure of a furnace and that the failure could not reasonably have been foreseen, or, if foreseen, could not reasonably have been provided against; and the alleged emission could not reasonably have been prevented by action taken after the failure occurred.

6.3 Sanctions

Fines are the most common sanctions for environmental criminal offences. The other types of sanction, such as custodial sentences, are rarely applied. In recent years, there has been a growing sense of dissatisfaction with the sanctions imposed upon those who commit environmental crime. The majority of environmental offences that are taken to court are punished with relatively small fines and a very low level of prosecutions is being detected (Bell et al. 2013, p. 301).

Concern about the level of sentences handed down for environmental crimes has recently (in 2014) led to the introduction by the Sentencing Council of the Sentencing Guidelines for Environmental Offences[64] in order to assist magistrates in dealing with environmental cases. For example, the Guidelines provide that fines are unlimited for the most serious environmental offences and that custodial sentences are available. However, experts have expressed concern about the fact that these sanctions are too often and too easily internalized by deviant actors, especially corporations.

While it is true that there has been an increase in public perception of the importance of environmental interests, a general perception still persists that environmental crimes are somehow less serious compared to other crimes, such as murder or burglary (Bell et al. 2013, pp. 177–278). In addition, a lack of sufficient training and expertise on the part of judicial operators when dealing with the specificities of environmental crimes has been detected. It follows that if the phenomena of pollution are conceived as less important than others, they are not placed at the top of the political agenda, therefore leading to weak enforcement action.

This level of dissatisfaction has led the regulatory bodies, such as the Environment Agency, to start imposing administrative sanctions outside the criminal law process, with specific regard to civil law sanctions. The view was expressed, in fact, that the latter types of sanctions could have a more dissuasive effect. The sanctioning regime that already introduced civil sanctions in 2008 has been conceived as an alternative to sanctions imposed through criminal prosecution.[65] Although some critics have pointed out that civil sanctions may confer too much power on the regulatory agencies at the expense of the courts, in some areas they can have greater effect than criminal offences.

---

[64]https://www.sentencingcouncil.org.uk/wp-content/uploads/Final_Environmental_Offences_Definitive_Guideline_web1.pdf.

[65]*Regulatory Enforcement and Sanctions Act 2008* (Part 3). On civil sanctions, see Macrory and Woods 2003.

6.4 Corporate Environmental Liability

Because of the scale of their industrial activities, legal persons cause the most serious breaches of pollution control legislation.[66] The general rule of corporate environmental liability is that when companies commit environmental crimes, prosecution is brought against them.

The main difference between an environmental crime committed by a major company and that committed by an individual is that in the second case criminal intent or negligence is more likely to be proved, whereas companies usually tend to "cover" their blameworthiness by reference to factors outside their control (Bell, McGillivray, and Pedersen, p. 282).[67] This difficulty does not arise for strict liability offences, so that the actions of the agents of the company can, without difficulty, be attributed to the company.

Corporate liability for acts against the environment can be thus attached to companies, for example, for having breached a licence held in the name of the company, or to individuals working for the company when they breached relevant pollution control legislation. In this second case, in order to establish personal liability, the offence has to be committed with their consent or connivance or must be attributable to their neglect.

According to the Environment Agency's report *Spotlight on Business Environmental Performance 2003*, which was published in July 2004, eleven company directors were personally fined sums up to £20,000. Data from the Environment Agency show that small- and medium-sized companies are responsible for up to 80% of all pollution incidents and for more than 60% of the commercial and industrial waste produced in England and Wales. In 2004, the subcommittee on environmental crime on the issue of corporate offences concluded that

> [e]ven putting aside the often extreme difficulty of finding and successfully prosecuting the individual behind the decision that led to an environmental crime, *the current sentencing system is just not flexible and imaginative enough adequately to punish corporate bodies or those in senior managerial positions within them. It is disgraceful that some companies openly boast about their crimes as though they manifested some sort of commercial talent or marketing genius. The Government must adopt a much tougher stance with businesses—regardless of their size and nationality—which flagrantly flout the law.*[68]

However, more recent figures show that pollution incidents by companies are slowly decreasing. By way of example, the *2012 Sustainable Business Report* by the Environment Agency reports that there have been 504 serious and significant

---

[66]For a detailed description of corporate liability in England and Wales, see Wells 2001, pp. 84 ff.

[67]See also (De Pret 2000, pp. 65–77).

[68]Emphasis in the original. House of Commons Environmental Audit Committee, *Environmental Crimes and the Courts, Sixth Report of Session 2003–04*, p. 16.

pollution incidents in England in 2012, down by 8% compared to 2011, and this is the lowest recorded level to date.[69] As to the level of fines, the *2012 Sustainable Business Report* states that

> 178 separate companies were fined for environmental offences in 2011, compared with 179 in 2010 and 317 in 2005. Total fines came to over £3.8 million. This is lower than the total of £4.8 million in 2010. Total fines in 2005 came to £2.7 million. The average company fine in 2011 was £21,600, compared with £26,900 in 2010 and £8,700 in 2005. (*2012 Sustainable Business Report*, p. 16)

This decrease seems to be linked, among other factors, to a growing tendency of the courts to impose much higher fines for environmental offences. As anticipated above, in 2014 the Sentencing Guidelines for Environmental Offences were introduced, including for corporate environmental crimes. The Guidelines introduced a step-by-step approach to calculation of a fine based upon the degree of culpability of the offender, the harm caused by the offence, and the size of the offending organization, to be assessed by reference to its turnover. Companies are divided into four categories, micro, small, medium, and large. By way of example, large organizations are identified as those with a turnover or equivalent of "£50 million and over" and the starting point for a fine under the Guidelines would be £60,000 with a range of £35,000–£150,000 before taking account of any aggravating and mitigating features.

The Guidelines, however, made it clear that the starting points and range of fines suggested did not apply to very large companies by stating that "where a defendant company's turnover or equivalent very greatly exceeds the threshold for large companies, it may be necessary to move outside the suggested range to achieve a proportionate sentence" (Guidelines, p. 7). By way of example, in a recent case (*R v Thames Water Utilities Ltd* [2015] EWCA Crim 960) the Court of Appeal was faced with the issue of a company's turnover that fell well outside the scale set down in the Guidelines. Thames Water, with a turnover of £1.9 billion, was found liable for discharging untreated sewage into a nature reserve, thus breaching regulations 38(1)(a) and 39(1) of the 2010 Environmental Permitting (England and Wales) Regulations. The difficulty in assessing the level of fine is well described in the following passage

> the need to impose a just and proportionate penalty will necessitate a focus on the whole of the financial circumstances of the company. We have already outlined the approach by reference to the guideline—starting with turnover, but having regard to all the financial circumstances, including profitability. In such a case, the objectives of punishment, deterrence and the removal of gain (for example by the decision of the management not to expend sufficient resources in modernisation and improvement) must be achieved by the level of penalty imposed. This may well result in a fine equal to a substantial percentage, up to 100%, of the company's pre-tax net profit for the year in question (or an average if there is more than one year involved), even if this results in fines in excess of £100 million. Fines of such magnitude are imposed in the financial services market for breach of regulations.

---

[69]https://www.gov.uk/government/uploads/system/uploads/attachment_data/file/297237/LIT_8546_b08a53.pdf.

> In a Category 1 harm case, the imposition of such a fine is a necessary and proper consequence of the importance to be attached to environmental protection. (para. 40)[70]

Rejecting Thames Water's appeal against a £250,000 fine, the Court eventually confirmed that amount and stated that fines against very large companies "had to bring home the appropriate message to the directors and shareholders of the company." This judgment follows another case (*R v Sellafield Ltd* [2014] EWCA Crim 49) where the Court of Appeal already gave guidance as to the approach to fines to be imposed on companies of very significant size. In that case Sellafield Ltd., with a turnover of £1.6 billion, was fined £700,000 for offences arising from the disposal of radioactive waste. Rejecting Sellafield's appeal, in the relevant passage the Court interestingly stated that

> [a] fine of the size imposed, even though only a little more than a week's profit and about 2% of its weekly income, would, in our view, in the circumstances achieve the statutory purposes of sentencing by bringing home to the directors of Sellafield Ltd and its professional shareholders the seriousness of the offences committed and provide a real incentive to the directors and shareholders to remedy the failures which the judge found existed. (para. 65)[71]

Against the background of the analysis conducted above, the following considerations can be made. It is clear that the UK (E&W) legal system widely accepts the criminalization of conducts detrimental to the environment, including corporate environmental liability. However, the relevant provisions have never been applied to cases of historical pollution. Criminal law plays only a minor role in the context of the contaminated land regime when there is a failure to comply with the remediation notice. Even in that case, though, the system seeks voluntary remediation first and, most importantly, no cases of criminal prosecution have been detected so far.

Thus, the relevance of criminal law in tackling phenomena of historical pollution is very limited. Polluting activities that in the past were (presumably) considered lawful are in practice not treated under the criminal law system. This is because, as demonstrated in the previous sections, the UK (E&W) system favours a non-criminal, more "voluntarist" and pragmatic approach. It could at least be argued, though, that the limited practical application of criminal law in this area does not completely rule out its applicability—at least in principle. In fact there are

---

[70]See also the case R v Day (2014) EWCA Crim 2683, para. 46.

[71]See also the decisions in January 2016 of Leeds Crown Court to fine Yorkshire Water £600,000 for unauthorized sewage pollution (available at https://www.gov.uk/government/news/yorkshire-water-fined-600000-after-sewage-pollution-hits-wakefield-fishery), and of St Albans Crown Court to fine Thames Water £1 million for repeated discharges of polluting matter from Tring STW (Sewage Treatment Works) into the Wendover Arm of the Grand Union Canal in Hertfordshire (available at https://www.gov.uk/government/news/thames-water-fined-1-million-for-pollution-to-grand-union-canal).

no legal obstacles to use, for example, corporate criminal liability also for cases of historical pollution. Therefore, should the policy choices on this matter change, it cannot be excluded that companies could face increasingly high fines as a result of past polluting activities.

7 Concluding Remarks

For many years pollution in the UK was seen as a natural consequence of the industrial activity associated with economic development. Only in the 1990s was the issue of redeveloping sites previously used for industrial activities recognized as a major problem needing regulatory intervention.

The contaminated land regime (Pt2A of 1990 Environmental Protection Act) is the regulatory tool that in the 1990s was given to enforcing authorities in order to deal with the potentially hazardous industrial legacy. It came into play as an administrative procedure, mainly based on remediation notices and application of strict, retrospective, joint, and proportionate liability. In practice, it focuses on existing uses of land that is affected by past polluting activities and that can constitute a risk of harm to human health (in this context, **Hypothetical Scenarios 1 to 4** have been answered, in the sense that only if damage or threats to human health exist as a consequence of the pollution would the relevant authorities be required to take action).

However, data have demonstrated that the majority of the sites identified as contaminated by the relevant authority have been remediated through voluntary clean-up via the planning system, which, instead, looks at the future or proposed use of the lands.

Overall, the system favours a pragmatic and "voluntarist" approach, whose main challenge is to find a balance between the need to pay for the restoration of the environment (and for the health effects) and the difficulty in considering someone responsible for an activity that they did not contribute to causing.

In addition, the research has demonstrated that phenomena of historical pollution can also be tackled through private law remedies (such as negligence and private or public nuisance) triggered by individuals claiming damages for the injuries suffered as a consequence of the contamination (in this context, **Hypothetical Scenario 5** has been answered in the sense that in the presence of a causal link between the pollution and the death or injury of people, the latter would be allowed to be awarded damages).

Criminal law plays a very limited role. Only failing to comply with a remediation notice issued by the relevant authority under the contaminated land regime constitutes a criminal offence. However, even in that case the system first seeks a voluntary remediation and no cases of prosecution have been reported so far. The non-criminal nature of the UK's response has been clearly pictured by Lees in the following terms:

> environmental regulation in the UK is primarily pragmatic. Law and guidance are drafted "practically" and the policy of regulatory agencies is one of mutual co-operation, dialogue and negotiation. Only rarely will the regulator resort to imposing criminal penalties onto those who damage, or risk damaging, the environment. (2015, p. 1)

Be that as it may, one cannot completely rule out the possibility, at least in principle, that criminal law could be also used to tackle phenomena of historical pollution. In fact, the UK (E&W) legal system widely accepts the criminalization of conducts detrimental to the environment, including corporate environmental liability, and there are no legal obstacles to applying the relevant offences also to phenomena of past pollution. It follows that, should policy choices on the applicability of the criminal law system to these types of phenomena change in the future, individuals and companies could face increasingly high fines as a result of criminal prosecution.

As a final consideration, a general lack of public awareness of the risks deriving from historical pollution has been noted. This could be due to the complexity of the matter. In this respect, the lack of sufficient resources available to enforcing authorities constitutes a major problem, as it undermines the possibility that they could take a more proactive role in informing the public about the serious risks of contamination.

References

Ashworth, A. (2013). *Positive obligations in criminal law*. Oxford: Hart.

Ashworth, A., & Horder, J. (2013). *Principles of criminal law*. Oxford: Oxford University Press.

Ashworth, A., & Redmayne, M. (2010). *The criminal process* (4th ed.). Oxford: Oxford University Press.

Ashworth, A., Zedner, L., & Tomlin, P. (2013). *Prevention and the limits of the criminal law*. Oxford: Oxford University Press.

Bishop, P., & Jenkins, V. (2011). Planning and nuisance: Revisiting the balance of public and private interests in land use development. *Journal of Environmental Law*, 285–310.

Carriage, R., & Ennis, O. (1990). Environmental Protection Act 1990 Part 2. Where does the *National Grid* case leave practitioners? *Journal of Planning and Environmental Law*, 1557–1563.

Catney, P., Henneberry, J., Meadowcroft, J., & Eiser, R. (2006). Dealing with contaminated land in the UK through 'development managerialism. *Journal of Environmental Policy & Planning, 6,* 331–356.

Brandon, E. (2013). *Global approaches to site contamination law*. Dordrecht et al.: Springer.

Bell, S., McGillivray, D., & Pedersen, O. (2013). *Environmental law*. Oxford: Oxford University Press.

De Pret, P. (2000). Excuses, excuses: The ritual trivialisation of environmental prosecution. *Journal of Environmental Law, 12,* 65–77.

Eisen, J. B. (2012). Finality in brownfields remediation and reuse. *Southwestern Law Review, 41,* 773–795.

Etherington, L. M. (2000). *Environmental rule-making and public consultation: A case study of the development of a new legal regime to clean up contaminated land*. Online thesis from British Library, Nottingham Trent University.

Faure, M., & Heine, G. (2000). *Environmental criminal law in the European Union*. Freiburg im Breisgau.

Fisher, E., Lange, B., & Scotford, E. (2013). *Environmental law: Text, cases, and materials*. Oxford: Oxford University Press.

Fogleman, V. (2005). *Environmental liabilities and insurance in England and the United States*. London: Witherby.

Fogleman, V. (2014). The contaminated land regime: Time for a regime that is fit for purpose (Part 2). *International Journal of Law in the Built Environment, 6,* 129–151.

Hamer, K. (2009). Environmental claims: A claimant's perspective. http://www.hendersonchambers.co.uk/wp-content/uploads/pdf/environmental-claims.pdf. Accessed May 2015.

Hamlin, C. (1990). *A science of impurity: Water analysis in nineteenth century Britain*. Berkeley: University of California Press.

Harris, M. R., & Denner, J. (1997). UK government policy and control. In R. E. Hester et al. (Eds.), *Contaminated land and its reclamation*. London: Thomas Telford.

Johnstone, D. W. M., & Horan, N. J. (1996). Institutional developments, standards and river quality: A UK history and some lessons for industrialising countries. *Water Science and Technology*, 212.

Lawrence, D., & Lee, R. (2003). Source permitting uncertainty: Owners, occupiers and responsibility for remediation. *The Modern Law Review, 66,* 261–276.

Lee, M. (2009). Personal injury, public nuisance and environmental regulation. *King's Law Journal, 20,* 129.

Lee, M. (2011). Tort law and regulation: Planning and nuisance. *Journal of Planning Law*, 8, 986–993.

Lees, E. (2012). The contaminated land regime—New guidance and a new philosophy? *Environmental Law Review*, 14, 267–278.

Lees, E. (2015). *Interpreting environmental offences, the need for certainty*. Oxford and Portland, Oregon: Hart Publishing.

Lewis, O. (2012). Contaminated land and regulation. In K. E. Makuch & R. Pereira (Eds.), *Environmental and energy law*. Wiley-Blackwell Publishing.

Luckin, B. (1986). *Pollution and control*. Bristol and Boston: Hilger.

Luo, Q., Catney, P., & Lerner, D. (2009). Risk-based management of contaminated land in the UK: Lessons for China? *Journal of Environmental Management, 90,* 1123–1134.

Markham, A. (1994). *A brief history of pollution*. New York: St. Martin's Press.

McCullough, A. (2010). The year's hottest cases reviewed—Part 3 access to justice and costs, nature conservation, nuisance, waste. *Environmental Law and Management, 2,* 262–273.

McLaren, J. P. S. (1983). Nuisance law and the industrial revolution—Some lessons from social history. *Oxford Journal of Legal Studies, 3,* 163.

Mosley, S. (2014). Environmental history of air pollution and protection. In M. Agnoletti & S. N. Serneri (Eds.), *The basic environmental history*. Dordrecht: Springer.

Mulheron, R. (2008). *Reform of collective redress in England and Wales, A perspective of need, a research paper for submission to the Civil Justice Council of England and Wales* (on file with the Author).

Norrie, A. (2014). *Crime, reason and history, a critical introduction to criminal law*. Cambridge: Cambridge University Press.

Pointing, J. (2010). Public nuisance re-visited? Paper presented to COBRA conference, Paris, available at http://eprints.kingston.ac.uk/17671/1/Pointing-J-17671.pdf. Accessed May 2015.

Potter, G. (2014). The Criminogenic effects of environmental harm: Bringing a 'green' perspective to mainstream criminology. In T. Spapens, R. White & M. Kluin (Eds.), *Environmental crime and its victims, perspectives within green criminology*. Farnham: Ashgate.

Sheau, J. (1997). The sustainable management of industrial watercourses: An English historical perspective. *Environmental History*, 197–215.

Spapens, T., White, R., & Kluin, M. (Eds.). (2014). *Environmental crime and its victims, perspectives within green criminology*. Farnham: Ashgate.

Thornton, J. (2007). The year's hottest cases: Part 2, 250–262 at http://www.ukela.org/content/doclib/99.pdf. Accessed May 2015.
Thorsheim, P. (2002). The paradox of smokeless fuels: Gas, coke, and the environment in Britain, 1813–1949. *Environment and History*, 381–401.
Tromans, S., & Turrall-Clarke, R. (2009). *Contaminated land*. London: Sweet and Maxwell.
Wald, R. (2009). The year's hottest cases reviewed—Part 2 nuisance, costs, nature conservation, 158–169 at http://www.ukela.org/content/page/2277/10-ELM-21-3%20Wald.pdf. Accessed May 2015.
Wells, C. (2001). *Corporations and criminal responsibility*. Oxford: Oxford University Press.
Wohl, A. S. (1983). *Endangered lives: Public health in Victorian Britain*. London: Dent.
Wolf, S., & Stanley, N. (2014). *Wolf and Stanley on environmental law*. London and New York: Routledge.
Zhao, X. (2013). *Developing an appropriate contaminated land regime in China: Lessons learned from the US and UK*. Heidelberg et al.: Springer.

Case Law

Alphacell Ltd v Woodward [1972] AC 824.
Corby Group Litigation, Re [2009] EWHC 1944 *(Technology and Construction Court).*
Corby Group Litigation v Corby Borough Council [2008] EWCA 463.
House of Lords: Judgments—R (on the application of National Grid Gas plc (formerly Transco plc)) (Appellants) v Environment Agency (Respondents) (Civil Appeal from Her Majesty's High Court of Justice).
R (National Grid Gas plc (previously Transco plc)) v Environment Agency [2006] EWHC 1083 (Admin).
R v Sellafield Ltd [2014] EWCA Crim 49.
R v Thames Water Utilities Ltd [2015] EWCA Crim 960.
The Queen on the Application of Redland Minerals Ltd. v Secretary of State for Environment, Food and Rural Affairs; Environmental Agency; Crest Nicholson Residential Ltd.; Thames Water Utilities Ltd.; Veolia Water Central Ltd. [2010] EWHC 913 (Admin); [2011] Env LR 2.

Reports and Guidance Documents

A Consultation Paper from the Department of the Environment and the Welsh Office (1994).
Clarinet, Sustainable Management of Contaminated Land: An Overview (2002).
Communities and Local Government: London, Planning Policy Statement 3(PPS3) (2011).
Contaminated Land Strategy of London Borough of Hounslow (2008).
DEFRA, Public Consultation on Changes to the Contaminated Land Regime under Part 2A of the Environmental Protection Act 1990 (2010).
DEFRA, Environmental Protection Act 1990: Part 2A Contaminated Land Statutory Guidance (2012).
DEFRA, Guidance on the Legal Definition of Contaminated Land (2008).
DEFRA and The Chartered Institute of Environmental Health, "Local Authority Guide to the Application of Part 2A of the Environmental Protection Act of 1990"—Extended to Cover Radioactive Contaminated Land (2007).
Evaluation Study on the Implementation of Directive 2008/99/EC on the Protection of the Environment through Criminal Law by Member States (2015).
Environment Agency, Environmental Management—Collection Land Contamination: Technical Guidance (2014).

Environment Agency, Model Procedures for the Management of Land Contamination, Contaminated Land Report 11 (CLR11) (2004) (updated 2014).
European Commission, DG Environment, 'Annex—Part A: Legal Analysis of the National Transposing Legislation', Implementation Challenges and Obstacles of the Environmental Liability Directive (2013).
Environmental Agency, National Planning Policy Framework: Planning and Contaminated Land (2012).
Environmental Agency, Reporting the Evidence: Dealing with Contaminated Land in England and Wales—A Review of Progress from 2000–2007 with Part 2A of the Environmental Protection Act (Bristol, Environmental Agency, 2009).
Environment Agency, Greener Business is Good Business, Spotlight on Business, Environmental Performance in 2003.
House of Commons Environmental Audit Committee, Corporate Environmental Crime Second Report of Session 2004-05 (2005).
Land Quality Management Ltd, SP1004 International Processes for Identification and Remediation of Contaminated Land Report No.: 1023-0 S (2013).
London Borough of Hillingdon, Contaminated Land Inspection Strategy (2013–2018).
Marion, H. (2010). The UK Regulatory Framework for Contaminated Land on Nuclear-Licensed Sites and Defence Sites Information Paper for the SAFEGROUNDS Learning Network (2010).
Planning Practice Guidance on Contaminated Land (updated 2014).
Planning Policy Statement 23, Planning and Pollution Control: Development of Land Affected by Contamination (2004).
The Environmental Audit Committee, Environmental Crime and the Courts: Sixth Report of Session 2003–04 (2004).
The Royal Commission on Environmental Pollution: Setting Environmental Standards, 21st Report (1998).
The Royal Commission on Environmental Pollution: A Review of the Royal Commission on Environmental Pollution's Nineteenth Report: Sustainable Use of Soil, 19th Report (1996).

Chapter 10
Historical Pollution in Spain: A Powerful Legal Framework for a Few Past Cases

Luis Arroyo Zapatero, Alejandro Luis de Pablo Serrano and Jordi Gimeno Beviá

Contents

L. Arroyo Zapatero (✉) · J. Gimeno Beviá
Universidad de Castilla-La Mancha, Ciudad Real, Spain
e-mail: Luis.ArroyoZapatero@uclm.es

J. Gimeno Beviá
e-mail: jordi.gimeno@uclm.es

A.L. de Pablo Serrano
Universidad de Valladolid, Valladolid, Spain
e-mail: alejandro.pablo@der.uva.es

F. Centonze and S. Manacorda (eds.), *Historical Pollution*,
DOI 10.1007/978-3-319-56937-6_10

1 Introduction: Historical Pollution Cases and Research Approach

The recent industrial history of Spain has seen significant cases of historical pollution as a result of economic and industrial activities which, once concluded, were found to have caused considerable damage to the environment or natural resources and human health, in some cases resulting in deaths. Our concept of historical pollution requires turning our attention to cases such as the pollution in the Ría de Huelva or the use of asbestos in the construction, textile and paint industries (notably in the *Unión Naval* case). At the same time, it excludes well-known environmental tragedies such as the *Prestige* or the *Aznalcóllar Mine* cases. Despite having had an indelible impact on the collective memory of both Spain and Europe, such incidents cannot be considered historical contamination, as they were accidents, and the damage to the environment or human health did not result from economic and industrial activity that caused harmful effects unknown at the time the activity took place. Similarly, in the *Ría Pontevedra* case, the dangerous activity is still ongoing, so the economic or industrial activities cannot be viewed as concluded. Nevertheless, we will analyse both types of cases in Sect. 2: those that precisely match the criteria for historical pollution, and those with close parallels to historical pollution cases. Looking at these incidents will allow us to draw out the characteristics of the concept of historical pollution on which our work is based, which we will define in accordance with most important contributions from the literature and the most important case law (Sect. 3).

We will then examine in detail the Spanish legal instruments available to sanction those responsible for polluting activities that are harmful to the environment and human health (Sect. 4). Updated largely based on EU directives, Spanish law is equipped with mechanisms for exacting broad and multidisciplinary liability, both criminal and non-criminal. On the one hand, there are traditional civil actions for damage, aimed at getting compensation, together with other administrative fines and penalties that pursue a more punitive effect; the aim of these, in addition to the effective legal implementation of the cleanup procedures, is reparation and rehabilitation (Sect. 5). On the other hand, there are different ecological crimes: the most important is set out in Article 325 of the Criminal Code, which covers an extensive list of polluting activities and all manner of harmful effects to the environment and human health, owing both to actions taken and to failure to act, either intentionally or negligently. The severity of the Spanish legal system is also reflected in how liability can be attributed to actors. Since 2010, Spanish criminal law has allowed the attribution of criminal liability to corporate actors, which in environmental law constitutes a vitally important instrument, as the majority of industrial and economic activities that harm the environment and human health are performed by companies (Sect. 7).

We conclude the chapter by drawing some conclusions regarding the phenomenon of historical pollution in Spain (Sect. 8). These conclusions connect with our initial hypothesis as stated in the title: that Spain has a powerful and effective legal framework oriented to compensation, reparation and penalization. Unlike other European countries and the USA, Spain has fortunately had few cases of historical contamination. It appears that Spanish law is sufficiently equipped to deal with them.

2 Sociological and Economic Context

Environmental pollution in Spain has resulted from lax and permissive industrial, waste and transport policies. This has allowed companies to obtain large profits from their operations because the costs associated with the pollution and the deterioration of public health were not included in their balance sheets. On the contrary, pollution costs have generally been assumed by citizens via the deterioration of the environment and public health, and the use of public funds to repair, to the extent possible, the polluting behaviour of companies (Greenpeace 2008, p. 15).

2.1 The Late Industrialization of Spain: The Asbestos Case

The unique historical circumstances of Spain explain the delay in tackling the most serious problems and crises regarding phenomena of historical pollution. The forty years of the Franco dictatorship delayed industrialization, and Spain was a late starter in the economic growth and industrial processes in Europe. As a result, the appearance of historical pollution phenomena, the adoption of legal mechanisms to combat such phenomena, and the awakening of a general environmental awareness of the polluting behaviour of companies also emerged much later on.

The best example of this historical delay is the case of asbestos. As in other European countries, asbestos was used extensively in Spain in the mid-twentieth century in the construction, textile and paint industries. It came from Canada and also from Spanish mines in Serranía de Ronda in Málaga and was used most heavily in Spain in the 1970s, following the industrial takeoff, with the import of asbestos reaching its height between 1960 and 1985. Concern over the safety of asbestos grew over the course of the twentieth century, and medical authorities demonstrated at the end of the 1990s that asbestos-related products caused cancer with a high mortality rate. In Spain, the widespread use of asbestos occurred later than in other countries. For this reason, Spain was over twenty years behind countries such as Sweden, Denmark, the UK, Holland, and Germany in tackling the problem. The use and commercialization of asbestos was prohibited in Spain in 2002. However, it is difficult to calculate the number of people exposed to asbestos and the effects on and costs to health programmes that will manifest themselves in the next thirty years as a result. Indeed, in Spain, asbestos-related deaths are expected to occur until 2040 (Cárcoba 2001, pp. 59, 60; Bernardo Reyes et al. 2013, pp. 1 ff.; Puche 2014, pp. 1–8).

In Spain, there have been several multi-million-euro judgments in asbestos-related cases. The company Uralita, which monopolized the Spanish asbestos sector in the twentieth century, was sentenced to pay €1.7 million to twenty-three workers at its factory in Getafe, Madrid, and €4 million to fifty residents in Ripollet and Cerdanyola, in Catalonia. But these are isolated rulings and a far cry from other collective actions in Europe for cases similar to those in Spain. Legal cases are based principally on three claims: (1) for a ruling that an illness is work-related; (2) for an order for the payment of benefits; and (3) for compensation for injury or harm, both physical and moral. On very few occasions have criminal complaints been pursued.

2.2 *The* Unión Naval *Case: How Compensation Hides a Huge Scandal*

In relation to asbestos, the judgment that has created the biggest impact in Spain is the *Unión Naval* case, from Valencia. The company Unión Naval de Levante, now called Unión Naval de Valencia, used asbestos in the construction of large ships. To date, more than fifty workers have died and around two hundred are affected by respiratory diseases caused by the inhalation of asbestos fibres.

Procedurally, the question has proved to be very complicated, mixing criminal and labour proceedings. The case was initially closed by the Criminal Court because the parties reached an agreement in 2009: compensation of €11 million to some of those affected, and prison for the legal representative of the company and three occupational safety managers.[1] However, another group affected did not agree and requested the case be reopened. The company considered the facts to be time-barred, and the prosecutor also claimed that they had been judged, but the Provincial Court reopened the case. At present, the company and the others affected are trying to reach another economic agreement to finally close the issue through criminal proceedings. On the other hand, the matter has also been the subject of

[1]The public prosecutor's initial requests for sentences of thirty-four and fifty-four years were, by agreement, reduced to less than two years to avoid prison terms, and to damages totalling €11 million for all claimants. A summary of the ruling can be found at http://www.porexperiencia.com/articulo.asp?num=46&pag=08&titulo=Sentencia-sobre-amianto-con-penas-de-prision.

Labour Court proceedings in 2015,[2] and more recently in 2016, according to which the company must pay compensation to four employees.[3] Aside from the legal proceedings, victims' organizations demand a compensation fund like those existing in other European countries. Such a fund would not preclude filing legal complaints for other claims, nor would it prevent the government from legally seeking reimbursement for paid-out benefits. It would mean, however, that the State rather than the victims would pursue the legal action, as it happens in Italy (Cárcoba et al. 2011, pp. 9–10).

This case may be considered as one of historical pollution because both the deceased and those affected have medical symptoms caused by asbestos, a material that was present from 1960 until 1990, whose consequences persist over time, in addition to being a pollutant. The majority of physical ailments resulted in cancer, and even the descendants of the victims suffer from the effects of asbestos.[4]

Either way, it is an issue that has not had great significance in Spain due to the procedural route. Because it has been addressed in different jurisdictions, the parties tried to reach an economic settlement, and the public prosecutor decided not to pursue a criminal claim, the *Unión Naval* case did not receive as much media coverage as other cases of pollution such as *Prestige* or *Aznalcóllar*.

2.3 Lack of Environmental Awareness: The Ría de Huelva *Case*

Criminological studies on the social perception of the risk of environmental pollution, awareness-raising regarding polluting activities, the demand for criminal liability, and other such tendencies show that in the field of historical and legacy pollution there is still much to do. There is awareness that environmental criminality is associated with companies and even large multinationals that cause serious damage to the environment through their economic activity.[5] However, the fact that such damage is caused by economic activity—essential for economic growth and social and technical progress [the result of a mistaken "predominantly developmentalist economic conception" (Paredes Castañón 2013, p. 755)]—means that the danger and risk entailed in such activities have no impact on the social

[2]There is a ruling of the Labour Court of Valencia on this case; it relates not only to criminal law, but also to labour law: see http://www.aepsal.com/wp-content/uploads/2015/05/TSasbestoMalRecurso.pdf.

[3]This ruling is from March 2016: http://www.abc.es/espana/comunidad-valenciana/abci-condenan-union-naval-valencia-indemnizar-cuatro-trabajadores-expuestos-amianto-201603231923_noticia.html.

[4]See http://www.lasprovincias.es/comunitat/201407/03/union-naval-victimas-amianto-20140703000310-v.html.

[5]Concerning the phenomenon of pollution caused by multinational companies, some authors focus their attention on the need for an international criminal environmental law, e.g. Nieto Martín (2002, pp. 136 ff).

consciousness, and they are seen as tolerable collateral damage. This insufficient social perception is further worsened by the scant attention paid by politicians to environmental pollution matters. This situation is clear in the insufficient budgetary allocation given to the courts to prosecute environmental crimes and in the existence of long and complicated procedures that require human and technical resources. The lack of initiative or energy on the part of the government is also partially due to the influence of lobby groups, which aim to guarantee the broadest scope of impunity for their activities and seek permissive regulations and high tolerance towards elevated levels of risk in hazardous activities (Paredes Castañón 2013, pp. 755–761). Indeed, on occasion, the Spanish authorities have underestimated the danger of certain polluting activities and have given precedence to the interests of particular companies in key sectors and to maintaining jobs (Ruggiero 2014, p. 118).

Some of the above circumstances—the lack of citizen consensus on the negative impact of polluting activities in the medium and long term, and the lack of interest of public powers in fighting such practices—can be seen in the *Ría de Huelva* case. The area of Ría de Huelva had been heavily polluted since the mid-twentieth century (1960) by companies dumping chemical waste from their production processes in the rivers, with adverse health consequences of which the most serious was cancer (Greenpeace 2015, p. 41). In an extensive study of this problem, Natali (2014, pp. 8 ff.) observes that, despite the magnitude of the pollution, there has been no collective citizen action and no vigorous response on the part of the authorities, which have acted as if nothing had happened.

The *Ría de Huelva* case is important because it faithfully reflects how little awareness there is on environmental matters in Spain. The worst circumstances were present in Huelva. (a) The polluting companies took advantage of the passage of time, how quickly people forget, and how difficult it is to specifically attribute liability. All of this contributed to the consolidation of the image and perception that such environmental crimes are not pursued, but go unpunished and enrich the offender. (b) The companies used sophisticated techniques like "greenwashing", publicity, and communications strategies used by polluters to paint an activity that inevitably causes damage to the environment and people in a pleasant and positive light.[6] (c) The apparently positive effect of the economic growth that the activity of these companies produced in the area also played in their favour, without taking into account the devastating environmental and health effects they generated (Natali 2014, p. 16; Ruggiero 2014, p. 126). (d) Over the years, there was an "amnesty" on environmental destruction because, in the end, it appeared to be less harmful and

[6]For example, in the *Huelva* case, the company Endesa built a new industrial plant that was aesthetically pleasing and had verses from the Spanish poet Juan Ramón Jiménez inscribed on its walls that paid homage to the pleasure one gets from walking the clean streets of Huelva (Natali 2014, p. 15). These companies thus send out the message that they are assuming their social responsibility to protect the environment while flouting the regulations or codes of responsible conduct. This is, indeed, a false display of "green credentials" in order to conceal irresponsible behaviour towards the environment (Nurse 2014a, p. 90).

more acceptable than it actually was, which helped to defuse the perception of the risk (Natali 2014, p. 17). The strict connection of the principle of legality to the concept of the environmental crime from a legal-procedural perspective means that events such as these do not receive the punishment they deserve. Thus, in the framework of the new green criminology, a socio-legal approach that classifies the damage as a crime, regardless of its legal definition, is gaining momentum (Ruggiero 2014, p. 119; Natali 2014, p. 9; Nurse 2014b, p. 7).

For years, the devastating effects of the pollution from Huelva's chemical industries went unnoticed, owing to the government's leniency and inaction on the part of the inhabitants. The situation continued until 2002, when the citizens' association Mesa de la Ría de Huelva was created. This association is a valuable example of citizen collaboration and involvement in environmental matters. It has worked tirelessly in denouncing pollution in the city, pressurizing public institutions at the local, regional, national, and even European level. Since it was founded, this association has organized awareness campaigns, demonstrations, symbolic protests, manifestos and appeals, recovery plans for the polluted area, and air quality studies.[7] Mesa de la Ría de Huelva has fought against the construction of new thermal power stations in the city. The association has focused most of its efforts on recovering the marshes of the Tinto River, where tonnes of the pollutant phosphogypsum (in Spanish, *fosfoyesos*) generated by certain chemical industries were dumped in ponds located ten kilometres from the city of Huelva.

In its judgment of 27 June 2007, Spain's National Court (Audiencia Nacional) confirmed that the administrative concession authorizing the company Fertiberia S. A. to deposit phosphogypsum waste in a plot of land in Huelva had expired. Thus, the court found that the company had illegally and in an uncontrolled manner dumped industrial waste, violating the terms of the concession.[8] The National Court prohibited opening new dumping ponds, ordered the dumps to be closed as of 31

---

[7]http://mesadelaria.es/huelva/?cat=7. On 30 October 2014, the Mesa de la Ría de Huelva association won the inaugural "Concurso Vigila tu Aire" [Look after Your Air Competition] for its proposal for a study on the air quality of the affected area, based on which the authorities would be able to verify air pollution levels and adopt the necessary measures to normalize them. For more information on this prize, see http://mesadelaria.es/huelva/?p=1924. The study that was subsequently conducted is available at http://www.biovia.es/concurso/.

[8]A number of directors of this company had already been convicted by the Huelva Criminal Court of First Instance No. 5 on 10 July 2002 for committing an environmental crime under Art. 325 of the Criminal Code, as a result of negligently handling the waste generated from the industrial activity, which was left in open-air dumps in inadequate conditions of humidity. Furthermore, the company violated rules of due care when transporting the waste to the dump sites, causing circumstances that could give rise to the airborne dispersion of particles and the spillage of these substances into a nearby river. Thus, the company directors were found guilty of an environmental crime for putting in serious harm both human health and the conditions for life in surface and underground waters. They received an eighteen-month prison penalty and fines of twelve months. In addition, owing to the civil liability arising from the crime (Art. 112 of the Criminal Code), they were ordered to adopt "adequate measures for avoiding the airborne dispersion of particles and spillage into the Ría de Huelva".

December 2010, and required the company responsible to regenerate the environment of the land affected by the waste dumps.

On 2 October 2014, the company Fertiberia S.A. submitted a proposal to cover the affected areas with topsoil and reforest the area by planting native species. Mesa de la Ría de Huelva opposed this proposal on the grounds that such a solution would not help to regenerate the area and that it violated environmental and urban planning legislation. Instead, the association proposed transporting the phosphogypsum to another site or injecting it deeper into the earth to return the marshes to their original condition. Mesa de la Ría de Huelva filed a complaint with the European Commission against the Fertiberia S.A. proposal. This complaint was dismissed. In its communication of 4 February 2015, the European Commission stated that the company and the Spanish authorities must be given the chance to respond before any measures for non-compliance could be taken, leaving open the possibility for a new action to be filed if the project selected by the authorities proved insufficient to completely regenerate the area. Recently, on 2 March 2016, the Spanish Ministry for Agriculture, Food and the Environment gave the green light to the project for covering the phosphogypsum ponds presented by Fertiberia S.A. Mesa de la Ría de Huelva and numerous environmental organizations, including Greenpeace, strongly objected to this decision, saying that "this is not a solution for the serious chemical and radioactive pollution generated by decades of industrial dumping in the Río Tinto marshes", but a cheap solution for the company responsible.[9] In fact, on 3 March 2016, Mesa de la Ría de Huelva filed ancillary proceedings for enforcement with the National Court. The association claimed that before the first measures of the plan were carried out, the national and regional governments should undertake the corresponding health and environmental impact studies to determine the viability and impact of the Fertiberia S.A. proposal from the legal, urban planning, environmental, and human health perspectives.[10]

2.4 *Other Cases Close to Historical Pollution: The* **Aznalcóllar** *Case, the* **Prestige** *Case, and the* **Ría Pontevedra** *Case*

On 25 April 1998, nearly six million cubic metres of polluted mud was spilled into the Guadiamar River owing to a breach in the dam of a waste-tip at a pyrite mine

[9]See http://www.europapress.es/andalucia/huelva-00354/noticia-greenpeace-ve-inadmisible-ministerio-fije-idoneo-proyecto-fertiberia-20160314144921.html. The situation became even more complicated following Greenpeace's claim that the government minister responsible was a former director of Fertiberia S.A., and therefore, there may have been a serious conflict of interest.

[10]In the press: http://mesadelaria.es/huelva/?p=2681. The ancillary proceedings for enforcement are available here: http://mesadelaria.es/huelva/wp-content/uploads/2016/03/incidente-de-ejecucion-MRH.pdf.

run by the company Boliden Apirsa in the municipality of Aznalcóllar (Seville). This spillage affected 4402 ha of agricultural and grazing land on the banks of the Guadiamar and Agrio rivers, close to the Doñana National Park. A retaining wall prevented the spillage from entering into the national park, although the polluted waters entered into the Guadalquivir River. The affected territory was covered with heavy-metal mud sediment which, at the points closest to the tip basin, reached a height of 1.5 m (Gómez Ligüere 2012, p. 3). Seventeen years after the disaster, civil liability is yet to be correctly determined. There was no criminal penalty. Following an "undesirable pilgrimage", in the words of the Supreme Court in its decision of 26 April 2012, the proceedings returned to where they started in terms of damage compensation in the Seville Court of First Instance.

The *Aznalcóllar* case does not constitute a typical case of historical pollution, at least not according to our definition. Even though serious damage was caused to ecosystems and natural resources and there was a threat to human health, these dangers were not caused by economic and industrial activity that appeared harmless at the time of the facts, but subsequently proved to be dangerous and harmful years later, after the activity had finished. In the case of the Aznalcóllar Mine, the harm to the environment was caused by an accident—the breaking of the dam containing the contaminated waste—not the economic activity itself, which means it is not a case of historical pollution.[11] However, given its environmental impact and because the pollution has not been cleaned up, it has become, over time, "historical"; thus, as it is a closely related case, it deserves a mention here. Furthermore, it has an aspect in common with a true of historical contamination, the *Ría de Huelva* case: the lack of social awareness of the pollution.

In the field of green criminology, the term "dirty collar crime" refers to environmental crimes that end up representing a source of income for the polluter, normally a company (Ruggiero and South 2010, pp. 257–260; Walters 2012, p. 103). In Spain, the *Aznalcóllar* case is not exactly an example of a dirty collar crime as it cannot be said that the company responsible, Boliden, benefited from the pollution. However, this case does demonstrate that the regulations for determining liability, and especially for ordering the payment of damages and compensation for the pollution, are too complicated and do not work. In the end, the obligation for repairing the damage is not met by the company responsible, and impunity triumphs in society.

The sinking of the *Prestige* is not a historical pollution case either, but it represents, without any doubt, one of the worst environmental tragedies that have taken place in Europe, on the Spanish (Galician) and French coasts. The incident occurred in November 2002, following the breakdown and subsequent sinking of the *Prestige* oil tanker off the Galician coast. The decision to send the boat away from the Spanish coast once it had broken down caused an oil slick that ran up the coast of Portugal and northern Spain, went on to France, and even reached Great Britain. The most affected region was Galicia, an autonomous community in northeast

[11] On the accidental circumstances of the Aznalcóllar case, see Conde Antequera (2004, pp. 299–300).

Spain. The oil spill had a major impact on waters, marine life, birds, fishing, and public health. Socially, it gave rise to a citizen movement known as "Nunca mais" (never again). Thousands of citizens from around Spain took part as volunteers in cleaning up the Galician beaches.

The sinking of the ship and the oil spill on the coast cannot be considered historical pollution because the pollution was not caused by industrial and economic activity, but rather a maritime accident owing to the poor condition of the vessel and the adverse sailing conditions.[12] However, the Galician coastline is still polluted as a result of the spill, and this pollution is becoming permanent and starting to appear "historical". In its judgment of 14 January 2016,[13] the Spanish Supreme Court convicted the ship's captain of environmental negligence (Art. 331 of the Criminal Code), with the aggravating circumstance of a "risk of irreversible or catastrophic deterioration" (Art. 326.e),[14] and gave him a two-year prison penalty, which modified the original sentence given for serious disobedience by the Provincial Court of A Coruña.[15] This conviction for environmental negligence also opens the door for a civil liability action under Article 110 of the Criminal Code for restitution and reparation of the damage, including environmental damage and compensation for economic and non-economic damage. This civil liability would be attributed to the ship's captain, the ship's insurer (the London Steamship Owners' Mutual Insurance Association), the ship owner (Mare Shipping), and even the FIDEC (an international fund that provides compensation for this type of catastrophe).[16]

The *Ría de Pontevedra* case refers to pollution caused by the EMCE cellulose factory in Lourizán, a town in Pontevedra. The issue is surrounded by controversy because the company had permission to carry out its activities until 2018, but the incumbent government has extended its licence until 2060. All these circumstances explain why we cannot consider this situation a historical pollution case, according to our concept: the industrial activity that causes harmful effects to the environment and to human health has not concluded, but is ongoing. The estuary is contaminated with high levels of mercury because the cellulose factory traditionally treated paper with chlorine bleach, containing traces of mercury. However, the levels of mercury in the estuary are gradually decreasing because the company no longer uses

---

[12]On the accidental circumstances in the *Prestige* case, Conde Antequera (2004, pp. 308–309).

[13]STS 865/2015, 14 January 2016.

[14]According to previous case law (STS 7/2002, 19 January), irreversible deterioration occurs "when the environmental damage that may be caused by the polluting emissions or spills is of such a magnitude that it cannot be remedied by nature's own regenerative capacity, and human intervention is required", and catastrophic deterioration occurs "when the damage is of considerable magnitude and extension according to the number of natural elements destroyed, the human population affected, and the duration of the effects of the polluting activity".

[15]For more on the criminal consequences in the *Prestige* case, see Martínez-Buján Pérez (2005), Alcántara González (2014), and Álvarez Rubio (2011).

[16]For more on civil liability in the *Prestige* case, see García Rubio and Álvarez González (2007). On Spanish regulations on environmental liability insurance, see Hebrero Álvarez (2002).

chlorine, but other less harmful materials, such as ozone. The battle is always the same: jobs (in an area with high levels of unemployment) versus the environment. The truth is that even if pollution levels are currently diminishing, it will be many years until the estuary becomes totally clean, because metals are persistent pollutants that hardly degrade.

3 Definition of Historical Pollution: The Question of "Late Results"

The definition of historical pollution emerges from three sources: legislation, scholars and case law. There are few useful references to historical pollution in Spanish legislation. On the other hand, valuable contributions about this issue by scholars can be found, along with some relevant case law statements.

3.1 Spanish Legislation

The Spanish legal system has no regulatory definition of historical pollution. However, experts and academics continue associating the characteristics that the international scientific community and other national legal systems attribute to historical pollution to long-term environmental damage. Likewise, we could find no references to related terms such as "legacy pollution" or "ongoing contamination". Some sources refer to "diffuse pollution", which is pollution from some vague origin, on a large scale, difficult to control and closely related to agriculture,[17] although this is not a common concept in the doctrine.

On this basis, references similar to historical pollution or other associated concepts in the Spanish legislation should be found. However, Spanish domestic law lacks a concept equivalent to the phenomenon of historical pollution. The offences classified in the environmental criminal law do not use this concept; such is the case of Article 325 of the Spanish Criminal Code. The administrative regulation also fails to make mention of such a concept. In Law 26/2007 on Environmental Liability (*Ley de Responsabilidad Ambiental*), historical pollution is not defined, nor is any other similar concept, although characteristics for defining environmental damage are established.[18]

[17]See for example "La otra contaminación", *Páginas de Información Ambiental*, 1 (1999).

[18]We will study the administrative regulations in greater depth below: see Sect. 4, General Legal Framework, and Sect. 5, Administrative and Private Law.

3.2 Spanish Scholars

Criminal scholars do not make any mention of historical pollution, whether affecting the environment or human health. A new concept of historical pollution should be developed through the various scholarly contributions of the last decades in the field of environmental law, whether criminal, civil, or administrative.

In Spanish environmental law, there is a fully consolidated doctrine on the distinction between "environmental damage" and "damage derived from environmental damage". Environmental damage (also known as "ecological damage" or "strict environmental damage") encompasses harm caused in any element integral to the environment (water, atmosphere, soil, flora and fauna, etc.), even if it affects a private property, where that is considered part of the environment (e.g. privately owned mountains). Damage derived from environmental damage, on the other hand, is the harm caused to people or their properties through the environment, due to contamination or pollution (e.g. diseases) (Conde Antequera 2004, pp. 23–26; Díez Picazo Giménez 1996, p. 1419; Valencia Martín 2005, p. 112[19]). As we have stated, this distinction is currently well accepted in the Spanish doctrine. For the purposes of our research on historical pollution, we would consider both types of damage to be the ordinary consequences of historical pollution; that is, historical pollution produces damage to the environment and/or damage to human health derived from industrial and economic activities.

Persistence and accumulation have traditionally been taken into account in environmental criminal law. This provides a legal technique for dealing with insignificant behaviours—by themselves inoffensive or not dangerous—that in isolation are not legally significant but become so if they are repeated. Environmental damage is often characterized by an accumulation of individual or isolated actions of damage, which, although harmless in isolation, on a massive scale can lead to large-scale environmental damage. Therefore, accumulation attributes to environmental damage, the level of harm required to justify the protection of the environmental criminal law, under the category of offences by accumulation. This directly protects the environment and, with it, the living conditions of future generations, establishing a criterion of minimum solidarity as the basis for the criminal charge, rather than personal autonomy (Alcácer Guirao 2002, pp. 8, 9, 30).

In addition, the criminal doctrine distinguishes between *permanent damage* (following an initial injury, permanent damage occurs, giving rise to subsequent harmful consequences), *continued damage* (the injury is caused both by the persistence of an initial uncorrected harm and by an external causal factor), and *late results* (the result occurs after the action or intensifies after a period of time, where there is a break in the action–result sequence) (Silva Sánchez 1989, pp. 678 ff.).

---

[19]Although Yanguas Montero (2009, pp. 36 ff.) uses different terminology, she also distinguishes between ecological damage (to the environment) and environmental damage or damage by contamination (to private interests).

In accordance with the above classification, the historical pollution examined in this chapter is not associated with permanent or continued damage but rather with late results, as the polluting activities generate unexpected harmful consequences, triggered or discovered in the future, delayed, or deferred,[20] and all while assuming a certain degree of scientific development characterized by the incapacity to determine with complete certainty the harmful consequences that may result.

3.3 High Court Case Law

The criminal case law on environmental protection has focused mainly on ecological crime under Article 325 of the Criminal Code.[21] Even in these cases, scarcely any formulation of the concept of historical pollution has been found.[22] And while the polluting activities investigated in these trials persisted over time and produced long-term effects, these circumstances have not led the judges to establish any concept or definition of the phenomenon that might be of interest for the purposes of our research. Nor for spillage pollution has criminal case law provided any concept that might prove useful, not even in cases such as Supreme Court Decision STS 776/2012, October 18, in which the polluting activity on trial (the extraction and treatment of aggregates) took place over a period of ten years.[23]

---

[20]We will analyse the imputation of late results below in Sect. 6, Criminal Law.

[21]Art. 325.1 of the Criminal Code. "Anyone who breaks the laws or other provisions of a general nature that protect the environment, directly or indirectly causes or creates emissions, spillages, radiation, extractions or excavations, landfills, noises, vibrations, injections or deposits, in the atmosphere, the ground, the subsoil or the surface water, groundwater, or sea water, including the high seas, even those affecting cross-border spaces, as well as the water catchment basins that, by themselves or together with others, cause or may cause substantial damage to the quality of air, the quality of soil or the quality of water, or to animals or plants, will be punished with imprisonment from six months to two years, a fine from ten to fourteen months, and with special barring from his profession or trade for a period from one to two years".

Art. 325.2, Para. 1 of the Criminal Code: "if the conducts mentioned in article 325.1, by themselves or together with others, may seriously damage the balance of the natural systems, shall be imposed a sentence of imprisonment from two to five years, fine from eight to twenty-four months and special barring for his profession or trade from one to three years".

Art. 325.2, Para. 2 of the Criminal Code: "if there has been a risk of a serious damage for human health, a punishment in the upper half shall be imposed, even the higher degree punishment"

[22]Among others, STS 141/2008, April 8; STS 708/2009, June 16; STS 1307/2009, November 5; STS 1112/2009, November 16; STS 1317/2011, December 2; STS 289/2010, April 19; STS 823/2010, September 18; STS 983/2010, November 23; STS 261/2011, April 14; STS 47/2011, February 1; STS 1162/2011, November 8; STS 1169/2011, November 11; STS 838/2012, October 23; STS 916/2012, November 28; STS 89/2013, February 11; and STS 410/2013, May 13.

[23]Also on environmental pollution by spillages, STS 722/2009, July 1; STS 411/2012, May 18; STS 776/2012, October 18; and STS 840/2013, November 11.

The aforementioned doctrinal distinction between continued damage, permanent damage, and late results has also been found in the case law. In its judgment of 19 November 2013 on the civil liability arising from the damage caused by thalidomide, the Madrid Court of First Instance No. 90 makes a distinction between *continued damage* (produced continuously and uninterruptedly), which, for the purposes of calculating the limitation period, starts when the definitive result is produced (see also STS 227/207, March 13); *permanent damage*, whereby "the damage occurs at a given moment but persists over time, with the possibility of becoming even worse owing to factors totally independent of the action or omission of the defendant" (STS 445/2010, July 14, and STS 512/2009, June 30); and *late result* (or *late-appearing damage*), which becomes apparent after the action occurs because previously it could not be detected owing to the state of scientific knowledge (Salvador Coderch et al. 2014, pp. 8–16).

3.4 The Concept of Historical Pollution: Conclusions

In the light of all these concepts and discussions, we define historical pollution, in accordance with the comparative research performed by other international scholars and experts, and taking into consideration the historical pollution cases described in Sect. 2, as (a) *contamination*, (b) *resulting from economic and industrial activities which took place in the past and are already finished*, but (c) *whose effects or impact on the environment (environmental damage) and/or human health (damage derived from environmental damage) are harmful over the medium to long term, so they can be observed in the present day.*

From the Spanish perspective, we believe that the concept of historical pollution should incorporate the distinction between *environmental damage* and *damage derived from environmental damage*, as we have done in the above definition. In this sense, historical pollution would produce both types of damage. This distinction is also made elsewhere, for example, in the French doctrine (Conde Antequera 2004, pp. 23–24). Secondly, we should pay attention to the "late results" element, as it has been noted by Spanish scholars and case law: the polluting activities generate unexpected harmful consequences, triggered or discovered in the future, delayed, or deferred, and all while assuming a certain degree of scientific development characterized by the incapacity to determine with complete certainty the harmful consequences that may result.

4 General Legal Background: Three Axes

If we describe the Spanish general legal framework in the field of historical pollution, we should refer, firstly, to the International Treaties and Conventions ratified by Spain that are immediately incorporated into Spanish domestic law. The starting point is the Stockholm Convention on Persistent Organic Pollutants of 2001.[24] This convention was ratified by Spain and has been incorporated into EU law in various regulations. These include Regulation (EC) No 850/2004 of the European Parliament and of the Council of 29 April 2004 on Persistent Organic Pollutants, amended by Directive 79/117/EEC, and the significant Directive 2004/35/EC of the European Parliament and of the Council of 21 April 2004 on environmental liability with regard to the prevention and remedying of environmental damage. Although Directive 2004/35/EC was meant to be transposed to the domestic legal systems prior to 31 April 2007, Spain did so after this date through Law 26/2007, of 23 October, on Environmental Liability (*Ley de Responsabilidad Medioambiental*).[25]

In Law 26/2007 on Environmental Liability, historical pollution is not defined, nor is any other similar concept, although characteristics for defining environmental damage are established.[26] This statute, taking concepts from Regulation (EC) 850/2004, defines "damage" as "any adverse and measurable change in a natural resource or injury to a natural resource service, whether produced directly or indirectly" (Art. 2.2) and distinguishes various types of environmental damage: (a) damage to wildlife and natural habitats; (b) damage to waters; (c) damage to coastline and riverbanks; (d) land damage. Lastly, again following the lead of the EU regulations, this statute also limits the punishable environmental damage to damage produced as a result of an emission, event, or incident occurring up to a maximum of thirty years ago, starting from the last day of the emission, event, or incident causing the damage occurred (Art. 4). The period of thirty years, following

---

[24]Prior to the adoption of this convention, the international community had made significant steps in the fight against persistent or historical pollution, and Spain participated as a committed member. The most significant international regulatory instruments signed by Spain in this area are as follows: (a) the 1979 Geneva Convention on Long-range Transboundary Air Pollution; (b) the Basel Convention on the Control of Transboundary Movements of Hazardous Wastes and their Disposal, adopted 22 March 1989; (c) the Convention for the Protection of the Marine Environment and the Coastal Region of the Mediterranean; (d) the Rotterdam Convention on the Prior Informed Consent Procedure for Certain Hazardous Chemicals and Pesticides in International Trade of 1998; (e) the Convention for the Protection of the Marine Environment of the North-East Atlantic (OSPAR Convention); and (f) the 1972 Convention on the Prevention of Marine Pollution by Dumping of Wastes and Other Matter (known as the London Convention), and the subsequent London Protocol of 1996.

As regards bilateral agreements and conventions, in 1998 Spain and Portugal adopted the Convention on Cooperation for the Protection and Sustainable Use of the Waters of the Spanish–Portuguese Hydrographic Basins (commonly referred to as the Albufeira Convention), in force since 17 January 2000 (BOE no. 37, 12 February 2000). This Convention is important for pollution as it aims at the collaboration between two countries to prevent, eliminate, mitigate, or control the effects of accidental pollution incidents against individuals or the environment (Arts. 10 and 11). On 4 April 2008, Spain and Portugal adopted the Protocol for Amending the Convention (BOE no. 14, 16 January 2010) to redefine concepts in the original text.

[25]See Jordano Fraga et al. (2005) for a deeper study of this European regulation.

[26]We will study the administrative regulation in greater detail below: see Sect. 5, Administrative and Private Law.

international and EU law, is undoubtedly the right criterion, given the difficulty in assessing environmental damage when it occurs, as it usually becomes apparent over time and remains for a long period, becoming increasingly visible with the passage of time (Borrillo 2011, p. 7).

Regulations on criminal law form an important part of the Spanish legal framework. Article 325 of the Spanish Criminal Code punishes causing or carrying out "emissions, spillages, radiations, extractions or excavations, landfills, noises, vibrations, injections, or deposits […] that cause or may cause substantial damage to air, soil, or water quality, or to animals or plants", or damage to the balance of natural systems and to human health;[27] this article has been recently modified, by Organic Law 1/2015, in order to introduce some recommendations from Directive 2008/99/EC.

And finally, from this general perspective, the Spanish legal framework also provides the possibility to exercise civil action, described in Articles 1902 and 1908 of the Spanish Civil Code, to compensate for and repair the damage. Civil liability will be described in more detail in the next section.

5 Administrative and Private Law: An Effective Tool Against Historical Pollution

5.1 Administrative Law and Liability: Law 22/2011, of 28 July, on Waste and Polluted Soils

In Spain, the most important administrative regulation on historical pollution and contaminated sites is Law 22/2011, of 28 July, on Waste and Polluted Soils (*Ley de residuos y suelos contaminados*).[28] There are other administrative regulations that tangentially affect the phenomenon of historical pollution and that contain their respective catalogue of prohibited behaviours, sanctions, fines and liable persons. These are Law 22/1988 on the Coastline (*Ley de Costas*; Arts. 90 onwards), Law 1/2001 on Waters (*Ley de Aguas*; Arts. 116 onwards), and Law 42/2007 on National Heritage and Biodiversity (*Patrimonio Natural y de la Biodiversidad*; Arts. 79 onwards).

5.1.1 Concepts

First of all, Law 22/2011 on Waste and Polluted Soils, under Article 3.x, defines "brownfield" (a subcategory within the broadest concept of historical pollution) as

---

[27]Below, in Sect. 6, criminal law, we will analyse in greater detail the content of Art. 325 of the Criminal Code and the problems connected with it.

[28]Published in BOE no. 181, 29 July 2011.

land “whose characteristics have been negatively altered by the presence of chemical components of a dangerous nature, that come from human activity, in such concentrations that it entails an unacceptable risk to human health and the environment, in accordance with the criteria and standards determined by the Government, and that has been declared as such in a specific decision”.

5.1.2 Liable Persons

Secondly, the regulation clearly defines the persons liable for the polluting behaviours. In general, any “physical or legal person committing the administrative offences set out in this statute” (Art. 45.1) is considered an offender and is liable. When liability jointly corresponds to several persons, “they are jointly and severally liable for the fines” (Art. 45.2). Specifically, liability is joint and several: (a) when the producer, the initial possessor, or the handler of the waste delivers it to a physical or legal person other than those set out in this statute, or (b) when several persons are liable and the degree of participation of each in committing the offence cannot be determined (Art. 45.3). Lastly, in accordance with Article 45.4, “when the harm caused to the environment occurs through an accumulation of activities by different persons, the competent body may individually attribute this liability and the economic effects”.

5.1.3 Sanctioning Polluting Activities

The legislation further offers a complete framework of sanctions for polluting activities, establishing three types of infringements in accordance with their level of seriousness. Among the very serious infringements (Art. 46.2), we find: (a) the abandonment, spillage, or uncontrolled elimination of dangerous waste; (b) the abandonment, spillage, or uncontrolled elimination of other kinds of waste, creating danger for human health or damaging the environment; (c) the preparation, import, or intra-communitarian acquisition of products with substances or prohibited preparations, because of the danger associated with the waste products that they generate; (d) the non-completion of cleaning and recovery operations when the ground has been declared polluted, following the corresponding requirement from the autonomous community or, if applicable, incompliance; and (e) the mixture of different categories of dangerous waste or of mixing these with other non-hazardous waste, with the caveat that public health has been seriously damaged or serious environmental damage or degradation has occurred as a consequence of doing so. Subsequently, similar behaviours to those described above are foreseen under Articles 46.3 and 46.4, as serious and minor infringements, although with a reduced scope and with fewer adverse effects on public health or environmental damage on a lower scale.

In this law, as in other administrative regulations, there are three levels depending on the seriousness of the infraction. These different levels of infraction can be considered in **Hypothetical Scenario 3**: from an administrative liability perspective, A would be sanctioned under Article 116.a of Law 1/2001 on Waters, which punishes activities causing damage to public water resources such as groundwater. A would also be punished for having diminished the quality of the water in accordance with Article 116.f of the same law. A would be fined between €10,000 and €1,000,000, depending on whether the crime was classified as minor, less serious, serious, or very serious. To determine the amount of the fine, various circumstances are taken into account: the significance of the behaviour for the safety of people and resources; the degree of malice, participation, and profit obtained; the deterioration caused in the quality of the resource; the economic value of the waters polluted; and the cost of treating and cleaning the waters (Art. 117). Lastly, A must also repair the damage caused to the public waters and restore the waters to their previous state (Art. 118.1).

In other administrative regulations, there are sanctions related to the information given by the landowner to the public administration in order to develop an industrial or economic activity, or to another person when the owner transfers the land. That is the case in **Hypothetical Scenario 2**: from the perspective of administrative liability, A has violated the obligation to state that polluting activities had occurred on the land when transferring it to another person (Art. 33 of Law 22/2011 on Waste and Polluted Soils). Law 1/2001 on Waters classifies intentionally hiding or changing this type of information as a very serious offence. As a result, A would be punished with a fine between €45,000 and €1,750,000.

The statute of limitations starts on the day on which the offence takes place. For continued offences, this period starts when the activity finishes or at the time of the last act entailing the offence. Where the acts or activities entailing an offence were unknown because there were no external indicators, this period starts when these indicators become apparent (Arts. 51.1 and 51.2 of Law 22/2011 on Waste and Polluted Soils). These circumstances are particularly important in cases of historical pollution, in which, as we set out at the beginning of this chapter, there is polluting activity that began in the past and continues producing the polluting effects in the present (as a point or moment of observation).

5.1.4 Fines

Under Articles 47 and 48 of Law 22/2011 on Waste and Polluted Soils, a wide catalogue of sanctions can be found that correspond to the offending person as defined in Article 45. These may be fines, up to a maximum of €1,750,000. As Castañón del Valle et al. (2003, p. 233) state, the most common sanctions in the administrative regime are fines or economic sanctions. However, sanctions may also take the form of barring for professions, trades, or activities foreseen in the law for a set period of time; the closure of facilities (temporary or permanent, total or partial); and the revocation or suspension of an authorization for a given period of time. In order to punish a polluting activity, from the administrative perspective, the legislator must follow some general principles: legality, culpability, proportionality, non-retroactivity and *ne bis in idem* (Alenza García 2009, pp. 424–431; Castañón del Valle et al. 2003, pp. 209 ff.; Conde Antequera 2004, pp. 221–227). The role of the *ne bis in idem* principle is quite important in analysing **Hypothetical Scenarios 4 and 5**. In **scenario 4,** criminal conviction under Article 325 of the Criminal Code (ecological crime) excludes administrative sanctions in virtue of the *ne bis in idem* principle and Article 120 of Law 1/2001 on Waters. In **scenario 5**, criminal conviction under Article 325 (ecological crime), Article 147 (bodily harm), and Article 138 (homicide) again excludes administrative sanction in virtue of the *ne bis in idem* principle and Article 120 of Law 1/2001 on Waters.

Legislators must ensure that the penalty matches the offence. In particular, they must take into account the impact of these offences and their importance for the health and safety of the public and the environment or the resources protected by the statute; the circumstances of the offender and the degree of intentionality, participation and benefit gained; whether it is a repeat offence, committed within one year of an offence of the same type as ruled by final judgment; and the irreversibility of the harm or deterioration caused. As stated in Article 52.3 of Law 22/2011 on Waste and Polluted Soils, "when one single act constitutes two or more offences under this or any other applicable statute, the more serious penalty will be imposed on the offender".

5.1.5 Compensation for Damage

Without prejudice to the penalty that may be imposed, the offender must restore the situation altered by him to its original state (Art. 19). In this sense, restitution (to the original conditions) and compensation, in that order and notwithstanding fines, are the common sanctions in every administrative regulation in cases of environmental damage (Alenza García 2009, p. 443; Castañón del Valle et al. 2003, p. 235). In the case of environmental damage, the offender must make such repairs in accordance with Spanish Law 26/2007, of 23 October, on Environmental Liability (Art. 20).

5.2 Cleanup Procedures

5.2.1 Concepts

The Spanish legal order provides for cleanup and/or reparation of the environmental damage in Law 26/2007, of 23 October, on Environmental Liability. The source of this law is found in the transposition of Directive 2004/35/EC on environmental liability in relation to the prevention and reparation of environmental damage. The transposition of the Directive has been well received in the Spanish doctrine, as it is generally accepted that the existing system was chaotic and incoherent (Valencia Martín 2005, pp. 110–111). The Directive incorporates an administrative regime of environmental liability into Spanish domestic law, objective and unlimited in nature, based on prevention and the "polluter pays principle" (PPP). Its scope of application (Art. 4) extends to historical pollution or persistent pollution, except pollution that is a result of armed conflict, war, national defence or a natural event of an exceptional, inevitable and unavoidable character (Art. 3.4).

Law 26/2007 establishes an objective liability system: the operator should adopt preventive measures and pay any sanction, whether he acted with purpose or negligently. In this sense, the preamble says that

> environmental liability is a liability of an objective nature in which the obligations to act are imposed on the operator apart from any guilt, *mens rea*, or negligence that may have existed in his or her behaviour. The legal framework for the protection of natural resources is thereby completed, as environmental damage originating in the commission of administrative or criminal infringements is already defined in the different sectorial regulations, which usually stipulate the obligation to make redress for damages arising from such illegal actions. In addition, in this way the polluter-pays principle is applied by passing the costs arising from the reparation of environmental damage away from society and onto the operators who benefit economically from exploiting natural resources. (Preamble, I, Para. 5)

Liability is accordingly provided for in Article 3.2:

> This law will also be applied to environmental damage and to imminent threats that such damage might occur, when they have been caused by economic or professional activities other than those listed in Annex III, in the following terms: (a) When criminal intent, blame, or negligence are applicable, measures for prevention, avoidance, and reparation will be required; (b) When criminal intent, blame, or negligence are not attributed, measures for both prevention and avoidance may be required.

This provision is complemented by Article 14.2:

> The operator will not be obliged to carry the costs of reparation when he or she demonstrates that he or she has not incurred blame, *mens rea*, or negligence, and that some of the following circumstances concur: (a) the discharge or fact that is the direct cause of the environmental damage constitutes the express and specific object of an administrative authorization, granted in accordance with the norms applicable to the activities listed under Annex III; (b) the operator proves that the environmental damage was caused by an activity, a discharge, or the use of a product that, at the time of its development or use, was not considered potentially harmful to the environment according to the existing scientific and technical knowledge at that time.

5.2.2 Liable Persons

Liability, regulated under Chapter III of the aforementioned law, is attributed to operators who have caused environmental damage, whether they are physical or legal persons or groups of companies (Arts. 9 and 10). Where the damage has been caused by more than one operator, joint liability will apply (Art. 11). Likewise, if the physical person holding liability was to die and/or the legal person was to enter receivership, liability would pass to heirs or subsequent owners, as occurs with tributary liabilities (Art. 12).

In Spain—unlike other countries, such as Italy—the trial *in absentia* of the accused is limited to misdemeanours, with an abbreviated process for certain crimes punishable by a term of imprisonment of one year (or six years if it is a question of another sanction of a different nature), as well as in criminal proceedings for slander and libel against individual parties. With regard to historical pollution, it can be difficult to determine liability, since on many occasions international elements coincide, complicating the attribution of liability. For example, in the case of the *Prestige*, its captain was Greek, the boat was awarded its certificate of seaworthiness in the USA, it was the property of a firm from Liberia, it was built in Greece, and the person who ordered it to sail away from the Spanish coastline was Spanish. Precise attribution of liability is therefore a complex issue, such that reparation is assumed by the authorities, although not under the heading of judicial compensation, but at the request of the administration.

Joint and several liability may also apply to managers and/or administrators in charge of the legal person who is the perpetrator of the environmental damage, even when they are no longer employed, on account of outstanding duties and obligations since the time of the closure of the operation. Liability may also apply to all those who have followed the manager in the post or in the management of the activity that caused the damage, along with any members of the boards of administration and liquidators of legal persons who have not done what may have been necessary earlier in order to comply with the duties and the obligations arising from such situations (Art. 13). Nevertheless, the operator will not pay the damages when these have been caused by a third party outside of the organization, or when they are caused by a binding order or instruction dictated by the competent authority—for example, in the case of the *Prestige*, part of the damage was linked to the decision taken at that time by the Director of the Merchant Navy to order the tanker to sail away from the coast. Likewise, the law also provides for the establishment of a State Fund for the reparation of environmental damage payable by the General Budget of the State (Art. 34).

5.2.3 Administrative Procedure for Environmental Liability

The administrative procedure for environmental liability is foreseen in Chapter VI of the Law 26/2007 (Arts. 41 and 47). There are different means to initiate the procedure: *ex officio* by the competent authority, by a complaint from that authority, by the operator, and even *at the request of the interested party*. In this last case, the action or omission and the damage caused should be mentioned in the request to institute the proceedings. This *general standing* allows any person harmed or threatened by environmental damage to act (López Sánchez 2008, pp. 390–391). However, not everybody satisfies the condition of an interested party: this only applies to those physical or legal persons recognized as such in Article 31 of Law 30/1992 or to not-for-profit legal persons in whose statutes environmental protection appears as an objective, whose constitutions are registered at least two years before the action (thereby avoiding personifications of associations for unfair or spurious ends), and whose territorial scope of action is the place where the damage has occurred. With this kind of limitation, it is understandable that we cannot talk of an absolute general standing or *actio popularis* (López Sánchez 2008, pp. 416–417). Information about the procedure, in view of the contents of Article 43, is public.

As in other legal orders, precautionary measures may be taken; the legislator foresees the adoption of provisional measures—*numerus apertus*—to prevent new damages ("remedial measures"), so as not to aggravate or worsen the situation. Given the implicit prejudice that the adoption of this type of measure involves for the operator, it is necessary to adopt these measures on the basis of a certain degree of proportionality. The decision should be reasoned and should at least contain the description of the threat or damage to clean up, the assessment of the environmental threat or damage to counter, the specific definition of the provisional measures (or "remedial measures")—both for prevention and for reparative compensation—the identification of the subject that should apply them, the time frame for their application, the quantity and the obligation for payment in cases where the measures have previously been adopted by the competent authority, and the procedures that the public administration should, if necessary, carry out.

The time frame for the notification of a decision is six months, with a single extension of three additional months. It is worth pointing out that although the deadline is an expiry date, it may nevertheless be suspended over the period in which the operator is required to determine the reparative measures. Likewise, finalization of the conventional proceedings is also possible, by means of a binding agreement between the parties that, as well as containing the requirements mentioned in the aforementioned decision, cannot be *contra legem*. Execution may be forced to ensure compliance with the agreement or decision, ordering the operator to comply and even imposing pecuniary fines, provided the amount of the sanction implies no delays in the environmental damage reparation.

5.2.4 Reparation of Environmental Damage

Once environmental damage has occurred, Law 26/2007 also regulates the regime for reparation (Arts. 19–23). The constitutional recognition of the environment as an economic and social principle, in Article 45 of the Spanish Constitution, makes the reparation a direct consequence of the environmental damage (and not only in the administrative regime, as we will see later) (Conde Antequera 2004, pp. 112–113).

The operator of economic or professional activities that cause environmental damage is obliged to bring the damage forthwith to the attention of the competent authority and to adopt the necessary remedial measures that may apply in accordance with the provisions of this law, even though criminal intent, blame, or negligence may not be attributed to the operator (Art. 19.1). Subsequently, the operator will submit a proposal to the competent authority, in accordance with Chapter VI, for measures to make good the environmental damage that has been caused (in relation to Annex II), notwithstanding the additional criteria that the autonomous communities may establish with the same objective in mind (Art. 20.1. b). On this point, the competent authority will have to take into account, among other aspects, the nature, scope, and seriousness of the environmental damage in each incident, as well as the possibilities of natural recovery.

In any case, the measures aimed at minimizing public health risks (Arts. 20.2 and 20.3) will take priority. Law 26/2007 states that environmental liability is unlimited, as the obligation for reparation (or, if applicable, for prevention) assumed by the liable operator consists in returning the damaged natural resources to their original state, supporting all the accumulated costs of the corresponding preventive and reparative actions. When placing emphasis on total restoration of the natural resources and the services that they offer, the environmental factor is paramount, which is hardly satisfied merely through financial compensation.

If the above refers to actions the operator will take under his or her own initiative (although supervised by the competent authority) to repair the damage, the law also attends to the measures the competent authority can impose under its own initiative for the reparation of environmental damage. Thus, the competent authority, faced with circumstances involving environmental damage, may at any time in a reasoned decision adopt various measures: (a) require the operator to supply additional information relating to the damage that has occurred; (b) adopt, require that the operator adopt, or give instructions to the operator with regard to all possible measures of an urgent nature so as to control, contain, neutralize and deal with the pollutants involved and any other harmful factors forthwith, to limit or prevent greater environmental damage, adverse effects on human health and greater damage to services; and (c) require the operator to adopt the necessary remedial measures (Art. 21). In case the operator were to fail to comply with the obligation to adopt the aforementioned measures, he or she would be subject to the disciplinary and sanctioning regime foreseen in Article 37.

5.2.5 Statute of Limitations

Environmental liability has a time limit of thirty years from the time of the discharge, event, or incident that caused the damage. The time is measured from the day on which it completely ceased, or the last time the discharge, event, or incident causing the damage to took place (Art. 4). From the criminal standpoint, liability for the criminal behaviours contained in Chapter III of Title XVI of the Criminal Code ceases after five years (Art. 131 of the Criminal Code).[29]

---

[29]See below, Sect. 6, Criminal Law.

5.3 *Civil Law and Liability*

Civil law plays a secondary role in protecting the environment: that is more the preserve of administrative and criminal law. Civil law does not in fact protect the environment, but provides private recourse to people who have suffered as a result of pollution or contamination (Martínez Vázquez de Castro 1996, 2013; Jordano Fraga 1995). As the High Court has stated: "Spanish civil law is the expression of the relations of a society conceived in an individualist form". It is a common opinion in the doctrine that civil law has many limitations in repairing environmental damage where it does not affect people, but only natural resources.

5.3.1 Civil Actions

From a general reading of the Spanish Civil Code, Article 1902 says that "who, by an act or an omission causes harm to another, intervening fault, or negligence, is obliged to repair the damage".[30] Article 1902 defines the scope of non-contractual civil liability, and it represents the most important instrument in seeking reparation for environmental damage in Spanish civil law (Valencia Martín 2005, p. 112). As civil law, it can only be used to compensate damages caused to people, and there is only standing for those whose property has been affected; that is, there is no general standing for civil actions arguing the environment has been damaged, because the environment and natural resources are not private property.

That is precisely the case in **Hypothetical Scenario 1**: there has not been any damage to human health, so there cannot be any civil liability. In **Hypothetical Scenarios 2 and 3**, however, A has sold the contaminated land to B, who has purchased private contaminated property, and because of that, B has suffered damage to his private assets. In this case, with regard to civil liability, B could sue A under Article 1902 of the Criminal Code to claim reparation (*in natura*—to the previous state of the land—or through payment of the costs of restoring the land to its state prior to the pollution) and damages. In **Hypothetical Scenarios 4 and 5**, risks and damages to human health have been created, as the polluted waters were used for human nutrition (**scenario 4**) and they even caused deaths (**scenario 5**). If it were possible for those harmed to prove that the polluted groundwater affected them in some way, or that these waters were the cause of the deaths, in accordance with Article 1902 they could sue B as landowner (or A, if B can prove that he purchased the land without any knowledge of the previous contamination).

More concretely, in regard to environmental liability, Article 1908.2° obliges an owner to repair damage caused by excessive fumes that are harmful to people or properties, and Article 1908.4° determines liability for emissions from sewers or

[30]Some experts, such as Valencia Martín (2005, p. 114), think there would not be any problem if we understand the legal clause "harm to another" as referring to the general interests of society. In this case, civil law would be also an adequate way to repair damages caused to the environment.

storage facilities for infectious materials that are built without adequate precautions. In Article 1908.4°, there is no strict or objective liability, because it requires negligent behaviour in the construction of the facilities. In this sense, Articles 1902 and 1908.4° are similar: both require fault or negligence, while Article 1908.2° contains a regime of objective liability (Ruda González 2009, p. 464; Díez-Picazo Giménez 2003, p. 136). In practice, the polluter should repair the harm or, if this proves impossible, pay economic compensation. Article 1908 also offers another valuable instrument to repair damages: a prohibitory action for any future activity that may produce damage.

In many cases, however, civil liability emerges from a crime, so reparation is frequently ordered in criminal proceedings. In accordance with Criminal Procedure Law (*Ley de Enjuiciamiento Criminal*, henceforth LECrim), civil actions may take place alongside criminal actions, or they may be set aside. Therefore, although Article 108 LECrim states that "civil actions have to be instituted together with criminal actions by the Public Prosecutor, whether or not there is an individual plaintiff; but if the offended party were to expressly renounce his or her right to redress, reparation, or compensation, the Public Prosecutor will limit itself to seeking the punishment of the guilty parties", Article 112 establishes that "having instituted only the criminal action, the civil action will also be understood to be underway, unless the harmed or prejudiced party were to renounce it or expressly to hold it in reserve to bring it after the criminal trial is over, if it were to take place".

5.3.2 Objective Liability: Causality

Briefly, we should also note that in civil law, liability for environmental damage requires the guilt of the liable person (Valencia Martín 2005, p. 112), according to Article 1902: "intervening fault or negligence". The causing of the damage reveals a lack of diligence on the part of the liable person, and the onus of proving they acted diligently lies upon them. The reason is simple: those who create risk through economic or industrial activities must be able to afford the obligation to compensate for the possible damage (Díez-Picazo Giménez 2003, pp. 151–157). Escaping liability requires more than just proving reasonable care; the level of diligence must be sufficient for the circumstances of the harmful activity (taking into consideration the circumstances of the person, the case, time and place—STS 1131/2006, 17 November), which means that if the activity entails great risk, greater diligence is required on the part of the operator (Cordero Lobato 2013, pp. 331; Ruda González 2009, p. 463).

A causal relationship between the polluting activity and the environmental damage must be proven. Thus, it is not enough simply to demonstrate that polluting behaviour occurred; there must be proof that the civil defendant's behaviour caused the damage. In other words, the activity must be the origin, source or cause of the damage. This question is especially complicated in cases of environmental pollution, because the origin could come from many factors. This problem is made worse when historical pollution is involved, since the harmful effects appear or become apparent long after the polluting activity (Ruda González 2009, p. 481). Given that environmental damage can be the consequence of diverse factors or polluting activities, the claimant needs to prove that, in light of the circumstances of the case and the technical reports (together with the commonly accepted technical knowledge), the polluting activity was an apt or ideal means of producing the damage (that is, the activity was the proximate cause). The burden of proof falls on the claimant, in accordance with the "equivalence of conditions" theory: the claimant must prove that the defendant's behaviour was one of the causes of the damage in such a way that without this conduct, the damage would not have taken place (Díez-Picazo Giménez 2003, p. 140; Ruda González 2009, pp. 479–491). Once this has been established, the judge may presume, when there is a high probability (and not mere conjecture or supposition; STS 1163/2006, 16 November) that the defendant's polluting activity was the cause of the damage. In conclusion, in the field of civil law and environmental damage, there is space for presumptions. The causal connection can be considered proved by presumptions, although these presumptions can be destroyed through proof or evidence to the contrary (Yanguas Montero 2006, pp. 157–163).

5.3.3 Civil Liability Derived from the Crime (Civil Liability Ex Delicto)

As a specific aspect of the Spanish system, if the harmful activity constitutes a crime, civil liability matters are resolved within the criminal proceedings, and this is provided for in the Criminal Code, except when the action is reserved for a subsequent civil procedure. In all cases, civil liability ex delicto includes the repair of the damaged resource (reparation *in natura* or by paying the costs of repair and damage prevention) and the payment of damages (Cordero Lobato 2014, p. 334).[31] Once more, the obligation to repair and restore derives from the constitutional recognition of the environment as an economic and social principle in Article 45 of the Spanish Constitution (Conde Antequera 2004, pp. 112–113).

Unlike in civil liability, civil liability ex delicto has an advantage: it allows for repair of damage caused not only to people but also to the environment. As stated in Article 109 of the Criminal Code, "committing an act classified by law as a crime requires repair, in the legally applicable terms, of the harm and injury caused by it". Article 110 of the Criminal Code states that this liability can imply:

(a) *restitution*. In accordance with Article 111 of the Criminal Code, "wherever possible, the same asset must be restored through payment for the deterioration and damage that the judge or court determines". In cases of crimes against natural resources, restitution seems to be the most appropriate solution in the field of civil liability; after that, reparation would be acceptable; and finally,

[31] For more detail, see Roig Torres (2000).

when there is no option for restitution or reparation, the liable person should pay compensation (STS 1073/2003, 25 September).

(b) *reparation*. Under Article 112 of the Criminal Code, "reparation of the damage may constitute obligations to give, to do, or not to do what the judge or court establishes, in accordance with the nature of the damage and the offender's personal conditions and assets".

(c) *compensation* for physical and moral damage, which, under Article 113 of the Criminal Code, "consists of damage caused to the victim and their relatives or third parties".

6 Criminal Law Pending on Article 325 of the Criminal Code: Ecological Crime

6.1 Prohibited Conducts

From the criminal perspective, we should refer to the environmental crimes contained in the Criminal Code. Chapter III of Title XVI includes crimes against natural resources and the environment along with serious damage to the equilibrium of natural systems and to public health (*ecological crime*), with this wording: "Anyone who breaks the laws or other provisions of a general nature that protect the environment, directly or indirectly causes or creates emissions, spillages, radiation, extractions or excavations, landfills, noises, vibrations, injections or deposits, in the atmosphere, the ground, the subsoil or the surface water, groundwater or sea water, including the high seas, even those affecting cross-border spaces, as well as the water catchment basins that, by themselves or together with others, cause or may cause substantial damage to the quality of air, the quality of soil or the quality of water, or to animals or plants" (Art. 325.1); these behaviours should be punished with imprisonment from six months to two years, a fine from ten to fourteen months, and with special barring from a profession or trade for a period from one to two years. The second paragraph provides for an aggravated punishment (imprisonment from two to five years, a fine from eight to twenty-four months, and special barring from a profession or trade from one to three years) when these conducts, by themselves or together with others, may seriously damage the balance of natural systems (Art. 325.2, Para. 1). If the behaviour may create a risk of a serious damage to human health, the upper half of the punishment may be applied, or even the higher degree punishment (Art. 325.2, Para. 2).

As De Vicente Martínez (2013, p. 294) points out, "emission" refers to the launching of material into the air, water, or earth; "spillage" is the tipping of waste into the air, waterways, or the earth[32]; "radiation" means the emission and

[32]For more detail about "spillages," see Torres Vela (2008, pp. 180–182).

propagation of energy by waves; "extract" means removing and "excavate" refers to digging wells, ditches, or galleries; "landfill" entails dumping earth into water to cause the drying out of a wet area; "noises" are emitted sounds propagated by waves; "vibrations" are oscillating sounds propagated by waves; "injection" is the introduction of a fluid into a body via an object; and "deposit" is the action of burying or containing something.[33]

According to the above, under Spanish criminal law, if there is no damage to the environment or human health, there can be no criminal liability or liability for not complying with administrative regulations. For this reason, there is no criminal liability in **Hypothetical Scenario 1** and **Hypothetical Scenario 2**.

Serious harm to the balance of natural systems or human health must be determined depending on whether and how various circumstances combine: the characteristics of the spillages; the condition of the animal, human, or plant life in the area; the intensity of the spillage; its duration over time; the repetition of the behaviour; the proximity to people and elements for consumption; and the irreversibility or difficulty of restoring the equilibrium to the systems (Landa Gorostiza 2004, p. 254). So, the key for courts in deciding whether there is a case for criminal liability is the level of danger caused by the polluting activity. If the polluting conduct infringes environmental regulations (the Law on Waters, for example) but does not create a situation of serious danger for natural systems, such conduct is not a criminal offence and merely constitutes a breach of the regulations. If the public authority believes there is criminal liability, it informs the competing jurisdiction court, the criminal procedure is initiated, and the sanctioning procedure is suspended. If the criminal procedure results in a conviction, there can be no administrative sanctions. If there is no criminal conviction, the public authority can proceed with the administrative procedure for imposing sanctions (Art. 120 of the Law on Waters).

In this sense, in relation to **Hypothetical Scenario 3**, if the Court states that a serious danger for causing harm to natural resources has been caused, Article 325.2, Paragraph 1 of the Criminal Code is applicable, and A may be punished with a prison sentence from two to five years, a fine from eight to twenty-four months, and a special barring from his profession or trade from one to three years. Furthermore, the aggravating circumstance of "falsifying or hiding information on environmental aspects of the activity" set out in Article 327.c of the Criminal Code is present. Therefore, the upper half of the punishment should be applied (as stated in Art. 66.1.3º): A would be punished with a prison sentence from three and a half to five years, a fine from sixteen to twenty-four months, and barring from his profession from two to three years. If A is a physical person, he is liable under Article 325 of the Criminal Code; if a legal person, then A is liable also in virtue of Articles 31 and 327 of the Criminal Code, which punish the environment crimes committed by legal persons. The obligation to repair the damage is applicable in accordance with the regulations of the Criminal Code (Arts. 109 onwards). **Hypothetical Scenario 4**

[33]Further definitions may also be found in Corcoy Bidasolo (2015, pp. 1171–1172).

can be analysed in a similar manner. From a criminal liability perspective, as the groundwater is used for human consumption there is danger to human health, and criminal proceedings are likely given that the behaviour is of sufficient magnitude to involve “serious harm for human health” (Art. 325.2, Para 2): A would receive a sentence of imprisonment from three and a half to seven and a half years, a fine from sixteen to thirty-six months, and special barring from his profession from two to four and a half years.

In the same chapter of the Spanish Criminal Code, we could also find the crimes of creating deposits and dumps for toxic or dangerous substances (Art. 328.1), exploiting dangerous installations (Art. 328.2), carrying out activities with hazardous waste (Art. 328.3), the illegal transfer of waste (Art. 328.4), and serious damage to a part of a natural protected area (Arts. 330 and 338).

6.2 *Blank Criminal Laws*

The confluence of the sanctioning regulations (criminal and administrative) described above creates certain problems. One such problem is the *accessoriness of administrative Law with regard to criminal Law* (known as the accessoriness or dependency principle), which is especially complex when the accessoriness is an act (administrative authorization) rather than a law (administrative legislation; de la Cuesta Arzamendi 1999, pp. 30–41).[34] Many of the prohibited behaviours mentioned revolve around the breach of other laws to which the criminal or administrative offence refers. This technique is known as *blank criminal law*. The doctrine usually warns of this problem, as conducts that seriously threaten the environment can go unpunished simply because they are performed without breaching environmental regulations (Alonso Álamo 2014a, b, pp. 353–355; Santa Cecilia García 2013, p. 912). The abuse of blank criminal law and referral to administrative regulations seriously contradicts the principle of legality and gives rise to numerous technical problems due to the unlimited nature of the referral, the complexity (and many reforms) of the administrative regulations in this and in other matters (de la Cuesta Arzamendi 1999, pp. 30–41; de Vicente Martínez 2013, p. 293; Gómez Rivero 2008, pp. 9 ff.),[35] and the varied sources of these blank laws (both in the

[34]On the differences between accessoriness in law (or general accessoriness) and accessoriness as an act, Gómez Rivero (2000, pp. 23 ff).

[35]On blank criminal laws and self-regulation, see Montaner Fernández (2015, pp. 13 ff).

autonomous communities and in European law[36]). Constitutional case law (STC 127/1990 and STC 53/1994) has declared that blank criminal laws will be considered legitimate with the concurrence of three requirements: a specific and justified referral; when the precept contains the essential core of the prohibition; and when the principle of certainty is satisfied (Mata y Martín 2010, p. 1272).

6.3 The Precautionary Principle and Crimes of Endangerment

The use of the blank criminal law technique and the presence of administrative norms in the scope of environmental criminal law are due to the growing influence of *the precautionary principle* as the guiding principle in this matter. This requires, first, a context of scientific uncertainty about a particular human activity, and, second, the likelihood of serious and irreversible harm, both to people and to the environment (Romeo Casabona 2010, pp. 133, 134). The feature of scientific uncertainty has great importance in this field, given that the resulting imputation will depend on the state of technical and scientific progress, according to how likely this result may be and in order to assess the harmfulness of the action (Gómez Rivero 2008, p. 12). But there is a problem: the growing incrimination of behaviours that could eventually cause damage is turning this broad spectrum of crimes of endangerment into crimes of risk, and some scholars have warned of the expansive effects this entails for criminal law and about the entry of administrative law into the criminal framework (Alonso Álamo 2014a, b, pp. 382, 389–391).[37]

In connection with the precautionary principle, another dogmatic problem of environmental criminal law refers to the incrimination of conducts using *crimes of endangerment*, which, rather than requiring the occurrence of actual damage to a legal interest, sanction its mere endangerment or threats against it. In this type of crimes, it is not required the confirmation of the causality of the damage, but rather the dangerous nature of the act of polluting as a prognosis of causality, something what is more in accordance with the nature of the polluting activity. (De Vicente Martínez 2013, p. 293; Corcoy Bidasolo 2015, p. 1168; Mata y Martín 2010,

---

[36]The most recent case law has made quite clear that the regulations used to complement blank criminal laws can even come from European Law and from the laws of the autonomous communities. In relation to European Law, there is solid case-law doctrine in this respect, ratified by the Decision of the Constitutional Court 120/1998, June 15 (STC 120/1998), which specifies that "in view of that doctrine, obviously, the conclusion is that an original or derived regulation of European Law, with regard to its primacy in the domestic law, is likely to integrate the factual supposition of a criminal norm, even if this criminal norm requires for its complement a legal status".

[37]On the debate related to the compatibility of the precautionary principle with basic requirements of criminal law (harm, minimum intervention, and subsidiarity principles), see Romeo Casabona (2010, pp. 135–138).

p. 1274; de la Cuesta Aguado 2007, p. 732). In any case, after some divergent opinions, the concept of endangerment as a *hypothetical or potential danger* was finally accepted, in order to add to the infringement of the administrative regulation, the necessity for the polluting behaviour to be sufficient and likely to endanger the environmental equilibrium (STS 3117/2014, 16 June; STS 410/2013, 13 May). The relation of causality was thereby linked to the congruity and likelihood of the prohibited conduct to threaten the environment. In summary, potential damage is required of a minimum level of harmfulness that may be specifically valued and that leaves insignificant behaviours out of the equation (De Vicente Martínez 2013, p. 296; Alonso Álamo 2014a, b, p. 350; de la Cuesta Arzamendi 1999, pp. 30–41).[38]

6.4 Perpetrators

Determining the person liable can be difficult in the case of complex companies, as the author is not usually the person who performs the last act of the prohibited conduct. In such cases, the general rules for offences in which someone is commanded (Art. 28.1) can be applied: persons who command another person to act and use them as instruments are considered the perpetrators (Del Moral García 2004, pp. 150–152). However, there are some people who cannot be considered liable in any way, although they form part of the company and work for it: directors of the company whose responsibilities do not involve conforming to the regulations in question, and directors whose tasks are related to the criminal regulations, but who the evidence shows to have acted correctly (Torres Vela 2008, p. 189).

The *criterion of commission by omission* (Art. 11 of the Criminal Code) is also applicable: the active causation of the result is equivalent to not avoiding the result when the person had a special duty to act (STC 62/1994, 28 February, STS 549/2002, 30 January). According to STS 600/2009, 5 July (Alastuey Dobón 2010a, b, p. 422), the requisites of an ecological crime committed by omission are as follows: the legal interest is in a position of guarantor, there is a legal obligation to act, and there is the capacity to avoid the contamination (see also Corcoy Bidasolo 2015, p. 1171; Del Moral García 2004, p. 157).

6.5 Causality: The Theory of Objective Imputation

The design of these criminal offences as crimes of endangerment and the predominance of the precautionary principle in view of scientific uncertainty mean that

[38]Considering the ecological crime in Art. 325.1 of the Criminal Code, an abstract endangerment crime, see Romeo Casabona et al. (2010, p. 530).

juridical-criminal liability is imposed in situations of uncertain danger, in which knowledge on whether the action may eventually endanger a legal interest is lacking. At this point, the risk, a question of the statistical probability that a result may occur, is turned into a normative criterion that affirms the relation of causality between the action (polluting activity) and the result (Alonso Álamo 2014a, b, p. 389). This can represent a problem for the basic safeguards of criminal law and procedural criminal law, such as the *presumption of innocence*, the *principle of in dubio pro reo*, and the *principle of minimum intervention.* Despite all this, it has already been accepted in the *Rapeseed Cooking Oil* case (STS 3654/1992, 23 April 1992) that there is no requirement in law for the scientific demonstration of a causal connection between the action and the result to the same exacting standards as in the natural sciences;[39] in law, *general (and not specific) causality* is sufficient to uphold a judgment of causality, that is, in an incomplete causal law that affirms the causal connection between the action and the result on the basis of the statistical judgment of probability (Gómez Rivero 2008, pp. 28–37, especially 30, 31).

Having affirmed the causality, it is then necessary to solve the problem of *imputation of results.* Spanish scholars and case law follow the theory of *objective imputation*, which consists of three criteria to affirm a causal connection: (a) that an unpermitted risk has been created; (b) that the danger or risk has had a result; (c) that the result is within the protected scope of the regulations or within the scope of the criminal definition. In the framework of historical pollution, specific problems can arise with objective imputation: the temporal question and the *imputation of late results.*

6.6 Statute of Limitations

From a *lege lata* perspective, as historical pollution refers to permanent crimes, we should take into consideration the problem of the statute of limitations. Historical pollution should be punished as an ecological crime, foreseen in Article 325 of the Criminal Code, as we have seen. According to the limitation periods set out in Article 131.1 of the Criminal Code, this type of offence cannot be punished after five years, as the maximum punishment set by law is imprisonment for less than five years in the case of ecological crime (Art. 325). Later on, Article 131.2 states that “in cases of continued crimes, sustained crimes, as well as crimes requiring assiduity, the statute of limitations terms shall be calculated, respectively, from the day on which the last infraction took place, from when the unlawful situation or the conduct ceased”.[40]

---

[39]STS 2298/2001, 4 December.

[40]However, recent case law has considered Art. 325 not as a permanent crime, but an only crime composed of many polluting activities (STS 215/2003, STS 89/2013, STS 713/2014). In this sense, see Corcoy Bodasolo (2015, pp. 1177–1178), Del Moral García (2004, pp. 162–163), Torres Vela (2008, pp. 193–194).

This regulation creates serious problems in historical pollution cases, as the polluting activity took place many years before being observed (the present day), so there is a danger that the crime could go unpunished. The statute of limitations in the Spanish Criminal Code does not allow polluting activities that date back more than five years to be punished (Art. 131.1). If the activity ceased before, its harmful effects could go unpunished; if the activity continues nowadays, there is no problem of limitations because the crime has not elapsed (Art. 131.2); but, this last case would not constitute historical pollution, which requires a ceased polluting activity. The problem arises when a long time passes between the action (polluting activity) and the effect (danger to the balance of natural systems or human health). At the same time, it is worth asking whether, after so much time between the action and the late result, one can continue talking of a "unit of meaning between the action and the late result" (Vásquez Shimajuko 2013, pp. 287–293).[41]

From a *lege ferenda* perspective, different possibilities have been tested to take into consideration the passing of time and the diminished need to punish by virtue of the principle of legal certainty. Some authors have proposed maintaining the temporal limits for the imputation of late results, because the behaviour is no longer in need of punishment—that is, the offence has been surpassed with the passing of time (Gómez Rivero 1998, pp. 65 ff.; Vásquez Shimajuko 2013, pp. 101–131, 303 ff.). Such a solution would be unsatisfactory in cases of historical pollution because most of these polluting activities concern facts that date excessively back in time. Other authors have proposed introducing corrective measures such as a *penalty adjustment* to reflect the lower value of the conduct; this proposal allows historical pollution crimes to be punished, but it would be contrary to the principle of legality and to the statute of limitations rules contained in the Criminal Code, in cases where more than five years have passed between the activity and the effect.

6.7 Intent

The crime against natural resources requires *dolus directus*, even *dolus eventualis*. Someone who is aware of the specific danger inherent in actions that cause a risk to others, but continues performing them anyway, is acting with intent (according to SSTS 3480/1999, 19 May, SSTS 442/2000, 13 March, and SSTS 96/2002, 30 January; de la Cuesta Aguado 2007, p. 732; Torres Vela 2008, p. 188). This intent

---

[41]There are also other specific difficulties in the procedural order. The LECrim imposes a temporal limit on the exercise of the criminal action by the production of a late result in one case: when previously a process has been initiated by the acts that evolve up until that result, in such a way that the change in the legal qualification of the facts can only take place before the judgment is handed down, at the time of presenting the definitive conclusions (Art. 793.3 LECrim). This regulation is a consequence of the principle of *res iudicata* and without prejudice to the scarce possibilities that open up in relation to the review appeal (Art. 954.4 LECrim). See also Gómez Rivero (1998, pp. 65–67), Gómez Rivero (2008, p. 42), Romeo Casabona (2010, p. 143), Vásquez Shimajuko (2013, pp. 335 ff).

must apply to act of causing spillages, emissions, etc., and their potential danger to the environment (Conde-Pumpido Tourón and López Barja de Quiroga 2007, p. 2509).

The prohibited behaviours foreseen in the chapter of the Criminal Code that we are analysing may also be committed through negligence, although this may only be punishable when the negligence is considered serious, that is, a serious crime that violates the duty of care required of the owner of the company or the person responsible for the source of danger, and to the technicians and specialists participating in the daily management of the dangerous activity (de la Cuesta Aguado 2007, p. 742; Conde-Pumpido Tourón and López Barja de Quiroga 2007, p. 2541; STS 1562/2000, 9 October). If the negligence is not serious, there is no crime (STS 472/2013, 8 February). That is a consequence of Article 331 of the Criminal Code: "the acts foreseen in this Chapter shall be punished, as appropriate, by the lower degree punishment, in their respective cases, when committed through serious negligence". In case law, negligence has been accepted for crimes against natural resources (Art. 325), waste management (Art. 328), and serious damage for any element of a protected space (Art. 330).

The importance of the legal duty of care violated depends on the seriousness of the prohibited risk generated by the polluting behaviour, the social utility of the behaviour (greater risk is permitted for greater utility), and the importance of the protected legal right (the greater the importance of the protected legal right, the lower the risk permitted).

This is what should be considered in **Hypothetical Scenario 3**, if it is demonstrated that B behaved negligently by not observing his duty of care. In such a case, B will be convicted of an environmental negligence crime under Article 331 of the Criminal Code and punished with imprisonment from one to two years, a fine from four to eight months, and special barring from his profession from six months to one year. A similar situation could arise in **Hypothetical Scenarios 4 and 5** if A had been negligent, applying the corresponding punishment in its lower degree.

6.8 Concurrent Crimes

In cases of historical pollution, situations of concurrent crimes are possible: for example, ecological crime as a consequence of a serious danger to human health (Art. 325) and the crimes of bodily harm (Art. 147) or homicide (Art. 138) (De Vicente Martínez 2013, p. 293; Corcoy Bodasolo 2015, p. 1168). For such cases, Article 77 states that "in the event of a sole fact constituting two or more crimes, or when one of them is the necessary means to commit the other [...] the upper half of the punishment foreseen for the most serious crime shall be applied, without exceeding the aggregate punishment that would be applicable if the crimes were punished separately" (Arts. 77.1, 77.3).

That would be the situation in **Hypothetical Scenario 5**. From a criminal perspective, there has been a serious danger for human health, which constitutes an environmental offence under Article 325.2, Paragraph 2 of the Criminal Code.

Furthermore, there is a causal relationship between the polluting activity and the injuries and deaths of several people. The court has to apply the theory of proximate cause to connect the damage to the polluting action. A would be convicted for causing bodily harm under Article 147 of the Criminal Code and sentenced to a prison term from three months to three years or a fine from six to twelve months if the injury required medical treatment; if such treatment was not required, the sentence would be a fine from one to three months. For the crime of homicide, under Article 138 of the Criminal Code, A would receive a prison sentence of between ten and fifteen years. As we have a situation of concurrent crimes in which one act has caused two or more crimes, the upper half of the punishment for the most serious offence is applied (Art. 77.2 of the Criminal Code). As the most serious crime is homicide, A would be sentenced to prison for twelve and a half to fifteen years.

7 Corporate Liability: A Wide Range of Sanctions for Companies

Since 2010, through reforms to the Criminal Code introduced by Organic Law 5/2010, legal persons may be held criminally liable. However, criminal liability is not attached to all crimes, but the legislator has previously established a *numerus clausus* list. According to Article 31bis of the Criminal Code, legal persons may hold criminal liability for crimes committed in their name or on their behalf, and to their benefit, by their legal representatives and receivers or legal administrators, or because of crimes committed by people under their charge, when failing to exercise due control over them. It should be understood that the legal person will indeed be held criminally liable for the commission of crimes against natural resources and the environment (Arts. 327 and 328.6).

Legal persons can be held both criminally and administratively liable, in accordance with whether the action committed constitutes an administrative infringement or a crime. Historical pollution, given its seriousness, could certainly engender criminal liability for a legal person (provided, as it has been said, the crime was committed after 2010). Prior imputation, formal accusation, and subsequent conviction are necessary for the determination of criminal liability. Therefore, as a passive part of the criminal process, it should be recognized as having the same rights as a physical person. General principles, therefore, such as *ne bis in idem*, the presumption of innocence, and the right to due process with full guarantees, are fully valid in the criminal proceedings (Gimeno Beviá 2014, pp. 56).

In reference to the sanctions available, the legislator has established a wide category, under Article 33.7 of the Criminal Code. Liquidation would produce the definitive loss of the company's legal personality; suspension of its activities (for a time that may not exceed five years); closure of its branches and establishments (for a time that may not exceed five years); future prohibition against conducting such activities in the exercise of which the accused may have committed, assisted, or concealed the crime (although this prohibition could be temporal or definitive, and if temporal, the duration should not exceed fifteen years); disqualification from obtaining compensation and public funds, from contracting with the public sector, and from benefiting from fiscal benefits and incentives or social security (for a time that may not exceed fifteen years); legal intervention to safeguard the rights of workers and creditors for the time that is thought necessary, which may not exceed five years (Nieto Martín 2008, pp. 263 ff.). Reading the *numerus clausus* list of sanctions, all of which are considered "serious" sanctions, throws up a conclusive result: the legal person, given that it cannot be subject to imprisonment, is likely to be sanctioned by pecuniary fines or interdictions.

Faced with the absence of procedural reform that might accompany Organic Law 5/2010, the legislator approved measures in Law 37/2011 to speed up certain procedural modifications of the LECrim that were unfortunately insufficient. There are a several questions—from the determination of the competent organ to the enforcement of the sentence—raised by criminal procedure against the legal person.

With regard to comparative law, it greatly warrants our attention that associations in Spain, in this case with an environmental mission, are not only instituting procedures as the civil party, but also exercising the right to act as the prosecuting party, by instituting the so-called popular action or *actio popularis*.[42] Likewise, holding large-scale mega-trials has also been questioned for crimes in which there are a group of accused or victims, as is usually the case in crimes of historical pollution; for this reason, a reform of the LECrim is planned to modify the rules of connection and allow the court to speed up the trial through the organization of separate proceedings. The burden of proof, unlike in other countries, cannot be shifted to the accused party, because in Spain the inversion of the burden of proof is unconstitutional, bearing in mind that nobody may be obliged to undertake a *probatio diabolica* of negative facts (Gimeno Beviá 2014, p. 288). The object of proof falls on the assessment of the compliance programme—where there is one—and if not, on the other evidence that may have been legally brought before the court. With regard to the standard of proof, it should be sufficient to weaken the presumption of innocence.

Article 130.2 of the Criminal Code establishes that criminal liability is not extinguished by the mere fact that a legal person merges, is taken over, or is the result of a demerger. Likewise, in cases in which the liquidation is apparent and not real (in other words, where the legal person subsequently continues with its

---

[42]This general standing (*actio popularis*) comes from Art. 45 of the Spanish Constitution, which lends the environment a constitutional dimension. Its position in the constitutional system in Chap. III of Title I of the fundamental norm means environmental protection is considered an economic and social principle that guides legislation, judicial practice, and the actions of public authorities, as set out in Art. 53.3 of the Spanish Constitution. The constitutional provisions on fundamental rights and duties establish the environment as a subjective right: "everyone has the right to enjoy an adequate environment for personal development and also the duty to protect it". It is, therefore, a collective subjective right, which implies the collaboration of social agents in public environmental management (Ortega Álvarez 2013, pp. 224–227; Ríos Angulo 2014, pp. 61–72; López Sánchez 2008, pp. 402–403).

economic activity and maintains a substantial identity through clients, suppliers, employers, etc.), its criminal liability will not be extinguished, unlike when the liquidation is real. Even when the Law on Swifter Procedural Measures has not expressly mentioned the circumstances of group business contracts, the solution appears clear. If the Criminal Code establishes that in such cases liability is not extinguished, neither should the accusation to decide whether or not such liability exists. With regard to criminal proceedings, the modifications that the legal person may undergo should not present a barrier: the proceedings should continue with the resulting company in cases of merger, takeover, or concealed dissolution. Faced with the demerger of the legal person, as established by the Italian legislator, "the one who benefits from it"—in other words, the one left with the larger part of the assets and social capital—will be the one against which the proceedings continue, and when the company that has benefited cannot be distinguished, the accusation will be transferred to all those who have emerged from that demerger.

Reparation of the damage caused, regardless of the criminal sanction, may be required of legal persons—cleaning up the pollution, in this case—or, where this is not possible, compensation, via civil proceedings. In Spain, discussion and decision-making over the criminal and civil action is allowed in a single proceeding (Art. 108 LECrim). When civil liability is attached to a crime, it may be joint and several for cases in which both the legal and the physical person/s are jointly convicted under circumstances in which the legal person is not convicted, but its employees or administrators are. It may therefore be considered a sort of civil liability of the legal person for guilt *in eligendo* or *in vigilando*, but it is not an objective liability, as it does not arise automatically, but is justified by shortcomings in the management of the company.

8 Conclusions: Ready to Deal with More Historical Pollution Cases

The application of technological developments to industrial and economic activities could be a source of danger in the future. While such activities may not initially cause noticeable damage, over time we may find them to be the cause of environmental pollution or harm to human health. This is precisely the problem entailed in the concept of historical pollution we have examined in this chapter. In Spain, cases such as the pollution of Ría de Huelva and the use of asbestos in the construction, textile, and paint industries required the public authorities to file civil liability actions against those responsible for activities that, once concluded, proved to be dangerous for the environment and human health. At the same time, concern in Spanish society grew as it witnessed the serious and permanent nature of the consequences and damage, and the impunity of the actors responsible. Environmental awareness developed late in Spain, as a result of great natural tragedies, but it has become a forceful element in spurring public authorities into action.

The Spanish legal system is equipped to handle cases of historical pollution on three fronts. In terms of penalties, criminal law and administrative law constitute two powerful instruments.

With regard to administrative law, Spanish legislation has a wide range of regulations that punish polluting activities when these do not entail a level of danger that make them crimes. Furthermore, taking the lead of EU law, specifically Directive 2004/35/EC of the European Parliament and of the Council of 21 April 2004 on environmental liability with regard to the prevention and remedying of environmental damage, Spanish administrative law now features a range of strong and advanced regulations. We can say, for the first time, that Spanish domestic law finally has a complete and real system for dealing with environmental damage. The best example of this is Law 26/2007 on environmental liability, which implements cleanup procedures and imposes on polluters, via a model of strict liability, the obligation to repair the damage caused and the duty to adopt whatever measures are necessary to return the natural resources affected to their original condition. Additionally, because of the problem of the criminal statute of limitations for historical pollution activities, Law 26/2007 establishes a longer administrative statute of limitation, of thirty years from the moment when the polluting activity concludes, thus safeguarding most effectively against impunity in such cases.

In relation to criminal law, we understand that the circumstances of historical pollution that arise in Spain may come under the ecological crime described in Article 325, whose literal meaning covers practically all types of imaginable pollution (emission, spillage, radiation, etc.) and any place where its effects occur (atmosphere, ground, subsoil, groundwater, etc.). Here, the problems that persist in case law and in doctrine refer to causality, objective imputation of late results, and the statute of limitations. With regard to *causality*, highly complex activities can produce polluting effects that existing scientific knowledge cannot explain; at this point, Spanish case law employs the criteria of *general causality* that connects actions and results on the basis of statistical probability. With regard to the *imputation of late results* and to the *statute of limitations*, because these activities can generate harmful long-term effects after the specific act, the problem arises of whether these late results may be sufficient grounds to impute the harmful act to the responsible person. Spanish case law follows a dual approach here: first, it verifies whether the criteria of the theory of objective imputation have been followed (a prohibited risk is created, the risk is turned into a fact, and the result is described in law); having attributed the act, if this is a late result, Spanish criminal scholars propose the penalty adjustment or subject the possibility of the imputation of late results to the statute of limitations of the offence in question.

In terms of compensatory measures, the main course of action in the Spanish legal system continues to be the filing of civil actions. At this point, it is important to show that Spanish civil law in the field of environmental damage has an individual perspective: it does not repair or compensate damage to the environment, but damage caused to people. However, Spain domestic law offers another way past this problem, through civil liability ex delicto: it is also possible to pursue such an action in criminal proceedings when the crime has caused civil liability in addition to criminal liability. In such cases, ex delicto civil liability includes restitution, repair and compensation, not only when there is damage to people, but also when the natural resources suffer the harmful effects of the polluting activity.

References

Alastuey Dobón, C. (2010a). Derecho Penal: la emisión de ruidos como delito de contaminación ambiental. In F. López Ramón (Ed.), *Observatorio de Políticas Ambientales 2010*. Pamplona: Aranzadi.

Alastuey Dobón, C. (2010b). Jurisprudencia penal: delitos contra el medio ambiente 2009–2013 In F. López Ramón (Ed.), *Observatorio de Políticas Ambientales 2010*. Pamplona: Aranzadi.

Alcántara González, J. M. (2014). El Prestige aún? Por qué no estamos preparados? *Diario La Ley*, nº8244.

Alcácer Guirao, R. (2002). La protección del futuro y los daños cumulativos. *Revista Electrónica de Ciencia Penal y Criminología*, 04–08.

Alenza García, J. F. (2009). Responsabilidades empresariales en materia de ambiente natural (I): responsabildad administrativa. In G. L. Barrios Baudor (Ed.), *Ambiente Natural, Empresa y Relaciones Laborales*. Thomson Reuters/Aranzadi: Navarra.

Alonso Álamo, M. (2014a). Riesgos no permitidos? Observaciones sobre la incidencia del principio de precaución en el Derecho penal. *Bien jurídico penal y Derecho Penal mínimo de los derechos humanos. Estudios*, Universidad de Valladolid, Valladolid, 2014.

Alonso Álamo, M. (2014b). Trama de la vida y protección penal del ambiente. *Bien jurídico penal y Derecho Penal mínimo de los derechos humanos. Estudios*, Valladolid: Universidad de Valladolid.

Álvarez Rubio, J. J. (2011). *Las lecciones jurídicas del caso Prestige*. Pamplona: Aranzadi, Pamplona.

Bernardo Reyes, A., Báez Baquet, P., & Puche, P. (2013). Fiebre del "oro blanco" en la Costa del Sol y en la serranía de Ronda. http://www.rebelion.org/docs/177954.pdf

Borrilo, D. (2011). Delitos ecológicos y derecho represivo del medio ambiente: reflexiones sobre el derecho penal ambiental en la Unión Europea. *Revista de Estudos Constitucionais, Hermenêutica e Teoria do Direito (RECHTD)*, 3(1).

Cárcoba, A. C. (2001). El amianto en España: situación actual y perspectivas. *Archivos de Prevención de Riesgos Laborales*, 4. http://www.scsmt.cat/Upload/TextComplet/2/1/214.pdf

Cárcoba, A. C., Báez Baquet, P., & Puche, P. (2011). El amianto en España: estado de la cuestión. http://www.rebelion.org/docs/136931.pdf

Conde Antequera, J. (2004). *El deber jurídico de restauración ambiental*. Comares: Granada.

Conde-Pumpido Tourón, C., & López Barja De Quiroga, J. (2007). De los delitos contra los recursos naturales y el medio ambiente. In C. Conde-Pumpido Tourón, (dir.) & J. López Barja De Quiroga (coord.), *Comentarios al Código Penal (Vol. 4)*. Barcelona: Bosch.

Corcoy Bodasolo, M. (2015). De los delitos contra los recursos naturales y el medio ambiente. In M. Corcoy Bidasolo & S. Mir Puig (Eds.), *Comentarios al Código Penal. Reforma LO 1/2015 y LO 2/2015*. Valencia: Tirant lo Blanch, Valencia.

Cordero Lobato, E. (2013). Daños a particulares y medio ambiente. In L. Ortega Álvarez & M. C. Alonso García (Eds.) & R. De Vicente Martínez (coord.), *Tratado de Derecho Ambiental*. Valencia: Tirant lo Blanch.

de la Cuesta Aguado, M. P. (2007). De los delitos contra los recursos naturales y el medio ambiente. In L. Arroyo Zapatero, I. Berdugo Gómez De La Torre, J.C. Ferré Olivé et. al (Eds.), *Comentarios al Código Penal*. Madrid: Iustel, Madrid.

de la Cuesta Arzamendi, J. L. (1999). Cuestiones dogmáticas relativas al delito de contaminación ambiental. *Revista Penal*, 4.

De Vicente Martínez, R. (2013). Derecho penal del medio ambiente. In L. Ortega Álvarez, M. C. Alonso García (Eds.) & R. De Vicente Martínez (coord.), *Tratado de Derecho Ambiental*. Valencia: Tirant lo Blanch.
Del Moral García, A. (2004). Aspectos problemáticos en los delitos contra el medio ambiente. In C. Granados Pérez, *Problemas derivados de la delincuencia medioambiental*. Madrid: Consejo General del Poder Judicial.
Díez-Picazo Giménez, G. (1996) Responsabilidad civil derivada de daños al medio ambiente (una duda acerca de su existencia). *Revista La Ley*.
Díez-Picazo Giménez, G. (2003). Responsabilidad civil ambiental. In A. Vercher Noguera, G. Díez Picazo Giménez, & M. Castañón Del Valle, *Responsabilidad Ambiental. Penal, civil y administrativa*. Madrid: Ecoiuris.
García Rubio, M. P., & Álvarez González, S. (2007). *La responsabilidad por los daños causados por el hundimiento del Prestige*. Iustel, Madrid.
Gimeno Beviá, J. (2014). *El proceso penal de las personas jurídicas*. Pamplona: Aranzadi.
Gómez Ligüere, C. (2012). Quien contamina paga ... si alguna jurisdicción es competente para condenar. Conflictos de jurisdicción en el caso Boliden. *InDret*, 2.
Gómez Rivero, C. (1998). *La imputación de los resultados producidos a largo plazo*. Valencia: Tirant lo blanch.
Gómez Rivero, C. (2000). *El régimen de autorizaciones en los delitos relativos a la protección del medio ambiente y ordenación del territorio*. Valencia: Tirant lo Blanch.
Gómez Rivero, C. (2008). Causalidad, incertidumbre científica y resultados a largo plazo. *Revista General de Derecho Penal*, 9.
Greenpeace. (2008). Informe contaminación en España. Febrero, 2008. http://www.greenpeace.org/espana/es/reports/contaminaci-n-en-espa-a/
Greenpeace. (2015). Radiografía del medio ambiente en España. http://www.greenpeace.org/espana/es/Informes-2015/Mayo/Radiografia-social-del-medio-ambiente-en-Espana/
Hebrero Álvarez, J. I. (2002). *El aseguramiento de la Responsabilidad Civil por daños al medio ambiente*. Madrid: Dykinson.
Jordano Fraga, J. (1995). *La protección del derecho a un medio ambiente adecuado*. Barcelona: Bosch.
Jordano Fraga et al., (coords.). *Estudios sobre la Directiva 2004/35/CE de responsabilidad por daños ambientales y su incidencia en el ordenamiento español*. Navarra: Aranzadi.
Landa Gorostiza, J.-M. (2004). Evolución jurisprudencial en los delitos contra los recursos naturales y el medio ambiente: una crónica (SSTS 2004). http://www.google.es/url?sa=t&rct=j&q=&esrc=s&source=web&cd=2&ved=0CCcQFjABahUKEwjaqpX00_bIAhVDc-hQKHcRnBl0&url=http%3A%2F%2Fwww.eitelkartea.com%2Fdokumentuak%2F12.kapitulua4.pdf&usg=AFQjCNH07wlDDNpD2GCLjFaHKDpSTdDqfQ&sig2=jcyCobivF2gMm_jkHV3mpg
López Sánchez, J. (2008). La legitimación procesal en mataría de medio ambiente. In A. Embid Irujo, *El derecho a un medio ambiente adecuado*. Madrid: Iustel.
Martinez Vazquez De Castro, L. (1996). La protección civil del medio ambiente. In Cabanillas Sánchez, *La reparación de los daños al medio ambiente*, Pamplona: Aranzadi.
Martínez-Buján Pérez, C. (2005). *Crónica Penal (Del Prestige y de otros relatos jurídico-penales)*. Valencia: Tirant lo Blanch.
Mata Y Martín, R. (2010). Artículo 325. De los delitos contra los recursos naturales y el medio ambiente. In M. Gómez Tomillo (Ed.), *Comentarios al Código Penal* (2 ed.), Valladolid: Lex Nova.
Montaner Fernández, R. (2015). *Es posible configurar la tipicidad del delito ecológico a través de las formas de autorregulación?* (p. 7418). No: Diario La Ley.
Natali, L. (2014). Green criminology, victimización medioambiental y social harm. El caso de Huelva (España). *Revista Crítica Penal y Poder*, nº 7 septiembre.
Nieto Martín, A. (2002). Bases para un futuro Derecho Penal Internacional del Medio Ambiente. *Anuario de la Facultad de Derecho de la Universidad Autónoma de Madrid*, 16.

Nieto Martín, A. (2008). *La responsabilidad penal de las personas jurídicas*. Iustel, Madrid: Un modelo legislativo.
Nurse, A. (2014a). Cleaning Up Greenwash: The Case for Enforcing Corporate Environmental Responsibility. In A. Nurse (Ed.), *Critical perspectives on green criminology, internet journal of criminology*. http://www.internetjournalofcriminology.com/Critical_Perspectives_On_Green_Criminology_June_2014.pdf
Nurse, A. (2014b). An Introduction. In Nurse. A. (Ed.), *Critical Perspectives on green criminology, internet journal of criminology*. http://www.internetjournalofcriminology.com/Critical_Perspectives_On_Green_Criminology_June_2014.pdf
Ortega Álvarez, L. (2013). Concepto de Medio Ambiente. In L. Ortega Álvarez, M. C. Alonso García (Eds) & R. De Vicente Martínez (coord.). *Tratado de Derecho Ambiental*. Valencia: Tirant lo Blanch.
Paredes Castañón, J. M. (2013). Derecho Penal español del medio ambiente: una evaluación crítica. In A. Álvarez García et al. (Eds.): *Libro homenaje al prof. Luis Rodríguez Ramos*. Valencia: Tirant lo Blanch.
Puche, P. (2014). Hacia el fin de la vida útil del amianto (uralita) instalado en España. http://docplayer.es/2264098-Hacia-el-fin-de-la-vida-util-del-amianto-uralita-instalado-en-espana-paco-puche-mayo-de-2014.html
Ríos Angulo, M. J. (2014). La configuración del derecho al medio ambiente adecuado como derecho procedimental. In A. Pigrau Solé & S. Borrás Pentinat (Eds.), & R. Clemente Oliveira Do Prado (coord). *Derecho internacional y Comparado del Medio Ambiente: temas actuales*, Barcelona: Huygens Editorial.
Roig Torres, M. (2000). *La reparación del daño causado por el delito: aspectos civiles y penales*. Valencia: Tirant lo Blanch.
Romeo Casabona, C. M. (2010). Conocimientos científicos y causalidad en el Derecho Penal. In C. M. Romeo Casabona & F. Guanarteme Sánchez Lázaro (Eds.), *La adaptación del Derecho Penal al desarrollo social y tecnológico*. Granada: Comares.
Romeo Casabona, C. M., Sola Reche, E., Hernández Plasencia, J. U., et al. (2010). Informe sobre los intentos de adaptación del Derecho Penal al desarrollo social y tecnológico: líneas de investigación y conclusiones. In C. M. Romeo Casabona & F. Guanarteme Sánchez Lázaro (Eds.), *La adaptación del Derecho Penal al desarrollo social y tecnológico*. Granada: Comares.
Ruda González, A. (2009) *Responsabilidades empresariales en materia de ambiente rural (II): Responsabilidad civil*. In G. L. Barrios Baudor (Ed.), *Ambiente Natural, Empresa y Relaciones Laborales*. Thomson Reuters/Aranzadi: Navarra.
Ruggiero, V. (2014). La destrucción creativa y la economía de los residuos. *Nova Criminis*, 8.
Ruggiero, V., & South, N. (2010). Green criminology and dirty collar crime. *Critical criminology, 18*(4).
Salvador Coderch, P., et al. (2014). Daños tardíos. *InDret* 1/2014.
Santa Cecilia García, F. (2013). Delitos contra los recursos naturales y medioambiente. In F. Álvarez García, A. Cobos Gómez De Linares, P. Gómez Pavón, et al. (coords.), *Libro homenaje al prof. Luis Rodríguez Ramo*. Valencia: Tirant lo Blanch.
Silva Sánchez, J. M. (1989). Sobre la relevancia jurídico penal de la no inmediatez en la producción del resultado. In *Estudios penales en Memoria del Profesor Agustín Fernández-Albor*. Santiago de Compostela: Servicio de Publicacións e Intercambio científico da Univesidade de Santiago de Compostela.
Torres Vela, M. (2008). Reseña jurisprudencial sobre la delincuencia medioambiental. Granados Pérez, C. *Problemas derivados de la delincuencia medioambiental*. Madrid: Consejo General del Poder Judicial.
Valencia Martín, G. (2005). El impacto (favorable) de la Directiva 2004/35/CE en el" sistema" español actual de responsabilidad por daños ambientales. In Jordano Fraga et al., (coords.), *Estudios sobre la Directiva 2004/35/CE de responsabilidad por daños ambientales y su incidencia en el ordenamiento español* (pp. 109–184). Editorial Aranzadi.

Vásquez Shimajuko, S. (2013). *La imputación de los resultados tardíos. Acerca de la dimensión temporal de la imputación objetiva*. Buenos Aires: BdeF.
Walters, R. (2012). Dirty collar crime and the environment. In *Proceedings of Crime, Justice and Social Democracy: an International Conference*. Brisbane: Queensland University of Technology QLD.
Yanguas Montero, G. (2006). *El daño no patrimonial en el derecho del medio ambiente*. Navarra: Thomson/Civitas.

Chapter 11
Confronting Historical Pollution in Germany: The Predominant Role of Administrative Law and Its Focus on Getting Results

Katharina Lienert

Contents

K. Lienert (✉)
Ludwig-Maximilians-Universität, Munich, Germany
e-mail: katharina.lienert@jura.uni-muenchen.de

F. Centonze and S. Manacorda (eds.), *Historical Pollution*,
DOI 10.1007/978-3-319-56937-6_11

1 Introduction: Historical Pollution in Germany

1.1 Altlasten *and* Kumulationseffekte*: Aspects of Historical Pollution Reflected in German Legal Doctrine and Discussion*

While there is no direct equivalent to the expression "historical pollution" in German law, and the term might even be misunderstood as a merely historical matter, the underlying concepts and problems have been discussed in German legal doctrine and addressed by German courts for several decades. In accordance with the definition employed in this volume, this report takes the term "historical pollution" to refer to pollution phenomena and challenges in counteracting pollution that are related to time. Regarding reactive measures, the situation raises two questions. The first is the question of how to determine who is responsible for pollution. The second is how to approach and regulate situations where no individual responsible can be made liable, either because the conduct was legal at the time, because the person responsible cannot be found, or because he is insolvent. The two questions often go hand in hand, for instance with regard to clean-up procedures of contaminated sites, an issue that in Germany is commonly known under the term *Altlasten* (legacy contamination)—literally translated as "old burdens" or burdens from the past. This term is currently defined in § 2 of the Federal Soil Protection Act (*Bundesbodenschutzgesetz*, BBodSchG), which came into force in 1999:

> (5) Contaminated sites (*Altlasten*) within the meaning of this Act are
>
> 1. closed-down waste management installations, and other real properties, in/on which waste has been treated, stored or landfilled (former waste disposal sites—*Altablagerungen*), and
>
> 2. real properties that house closed-down installations, and other real properties, on which environmentally harmful substances have been handled, except for installations that can be

closed down only under a licence pursuant to the Atomic Energy Act (former industrial sites—*Altstandorte*),

that cause harmful changes (to the soil) or other hazards for individuals or the general public.

In § 2, the Federal Soil Protection Act also provides further definitions for the different terms used in the statute, such as soil, soil functions, negative changes to the soil, suspected contamination sites, and clean-up measures: for example, negative changes to the state of the soil—the expression used in the act to describe soil contamination—are defined as changes which interfere with the soil's functions in a manner likely to cause danger, considerable disadvantage, or substantial nuisance to individuals or the general public.

Although the Federal Soil Protection Act is relatively recent, the definition of *Altlasten* reflects the term as used during the preceding decades (e.g. in BReg 1990, 1995). Nonetheless, the statute does not contain an exact definition of the temporal dimension of the term, otherwise often used to define contamination resulting from the time before any of the special environmental statutes were in force (e.g. Vierhaus 1998, p. 1263).[1] However, most rules are also applicable to recent soil contamination.

With the passage of time, different legal and factual problems arise: the regulatory framework changes, reflecting the changing attitude towards pollution and the increasing understanding of harmful substances. Claims and prosecution are time-limited or statute-barred. The availability of evidence deteriorates. Contamination lingers and is often the result of many different actions by a multitude of persons spread over the entire globe as well as over time, a phenomenon referred to as cumulative or diffuse pollution (Ronzani 1992, pp. 41 ff.). Although there are no separate express terms for cumulative and diffuse pollution in German law, it is possible to distinguish between the two concepts. Cumulative pollution, the more generic term, refers to any kind of pollution caused by more than one polluter, while diffuse pollution addresses the situation of a multitude of unmanageable and even unidentifiable polluting factors. The causation and objective imputation of cumulative pollution have not only long been the object of fierce discussion in German criminal law under the heading *Kumulationseffekte*, i.e. cumulative effects (Samson 1987; Ronzani 1992; Daxenberger 1997); cumulative pollution is also relevant in ascertaining civil liability or responsibility for conduct in administrative law (Daxenberger 1997, pp. 34 ff.; Marburger 1992, p. 23) and constitutes one of the main motives and challenges for preventative environmental legislation and regulation (cf. Hermann and Wagner 2005; Roßnagel in Führ 2016; § 5 Rn. 172 ff., 458 ff.).

[1]With regard to the definition of "*Altlast*" before the Federal Soil Protection Act, see Müggenborg (1992, pp. 845 f).

1.2 *The Downside of Industrialization and Economic Prosperity: Historical Pollution in Twentieth-Century Germany*

Whereas the purity of potable water and waste disposal have been issues throughout centuries (Heine 1989; Kloepfer 1994), the quantity of polluting activity and potentially harmful behaviour has significantly increased in tandem with industrial and scientific progress at the end of the last millennium. In the mid-twentieth century, polluted rivers, air pollution, and industrial sites contaminated with hazardous substances that had escaped due to ignorance of their harmful potential or careless handling were common occurrences in Germany (UBA 2015, pp. 6 ff.). Waste disposal sites endangered the health of inhabitants and neighbours (UBA 2015, pp. 73 ff.; Breuer 1987). An especially noteworthy example is the Ruhrgebiet, the industrial area around the river Ruhr in northwest Germany, home to coal mines, power plants, and factories (Asendorph 2013). In the 1960s, emissions were discharged into the air unfiltered, causing such high levels of pollutants that special weather conditions in December 1962 led to a smog that killed more than 150 people (Knebelspieß 2014).[2] The river Emscher served to collect industrial wastewater, along with the household wastewater of 2.3 million people, making it the biggest open sewer in Europe (Asendorph 2013).

From the late 1960s to the 1980s—in particular at the turning point of the 1973 oil crisis—pollution as the downside of industrial progress and economic prosperity became the focus of public awareness and politics, and the alarming consequences to human health became apparent (Saliger 2012, Rn. 19; UBA 2015, pp. 12 ff., 111 ff.; BT-Drs. VI/2710; zu BT-Drs. VI/2710). This situation gave rise to several acts of environmental legislation, both administrative and criminal, as well as administrative action and court proceedings over the following decades. The question of clean-up was raised again by the reunification of Germany, which confronted the federal republic with the immense contamination left by the socialist regime (Schindler 2014). The uranium mines in Saxony provide a striking example of contamination so severe that the region was nicknamed the "valley of death" (Tarli 2015).[3]

The legal landscape has continuously evolved, including the introduction of legislation on civil liability in environmental matters in later years. Still, new environmental challenges continue to emerge, including explosions in nuclear reactors and the question of permanent storage of radioactive waste, as well as the depletion of the ozone layer and climate change. Nevertheless, the number of registered environmental crimes has continued to decrease since the 1990s, reflecting a declining interest on the part of the public (Saliger 2004, with further

[2]A similar situation in London in December 1952 claimed as many as 12,000 lives: 4000 persons suffocated directly, and 8000 more succumbed to their injuries during the subsequent months.

[3]The uranium mining caused both environmental damage and health damage (cancer) to the employees of the mines. The environment has now been restored.

references), and polls show that the public considers the environmental situation in Germany to be very positive (BMU 2014, pp. 11 f.).

Due to the environment's natural ability to regenerate, the prevention and reduction of new pollution have been sufficient to improve the situation in many cases, especially as regards air contamination (cf. Knebelspieß 2014; Roßnagel in Führ 2016, § 5 Rn. 494 ff.).[4] In other cases, particularly those of soil contamination and contaminated rivers, clean-up has become necessary (Schindler 2014). The ongoing remediation of the river Emscher, already mentioned above, provides an example of a clean-up and development project of truly spectacular dimensions.[5] The entire wastewater is to be channelled into an underground canal fifty-one kilometres in length and 1.4–2.8 m in diameter, recreating the natural river above. This project, estimated at €4.5 billion, is being carried out by the Emschergenossenschaft—a cooperative entity of public law, formed of members from the industry, cities, and municipalities, and co-financed by the European Union and the German state.

In several cases of soil contamination, of hazardous waste disposal sites and sites of old factories, the polluters and in particular the owners of the properties could be held liable. Many clean-up procedures were, nevertheless, state-financed, indicating the difficulty in making the original polluters pay. Waste disposal sites were often state-owned and state-operated, making the state responsible for clean-up or shielding the environment from the resulting contamination and thereby shifting the expenses from the generators of the waste to the taxpayer.[6] The state's responsibility for the environment and the health of its citizens also becomes apparent in planning. When, after its closure, a waste disposal site in Bielefeld was designated an area for residential development in spite of being unfit for this purpose, the state was held liable for the damages of the residents because the planning authorities had not appropriately investigated and considered the consequences of the site's previous use (BGH, NJW 1989, 976).[7]

Although the clean-up of contaminated sites has been pursued methodically since the 1980s (Schindler 2014, pp. 113 ff.),[8] the number of contaminated sites and suspected contamination sites is still immense and has grown beyond previous expectations.[9] According to the statistics on site contamination published by the Working Group on Soil Protection (Bund/Länder-Arbeitsgruppe Bodenschutz,

---

[4]The prevention of new contamination is also aimed at restoration.

[5]Information on the project and the agents responsible may be found on the following homepages: http://www.abwasserkanal-emscher.de/; http://www.eglv.de/emschergenossenschaft.

[6]Hamburg spent DM33 million on one former waste disposal site, Georgswerder, alone (Grziwotz 1990, p. 285). See also Weber and Otting (2014) with regard to current burdens borne by municipalities as owners of properties.

[7]See Schink (1990) for further references.

[8]An important first step was the Concept for Soil Protection issued by the Federal Government in 1985 (BT-Drs. 10/2977).

[9]Estimated clean-up costs have increased more than fourfold between 1985 and 1990; see Grziwotz (1990, p. 285).

LABO) in 2014, there are 15,837 contaminated sites in Germany, and 317,036 listed as suspected contamination sites (ALA 2014, p. 2, with links to the state statistics in Annex 2). It is interesting to note that investigation measures leading to a final assessment of suspected contamination sites are limited to 93,911 cases. Accordingly, the investigation and clean-up costs lying in wait are significant.[10]

Notwithstanding the high number of pending cases, the clean-up in this context is technically possible, and the legal framework has already been resolved—even if changes naturally remain possible. There also exist scenarios, however, in which the clean-up is difficult or even technically impossible by today's standards, either because there are no feasible ways to isolate and remove contaminating substances —for example regarding the contamination of the high seas or ozone depletion—or because there is no safe way of disposing of them. Most notorious perhaps is the case of the salt mine Asse II (BMU 2012; König and Hoffmann 2009; Schneider 2010),[11] a former mine shaft that has been used to store radioactive waste in barrels. Water has now entered the mine, corroding the barrels and releasing radiation into the environment. This case represents a particularly clear example of the problem of changing regulatory frameworks. The site was built as a salt mine and was later used as a research site on the permanent disposal of nuclear waste under mining law; the question of whether the mine's operation and shut-down fell under the newly reformed Atomic Energy Act of 1976 became a matter of severe dispute for several decades, forcing the operation to continue on a series of temporary administrative acts (König and Hoffmann 2009, p. 55; Schneider 2010, pp. 24 ff.). The case is also an example of changing scientific knowledge and practice, as well as of a certain carelessness predominant in the treatment of the environment in the past. The risk of flooding, for example, might have been foreseen, as both neighbouring mine shafts had been closed due to the same problem—well in advance of the construction of the experimental disposal site. Furthermore, the barrels were transported into the shaft by dropping them into it, which caused them to break in some cases (König and Hoffmann 2009, p. 353). At present, the removal of the radioactive waste has been decided on, confronting the authorities with the challenge of safely retrieving the barrels and finding a more appropriate disposal site (see *Gesetz zur Beschleunigung der Rückholung radioaktiver Abfälle und der Stilllegung der Schachtanlage Asse II vom 20. April 2013*, BGBl. I p. 921; Gaßner and Buchholz 2013; BMU 2012).

---

[10]Even in 1995, the costs were estimated at €500 billion (Gaier 2005, p. 161).

[11]Current and further information may also be found at the homepage of the responsible ministry in Lower Saxony: http://www.umwelt.niedersachsen.de/atomaufsicht/endlager/asse/stilllegung-der-schachtanlage-asse-ii-93833.html.

2 Legal Sources and the Structure of Environmental Legislation in Germany: An Intricate Network

Germany is bound by international customary law and is signatory to several international treaties on specific environmental matters.[12] Environmental legislation has furthermore formed an important domain for European law almost from the beginning (cf. Epiney in Landmann and Rohmer 2015, Art. 191 AEUV Rn. 1 ff.): at present, EU regulations and directives form the basis of a large percentage of the administrative legislation on environmental matters in Germany (cf. Voßkuhle 2013, p. 2).[13]

At national level, the aim of protecting the wildlife and natural resources as the basis of human life for current and future generations has constitutional rank, being laid down in Art. 20a GG in 1994 (Voßkuhle 2013, pp. 4 f.; Gärditz in Landmann and Rohmer 2015, Art. 20a GG Rn. 1 ff., 24). However, a single specific and comprehensive regulation regarding the legislative power for enacting environmental legislation does not exist (Gärditz in Landmann and Rohmer 2015, Art. 20a GG Rn. 98 ff.; Voßkuhle 2013, pp. 3 f.). Legislative powers are shared between the federal level and the sixteen states, the *Bundesländer*. Some competence titles even allow the states to issue divergent laws, the more current legislation taking priority; see Art. 72 II, III GG (Gärditz in Landmann and Rohmer 2015, Art. 20a GG Rn. 117 f.).

In consequence, German environmental legislation consists of an intricate network of different regulations (cf. Schulze-Fielitz 2007), the attempt to systemize German environmental law by creating a unified environmental code (*Umweltgesetzbuch*, UGB) having failed twice (cf. Scheidler 2007). This chapter will focus on federal law and common aspects of the different state laws, providing examples from Baden-Württemberg when necessary for illustration.

Environmental law comprises both legislation on the protection of the environment itself and legislation on the protection of human health and property from pollution, which indirectly protects the environment. The strategy of environmental legislation is based on two pillars: the prevention or limitation of new contamination on the one hand, and reactive measures on the other (including compensation or damages for victims, punishment, the clean-up of existing contamination, and the restoration of the environment to prevent further damage); however, there obviously are interactions between the two. Given the multiple aims, relevant legislation forms part of all main areas of the law—administrative, civil, and criminal. The first pillar aims at preventing or limiting new contamination, for example by instituting maximum permissible values for the concentration of contaminants and technical

---

[12]A complete list may be found in the "Fundstellennachweis B", last version 2015/12/31, available at http://www.bgbl.de/xaver/bgbl/text.xav?SID=&tf=xaver.component.Text_0&tocf=&qmf=&hlf=xaver.component.Hitlist_0&bk=bgbl&start=%2F%2F*[%40node_id%3D%271197685%27]&skin=pdf.

[13]According to Voßkuhle (2013), it is as high as 65%.

requirements, both monitored and controlled by a system of permits. Environmental taxes or trading schemes (for carbon dioxide) form an essential element of modern preventative regulation by administrative law, especially with regard to diffuse pollution (see Sacksofsky 2000; Lübbe-Wolff 2001 with further references). In addition, the Aarhus Convention, and the European and German[14] legislation implementing it, attempts to increase the awareness and influence of the public and environmental organizations by providing a right to information, a right to public participation in environmental decision-making, and a right to challenge administrative decisions before the courts (see Hofmann in Landmann and Rohmer 2015, § 9 UVPG Rn. 1). However, in accordance with the aim of the investigation, this chapter will mainly focus on the regulation confronting the challenges inherited from the past.

3 Administrative Law: The Centerpiece of Environmental Regulation

3.1 The Starting Point: General Police and Regulatory Law as the German Headstart in Ordering Clean-up Procedures

The starting point and regulatory framework originally employed for the clean-up of contaminated sites or any sort of contamination was the general authorization under general police and regulatory law (*polizeirechtliche Generalklausel*) (Papier 1986; Breuer 1987, pp. 753 ff.; Müggenborg 1992). Although each state has its own law, it follows a common structure throughout Germany (see, for example, Pieroth et al. 2014, § 2 Rn. 35, 45 ff.). Its objective was (and is) the defence against dangers to public security and order. General police and regulatory law is not only important as a means of understanding the common structure; it also remains relevant in cases originating from before the specific sectorial law came into force. Responsibility under police law is not subject to time limitation (*polizeirechtliche Ewigkeitshaftung*) (see Ossenbühl 1995 [for potential restrictions], Hullmann and Zorn 2010).

The first prerequisite for action on the part of the regulatory agencies is a danger to or a disturbance of public security and order. The definition of public security and order is very extensive—not to be compared to the identical expression in Art. 36 AEUV (Kingreen in Callies and Ruffert 2011, Art. 36 AEUV Rn. 76)—and comprises not only all individual goods but also the entire legal system, among others (see, for example, Pieroth et al. 2014, § 8 Rn. 3 ff.). As a consequence, any

[14]Federal Act on Legal Remedies in Environmental Matters (*Umweltrechtsbehelfsgesetz*, UmwRG); Environmental Information Act (*Umweltinformationsgesetz*, UIG).

illegal activity, for example polluting activity exceeding permitted emission values, can be prevented by regulatory action.

Administrative acts will generally be addressed to the person responsible for the danger. Responsibility can be based on conduct (*Handlungsstörer*), § 6 PolG (*Polizeigesetz Baden-Württemberg*, Police Act) or on a person's control over the state of an object or property that is the source of the danger (*Zustandsstörer*), § 7 PolG. Responsibility for conduct is incurred by causing the danger. Administrative law states that an action is causal if the perpetrator has directly crossed the threshold to creating a danger (*Überschreiten der Gefahrenschwelle, Theorie der unmittelbaren Verursachung*) (see, for example, Pieroth et al. 2014, § 9 Rn. 5 ff.; Oerder 1992, pp. 1034 f.). The question is whether the action has sufficient gravity and is sufficiently close to the damage to justify attributing responsibility for the consequences. A legal action cannot therefore lead to responsibility under this provision. Consequently, the existence of authorization excludes in principle the responsibility for conduct. However, the extent of the legalizing effect is subject to controversy. While some authors want to exclude liability completely, it seems reasonable to limit the legalizing effect to dangers that were actually considered in the decision, i.e. dangers that were known at the time the authorization was issued (see Pieroth et al. 2014, § 9 Rn. 17, 63 f.; Oerder 1992, pp. 1034 f.; Papier 1986, pp. 257 ff.; Dombert in Landmann and Rohmer 2015, § 4 BBodSchG Rn. 49 ff., with further references). In cases of diffuse pollution, the individual contribution to the damage will, nevertheless, be too insignificant to cause responsibility.

Responsibility based on the control over a source of danger is incurred by the owners of a property and the holder of the actual control over it (e.g. Pieroth et al. 2014, § 9 Rn. 33 ff.). The responsibility of innocent owners—owners who did not cause the contamination—has long been subject to criticism and was limited by the Federal Constitutional Court to the value of the property after restoration, alleging that the social responsibility for the property cannot exceed its privileges (BVerfGE 102, 1).[15] This limitation is, however, restricted to owners who did not know and (exercising reasonable care) could not have known of the contamination when acquiring the property, and subject to exceptions with regard to the extent of admissible costs (cf. Müggenborg 2001, p. 41; Gaier 2005, pp. 162 f.).[16]

The competent administrative authority disposes of a certain measure of administrative discretion both with regard to acting at all (*Entschließungsermessen*) and with regard to selecting the appropriate action and addressee (*Auswahlermessen*) (e.g. Pieroth et al. 2014, § 10 Rn. 32 ff.). In accordance with the basic principle of police law, the effectiveness of preventing dangers (*Effektivität der Gefahrenabwehr*), the responsibility of the owner is not subsidiary to that of the polluter (Papier 1986, pp. 262 f; see also Gaier 2005). Because of the difficulty in

[15]A restriction had long been proposed in legal literature, for example, by Papier 1986, pp. 261 f. This is criticized by Pieroth et al. (2014, § 9 Rn. 71 ff.) for being incompatible with the objective character of police law.

[16]Müggenborg (2001) suggested that this may cause complications in practice.

identifying the polluter, clean-up orders are addressed to the owners in the majority of cases (Gaier 2005). To enforce the order, the competent authority can impose penalty payments but may also undertake the necessary measures itself or commission them to a third party, retrieving the necessary expenses from the responsible (*Selbstvornahme*, *Ersatzvornahme*) (see, for example, Pieroth et al. 2014, § 24 Rn. 1 ff., 10 f., § 25 Rn. 9 ff.). If damages are imminent, direct action by the authorities can be admissible (*unmittelbare Ausführung)*, § 8 I PolG.

3.2 *Making Clean-Up and Remediation Even More Effective: Specific Environmental Legislation from Later Decades*

3.2.1 The Federal Soil Protection Act: The Central Legislative Instrument for Clean-up

The Central Prerequisite for Action: Soil Contamination

The statute currently most important for clean-up procedures is the Federal Soil Protection Act, already mentioned above (see, for example, Vierhaus 1998, pp. 1264 f.). The Federal Soil Protection Act is applicable to legacy contamination (*Altlasten*), but also, with certain exceptions, to recent soil contamination.

The prerequisite for action under the Federal Soil Protection Act is a negative change to the soil or a danger thereof. The practical question is that of when potential consequences of a change to the state of the soil can be considered a substantial nuisance or danger to individual persons or the general public. Pursuant to § 4 IV of the Federal Soil Protection Ordinance (*Bundesbodenschutz- und Altlastenverordnung*, BBodSchV), issued under § 8 BBodSchG, this is determined in accordance with specific threshold values for contaminants (cf. Kobes 2000, pp. 264 f.; Dambert in Landmann and Rohmer § 8 BBodSchG Rn. 1 ff.). These threshold values, found in Annex 2 to the BBodSchV, depend on the designated use of the site, categorized by different contamination "pathways" or linkages—soil–human, soil–plant, or soil–water—which reflect the varying susceptibility to contamination. If there are no threshold values in the annex yet, the necessity for clean-up is to be determined in accordance with the Methods and Standards for the Derivation of Threshold Values (*Methoden und Maßstäbe für die Ableitung von Prüf- und Maßnahmenwerten*) (Bundesanzeiger Nr. 161a vom 28. August 1999), § 4 V BBodSchV.

If the concentration of contaminants is known to the regulatory agencies, determining the existence of a danger or disturbance will be relatively simple, provided there is sufficient scientific knowledge on the harmful potential of the substance. However, in many cases, the exact contamination will be unknown and the regulatory agencies can only suspect contamination based, for example, on previous industrial use (*Gefahrenverdacht*). It had long been a matter of controversy whether the regulatory agencies could call upon the suspected responsible person to perform the necessary investigation to confirm the existence of the danger (*Gefahrerforschungsmaßnahmen*) (Papier 1986, p. 257). The question has been decided in the Federal Soil Protection Act: pursuant to § 9 I BBodSchG, the regulatory agency has to undertake the initial investigations at its own cost. The owner of the property is obliged to tolerate them (Posser in Giesberts and Reinhardt 2015, § 9 BBodSchG Rn. 13 f.; Ewer in Landmann and Rohmer 2015, § 9 BBodSchG Rn. 42 ff., 85).

The prerequisite of a negative change to the soil or a danger thereof seems to be missing in **Hypothetical Scenario 1**[17] and **Hypothetical Scenario 2**. As has been stated above, the requirement for a clean-up order is a change to the soil that interferes with the soil's functions in a manner likely to cause danger, considerable disadvantage, or substantial nuisance to individual persons or the general public. This will be determined in accordance with threshold values that depend on the use of the site (§ 4 IV BBodSchG). Although pollution has taken place, it does not cause any damage or danger to human health. The legal assessment obviously depends on the exact circumstances and consequences originating from the pollution. It has to be emphasized that the Federal Soil Protection Act does not require a concrete danger to human health; a negative change to the soil or a danger thereof is sufficient. Furthermore, a negative change to the soil can already be presumed when the pollution may result in substantial nuisance. Nevertheless, further information would be required in order to determine whether this threshold is reached in the scenarios. Additionally, even if there is no present danger, it is possible that the competent authority has the power to forbid A's activity with regard to the future, that is if A exceeds threshold values for contamination. As these threshold values are oriented at the precautionary principle and are established at a level well below the point where concrete danger to human health is to be expected, it is likely that they will be exceeded before there are notable consequences. However, the information provided is yet again insufficient to draw a clear conclusion.

In **Hypothetical Scenarios 3, 4, and 5**, A's polluting activity had much more severe consequences, from contamination of groundwater used in agriculture to contamination of groundwater used as drinking water, and on to the causally linked death or injury of persons. In all three scenarios, the necessary negative change of the soil required for action by the competent authority is present. Although the consequences in **Hypothetical Scenario 5** are clearly more severe, the contamination in all scenarios has progressed significantly. Considering that the contamination in **Hypothetical Scenarios 3 and 4** has already leaked into the groundwater, which is used either directly or indirectly for human alimentation, it is highly likely that there exists at least an abstract danger to human health. The competent authority is therefore entitled to take action.

[17]See Chap. 1, footnote 7 in this volume.

Measures, Procedure, and Enforcement: From Preliminary Investigation to the Actual Clean-Up and Remediation

The procedure regarding the treatment of contaminated sites is divided into several different steps (see Kobes 2000, pp. 261 ff.). The first step is the preliminary investigation (*orientierende Untersuchung*; § 9 I BBodschG), to determine whether there is sufficient indication of contamination to categorize the site as a suspected contamination site. This will be done in accordance with the threshold values provided in Annex 2 of the BBodSchV, already mentioned above. If these values are exceeded, the competent authority generally can (and is supposed to) issue an administrative act ordering the responsible person to perform a detailed investigation of the contamination (*Detailuntersuchung*; § 9 II 1 BBodSchG), and, if the existence of contamination demanding remediation is confirmed, to draw up a detailed clean-up plan (*Sanierungsplan*; § 13 I, II BBodSchG). If there is no responsible person or if the addressee of the order does not comply, the competent authority can create the plan itself (§ 14 BBodSchG). Subsequently, the competent authority can issue an order imposing the necessary clean-up measures (§ 10 I BBodSchG), or declare the plan legally binding (§ 13 VI BBodSchG). Declaring the plan legally binding has the advantage that the plan then substitutes otherwise necessary permits (§ 13 V, VI BBodSchG), for example planning approval for the on-site treatment of contaminated soil under the Recycling Management Act, which, if applied for separately, would require time-consuming public participation (Dombert in Landmann and Rohmer 2015, § 13 BBodSchG Rn. 29 ff.).

The main potential clean-up measures are either the decontamination of the property (§ 2 VII Nr. 1 BBodSchG), for instance by excavating and disposing of the contaminated soil or by treating it on site, or the containment of the contamination (§ 2 VII Nr. 1 BBodSchG), for example by installing barriers between the contaminated soil and the groundwater or between the contaminated soil and the surface layer (Nies in Landmann and Rohmer 2015, § 2 BBodSchG Rn. 40 f.). The competent authority can also oblige regular surveys of the concentration of contaminants on the site (§ 15 II BBodSchG), as an alternative to or in order to monitor the success of the other measures.

Instead of imposing clean-up action by administrative order, the competent authorities can also cooperate with the responsible person (Vierhaus 1998, pp. 1267, 1268 f.; Frenz and Heßler 2001; Müllmann 1994, also in the following). In particular, it is possible to make the clean-up plan the subject of an administrative contract (§ 13 IV BBodSchG). A cooperative approach has the advantage of including the responsible person in the decision-making process and to enable compromises, especially when there are points of doubt. This not only facilitates the concrete planning, but will also prevent appeals and time-consuming and costly court proceedings. Furthermore, while the Federal Soil Protection Act, as a part of police and regulatory law, is limited to measures to combat dangers and prevent further damages, and may not therefore allow the imposition of measures aiming for the full restitution to the *status quo ante* (Breuer 1987, p. 755), the cooperative approach is more flexible (Vierhaus 1998, p. 1267). In addition, it is also possible to combine clean-up with urban planning (see Müllmann 1994, p. 877; Tomerius 2005, p. 18). Consequently, contracts between the administration and persons responsible for clean-up are frequently used. These contracts will often include a provision that guarantees the property's removal from the corresponding Register of Contaminated Sites, which will increase property value (Vierhaus 1998, pp. 1268 f.).

Administrative competences in environmental matters have long been divided between the general administrative agencies and special environmental agencies, the exact structures and organization being subject to state law. As Baden-Württemberg has recently adopted a reform integrating the special agencies into the general administration (see IMBW), measures under the Soil Protection Acts are administered by the general administrative agencies, in the first instance the administrative districts (UMBW 2014).

If the competent authority and the addressee cannot decide on a consensual approach, the competent authority disposes of the same methods of enforcement as under general regulatory and police law, particularly the possibility of performing the clean-up itself and recovering the costs from the responsible person (§ 24 I BBodSchG). In addition, the negligent or intentional non-compliance with a clean-up order or surveillance obligations constitutes a regulatory offence under § 26 BBodSchG.

Liable Persons and Compensation: Effectiveness Versus Polluter Pays

With regard to liable persons, the Federal Soil Protection Act still distinguishes between responsibility for conduct and responsibility for a state. The responsibility for conduct now expressly includes legal successors in addition to the person(s) causing the danger (§ 4 III BBodSchG). The responsibility for a state is incurred by the owner of a contaminated property, the holder of the actual control over it (for example a tenant), as well as the former owner of a derelict property (see Sanden 2014) and executives of legal persons owning a property (§ 4 III BBodSchG). In addition, vendors of properties are responsible if they have sold the property after the statute came into force, knew of the contamination or ought to have known it, and did not legitimately expect the absence of contamination when buying the property themselves (§ 4 VI BBodSchG).

This regulation of responsibility has been criticized for being very extensive, especially with regard to the sellers of properties but also to executive officers of companies owning a contaminated property (e.g. Müggenborg 2000). It was introduced to counteract attempts at bypassing responsibility for clean-up procedures by selling contaminated properties to insolvent companies or persons, but it effectively applies to a much larger range of cases (Giesberts and Hilf in Giesberts and Reinhardt 2015, § 4 BBodSchG Rn. 33 ff.).

The Federal Soil Protection Act provides for a claim for compensation by the addressee of the administrative act against other responsible persons (§ 24 II BBodSchG) (see Hellriegel and Schmitt 2009). Costs are shared in accordance with the proportion of responsibility and the liability of different persons responsible for conduct is joint and several (§ 24 II 2 BBodSchG). The claim is only directed at a person responsible for conduct, emphasizing the polluter pays principle (Vierhaus 1998, p. 1267; Dombert in Landmann and Rohmer 2015, § 24 BBodSchG Rn. 18 ff.). The principle was further strengthened by case law that allows the transfer of the presumptions and partial reversal of the burden of proof provided in §§ 6, 7 UHaftG (*Umwelthaftungsgesetz*, Environmental Liability Act) to determine responsibility for conduct that forms the base for the compensation claim (BGHZ 158, 354; Hellriegel and Schmitt 2009; see also below). The polluter, on the other hand, cannot claim compensation from the owner under this provision (Vierhaus 1998, p. 1267; Dombert in Landmann and Rohmer 2015, § 24 BBodSchG Rn. 18 ff.). The question of whether the provision also grants a claim for compensation between two owners is subject to controversy (in favour, see Sandner 2001, p. 2045 and Hilf in Giesberts and Reinhardt 2015, § 24 BBodSchG Rn. 24 f., with notable reasons; against, see Dombert in Landmann and Rohmer 2015, § 24 BBodSchG Rn. 18 ff.). The claim is time-barred three years after the claimant knew of the claim, and at the latest thirty years after the clean-up is finished (§ 24 II 3, 4, 5 BBodSchG).

In **Hypothetical Scenarios 3, 4, and 5**, the competent authority will most likely order first a preliminary and then a detailed investigation of the contaminated site, § 9 I, II 1 BBodSchG, and, ultimately, the submission of a clean-up plan and the clean-up itself, although in theory it also has the option, at its discretion, not to act. However, considering the fact that the contamination has already reached the groundwater used directly or indirectly for human alimentation in **Hypothetical Scenarios 3 and 4**, action seems to be urgent, and even more so in **Hypothetical Scenario 5**, where the contamination has already led to injuries and even the death of a person. In this context, it has to be noted that the administration has a duty to protect the life and health of the population, which will influence its decisions. The investigation and clean-up orders can either be addressed to A, as the perpetrator (§ 4 III BBodSchG), or to B, as the current owner of the site (§ 4 III BBodSchG). The choice of the addressee is at the discretion of the competent administrative authority, the effectiveness of the action constituting the deciding factor, which will make B the more likely choice. Yet again, considering the progression and the severe consequences, the case may even be sufficiently urgent to warrant direct action and clean-up by the competent authorities themselves, who can then retrieve the expenses from A and B under § 24 I BBodSchG. Direct action is possible when addressing the clean-up order to one of the liable persons will take too long to effectively counteract the danger.

If the administration addressed the clean-up order to B, he can retrieve the entire costs of clean-up from the perpetrator, A, under § 24 II BBodSchG, barring a different regulation in the sales contract. In addition, contamination also constitutes a defect to the property, which gives rise to contractual liability.

Retrospectivity: Effectively Regulating Old Cases

The Federal Soil Protection Act encompasses far-reaching regulations, especially with regard to liable persons, posing the question of whether they can also be applied to contamination resulting from activity before the statute came into force (Leitzke and Schmitt 2005). Unlike in criminal law, retrospective application is not generally inadmissible, but is still subject to constitutional restrictions (BVerfGE 30, 367, 385; Grzeszick in Maunz and Dürig, Art. 20 GG Rn. 72 ff.), the exact prerequisites depending on the kind of retrospectivity. True retrospective application (*Rückbewirkung von Rechtsfolgen, echte Rückwirkung*), where new regulations apply to situations that have already been concluded in the past, is only possible in very exceptional cases; quasi-retrospective application (*tatbestandliche Rückanknüpfung, unechte Rückwirkung*), which applies new regulations to situations that began in the past but are still ongoing, is generally admissible (BVerfGE 72, 201ff., 241f.; Grzeszick in Maunz and Dürig 2014, Art. 20 GG Rn. 78 f.).[18] The Federal Court applied both the new regulation on legal successors and the claim for compensation between different responsible persons retrospectively (BGHZ 158, 354; see also Gaier 2005, pp. 164 f.; Hilf in Giesberts and Reinhardt 2015, § 24 Rn. BBodSchG 22 ff.).

3.2.2 The Environmental Damage Act: Minimum Requirements of European Origin

In 2007, the Environmental Damage Act (*Umweltschadensgesetz*, USchadG) came into force, implementing the Directive on Environmental Liability[19] regarding the prevention and compensation of damages to the environment itself, that is, damages to biodiversity, water, and soil, defined in § 2 Nr, 1, 2 USchadG (cf. Becker 2007; Ruffert 2010; Scheidler 2007). Unlike general police and regulatory law and the Federal Soil Protection Act, the Directive and consequently the Environmental Damage Act impose liability only on certain professionals, not on the owners, emphasizing the polluter pays principle (BT-Drs. 16/3806, S. 13; Becker 2007). A number of conducts that are considered especially dangerous to the environment, enumerated in an annex to the statute, lead to fault-independent liability (§ 3 I Nr. 1 USchadG). Liability for damages resulting from other professional activity requires negligence or intent (§ 3 I Nr. 2 USchadG) (Beckmann and Wittmann in Landmann and Rohmer 2015, § 3 USchadG Rn. 9 f.; Ruffert 2010, p. 1181).

The Environmental Damage Act imposes several duties on the person responsible (see, for example, Becker 2007, pp. 1109 ff.): the duty to prevent imminent danger (§ 5 USchadG), the duty to report imminent danger and damages after they

[18]See Grzeszick in Maunz and Dürig (2014) for further references and clarification on the definitions.

[19]See Becker (2005) with regard to the Directive.

have occurred (§ 4 USchadG),[20] and the duty to contain and remedy any damage (§ 6 USchadG). This also includes the duty to investigate and plan the necessary remediation measures and to submit a clean-up plan to the competent authorities (§ 8 I USchadG). The duties are all defined in § 2 USchadG. Pursuant to §§ 2 Nr. 8, 8 USchadG, the applicable remediation measures are determined in accordance with the respective sectorial legislation (*Fachgesetze*) (Balensiefen 2012, § 2 Rn. 10; Seuser 2013 [2], p. 398). Annex II of the Environmental Liability Directive is referenced in § 19 IV of the Federal Nature Protection Act (*Bundesnaturschutzgesetz*, BNatSchG). With regard to damages to biodiversity or habitats, remediation is therefore based on a three-tier system (see Bruns et al. 2009, pp. 155 ff.; Gellermann in Landmann and Rohmer 2015, § 19 Rn. BNatSchG Rn. 28 ff.). The primary duty is to restore the environment to its exact former state. If this is not possible, the responsible person has to implement comparable restoration measures, for example creating an equal habitat on a neighbouring area if the contamination does not permit the continued survival of the damaged species on the affected site. The tertiary duty includes interim measures to compensate for damages occurring before the primary remediation is complete. These measures are also directed at compensation in kind but can include any projects of equal value that improve the situation of protected species and habitats (Bruns et al. 2009, p. 156).

On the other hand, § 90 WHG does not specify particular remediation measures with regard to water contamination. Nevertheless, remediation will aim at the restitution of the *status quo ante* and can be based on the same three-tier system (Seuser 2013 [2], pp. 398 f.). The costs incurred by complying with the duties under the Environmental Damages Act are borne by the responsible person (§ 9 I USchadG), although the Environmental Damages Act allows for state legislation exempting the responsible persons from liability in certain cases (see Becker 2007, pp. 1110 ff.).

The competent authority monitors the action of the responsible person and decides on necessary measures (§§ 7 I, 8 II USchadG). However, environmental associations and affected citizens have the right to apply for the initiation of investigation and remediation measures under the Environmental Damage Act and other sectorial environmental legislation that the Environmental Damage Act refers to (§§ 10, 11 II USchadG). Although it is not expressly specified in the statute, the authority can also undertake necessary measures itself and retrieve its expenses from the polluter (§ 9 I USchadG) (Schrader and Hellenbroich 2007; Wittmann and Beckmann in Landmann and Rohmer 2015, USchadG § 10 Rn. 1 ff.). If there is more than one responsible person, the authority can draw on one person for the entire remediation. By reiterating the necessity of proving individual causation, however, § 3 IV USchadG effectively excludes the application to cases of diffuse pollution (Balensiefen 2012, § 3 Rn. 4). The person tasked with the clean-up then

[20]A duty previously unknown in German administrative environmental law; see Becker (2007, p. 1111).

has a claim for compensation against the other responsible persons (§ 9 II USchadG). The costs are shared in accordance with the proportion of responsibility.

The Environmental Damage Act is only applicable to damages caused by incidents after 30 April 2007 or by activity that has not ended before that date (§ 13 I USchadG). Action under the Act is furthermore time-limited to thirty years after the damages occurred (§ 13 II USchadG), unlike action under traditional German regulatory law such as the Federal Soil Protection Act (see, for example, Becker 2007, p. 1112).

While clean-up with regard to the contamination of soil and water had already been possible in Germany, the responsibility for damages to biodiversity was significantly extended (Bruns et al. 2009). The existing specific sectorial legislation is still applicable if the protection it conveys is more extensive than that provided by the Environmental Damage Act (§ 1 II USchadG), which is only meant to institute minimum requirements (BR-Drs. 678/06, S. 36 f.; Scheidler 2007, p. 1114). In consequence, the Federal Soil Protection Act, which not only applies to older legacy contamination but also provides for more persons responsible, largely takes precedence in cases of soil contamination (Becker 2007). In the case of damages to water or the landscape, the Environmental Damage Act and the new provisions introduced in the Federal Water Act and the Federal Nature Conservation Act, respectively, closely interact, the Environmental Damage Act regulating duties and enforcement authorization and the sectorial law the content requirements (Becker 2007, pp. 1107 f.; see also Balensiefen 2012, § 2 Rn. 10; Seuser 2013 [2], p. 394).

3.3 The Planning Aim of Reducing Urban Sprawl: Brownfield Redevelopment

The English term "brownfield" may be translated as *Brachfläche* and refers to abandoned sites within towns or cities, whether contaminated or not. Although there are many industrial brownfields in Germany—due to globalization and a shift from an industrial to a service society—a significant number of building projects are still realized on new, the so-called green, land (Heinz 2005, p. 81; Tomerius 2005, pp. 14 f.). However, the redevelopment of brownfield sites, also referred to in Germany as the recycling of areas (*Flächenrecycling*), has become a main aim of urban development and planning. The Federal Government wants to reduce the use of new areas to thirty hectares per day until the year 2020 (in comparison with 130 ha in 2002) (BReg 2002, pp. 99 f.; BReg 2012, pp. 194; BMU 2013, pp. 14), and Baden-Württemberg is even pursuing the long-term aim of covering the entire demand for building land by reusing brownfield sites.[21] The aim of using brownfields instead of new undeveloped land is even laid down in § 1a II BauGB

---

[21]See MUVBW (2011); MVIBW provides an overview of brownfield redevelopment projects in Baden-Württemberg.

(*Baugesetzbuch*, Federal Building Code) as an aim of the planning process (Tomerius 2005, pp. 15 ff.; Krautzberger and Wagner in Ernst et al. 2015, § 1a Rn. 46 ff.)[22] and is facilitated by the new, less strict regulations for planning procedure found in § 13a BauGB (cf. Tomerius 2008).

Many brownfields are so severely contaminated that the owners are unable to finance the clean-up. Due to the large number of unused areas and the ongoing designation of new land for building purposes, the clean-up costs may even exceed the property value and make the sale impossible (Heinz 2005, p. 81; Mitschang and Roeper 2011). Although brownfields are already developed, thereby saving on service infrastructure costs, and centrally located (Tomerius 2005, p. 20), redevelopment through reselling the property remains challenging. Even if there is no known contamination, the far-reaching liability of owners under the Federal Soil Protection Act may discourage both sellers and investors, especially since the contractual arrangement has no influence on obligations under administrative law and a clean-up order can therefore severely affect the balance of the transaction (Gaier 2005, p. 166; Dombert in Landmann and Rohmer 2015, § 4 BBodSchG Rn. 67 ff.). With regard to the former GDR, the discouragement to investors was so strong that parliament saw fit to provide the possibility of the Government taking over liability for contamination, exempting investors from it (Art. 1 § 4 III UmwRG; Tettinger 1991; Dannecker and Klink 2003; BMU 2008).

Planning law may, nevertheless, be an effective tool to encourage redevelopment (Heinz 2005; Tomerius 2005). Not only can it favour the reuse of uncontaminated property by refusing to make new land available, it can also influence the clean-up of contaminated sites. Because planning and building law—unlike regulatory law—is not limited to measures preventing damages, but follows a precautionary approach, it can impose stricter requirements on clean-up and remediation. The required standard for remediation will, furthermore, change automatically with the change of the site's designated use, because safe levels of contaminants are determined in accordance with the use (Dombert in Landmann and Rohmer 2015, § 4 BBodSchG Rn. 67 ff.; Tomerius 2005, p. 18). Because the demand for industrial areas is lower than that for residential building land, this change to the permitted use will also increase the property value. It is therefore possible that investors will voluntarily submit to remediation measures in the course of new building projects (cf. Heinz 2005, p. 82).[23] However, it is also of essential importance to cut down on fiscal instruments that favour urban sprawl and the use of new land, first and foremost the commuting allowance but also first-home buyer allowances. The extra revenue generated could be redirected to promote brownfield redevelopment (Tomerius 2005, p. 19).

---

[22]Tomerius (2005) also provides an overview of different applicable planning instruments.

[23]It is also possible to combine a clean-up contract under § 13 IV BBodSchG with a planning contract under § 11 BauGB (Tomerius 2005, p. 18).

3.4 The Challenge of Sheer Numbers: Financing and Coordinating Clean-up and Redevelopment

The legislation on the clean-up of site contamination in Germany is far-reaching and provides the owner as a liable person, who is easy to find. Nevertheless, it is possible that no responsible person can be made liable. Uncontaminated brownfield sites, while not directly posing a threat to health or the environment, can affect the attractiveness and living quality of cities and may indirectly favour further use of new land with all its ecological consequences (Tomerius 2005, p. 14). If planning decisions alone are unable to attract investors, it is important to find other modes of financing clean-up and redevelopment. In larger industrial areas, the state has therefore implemented large-scale clean-up and redevelopment projects, the so-called lighthouse projects (Heinz 2005, p. 81). Nevertheless, contaminated sites and brownfields are widespread, and smaller sites also await remediation and redevelopment (Heinz 2005, p. 81).

Each state has a different strategy and different projects for implementing redevelopment.[24] In Baden-Württemberg, state aid for financing clean-up measures is currently subject to the Regulation on Financial Aid for the Treatment of Legacy Contamination (*Förderrichtlinien Altlasten*, FrAl), enacted in 2014, which focuses on providing funding for investigation and remediation measures for brownfields (UMBW 2014). The treatment of contaminated sites was facilitated by the institution of a fund for financial aid (*Kommunaler Altlastenfond*) in 1988. Since then, approximately €730 million have been granted to the municipalities to finance remediation measures. An overview of different financing possibilities in Germany may be found in the Database on Funding Programmes (*Förderdatenbank*) provided by the Federal Ministry for the Economy and Energy (Bundesministerium für Wirtschaft und Energie, BMWi).[25]

Financial aid by the government is mainly financed through taxes. To shift the burden back to the polluter, it has been proposed to impose special levies (*Sonderabgaben*) on certain branches of industry, for example to account for damages to forests resulting from acid rain (Marburger 1992, pp. 33 f.; v. Hippel 1990). However, the imposition of a special levy is subject to severe restrictions, as the constitution provides for general taxes as the main means of financing public spending (Kube in Epping and Hillgruber 2015, Art. 105 GG, Rn. 1 f., 16 ff., with further references; Breuer 1987, pp. 759 ff.). Because of this, and due to issues of federal competence, attempts at imposing special levies for compensation funds were declared unconstitutional by the Federal Constitutional Court (BVerfGE 113, 128 (Solidarfond Abfallrückführung); BVerfG, NJW 1998, 2346; cf. Bothe 1998).

[24]See Breuer (1987, pp. 756 ff). for an overview of different models.

[25]http://www.foerderdatenbank.de/Foerder-DB/Navigation/root.html.

Notwithstanding financing possibilities, the number of suspected contamination sites does not permit immediate intervention in all cases (UMBW 2014). The procedure for handling cases of soil contamination in Baden-Württemberg is based on the Concept for the Treatment of Contaminated Sites and Suspected Contaminated Sites, a ministerial decree from 1988 (*Konzeption zur Behandlung altlastverdächtiger Flächen und Altlasten in Baden-Württemberg*). The strategy encompasses a step-by-step plan (*Stufenplan*), providing for the state-wide identification of suspected contamination sites, the accurate and comparable assessment of the dangers resulting from the contamination, the establishment of registers of contamination sites and suspected contamination sites, and only finally the actual clean-up and redevelopment procedure. While the administrative districts are competent for the actual enforcement, the final responsibility for remediation lies with the Ministry of Environment, and financial aid is decided upon by the regional authorities (*Regierungspräsidien*). The State Institute for the Environment, Measurements, and Nature Conservation (*Landesanstalt für Umwelt, Messungen und Naturschutz*, LUBW) coordinates the clean-up strategy and aids and advises the authorities with regard to technical issues.[26]

On the federal level, responsibility lies with the Federal Ministry for the Environment (*Bundesministerium für Umwelt, Naturschutz, Bau und Reaktorsicherheit*, BMU) and the Federal Environment Agency (*Umweltbundesamt*, UBA), which enforces environmental legislation, undertakes research, advises the government on environmental issues, and informs the public.[27] The Agency is aided and advised by two expert committees, The Expert Council for Soil Investigation (*Fachbeirat Bodenuntersuchungen*, FBU) and the Commission for Soil Protection (*Kommission Bodenschutz*).[28]

Activities at the federal and state levels regarding soil protection are coordinated by the Working Group on Soil Protection, with regard to legacy contamination especially the Standing Committee on Legacy Contamination (*Ständiger Ausschuss "Altlasten"*, ALA).[29]

---

[26]Information on the strategy may be found at the homepage of the LUBW, http://www.lubw.baden-wuerttemberg.de/servlet/is/50997/.

[27]http://www.bmub.bund.de/bmub/aufgaben-und-struktur/.

[28]http://www.umweltbundesamt.de/themen/boden-landwirtschaft/kommissionen-beiraete/kommission-bodenschutz-0; http://www.umweltbundesamt.de/themen/boden-landwirtschaft/kommissionen-beiraete/fachbeirat-bodenuntersuchungen-fbu.

[29]https://www.labo-deutschland.de/Startseite.html; https://www.labo-deutschland.de/Ausschuesse-ALA.html.

4 Civil Law: Inadequate Responses to Injuries to Health and Property?

4.1 Tort Law: The Difficulties of Fault-Based Liability and the New Risk-Oriented Special Environmental Regulations

Civil law focuses on individual goods and rights—in this context compensation for injuries to health and property by way of the environment (cf. Marburger 1992, pp. 30 ff.; Wagner in Habersack 2013, § 823 Rn. 703 ff.). The environment itself can only be protected indirectly by civil law, i.e. if it forms part of private property (Spindeler in Bamberger and Roth 2015, § 823 BGB Rn. 568 f.). Claims are directed at restitution in kind, § 249 BGB (Schubert in Bamberger and Roth, §§ 249–252), i.e. at the re-establishment of the *status quo ante*. Only if the restitution is unacceptable to the claimant, if it proves impossible, or if its costs are disproportionate, does the claim provide monetary compensation (§ 251 BGB).

The basic principle underlying tort law is that damages have to be borne by the holder of goods and rights—as the downside to disposition and profits—unless there is a special reason to shift the risk (*casum sentit dominus*). The two main motives to shift the risk forming the basis of claims for damages are fault and dangerous conduct (Spindeler in Bamberger and Roth 2015, § 823 BGB Rn. 0.2). Before specific legislation on environmental liability came into force, general tort law, especially § 823 I BGB, formed the main basis for damages. Currently, claims under fault-oriented general tort law remain simultaneously applicable to claims under the new risk-oriented special bases of liability.

Under § 823 I BGB, the award of damages is tied to the infringement of an individual good or right such as health or property. The first problematic requisite for the claim is causation, both between the action or omission of the defendant and the infringement of the legal good (*haftungsbegründende Kausalität*), and between this infringement and the damages (*haftungsausfüllende Kausalität*). The general distribution of the burden of proof places it on the claimant (Wagner in Habersack 2013, § 823 Rn. 71), imposing severe difficulties in many environmental cases, especially as there is no pre-trial discovery in Germany (Marburger 1992, p. 27). However, there are specific rules of evidence that may facilitate the reasoning, and case law (BGHZ 92, 143 (*Kupolofen*); BGH, NJW 1997, 2748; Hager in Landmann and Rohmer 2015, § 6 UmweltHG Rn. 4 ff.) has introduced certain presumptions. Exceeding permitted immission[30] values will, for example, lead to the reversal of the burden of proof with regard to the causation of the infringement (Wagner in Habersack 2013, § 823 Rn. 74). Still, this rule will not help if lack of knowledge on

[30]Immission is a technical term used in German law referring to the environmental impact of a certain substance *on* a specific site, as opposed to the term emissions, which refers to substances that are discharged *from* a specific site (§ 2 II, III BImSchG). Exceeding emission values does not trigger the reversal of the burden of proof.

contaminants has already prevented the institution of immission values at the time of the polluting activity or in cases involving multiple polluters. As the threshold values that trigger this reversal must be immission levels, i.e. refer to measurements made at the claimant's property, the claimant still has to prove that the immission originated from the individual defendant's factory or other polluting activity (Spindeler in Bamberger and Roth 2015, § 823 Rn. 579 f.). Pursuant to § 830 I 1, II BGB, the defendant will, nevertheless, be jointly and severally liable for the entire damage in cases of alternative causation, i.e. cases in which it cannot be determined which one of the potential perpetrators or sources actually caused the harm, but where each would have been capable of doing so on its own (Wagner in Habersack 2013, § 830 Rn. 38). The norm is, however, generally regarded as inapplicable in cases of cumulative pollution, that is cases in which the different sources act together to cause the danger (BGHZ 67, 14; BGHZ 72, 355; criticized by Braun 1998, p. 2320; see also Hager in Landmann and Rohmer 2015, § 6 UmweltHG Rn. 12). Liability in the typical case of diffuse environmental pollution is therefore excluded, even if the emissions were illegal. A market share or pollution share liability, which bases liability for damages on the statistical proportion of the emissions, does not exist in German tort law (Spindeler in Bamberger and Roth 2015, § 830 Rn. 37).

The claims are, furthermore, dependent on fault, that is, intent or negligence. Because of the claimants' lack of knowledge of the internal organization and the production processes and substances used (Wagner in Habersack 2013, § 823 Rn. 56 ff.), the burden of proof for fault in environmental tort law was reversed by case law. The relevant duty of care (*Verkehrspflicht*) is determined in accordance with threshold values and obligations under administrative law, but may be stricter if the concrete situation demanded an elevated standard of care to avoid injuries and damages (Spindeler in Bamberger and Roth 2015, § 823 BGB Rn. 570). Observing threshold values and obligations under administrative law will, nevertheless, usually exclude negligence, because the owner may trust in their adequacy. The proof of observing emission values will therefore generally serve to refute the presumption of fault (Spindeler in Bamberger and Roth 2015, § 823 BGB Rn. 582). In many cases of historical pollution, a lack of knowledge of the danger of the substances employed at the time will make the proof of fault impossible.

The first federal norm recognizing polluting activity as sufficiently dangerous to warrant a special regulation of liability was ex-§ 22 WHG (*Wasserhaushaltsgesetz*, Federal Water Act) (now § 89 WHG) (see Seuser 2013 [1]), which was introduced in 1960 (Petersen in Landmann and Rohmer 2015, § 89 WHG Rn. 6); § 89 I and II WHG provide two separate claims for damages to individual goods, basing liability on passing contaminants into water or on being the owner or operator of a factory from which contaminants get into the water (Petersen in Landmann and Rohmer 2015, § 89 WHG Rn. 9, 35). This liability is independent of fault and also applies if the contamination was authorized under public law, but is excluded in cases of force majeure (Petersen in Landmann and Rohmer 2015, § 89 WHG Rn. 81). Joint and several liability not only applies in cases of alternative causation. The defendant is also liable for the entire damage if it was caused cumulatively and his own contribution was sufficiently high to have caused the resulting damage, alone or together with the other polluting actions (BGHZ 57, 257; Petersen in Landmann and Rohmer 2015, § 89 WHG Rn. 58; Schwendner in Sieder et al. 2014 § 89 Rn. 50). Liability for diffuse pollution is, nevertheless, excluded because the contributing factor would be too insignificant to fulfil this requirement.

A more general and far-reaching liability for polluting activity was introduced in 1990 by the Environmental Liability Act (*Umwelthaftungsgesetz*, UmweltHG (see Marburger 1992, pp. 2 ff., 17 ff.). The Environmental Liability Act is applicable to all damages that have been caused after it came into force (§ 23 UmweltHG).[31]

Liability under the Environmental Liability Act is based on the operation of certain plants or facilities, enumerated in Annex 1 to the act (§ 3 II UmweltHG). The damage must have been caused via environmental pollution (§ 3 I UmweltHG).[32] Liability under the Environmental Liability Act is independent of fault and even includes damages caused during the regular and authorized operation of the plant (cf. Hilf in Giesberts and Reinhardt 2015, § 89 WHG Rn. 43 f.; Rehbinder in Landmann and Rohmer 2015, § 1 UmweltHG Rn. 17 f.). However, liability is excluded if the damage was caused by force majeure (§ 4 UmweltHG) and is limited to €85 million per event (§ 15 UmweltHG), a limitation very common for risk-based claims (cf. Marburger 1992, p. 19; Braun 1998, p. 2322). Insurance is compulsory (§§ 19, 20 UmweltHG).

The burden of proof of causation is partially reversed by § 6 UmweltHG (see, for example, Marburger 1992, pp. 20 f.). This presumption is not applicable if the operation was in accordance with the usual obligations, in particular those resulting from administrative regulations and acts and applicable threshold values (§ 6 II, III UmweltHG). The Act's applicability to legacy contamination is therefore very limited. Yet again, the determination of liability is difficult in cases involving several polluters[33] where the pollution was cumulative (Hager in Landmann and Rohmer 2015, § 6 UmweltHG Rn. 29 ff.). The applicability of the presumption in this case is subject to controversy. Although the text of the norm suggests a restrictive interpretation, excluding this frequent and important constellation would severely reduce the value of the presumption for the victim (Hager in Landmann and Rohmer 2015, § 6 UmweltHG Rn. 35a). At any rate, diffuse polluting activities will yet again be insufficient to trigger the presumption, as the increase in risk to the protected good associated with each individual contribution is too minor to justify it. Liability for diffuse pollution is therefore mostly excluded under the Environmental Liability Act (Hager in Landmann and Rohmer 2015, § 6 UmweltHG Rn. 35a ff.; Ruffert 2010, p. 1178; Marburger 1992, p. 33).

---

[31]See Marburger (1992, p. 21) for a discussion of the applicability to cases of ongoing pollution.

[32]Marburger (1992, p. 18) considered the definition provided as ambiguous.

[33]See Hager in Landmann and Rohmer (2015, § 7 UmweltHG Rn. 11 ff) for an elaborate presentation of the different constellations and the controversy.

4.2 *Fault-Independent Claims for Omission, Removal, and Compensation: Clean-Up as Subject of Traditional Neighbour Law*

Under § 1004 BGB, a claim is provided for the omission and removal of any interference with property and—extended by case law—other individual goods and rights, such as personal health (*quasi-negatorische Ansprüche*) (Berger in Jauernig 2014, § 1004 Rn. 2). The claim entitles the owner not only to the removal of the offending substance (which is often impossible), but to the full restoration of the property to its previous state and use, in some cases making clean-up procedures the subject of neighbour law (BGH, NJW 1996, 845; BGHZ 135, 235 = NJW 1997, 2234). Unlike tort law, the claim is not fault-based (Fritzsche in Bamberger and Roth 2015, § 1004 BGB Rn. 13 ff.; Baldus in Habersack 2013, § 1004 Rn. 149 ff.). However, causation must be proven, raising the traditional difficulties with regard to cumulative pollution (see Fritzsche in Bamberger and Roth 2015, § 1004 Rn. 25; § 906 Rn. 49 ff., 90 ff.).[34]

The claim is excluded under several circumstances, for example if a legal obligation to tolerate the interference exists, such as when the interference was caused by an imponderable substance—which includes most typical pollutants—and is negligible (§ 906 I 1 BGB).[35] This exception is especially important in the environmental context because the interference is generally considered negligible if pollution levels keep within the threshold values instituted by administrative regulations (§ 906 I 2, 3 BGB; see Fritzsche in Bamberger and Roth 2015, § 906 Rn. 41 ff.). If the claim for omission and removal is excluded although the interference is significant, the owner is entitled to just compensation, pursuant to § 902 II 2 BGB or § 14 S. 2 BImSchG. The claim under § 902 II 2 BGB has been extended by case law to situations in which the owner had to tolerate the interference although it was illegal, for instance because the imminent danger had not been noticeable beforehand (see Fritzsche in Bamberger and Roth 2015, § 906 Rn. 74, 84 for references).

All of the claims described, both for damages as well as for omission, removal, and compensation, are time-barred from three years after the end of the year in which the claimant knew or ought to have known of the claim (§§ 195, 199 BGB), but at the latest from thirty years after their origin in the case of intentional injury to life and health or ten years otherwise (§ 197 BGB; Grothe in Habersack 2013, § 195 BGB Rn. 12), creating yet another obstacle to their application in cases of historical pollution.

---

[34]This is especially relevant for removal or compensation claims.

[35]Cumulative effects will be evaluated together in their entirety (Fritzsche in Bamberger and Roth 2015, § 906 Rn. 49 ff.). See, however, Rn. 90 ff. on the difficulty in determining the legal consequences. A distinction has to be made between claims for omission (normally possible until the threshold of acceptable immissions is crossed again) and claims for removal and compensation, which are subject to the familiar problem of allocating responsibility for diffuse pollution.

4.3 Related Administrative Law: Social Insurances Mitigating the Financial Burdens of Pollution-Related Sickness in the Population

Germany has an elaborate system of social security, originally introduced in the nineteenth century (Benkel 2002), which takes a significant amount of pressure from general tort law in cases of personal harm (cf. FAZ 2002). Social accident insurance (*gesetzliche Unfallversicherung*) provides compensation for work accidents and, since 1922, for occupational illnesses (*Berufskrankheiten*).[36] Employees of companies producing or handling hazardous substances often suffer from illnesses related to frequent exposure, for example cancer caused by breathing in asbestos.[37] While benefits do not depend on the fault of the employer, the insurance will only pay if the illness was caused by the claimant's work (Krasney 1993; Ricke in Körner et al. 2015, § 9 SGB VII Rn. 2, § 8 Rn. 4 ff.). As in most other cases of historical pollution, this proof of causation is challenging, especially because many diseases have a period of latency of several decades (Breuer 1995; Krasney 1993, pp. 94 f.; Radek 1990, p. 597). Even if significant exposure is proven, it is necessary to determine that there are no non-work-related factors that could have caused the disease (§ 9 III SGB VII), although it is not required to relate the disease to one individual employer (Ricke in Körner et al. 2015, § 9 SGB VII Rn. 29, 57a). A classic example: an employee claiming compensation for lung cancer worked with asbestos for several years, but has also smoked for most of his life (BSG, NZS 2007, 594).[38] Because of the difficulties presented by such cases, compensation for occupational illnesses is limited to illnesses enumerated in the annex to the Regulation on Occupational Diseases (*Berufskrankheitenverordnung*, BKV), instituted in accordance with § 9 I SGB VII, unless new scientific facts were discovered after the last revision of this list (§ 9 II SGB VII) (Ricke in Körner et al. 2015, § 9 SGB VII Rn. 7 f.). The Regulation on Occupational Diseases may also provide for specific presumptions in relation to individual illnesses, such as the presumption in Nr. 4114 of the annex to the Regulation that a lung tumour was caused by asbestos if a certain period of exposure to a specific concentration of asbestos can be proven (Ricke in Körner et al. 2015, § 9 SGB VII Rn. 27; Breuer 1995, p. 152). Nevertheless, many affected persons do not succeed in obtaining compensation.[39] A more general reversal to the burden of proof has therefore often been suggested, but has never been realized and has sometimes been described as incompatible with the character of the insurance (Radek 1990, p. 597; Breuer 1995;

---

[36]For the history of the social accident insurance, see Ricke in Körner et al. (2015), Vorbem. SGB VII Rn. 1. Statistics on reported and recognized cases in recent years may be found at BMAS (2013, pp. 247 ff).

[37]Asbestos is the most frequent cause of death with regard to occupational disease; cf. BMAS (2013, pp. 253 ff).

[38]The Court, nevertheless, affirmed causation.

[39]A fact that was particularly addressed by the general media: Engler (2014), Lenz (2011).

Ricke in Körner et al. 2015, Vorbem. SGB VII Rn. 10a, 10b, § 9 SGB VII Rn. 27, with further references).

Insurance benefits cover the costs of treatment, disability pensions, and pensions to surviving dependants. Membership is compulsory for every employer, who is also solely responsible for paying the premiums (§ 150 SGB VII). In return, the insurance benefits replace the employer's (and co-worker's) civil liability to the employee if the accident or illness was not caused intentionally (§§ 104, 105 SGB VII) (Ricke in Körner et al. 2015, Vorbem. SGB VII Rn. 3; Krasney 1993).

In addition to the special social accident insurance, the German social system comprises statutory health and nursing care insurance (*gesetzliche Kranken- und Pflegeversicherung*), public unemployment insurance (*Arbeitslosenversicherung*), and statutory pension insurance (*gesetzliche Rentenversicherung*), the costs of which are shared between employers and employees and are dependent on income, as well as income support and social benefits (*Sozialhilfe*), regulated in the different volumes of the Social Security Code (see Seewald in Körner et al. 2015, Vorbem. §§ 18-29 SGB I, Rn. 2.). None of these benefits depends on causation.[40] Treatment costs resulting from illnesses caused by pollution and living costs are therefore covered without further requirements.

4.4 State Liability: The State's Responsibility for Protecting Its Citizens

As the state is subject to obligations under general tort law, it may be liable, for example, if the operation of a public waste disposal site causes damages to neighbours (Papier in Habersack 2013, § 839 BGB Rn. 169 ff.). In addition, the state has a constitutional duty to protect its citizens' and their rights from harm (*Schutzpflicht*) that is reflected in more specific obligations such as the obligation to consider all health-relevant factors when designing a development plan (Runkel in Ernst et al. 2015, Vorbem. zum BauGB, Rn. 82). In exceptional cases, neglecting this duty also may result in state liability pursuant to § 839 BGB, Article 34 GG (Runkel in Ernst et al. 2015, Vorbem. zum BauGB, Rn. 56 f., 82, 94 ff.; Papier in Habersack 2013, § 839 Rn. 266).

Two examples gain special significance in the environmental context. The first example has already been illustrated above, regarding the plan to designate a contaminated site a residential area (see also Schink 1990). In addition, an attempt has also been made—this time unsuccessfully—to make the state liable for the consequences of diffuse legal pollution in the case of damage to forests by acid rain (BVerfG, NJW 1998, 3264; v. Hippel 1985). In this context, it has to be noted that parliament can—with certain exceptions based on European law—neither be held

[40] The requirement of causation is, therefore, a distinct and necessary characteristic of the social accident insurance (Radek 1990, p. 597).

liable for enacting illegal legislation, nor for the failure to adopt necessary legislation (Dörr in Bamberger and Roth 2015, § 839 BGB Rn. 643 ff., 287 ff.; Papier in Habersack 2013, § 839 Rn. 260 f.). State liability is therefore primarily applicable in cases of faulty individual administrative actions (Papier in Habersack 2013, § 839 Rn. 99 ff., 260 f.).

5 Criminal Law and Regulatory Offences: Backing up Administrative Law

5.1 The Long Tradition of German Environmental Criminal Law and the Role of the Offences for the Protection of Human Life and Human Health

Environmental criminal law was initially part of the various different environmental statutes but was transferred to the Criminal Code in 1980,[41] a legislative decision made to emphasize the relevance of environmentally harmful conducts (BT-Drs. 8/2382, p. 10; Saliger 2012, Rn. 18 ff.). Environmental criminal law has been subject to several reforms, the last of which implemented European Directive 2008/99/EC on the Protection of the Environment through Criminal Law.[42]

German environmental criminal law is not solely anthropocentric, but recognizes the environment itself—comprising its different media (soil, air, water) and the flora and fauna—as its object of protection (Saliger in Satzger et al. 2014, Vor 324 ff. Rn. 10 ff.; Kloepfer and Heger 2014, Rn. 35). The same twofold aim is seen in the Directive on the Protection of the Environment through Criminal Law, which makes sanctions conditional on the likelihood of severe harm to the environment itself or to human health (Saliger 2015, pp. 19 f.). Criminal liability is dependent on or accessory to administrative law (*Verwaltungsrechtsakzessorietät*) and administrative acts (*Verwaltungsaktakzessorietät*) (e.g. Saliger 2012, Rn. 67 ff.). Administrative decisions take priority, even if illegal, unless they are void or acquired by duress, collusion, or deceit (§ 330d I Nr. 5 StGB), a decision traditionally subject to fierce controversy (see, for example, Saliger in Satzger et al. 2014, Vor 324 ff. Rn. 78, with further references).

---

[41]By the First Act to Fight Environmental Crime (*Erstes Gesetz zur Bekämpfung der Umweltkriminalität*, 1. UKG), 18. StrÄndG vom 28. März 1980, BGBl. I, S. 373.

[42]Directive 2008/99/EC of the European Parliament and of the Council of 19 November 2008 on the protection of the environment through criminal law, Official Journal L 328, 6.12.2008, pp. 28–37. The most important of these reforms were the Second Act to Fight Environmental Crime (*Zweites Gesetz zur Bekämpfung der Umweltkriminalität*, 2. UKG), 31. StrÄndG vom 27. Juni 1994, BGBl. I, S. 1440, and the 45th Act to Amend the Criminal Law (Fünfundvierzigstes Strafrechtsänderungsgesetz zur Umsetzung der Richtlinie des Europäischen Parlaments und des Rates über den strafrechtlichen Schutz der Umwelt vom 6. Dezember 2011), BGBl. I, S. 2557; BT-Drs. 17/5391.

The section comprises several individual offences that either focus on the protection of one of the environmental media—§ 324 StGB (water), § 324a StGB (soil), § 325 StGB (air)—protected areas (§ 329 StGB), or relate to specific conducts, including the handling of waste (§ 326 StGB) or radioactive materials or chemicals (§ 328 StGB), the causing of noise or the release of ionized radiation (§ 325a StGB), or the operation of certain plants without authorization (§ 327 StGB) (Saliger 2012, Rn. 49). This structure also reflects the system of administrative environmental law.

Environmental offences follow different structures, some requiring actual harm or a concrete danger (*Erfolgsdelikte, konkrete Gefährdungsdelikte*, e.g. § 324 StGB), but the majority of which sanction certain behaviours (*abstrakte Gefährdungsdelikte*) or base criminal liability on the likelihood of causing severe harm (*Eignungsdelikte*), a structure typical of environmental criminal law that has also found its way into the European Directive (Saliger 2012, Rn. 51 ff.; Kloepfer and Heger 2014, Rn. 59).

Offences are typically punishable with fines or prison sentences of up to five years (with slight variations and lower sentences for negligence), resulting in a time limit for prosecution of five years (§ 78 III Nr. 4 StGB). Criminal liability for active perpetration in cases of historical pollution is thereby effectively excluded (Franzheim and Pfohl 2001, Rn. 172 ff.; Michalke 2000, Rn. 137).

In the gravest cases of polluting activity, offences relating to the protection of life and health can theoretically become relevant. These offences require a specific outcome (*Erfolgsdelikt*), the death or the injury of a person, meaning that both the outcome as well as causation between the act and the outcome and the objective imputation of the outcome have to be proven (*Kausalität und objective Zurechnung*). Perpetration by omission and—especially—negligent perpetration are significantly more relevant in practice (although this relevance is limited for all forms). In addition to offences that focus on the protection of human life and human health as an individual legal good, offences against public safety, such as the poisoning of drinking water (*Gemeingefährliche Vergiftung*, § 314 StGB), that base liability on an abstract danger to human health, may become relevant (Herzog and Kargl in Kindhäuser et al. 2013, § 314 Rn. 2 ff.).

With regard to **Hypothetical Scenarios 3, 4, and 5**, A's polluting behaviour is likely to be criminally relevant. In **Hypothetical Scenarios 3 and 4**, potential offences are § 324a StGB (contamination of soil) and § 324 StGB (contamination of water); § 324a StGB provides for criminal sanctions against soil contamination that is of significant magnitude or potentially harmful to humans, animals, plants, or the groundwater. Since the groundwater has already been affected by the contaminants that A introduced into the soil and is destined for human consumption, the objective requirements are fulfilled, provided the contamination is sufficiently concentrated. In the end, however, A's liability will depend on whether the subjective requirements can be proven: A has to have acted with intent or negligence, i.e. he must have been aware or potentially aware of the fact that the contaminants entered the soil and could be dangerous to human health, flora and fauna, or the groundwater. His act also fulfils the objective requirements of § 324 StGB, which censures the contamination of surface water and groundwater. Again A must have acted intentionally or negligently, i.e. he must have known or been capable of knowing that the contaminants would enter into the groundwater and change its state negatively, which requires a slightly further-reaching awareness on his part. Since § 324a StGB only requires potential harm to the groundwater, while § 324 StGB requires actual harm, § 324 StGB would take precedence over § 324a if the requisites of both offences are fulfilled. However, both offences are punishable by a fine or a prison sentence of up to five years.

In **Hypothetical Scenarios 1 and 2**, A also contaminated the soil, so that § 324a might theoretically apply. As German environmental law also recognizes the environment itself as its subject of protection, criminal liability is possible even without a concrete danger to human health. However, the contamination has to be of a significant extent or potentially harmful to animals, plants, or the groundwater, even if it is not potentially harmful to humans. Yet again, a final conclusion would require more detailed information. As there have been no damages and criminal law is stricter than administrative law, liability pursuant to § 324a StGB is unlikely in this scenario.

In **Hypothetical Scenario 5**, A could in addition be liable for murder (§ 212 StGB), negligent manslaughter (§ 222 StGB), or intentionally (§§ 223, 224 StGB) or negligently causing bodily harm (§ 229 StGB), in addition to the environmental offences already discussed with regard to **Hypothetical Scenarios 3 and 4**. Yet again, proving intent or negligence on the part of A will be the most severe challenge for the prosecution (this time even more so). While it is unlikely that A actively considered and accepted the possibility of harming humans, it is possible that he merely ignored the potential dangers arising from his activity, a behaviour which might constitute negligence. It has to be noted that §§ 324a, 324 and, for example, § 229 StGB do not only demand increasing objective requirements; proving the subjective component will get more difficult at the same time. While it will be relatively easy to allege that the owner of a production plant always has to keep the possibility of leakages in mind, proof of knowledge regarding the possible further dispersion of pollution through the soil and groundwater and finally into the reach of humans will require additional effort, even if, as in this case, the prosecution is not hindered by the cumulation of several possible contamination sources.

5.2 *The Challenges of Criminal Prosecution in Historical Pollution Cases*

5.2.1 The Problem of Determining Causation and Objective Imputation

In criminal law, causation is the mere factual relationship between the action (or omission) and the outcome (*Erfolg*) of the case. It requires that the outcome would not have occurred without the action (see, for example, Roxin 2006, § 11 Rn. 6 ff.). The objective imputation considers normative elements (Roxin 2006, § 11 Rn. 44 ff.). The question of how to treat cumulative pollution is relevant in all offences that require the actual harm or concrete danger to a protected good and in determining the likelihood of causing harm (Samson 1987, p. 620; Roxin 2006, § 11 Rn. 1). The first step is to determine whether the action or omission at hand was causal at all, either as the only cause or as one cause among many (see, for example, Saliger 2012, Rn. 230 ff.). Even this can pose severe difficulties in environmental cases because it is often possible that the harm was entirely caused by different circumstances. In this context, the complexity of human health and the environment, the number of different influences, and, especially, the lack of scientific knowledge come into play (Kühne 1997). The two most important cases decided in this context are the *Lederspray* case (BGHSt 37, 106; see, among many, Kuhlen 1990; Samson 1991; Hilgendorf 1994; Hoyer 1996) and the *Holzschutzmittel* case (BGHSt 41, 206; see Schmidt-Salzer 1996; Puppe 1996; Volk 1996), nicknamed after the harmful substances in question. In both cases, the products, a waterproofing spray for leather and a wood preservation agent used on the surfaces of dwellings, were alleged to have caused injuries to the exposed users or inhabitants, resulting in the charges of negligently causing bodily harm (*fahrlässige Körperverletzung*) and intentionally causing bodily harm by omission (*Körperverletzung durch Unterlassen*) being aimed at the directors of the companies. In the Lederspray case, the alleged symptoms were severe and consistent in all patients who had used the waterproofing spray, ranging from breathing difficulties to pulmonary embolisms. Nevertheless, the individual chemical substance in the spray that had actually caused the illnesses could not be identified in tests. The Court decided that the proof of causation does not require the exact identification of an individual substance if alternative causes other than the spray could be excluded. In the Holzschutzmittel case, the situation was even more complex. Symptoms were diffuse, ranging from cerebral damage in children to breathing difficulties and afflictions of the digestive system. The Court, nevertheless, decided that the exposure to the product caused the injuries, stating that it is sufficient to prove that the action at hand was one of several causes. This decision has been criticized for infringing the principle of *in dubio pro reo* by effectively letting the mere plausibility or probability of causation suffice (Samson 1991, p. 183; Daxenberger 1997, p. 45; Volk 1996, p. 108).

The question of whether the final result can be imputed to each individual pollutant is even more critical (Samson 1987, p. 618; Saliger 2012, Rn. 239 ff.). Even if it can be ascertained that the action contributed to the harm at all, it often was only one cause among several, the harm therefore being the result of a so-called cumulative effect (*Kumulationseffekt*). Cumulative effects can be divided into cases of mere summation of similar substances that cause the harm together (*Summationseffekt*), and cases of synergetic effects, where the effect of the combined substances has an entirely new quality because of chemical reactions (*synergetischer Effekt*) (Ronzani 1992, pp. 45 ff.; Daxenberger 1997, pp. 17 ff.). The principle of responsibility (*Verantwortungsprinzip*) states that each person is responsible only for the harm caused by his or her own actions (Roxin 2006, § 24 Rn. 21 ff.). An action that is harmless in itself cannot be criminalized just because it is possible that further conduct, which would then become illegal, causes harm (Saliger 2012, Rn. 247). Attributing the consequences of another person's actions is therefore generally dependent on further requirements of complicity or other forms of joint perpetration that are not present in most cases of pollution (Samson 1987, p. 627; Daxenberger 1997, pp. 87 ff.). This is one reason why the offences for the protection of human life and health have relatively little importance in pollution cases.

With regard to environmental offences, the situation is yet more complex. Unlike death, the necessary outcomes—for instance the negative change in water quality in § 324 StGB—in themselves provide a significant margin of interpretation (Samson 1987, pp. 619 ff.; Saliger 2012, Rn. 225 ff.). The change in water quality can be measured gradually, i.e. without any qualitative change such as death, even leading some authors to doubt causation for the entire harm in cases of summation by stating that each change in concentration has to be evaluated separately and each action is only causal for the direct change in concentration (Samson 1987, pp. 626 ff., 629 f.; contra Saliger 2012, Rn. 235).

With regard to water pollution, the courts have affirmed criminal liability, stating that any change to water quality likely to result in severe harm, individually or in combination with other changes, is sufficient to constitute the offence (OLG Stuttgart, ZfW 1977, 177, 179; strongly criticized by Samson 1987, p. 621. See also BGH, NStZ 1987, 323). They therefore redefined the necessary outcome, effectively changing the nature of the offence into an offence sanctioning abstract dangers rather than actual harm (Samson 1987, p. 621). The same line of argumentation can be applied to determining the likelihood of causing severe harm to the environment or human health.

To justify criminal liability in cases of cumulative pollution, legal doctrine has developed a special type of offence, the cumulative offence (*Kumulationsdelikt*) (Kuhlen 1986, pp. 389, 399; Kuhlen 1993, pp. 697, 716). This type of offence is viewed as appropriate to criminalize behaviour that is typically committed by a multitude of different persons, each making individually minor contributions that cause severe harm in combination. Liability is based on individual contribution, thereby avoiding the necessity to impute collective harm (Saliger 2012, Rn. 244 f.).

However, these approaches do not reflect the current structure of the legislation. Many offences provide for *de minimis* exceptions, which would be bypassed by focusing on the potential to cause harm together with other polluting actions (Samson 1987, pp. 624 ff., 634 f.; Saliger 2012, Rn. 225 ff., with further references). *De minimis* exceptions may also be found in the European Directive.[43]

It therefore remains necessary to apply the general principles of objective imputation to the collectively caused harm (Saliger 2012, Rn. 245). Some authors dismiss any liability if the individual conduct was in itself insufficient to cause

[43]With regard to the transport of waste, for example, in Art. 3 lit. c 2008/99/EC.

significant harm, thereby effectively excluding many cases of cumulative pollution from the scope of criminal law (Samson 1987, pp. 624, 635; Daxenberger 1997, p. 75). However, it appears preferable to determine the harm caused by the polluting act in accordance with the already polluted environment (Saliger 2012, Rn. 246 ff.). If previous actions had already raised the concentration of contaminants and the last minor contribution led to critical conditions, the entire harmful state has to be imputed to the last actor. On the other hand, previously harmless actions cannot be made to constitute an offence only because a successive contribution caused harm. Therefore, objective imputation should be possible with regard to simultaneous polluting actions (the final liability depending on intention or negligence, i.e. the foreseeability of the other contamination) and to subsequent polluting actions, but not to previous polluting actions (Saliger 2012, Rn. 247 ff.).

To avoid these difficulties, many environmental offences now base liability on the mere conduct, taking potentially severe harm by accumulation as the motive for criminalizing individually still relatively harmless conducts (Saliger 2012, Rn. 53). This has been criticized with regard to the principles of fault and proportionality (e.g. Hassemer 1992, pp. 382 f.).

5.2.2 Liability for Omission: Preventing Further Harm Originating from Sites of Historical Pollution

In most cases of legacy contamination, prosecution for actively causing harm to the environment is time-barred or impossible because the act was in accordance with the scientific knowledge of the time and therefore was neither intentional nor negligent, or it was in accordance with administrative law. If the contamination of the site spreads and causes further harm, the question of criminal liability (for example under § 324a StGB) for failure to clean up has to be raised (among many others, Lackner and Kühl 2014, § 324a Rn. 6; more restrictive, Ransiek in Kindhäuser et al. 2014, § 324a Rn. 14). In Germany, most offences can be committed either actively or by omission (*Unterlassen*) (§ 13 StGB)—by omission only if the perpetrator is responsible as guarantor, i.e. has a duty of care (*Garantenstellung*, *Garantenpflicht*) (Heine and Hecker in Schönke and Schröder 2014, § 324a Rn. 7). This duty of care mainly stems from responsibility for the protection of an endangered subject or object (*Beschützergarant*) or for preventing danger (*Überwachergarant*) resulting from former non-diligent actions (*Ingerenz*), or from sources of danger in the sphere and under the control of the perpetrator (Freund in Joecks and Miebach 2006, § 13 Rn. 106 ff.). Causes giving rise to a duty of care under criminal law are to a certain extent parallel to the responsibilities for clean-up under administrative law, but do not reflect the new extension of these responsibilities introduced in the Federal Soil Protection Act (Heine and Hecker in Schönke and Schröder 2014, § 324a Rn. 7). In the context of legacy pollution, the duty of care will most likely result from the responsibility for a source of danger, that is if the original polluting activity was in accordance with the due diligence of the time (Heine and Hecker in Schönke and Schröder 2014, § 324a Rn. 7).

The dependence on administrative law, furthermore, requires an action contrary to administrative law or administrative acts and therefore an enforceable clean-up order, for example under § 4 III BBodSchG (Heine and Hecker in Schönke and Schröder 2014, § 324a Rn. 7; Ransiek in Kindhäuser et al. 2014, § 324a Rn. 14). Mere non-compliance with a valid clean-up order that does not cause further harm only constitutes a regulatory offence under the Federal Soil Protection Act (§ 26 BBodSchG).

Liability for omission is also possible in cases of cumulative pollution if the accused, generally a public official, was responsible for the protection of the environment.

5.2.3 Public Officers: Guardians of the Environment?

It is possible for public officers not only to be criminally liable to the same extent as normal citizens, when the state operates plants or actively pollutes the environment in other ways (*Betreiberfälle*) (see, for example, Saliger 2012, Rn. 177 ff.), but also with regard to the state's function of regulating and monitoring polluting activities. A public officer can, for example, incur liability for active perpetration through an agent if he issues an authorization that prevents the actual polluter from being punished by legalizing his actions, provided the offence is not a special offence (see, for example, Steindorf in Odersky 2005, Vor § 324 Rn. 53 f.; Ransiek in Kindhäuser et al. 2014, § 324 Rn. 74, with further references). This criminal liability can even extend to specific authorities within the administration who do not directly issue permits but are involved in the proceedings because of their special knowledge. As an example, in the *Abfallmanager* case an employee of the waste authority who recommended the issuing of a permit for the relocation of mixed household and hazardous waste was convicted for unlawful disposal of waste through an agent after the waste in question was transferred to an unsuitable disposal site (BGHSt 39, 381).

If the public official and the recipient of the permit collude, the permit is regarded as invalid for the purposes of criminal law, resulting in criminal liability for the recipient and the public official's liability for instigation, aiding and abetting, or even complicity (Saliger 2012, Rn. 191 f.).

In the case of historical pollution, liability for omission is even more relevant. Liability for omission is conceivable for neglecting to revoke illegal permits—either illegal from the beginning or made so by a change in legislation—or for condoning illegal activity (see Saliger 2012, Rn. 202 ff., 213 ff., with further references). As has already been stated, criminal liability for omission depends on a duty of care. A duty of care can be drawn from previous behaviour, obliging public officers to revoke previously issued illegal permits, an obligation that was also extended to successors in office (Lackner and Kühl 2014, Vor § 324 Rn. 11;

Steindorf in Odersky 2005, Vor § 324 Rn. 57; see also, differing, Saliger 2012, Rn. 204 ff., Schmitz in Joecks and Miebach 2006, Vor §§ 324 ff. Rn. 112). It is also held that public officials of special environmental authorities (but not those of general administrative authorities) are guarantors for the protection of the environment and can therefore be liable for omission in many environmental cases (see Saliger 2012, Rn. 208 ff., with further references).

A notable case example for the conviction of a public official is that of the village mayor who neglected to intervene in the illegal discharge of raw sewage into a brook although ordered to do so by the competent water authority (*Bürgermeisterfall*) (BGHSt 38, 325; Pfohl 1994). This example is also interesting in the context of cumulative pollution, because the contamination and damage to the ecosystem of the brook was caused by the sum of several minor discharges that were individually insignificant and only resulted in fines for the perpetrators. It, furthermore, serves as a clear illustration of the proof of quasi-causation, exactly delineating the consequences of the mayor's intervention and taking into account possible appeals against administrative orders and the time required for court proceedings and enforcement action (see Saliger 2012, Rn. 213 ff.).

5.2.4 Criminal Procedure: Problems of Evidence and the Broken Ideal of Mandatory Persecution

German criminal procedure is regulated in the German Code of Criminal Procedure (*Strafprozessordnung*, StPO). Jurisdiction is laid down in the Courts Constitution Act (*Gerichtsverfassungsgesetz*, GVG). German criminal procedural law is based on the principle of mandatory prosecution (*Legalitätsprinzip*), forcing the authorities to investigate and bring charges if the perpetration of a criminal offence is sufficiently likely (§ 152 II StPO) (see Diemer in Hannich 2013, § 152 StPO, Rn. 1 ff.). However, many cases of environmental crime remain unrecorded and unknown to the prosecution as there are no victims who could report the offence (Saliger 2012, Rn. 527 ff.). Due to the complexity of environmental cases, investigation proceedings will often be concluded without bringing charges pursuant to § 170 II StPO or without opening the main trial (§ 203 StPO), even if they have been initiated (Saliger 2012, Rn. 527 ff.). In the case of misdemeanours, the prosecution (or later the court) may also dispense with court action if the offence was petty or if the suspect submits to certain conditions and instructions (*Einstellung wegen Geringfügigkeit*), for example the payment of a sum of money to a non-profit organization (§§ 153, 153a StPO). As most environmental offences are misdemeanours, this possibility is frequently used (Saliger 2012, Rn. 536 ff.).

6 Division of Labour and Division of Responsibility: Corporate Liability and Liability in Corporations

6.1 General Rules Modified in Favour of Effectiveness: Liability of Legal Persons in Private and Administrative Law

Legal persons and partnerships are subjects of both private and administrative law, acting through their organs or representatives. Liability follows the general rules of company law, especially important with regard to limitation of liability (Eidam 2014, Kapitel 5, Rn. 7 ff.). Acting representatives and leading executives can themselves be responsible under the Environmental Damage Act, however, thereby incurring high personal risks (Scheidler 2007, p. 1115; Balensiefen 2012, § 2 Rn. 5 ff.). The Federal Protection Act makes specific provisions in this context, enabling the administration to address clean-up orders or recover costs not only from the company that owns or owned a property, but under certain circumstances also from the representative, piercing the corporate veil (§ 4 III BBodSchG) (Giesberts and Hilf in Giesberts and Reinhardt 2015, § 4 BBodSchG Rn. 39 ff.; see also Becker and Fett 1999). This regulation was adopted to prevent the evasion of liability under the Federal Soil Protection Act by transferring contaminated properties to insolvent companies (Tiedemann 2008, pp. 257 f.).

In addition, under the Transformation Act (*Umwandlungsgesetz*, UmwG) the responsibility of conduct incurred by the legal successor may be applicable in cases of extraordinary business operations such as a merger of different companies (§ 2 UmwG) (Giesberts and Hilf in Giesberts and Reinhardt 2015, § 4 BBodSchG Rn. 27 ff.). In the case of the responsible company's insolvency, the responsibility under regulatory law can fall onto the administrator of the estate (*Insolvenzverwalter*), either as a person or as acting for the company. However, the insolvency administrator can release a contaminated property from the insolvency estate (*Freigabe aus der Insolvenzmasse*), thereby ending his and even the company's responsibility (BVerwGE 122, 75; Kebekus 2001).[44]

6.2 Criminal Liability in Corporations: Problems of Individual Responsibility and Imputation

A large part of environmentally harmful activity is of industrial origin. Corporations are often accused of being organized in a way that prevents the allocation of responsibility (*organisierte Unverantwortlichkeit*) (Joecks in Joecks and Miebach

[44]Severe criticism can be found in Schmidt (2010). See Schall (2006, p. 164) and Heine and Hecker in Schönke and Schröder (2014, § 324a Rn. 7) for the effect on criminal law.

2006, Vor § 25 Rn. 18 ff.; Dannecker in Wabnitz and Janovsky 2014, 1. Kapitel, Rn. 23). Division of labour also leads to a division of responsibility both horizontally—e.g. between different directors—and vertically (Schmidt-Salzer 1996; Saliger 2012, Rn. 162 ff.). In many cases, only the person directly acting—for example the person opening a valve to discharge chemicals into a river—can be held criminally liable, although only the company profits from the offence (Raum in Wabnitz and Janovsky 2014, 4. Kapital, Rn. 59; Saliger 2012, Rn. 158). In order to get hold of the perceived truly responsible person as a perpetrator, such as the director of the company or another influential company representative, the courts have focused on the legal concept of perpetration through an agent (Saliger 2012, Rn. 169). Even if the agent in this case is fully responsible for his actions, attributing authority to the director is held to be justified by his control over the organization (*Organisationsherrschaft*) (BGHSt 49, 147; BGHSt 48, 331; BGH, JR 2004, 245; BGH, NStZ 1998, 568; BGHSt 43, 219; BGHSt 40, 257; BGHSt 40, 218).[45]

Alternatively, if the harmful conduct results from the general production and operation of a firm, the allocation of responsibility focuses directly on the directors (the so-called top-down approach), either for negligence (which does not distinguish between perpetration and instigation) or for not preventing the crime or harmful conduct, i.e. for omission (Saliger 2012, Rn. 160 f.). The main question is that of the extent of the duty of care to prevent criminal offences from being committed by employees of the company (Schall 1999; see also Lackner and Kühl 2014, § 13 Rn. 14, with further references). An extensive interpretation would bypass the principle that each person is responsible only for his or her own actions (Schall 1999, pp. 271 ff.).

The Lederspray case delineated above provides a clear example of the challenges of this approach. The charges were, among others, the omission of a product recall. The relevant questions were that of the existence of a duty of care[46] and the responsibility of each member of the board of directors. The Court decided that while a division of responsibilities into different departments (e.g. technical and production, legal, and marketing and sales) is generally relevant, it is superseded by a general and joint responsibility of all directors in the case of an emergency. Each director, therefore, had a duty to work towards the necessary product recall. The second question was that of causation (or rather the so-called quasi-causation, the equivalent in the case of omission). Each individual director alleged that his vote alone could not have prevented the refusal of the recall because a majority decision would have been sufficient. The Court did not accept this argument, deciding that

[45]These decisions have been severely criticized by several authors, e.g. Roxin (2003, § 25 Rn. 130), who originally developed this legal concept (in Roxin 1963) to permit adequate treatment of leaders in unjust regimes, with a view on the Third Reich.

[46]If the danger was discovered after putting the product into circulation it is difficult to speak of previous behavior breaching a legal duty. See Roxin (2003, § 32, Rn. 195 ff.); Winkelbauer in Foerste and Graf von Westphalen (2012, § 81 Rn. 11 ff.); Freund in Joecks and Miebach (2006, § 13 Rn. 129).

the equal omission of the other directors could not exonerate the accused because the product recall would have been made if all directors had acted in accordance with their obligation. The passage of the judgment delineating the consequences of the recall is also notable. To refute the argument that a recall would not have been successful after distribution, the Court clearly emphasizes that the products in question were still in the possession of wholesalers and would have been kept from falling into the hands of customers.

Special personal characteristics are attributed to company directors or other employees with a position that entrusts them with sufficient responsibility pursuant to § 14 StGB, § 9 OWiG (see Schmid and Fridrich in Müller-Gugenberger 2015, § 30 Rn. 74 ff.).

6.3 *Criminal Liability of Companies?* Societas Delinquere Non Potest *Versus Company Fines Under the Law of Regulatory Offences*

The difficulty in identifying a natural person responsible for an offence has led to a demand for instituting a separate criminal liability of legal persons. Although its introduction has been fiercely discussed over recent years and still is, *de lege lata* only natural persons are subject to criminal liability in Germany (*societas delinquere non potest*) (see Müller-Gugenberger in Müller-Gugenberger 2015, § 30 Rn. 34 ff., 50 ff.; Eidam 2014, Kapitel 5 Rn. 1, 145). This situation was not changed by the European Directive on the Protection of the Environment through Criminal Law, which merely requires effective sanctions for corporations but does not specifically call for criminal liability (Saliger 2015, pp. 23 f.).

Legal persons can, however, be the addressees of forfeiture orders issued in criminal proceedings carried out against certain employees (Winkelbauer in Foerste and Graf von Westphalen 2012, § 80, Rn. 15) and can also be liable under the law of regulatory offences (Eidam 2014, Kapitel 5 Rn. 3 ff.; Müller-Gugenberger in Müller-Gugenberger 2015, § 30 Rn. 36 ff.). The possibility of imposing a fine on a corporation is provided for in § 30 OWiG, if one of its chief executive officers has committed a crime or a regulatory offence (see Raum in Wabnitz and Janovsky 2014, 4. Kapitel Rn. 213 ff.). While basing the fine on a crime committed by the employee is subject to the same difficulties in attributing responsibility within an organization as determining the presence of the individual criminal, § 30 OWiG provides for a far-reaching and effective sanctioning mechanism in connection with § 130 OWiG (Rogall in Senge 2014, § 30 OWiG Rn. 92). According to § 130 OWiG, the company owner's intentional or negligent omission of necessary supervision that would have prevented or impeded the commission of a criminal or regulatory offence by an employee is also a regulatory offence. As it is not necessary to prove that the appropriate supervisory measure would have prevented the offence, but is sufficient to show that it would have made the commission significantly more difficult, an employee's crime or regulatory offence can rather easily result in a company fine under § 30 OWiG if the owner failed to comply with ordinary care with regard to organization or supervision.[47]

7 Concluding Remarks

The challenges of pollution require forward-looking and precautionary legislation as well as the handling of existing historical contamination by cleaning up and redeveloping legacy contamination sites. Although the compensation mechanisms of private law for damages to health and property resulting from contamination sites have failed in cases of historical pollution, the characteristics identified have given rise to several new regulations that attempt to shift the risk back to the polluters by creating fault-independent liability for damages to health, property, and the environment itself and presumptions facilitating the proof of causation. These regulations are often of European origin, such as the Environmental Damage Act and the Environmental Liability Act. Similarly, criminal law is only partially applicable in cases of legacy contamination. However, the fundamental decision of basing responsibility on abstractly dangerous conducts or the likelihood of causing severe harm—made compulsory by the Directive on environmental criminal law—ensures the future applicability of criminal offences also in cases of cumulative pollution, thereby aiming to prevent the emergence of new contaminated sites. With regard to widespread diffuse pollution—especially the legal pollution that makes up the majority of the environmental burden—environmental criminal law is, nevertheless, inapplicable, even if severe danger arises in accumulation. Effective environmental protection, therefore, mainly depends on stricter preventative administrative requirements and consequently on political decisions. However, it is also important to prevent the evasion of restrictions and safety precautions. The illegal export of hazardous waste to developing countries is one such example, which has to be regulated and controlled so that further restrictions combat and prevent contamination instead of merely shifting it (cf. Compte 2006, p. 192).

With regard to reactive measures in historical pollution cases, the Federal Soil Protection Act is an effective instrument that enables the clean-up of historically contaminated sites. It comprises a far-reaching administrative regulation, oriented at effectiveness, that is even applicable to cases from before its entry into force and partially extends to new contamination. The sheer quantity of contaminated sites for which no private responsible person exists and whose financial burden must be borne by the general public is, nevertheless, challenging. Clean-up and remediation measures will accompany us for some time to come.

---

[47]See Rogall in Senge (2014, § 130 Rn. 113 ff.) for the exact requirements on causation and breach of duty.

References

Asendorph, D. (2013). Alles so sauber, Zeit online, 21. März 2013, http://www.zeit.de/2013/13/Ruhrgebiet-Umweltschutz

Balensiefen, G. A. (2012). *Umweltschadensgesetz—USchadG, Kommentar, Erläuterungen, Das deutsche Bundesrecht*. Baden-Baden: Nomos-Verlag.

Bamberger, H. G., & Roth, H. (2015). *Beck'scher Online-Kommentar BGB* (37th ed.). Munich: C.H. BECK.

Becker, B. (2005). Einführung in die Richtlinie über Umwelthaftung zur Vermeidung und Sanierung von Umweltschäden. *Neue Zeitschrift für Verwaltungsrecht, 2005,* 371–376.

Becker, F., & Fett, T. (1999). Verantwortlichkeit für Verunreinigungen nach dem neuen Bundes-Bodenschutzgesetz im Spannungsfeld von Umwelt- und Gesellschaftsrecht. *Neue Zeitschrift für Gesellschaftsrecht, 1999,* 1189–1198.

Benkel, J. (2002). Sozialrecht und Sozialgerichtsbarkeit im deutschen Rechtssystem. *Neue Zeitschrift für Sozialrecht, 2002,* 617–620.

Bothe, M. (1998). Zulässigkeit landesrechtlicher Abfallabgaben. *Neue Juristische Wochenschrift, 1998,* 2333–2335.

Breuer, J. (1995). Berufskrankheitenrecht ohne Kausalitätsbeweis? *Neue Zeitschrift für Sozialrecht, 1995,* 146–152.

Breuer, R. (1987). Rechtsprobleme der Altlasten. *Neue Zeitschrift für Verwaltungsrecht, 1987,* 751–761.

Bruns, E., et al. (2009). Anforderungen an die Erfassung, Bewertung und Sanierung von Biodiversitätsschäden nach dem Umweltschadensgesetz. *Natur und Recht, 2009,* 149–159.

Bund/Länder-Arbeitsgemeinschaft Bodenschutz (LABO), Ständiger Ausschuss "Altlasten" ALA. (2014). Bericht des ALA über „Bundesweite Kennzahlen zur Altlastenstatistik", S. 2, with links to the state statistics in Annex 2 S. https://www.labo-deutschland.de/documents/Bericht_Altlastenstatistik_2014_endgefasst_.pdf

Bundesministerium für Arbeit und Soziales (BMAS). (2013). Sicherheit und Gesundheit bei der Arbeit 2013, http://www.baua.de/de/Publikationen/Fachbeitraege/Suga-2013.pdf?__blob=publicationFile&v=16

Bundesministerium für Umwelt, Naturschutz, Bau und Reaktorsicherheit (BMU). (2008). Sanierung der ökologischen Altlasten in den neuen Bundesländern, Stand, November 24, 2008. www.bmub.bund.de/P673/

Bundesministerium für Umwelt, Naturschutz, Bau und Reaktorsicherheit (BMU). (2012). Schachtanlage Asse II, Stand, March 1, 2012. http://www.bmub.bund.de/themen/atomenergie-strahlenschutz/endlagerprojekte/schachtanlage-asse-ii/

Bundesministerium für Umwelt, Naturschutz, Bau und Reaktorsicherheit (BMU). (2013). Dritter Bodenschutzbericht der Bundesregierung—Beschluss des Bundeskabinetts vom 12. Juni2013, S. 14.

Bundesministerium für Umwelt, Naturschutz, Bau und Reaktorsicherheit (BMU). (2014). Umweltbewusstsein in Deutschland 2014—Ergebnisse einer repräsentativen Bevölkerungsumfrage, https://www.umweltbundesamt.de/sites/default/files/medien/378/publikationen/umweltbewusstsein_in_deutschland.pdf

Callies, C., & Ruffert, M. (2011). *EUV/AEUV: Das Verfassungsrecht der Europäischen Union mit Europäischer Grundrechtecharta - Kommentar* (4th ed.). Munich: C.H.BECK.

Compte, F. (2006). Environmental crime and the police in Europe: A panorama and possible paths for future action. *European Energy and Environmental Law Review, 2006,* 190–231.

Dannecker, M., & Klink, T. (2003). Beseitigung von DDR-Umweltschäden durch Investoren—Zwischenbilanz und neue Aspekte der Altlastenfreistellung nach Art. 1 § 4 III DDR-URG. *Landes- und Kommunalverwaltung, 2003,* 159–164.

Daxenberger, M. (1997). *Kumulationseffekte—Grenzen der Erfolgszurechnung im Umweltstrafrecht*. Baden-Baden: Nomos.

Deutscher Bundestag. (1990). Sondergutachten "Altlasten" des Rates von Sachverständigen für Umweltfragen BT-Drs. 11/6191.
Deutscher Bundestag (1995). Sondergutachten „Altlasten II“ des Rates von Sachverständigen für Umweltfragen, BT-Drs. 13/380.
Die Bundesregierung (2002). Perspektiven für Deutschland—Unsere Strategie für eine nachhaltige Entwicklung, http://www.bundesregierung.de/Content/DE/_Anlagen/Nachhaltigkeit-wiederhergestellt/perspektiven-fuer-deutschland-langfassung.pdf?__blob=publicationFile&v=3
Die Bundesregierung (2012). Nationale Nachhaltigkeitsstrategie—Fortschrittsbericht 2012, http://www.bundesregierung.de/Content/DE/_Anlagen/Nachhaltigkeit-wiederhergestellt/2012-05-21-fortschrittsbericht-2012-barrierefrei.pdf?__blob=publicationFile&v=1
Eidam, G. (2014). *Unternehmen und Strafe* (4th ed.). Cologne: Carl Heymanns Verlag.
Engler, M. (2014). Wer Hilfe will, muss durch die Bürokratie-Hölle, Zeit online, 25th April 2014, http://www.zeit.de/wissen/gesundheit/2014-03/asbest-krebs-berufskrankheit/komplettansicht
Epping, V., & Hillgruber, C. (2015). *Beck'scher Online-Kommentar Grundgesetz* (26th ed.). Munich: C.H. BECK.
Ernst, W., et al. (2015). *Baugesetzbuch* (116th ed.). Munich: C.H.BECK.
Foerste, U., & Graf von Westphalen, F. (2012). *Produkthaftungshandbuch* (3rd ed.). Munich: C.H. BECK.
Frankfurter Allgemeine Zeitung (FAZ). (2002). Finanzen, In Deutschland drohen kaum Asbest-Klagen, v., September 25, 2002. http://www.faz.net/aktuell/finanzen/aktien/schadensersatz-in-deutschland-drohen-kaum-asbest-klagen-173770.html. Last visited October 11, 2015.
Franzheim, H., & Pfohl, M. (2001). *Umweltstrafrecht: Eine Darstellung für die Praxis* (2nd ed.). Cologne: Carl Heymanns Verlag.
Frenz, P., & Heßler, W. (2001). Altlastensanierung und öffentlich-rechtlicher Sanierungsvertrag. *Neue Zeitschrift für Verwaltungsrecht, 2001,* 13–16.
Führ, M. (2016). *Gemeinschaftskommentar zum Bundes-Immissionsschutzgesetz*. GK-BImschG, Cologne: Carl Heymanns Verlag.
Gaier, R. (2005). Die zivilrechtliche Haftung für Altlasten und schädliche Bodenveränderungen. *Neue Zeitschrift für Miet- und Wohnungsrecht, 2005,* 161–167.
Gaßner, H., & Buchholz, G., Lex Asse—Gesetz zur Beschleunigung der Rückholung radioaktiver Abfälle und der Stilllegung der Schachtanlage Asse II, *Zeitschrift für Umweltrecht* 2013, 336–343.
Giesberts, L., & Reinhardt, M. (2015). *Beck'scher Online-Kommentar Umweltrecht* (37th ed.). Munich: C.H. BECK.
Grziwotz, H. (1990). "Altlasten" und Grundstücksverkehr. *Mitteilungen des Bayerischen Notarvereins, 1990,* 282–294.
Habersack, M. (2013). *Münchener Kommentar zum Bürgerlichen Gesetzbuch: BGB* (6th ed.). Munich: C.H.BECK.
Hannich, R. (2013). *Karlsruher Kommentar zur Strafprozessordnung: StPO* (7th ed.). Munich: C.H. BECK.
Hassemer, W. (1992). Kennzeichen und Krisen des modernen Strafrechts. *Zeitschrift für Rechtspolitik, 1992,* 378–383.
Heine, G. (1989). Ökologie und Recht—Zur historischen Entwicklung normativen Umweltschutzes. *Goltdammer's Archiv für Strafrecht, 1989,* 116–131.
Heinz, K. (2005). Flächenrecycling und Immobiliendevelopment: Die Industriebrache im Spannungsfeld zwischen Leuchtturmprojekt und Unveräußerlichkeit—Einige Anmerkungen zu Chancen und Risiken des Flächenrecyclings für kleinere und mittlere Areale. *Neue Zeitschrift für Baurecht und Vergaberecht, 2005,* 79–84.
Hellriegel, M., & Schmitt, T. (2009). Aufwertung des bodenschutzrechtlichen Ausgleichsanspruchs—Die BGH-Rechtsprechung zur Inanspruchnahme des Verursachers von Altlasten durch den Grundstückseigentümer. *Neue Juristische Wochenschrift, 2009,* 1118–1121.

Hermann, N., & Wagner, T. (2005). Grenzen der Summation und Kumulation im Umweltrecht. *Natur und Recht, 2005,* 20–27.
Hilgendorf, E. (1994). Fragen der Kausalität bei Gremienentscheidungen am Beispiel des Lederspray-Urteils. *Neue Zeitschrift für Strafrecht, 1994,* 561–566.
Hoyer, A. (1996). Die traditionelle Strafrechtsdogmatik vor neuen Herausforderungen—Probleme der strafrechtlichen Produkthaftung. *Goltdammer's Archiv für Strafrecht, 1996,* 160–178.
Hullmann, C., & Zorn, M. (2010). Zeitliche Grenzen der Altlastenhaftung. *Neue Zeitschrift für Verwaltungsrecht, 2010,* 1267–1271.
Innenministerium Baden-Württemberg (IMBW). Stabstelle für Verwaltungsreform, Die neue Verwaltungsorganisation in Baden-Württemberg, http://www.boa-bw.de/downloads/redbar// frei/bsz34970435x/0/verwaltungsreform_neue_verwaltungsorganisation_nr_24.html.pdf
Jauernig, O. (2014). *Bürgerliches Gesetzbuch: BGB* (15th ed.). Munich: C.H. BECK.
Joecks, W., & Miebach, K. (2006). *Münchener Kommentar zum Strafgesetzbuch.* Munich: C.H. BECK.
Kebekus, F. (2001). Altlasten in der Insolvenz—aus Verwaltersicht. *Neue Zeitschrift für Insolvenz- und Sanierungsrecht, 2001,* 63–69.
Kindhäuser, U., et al. (2013). *Nomos-Kommentar Strafgesetzbuch (StGB)* (4th ed.). Baden-Baden: Nomos.
Kloepfer, M., & Heger, M. (2014). *Umweltstrafrecht* (3rd ed.). Munich: C.H. BECK.
Kloepfer, M. (1994). *Zur Geschichte des deutschen Umweltrechts.* Berlin: Duncker & Humblot.
Knebelspieß, S. (2014). Das Umweltbundesamt und Meilensteine in der Geschichte der Luftreinhaltung, in: Umweltbundesamt, Schwerpunkte 2014, https://www.umweltbundesamt.de/sites/default/files/medien/378/publikationen/schwerpunkte_2014.pdf
Kobes, S. (2000). Die Untersuchung, Bewertung und Sanierung von Altlasten nach dem Bundes-Bodenschutzgesetz—Zugleich ein Beitrag zur Bundes-Bodenschutz- und Altlastenverordnung. *Neue Zeitschrift für Verwaltungsrecht, 2000,* 261–268.
König, W., & Hoffmann, M. (2009). Asse II: Der lange Weg vom» Forschungsbergwerk «zum» Endlager für radioaktive Abfälle. *Zeitschrift für Umweltrecht, 2009,* 353–358.
Körner, A., et al. (2015). *Kasseler Kommentar Sozialversicherungsrecht* (85th ed.). Munich: C.H. BECK.
Krasney, O. E. (1993). Die gesetzliche Unfallversicherung—Bestand und Wandel in mehr als 100 Jahren. *Neue Zeitschrift für Sozialrecht, 1993,* 89–95.
Kuhlen, L. (1986). Der Handlungserfolg der strafbaren Gewässerverunreinigung (StGB § 324). *Goltdammer's Archiv für Strafrecht, 1986,* 389–408.
Kuhlen, L. (1990). Strafhaftung bei unterlassenem Rückruf gesundheitsgefährdender Produkte. *Neue Zeitschrift für Strafrecht, 1990,* 566–570.
Kuhlen, L. (1993). Umweltstrafrecht—auf der Suche nach einer neuen Dogmatik. *Zeitschrift für die gesamte Strafrechtswissenschaft, 105*(1993), 697–726.
Lackner, K., & Kühl, K. (2014). *Strafgesetzbuch: StGB* (28th ed.). Munich: C.H. BECK.
Landmann, R., & Rohmer, G. (2015). *Umweltrecht: UmweltR* (75th ed.). Munich: C.H. BECK.
Leitzke, C., & Schmitt, T. (2005). Eine zulässige echte Rückwirkung im Rahmen von § 4 Abs. 3 Satz 1, 2. Alt BBodSchG? *Umwelt- und Planungsrecht, 2005,* 177–181.
Lenz, P. (2011). Der verzweifelte Kampf der Berufskranken, Spiegel online, 15. November 2011, http://www.spiegel.de/karriere/berufsleben/asbest-uran-der-verzweifelte-kampf-der-berufskranken-a-797734.html
Lübbe-Wolff, G. (2001). Instrumente des Umweltrechts—Leistungsfähigkeit und Leistungsgrenzen. *Neue Zeitschrift für Verwaltungsrecht, 2001,* 481–593.
Marburger, P. (1992). Grundsatzfragen des Haftungsrechts unter dem Einfluß der gesetzlichen Regelungen zur Produzenten- und zur Umwelthaftung. *Archiv für die civilistische Praxis, 192* (1992), 1–34.
Materialienband zum Umweltprogramm der Bundesregierung, zu BT-Drs. VI/2710.
Maunz, T., & Dürig, G. (2014). *Grundgesetz* (63rd ed.). Munich: C.H. BECK.
Michalke, R. (2000). *Umweltstrafsachen* (2nd ed.). Heidelberg: C.F. Müller.

Ministerium für Umwelt, Klima und Energiewirtschaft Baden-Württemberg (UMBW). (2014). Pressemitteilung vom, March 28, 2014: Neue Förderrichtlinien für die Behandlung von Altlasten. http://www.baden-wuerttemberg.de/de/service/presse/pressemitteilung/pid/neue-foerderrichtlinien-fuerdie-behandlung-von-altlasten-1/

Ministerium für Umwelt, Naturschutz und Verkehr Baden-Württemberg (MUVBW). (2011). Strategieprogramm zur Reduzierung des Flächenverbrauchs—Bilanz zur Umsetzung und Ausblick Januar 2011, http://www2.mvi.baden-wuerttemberg.de/servlet/is/49873/Eckpunkte%202011%20Strategieprogramm%20Flaechenverbrauch.pdf?command=downloadContent&filename=Eckpunkte%202011%20Strategieprogramm%20Flaechenverbrauch.pdf

Ministerium für Verkehr und Infrastruktur Baden-Württemberg (MVIBW). Die Flächenmanagement-Plattform, http://www2.mvi.baden-wuerttemberg.de/servlet/is/60481/

Mitschang, S., & Roeper, K. (2011). Stadtumbau auf Gewerbe- und Industriearealen—Gibt es Typologien? *Zeitschrift für deutsches und internationales Bau- und Vergaberecht, 2011,* 10–24.

Müggenborg, H. (1992). Rechtliche Aspekte der Altlastenproblematik und der Freistellungsklausel. *Neue Zeitschrift für Verwaltungsrecht, 1992,* 845–852.

Müggenborg, H. (2000). Die Haftung des früheren Eigentümers nach § 4 VI BBodSchG. *Neue Zeitschrift für Verwaltungsrecht, 2000,* 50–52.

Müggenborg, H. (2001). Zur Begrenzung der Zustandshaftung bei Altlasten. *Neue Zeitschrift für Verwaltungsrecht, 2001,* 39–41.

Müller-Gugenberger, C. (2015). *Wirtschaftsstrafrecht—Handbuch des Wirtschaftsstraf- und -ordnungswidrigkeitenrechts* (6th ed.). Cologne: Dr. Otto Schmidt.

Müllmann, C. (1994). Altlastensanierung und Kooperationsprinzip—der öffentlichrechtliche Vertrag als Alternative zur Ordnungsverfügung. *Neue Zeitschrift für Verwaltungsrecht, 1994,* 876–879.

Odersky, W. (2005). *Strafgesetzbuch. Leipziger Kommentar, 11th ed., Band 8: §§ 302a–335a,* Berlin: Walter de Gruyter.

Oerder, M. (1992). Ordnungspflichten und Altlasten. *Neue Zeitschrift für Verwaltungsrecht, 1992,* 1031–1039.

Ossenbühl, F. (1995). Verzicht, Verwirkung und Verjährung als Korrektive einer polizeilichen Ewigkeitshaftung. *Neue Zeitschrift für Verwaltungsrecht, 1995,* 547–550.

Papier, H. (1986). Die Verantwortlichkeit für Altlasten im öffentlichen Recht. *Neue Zeitschrift für Verwaltungsrecht, 1986,* 256–263.

Pfohl, M. (1994). Strafbarkeit von Amtsträgern wegen Duldung unzureichender Abwasserreinigungsanlagen. *Neue Juristische Wochenschrift, 1994,* 418–423.

Pieroth, B., et al. (2014). *Polizei- und Ordnungsrecht* (8th ed.). Munich: C.H.BECK.

Puppe, I. (1996). Zur Kompetenz des Tatrichters über die Gültigkeit eines zweifelhaften Kausalgesetzes zu entscheiden. *JuristenZeitung, 1996,* 318–320.

Radek, E. (1990). Aktuelle Probleme und Entwicklungen im Bereich des Berufskrankheitenrechts. *Neue Zeitschrift für Arbeitsrecht, 1990,* 592–597.

Ronzani, M. (1992). *Erfolg und individuelle Zurechnung im Umweltstrafrecht.* Freiburg: Max-Planck-Institut für ausländisches und internationales Strafrecht.

Roxin, C. (1963). Straftaten im Rahmen organisatorischer Machtapparate. *Goltdammer's Archiv für Strafrecht, 1963,* 193–207.

Roxin, C. (2003). *Strafrecht Allgemeiner Teil, Band II: Besondere Erscheinungsformen der Straftat.* Munich: C.H. BECK.

Roxin, C. (2006). *Strafrecht Allgemeiner Teil, Band I: Grundlagen. Der Aufbau der Verbrechenslehre.* Munich: C.H. BECK.

Ruffert, M. (2010). Verantwortung und Haftung für Umweltschäden. *Neue Zeitschrift für Verwaltungsrecht, 2010,* 1177–1183.

Sacksofsky, U. (2000). Verfolgung ökologischer und anderer öffentlicher Zwecke durch Instrumente des Abgabenrechts. *Neue Juristische Wochenschrift, 2000,* 2619–2626.

Saliger, F. (2012). *Umweltstrafrecht.* Munich: Franz Vahlen.

Saliger, F. (2015). Grundfragen des heutigen Umweltstrafrechts. In M. Heger & M. Kloepfer (Eds.)), *Das Umweltstrafrecht nach dem 45. Strafrechtsänderungsgesetz* (pp. 9–32). Berlin: Duncker & Humblot.

Samson, E. (1987). Kausalität- und Zurechnungsprobleme im Umweltstrafrecht. *Zeitschrift für die gesamte Strafrechtswissenschaft, 99*(1987), 617–636.

Samson, E. (1991). Probleme strafrechtlicher Produkthaftung. *Strafverteidiger, 1991,* 182–186.

Sanden, J. (2014). Altlastenverantwortlichkeit trotz Dereliktion. *Neue Zeitschrift für Verwaltungsrecht, 2014,* 1329–1333.

Sandner, W. (2001). Wer sind die Beteiligten des Anspruchs auf Ausgleich der Sanierungskosten nach § 24 II BBodSchG? *Neue Juristische Wochenschrift, 2001,* 2045–2051.

Satzger, H., et al. (2014). *StGB—Strafgesetzbuch* (2nd ed.). Köln: Carl Heymanns.

Scheidler, A. (2007). Umweltschutz durch Umweltverantwortung—Das neue Umweltschadensgesetz. *Neue Zeitschrift für Verwaltungsrecht, 2007,* 1113–1119.

Schindler, R. (2014). Von der Altlastensanierung über das Bundes-Bodenschutzgesetz zum internationalen Bodenschutz—40 Jahre Bodenschutz im Umweltbundesamt. *Umweltbundesamt, Schwerpunkte, 2014,* 112–121. http://www.umweltbundesamt.de/publikationen/schwerpunkte-2014

Schink, A. (1990). Konfliktbewältigung und Amtshaftung bei der Bauleitplanung auf Altlasten. *Neue Juristische Wochenschrift, 1990,* 351–357.

Schmidt, K. (2010). Keine Ordnungspflicht des Insolvenzverwalters?—Die Verwaltungsrechtsprechung als staatliche Insolvenzbeihilfe für Umweltkosten. *Neue Juristische Wochenschrift, 2010,* 1489–1493.

Schmidt-Salzer, J. (1996). Konkretisierungen der strafrechtlichen Produkt- und Umweltverantwortung. *Neue Juristische Wochenschrift, 1996,* 1–8.

Schneider, H. (2010). Übergang der Zuständigkeit des Salzbergwerks Asse II auf das Bundesamt für Strahlenschutz. *atw - Internationale Zeitschrift für Kernenergie* 2010, 23–34.

Schönke, A., & Schröder, H. (2014). *Strafgesetzbuch: StGB* (29th ed.). Munich: C.H.BECK.

Schulze-Fielitz, H. (2007). Umweltschutz im Föderalismus—Europa. *Bund und Länder, Neue Zeitschrift für Verwaltungsrecht, 2007,* 249–259.

Senge, L. (2014). *Karlsruher Kommentar zum Gesetz über Ordnungswidrigkeiten: OWiG* (2nd ed.). Munich: C.H.BECK.

Seuser, A. A. (2013a). Die Haftung für nachteilige Gewässerveränderungen und die Sanierung von Gewässern (§§ 89, 90 WHG), Teil 1—Die nachteiligen Gewässerveränderungen in sonstiger Weise (§ 89 Abs. 1 WHG). *Natur und Recht, 35*(2013), 248–257.

Seuser, A. A. (2013b). Die Haftung für nachteilige Gewässerveränderungen und die Sanierung von Gewässern (§§ 89, 90 WHG) Teil 2—Die nachteilige Gewässerveränderungen durch Anlagen (§ 89 Abs. 2 WHG) und die Sanierung von Gewässern (§ 90 WHG). *Natur und Recht, 2013* (391), 398.

Sieder, F., et al. (2014). *Wasserhaushaltsgesetz, Abwasserabgabengesetz: WHG* (47th ed.). Munich: C.H. BECK.

Tarli, R. (2015). Uranabbau in der DDR—Strahlende Landschaften, Neue Zürcher Zeitung, June 16, 2015. http://www.nzz.ch/international/deutschland-und-oesterreich/strahlende-landschaften-1.18562954

Tettinger, P. (1991). Risiken eines Flächenrecycling für Anlagenerwerber in der früheren DDR. *Deutsch-Deutsche Rechts-Zeitschrift, 1991,* 40–46.

Tiedemann, M. (2008). Die bodenschutzrechtliche Einstandsverantwortlichkeit und der existenzvernichtende Eingriff—Öffentlich-rechtliche Auswirkungen der Änderung der Rechtsprechung des BGH zur gesellschaftsrechtlichen Durchgriffshaftung. *Neue Zeitschrift für Verwaltungsrecht, 2008,* 257–261.

Tomerius, S. (2005). Flächenrecycling als Instrument nachhaltiger Stadtentwicklung—Strategien und Rechtsfragen in der kommunalen Praxis. *Natur und Recht, 2005,* 14–20.

Tomerius, S. (2008). Auswirkungen der Baugesetzbuch-Novelle 2007—Impulse für mehr Innenentwicklung und Flächensparen in den Gemeinden? *Zeitschrift für Umweltrecht, 2008,* 1–7.

Umweltbundesamt (UBA). (2015). 40 Jahre Umweltbundesamt, S. 6 ff. http://www.uba.de/40jahre
v. Hippel, E.(1985). Staatshaftung für Waldsterben? *Neue Juristische Wochenschrift, 1985*, 30–32.
v. Hippel, E. (1990). Rechtlose Umweltopfer—Das Verursacherprinzip steht weithin nur auf dem Papier, DIE ZEIT, October 19, 1990 Nr. 43. http://www.zeit.de/1990/43/rechtlose-umweltopfer/komplettansicht
Vierhaus, H. (1998). Das Bundes-Bodenschutzgesetz. *Neue Juristische Wochenschrift, 1998,* 1262–1269.
Volk, K. (1996). Kausalität im Strafrecht: Zur Holzschutzmittel-Entscheidung des BGH vom, August 2, 1995, NStZ 1995, 590. *Neue Zeitschrift für Strafrecht, 1996,* 105–110.
Voßkuhle, A. (2013). Umweltschutz und Grundgesetz. *Neue Zeitschrift für Verwaltungsrecht, 2013,* 1–8.
Wabnitz, H., & Janovsky, T. (2014). *Handbuch des Wirtschafts- und Steuerstrafrechts* (4th ed.). Munich: C.H. BECK.
Weber, F. C., & Otting, O. (2014). Grenzen der kommunalen Zustandsstörerhaftung nach dem Bundes-Bodenschutzgesetz. *Neue Zeitschrift für Verwaltungsrecht*, 1618–1623.

Chapter 12
The Approach to Historical Pollution in France: Remedy, Compensate, and Punish

Luca d'Ambrosio

Contents

L. d'Ambrosio (✉)
Collège de France, Paris, France
e-mail: luca.dambrosio@college-de-france.fr

F. Centonze and S. Manacorda (eds.), *Historical Pollution*,
DOI 10.1007/978-3-319-56937-6_12

1 Introduction: Historical Pollution, a Challenge for Legal Liability

Historical pollution may be considered as one of the biochemical signatures of human activity on the environment and human beings.[1] The juridical approach to this phenomenon certainly depends on *intrinsic factors* related to the physical and biochemical processes of pollution, such as the quantity and quality of pollutants or the type of human or environmental elements concerned.[2] But it also depends on *extrinsic factors*—such as technical progress in the detection of polluting substances, public awareness about the harmful effects of pollution, and political willingness to uphold environmental and social concerns about economic growth—which make the juridical approach to historical pollution a process propelled and criss-crossed by social conflicts. In effect, the assignment of liabilities for historical pollution contributes to the redistribution, within society, of the environmental and human "costs" of industrial development.

France has not been exempted from this social and juridical process. After two centuries of industrialization, essentially regulated through preventive measures, this country was witness in the last decades of the twentieth century to a growing societal demand for remedial and retributive actions with regard to environmental and social injustices originating from industrial development (Laurian 2008). In this period, legal liability for breaches of environmental regulations has been progressively reinforced. However, it has mainly been implemented in a forward-looking respect. When facing into the past, the assignment of liability for environmental and health harms has traditionally encountered the obstacles that *time* so often poses in the legal field: non-applicability of statutory limitations, retroactive rules, legal certainty, and legal activism (Ost 1999; Cour de Cassation 2014).

Moreover, historical pollution poses specific obstacles that make the assignment of liability especially complex. These obstacles are well known and concern the proof of the author's knowledge of the emissions' harmful effects, as well as proof of causation between past polluting activities and current damage to the environment or health. As an example of the first type of obstacles, which usually arise from the temporal gap existing between knowledge production and the production of regulations, we can quote a judgment delivered in 2015 by the Cour de Cassation (the French Supreme Court) in a criminal case concerning the exposure of workers to asbestos during the 1970s and 1980s. In this decision, the Supreme Court

[1]Many scientists investigating the climatic, biological, and geochemical markers of human activity (Crutzen 2002) are now using the term "Anthropocene" to suggest that the era since the Industrial Revolution should be recognized as a distinct geological epoch to the "Holocene". For a comprehensive panorama of the challenges that the concept of Anthropocene raises in the legal field, see Viñuales 2016.

[2]See the Preamble to the Stockholm Convention on Persistent Organic Pollutants.

dismissed the charges of involuntary manslaughter against public and private officers on the ground that negligence was not to be attributed prior to the legal banning of the lethal fibre in 1996.[3] According to the Court, *and despite fact that the carcinogenic effects of asbestos had been unanimously declared by scientists as early as 1973*, the accused could not "assess the particular gravity of the risk to which they exposed the victims".[4]

But the supposed ignorance of past generations about future damages related to current pollution may match with the effective powerlessness of the present generations when tracing causation between current damage and past pollution. As is well known, most of the pollution-related diseases are multifactorial. Moreover, several factual elements may be interposed between the acts of pollution and assessment of their harmful effects: one of the most common is the succession of natural or legal persons operating on the same contaminated site or, in case of a single corporation operating on the site, the succession of its officers and directors. In these cases, it is quite hard to identify which "portion" of the damage may be attributed to each operator or to each officer (d'Ambrosio 2016, p. 330). Finally, it may be impossible to attribute current damage to individual actors when pollution crosses both time and space: this is the case for pollution (especially agricultural) which may originate either from several collective and not individually identifiable sources (*diffuse pollution*), or from several sources that are only collectively harmful to the environment or to health (*cumulative pollution*).[5]

Dealing with historical pollution is therefore a significant challenge for legal liability. In order to discuss how this challenge was taken on in France, we describe in Sect. 2 the social and economic context that engendered public awareness on historical pollution. Section 3 presents some elements that may contribute to the legal definition of historical pollution, whereas Sect. 4 presents the legal background applying to historical pollution: this essentially relies on the "patchwork" of regulations composing the national "Contaminated Land Regime". The specific issue of liability for historical pollution is addressed in the subsequent sections: regulatory liability for remediation of historical pollution is addressed in Sect. 5, while civil and criminal liability is, respectively, addressed in Sects. 6 and 7. Section 8 focuses on corporate criminal liability. Section 9 provides some concluding remarks.

[3]The gap between the scientific assessment of the harmful effects of asbestos and its legal ban was essentially attributable in France to the lobbying of the Comité Permanente Amiante. This Committee was made up of scientists, representatives of firms in the industry, public officers, and trade unionists.

[4]Cour de Cassation, Chambre criminelle, 14 April 2015, case no. 14-85.333.

[5]See para. 13 of Directive 2004/35/CE of the European Parliament and of the Council of 21 April 2004 on environmental liability with regard to the prevention and remedying of environmental damage. This is also the case for CO_2 emissions, which are excluded from the discussion in this research.

2 Making Historical Pollution "Visible": Social and Judicial Activism in France

France may be inscribed among the European countries who embraced industrialization at its very beginning (Asselaim 2011): for over almost two centuries, from industry to agriculture, different kinds of productive activities contributed to the contamination of air, water, and soils.[6]

As in other developed countries, the first public denunciations of the harmful effects of industrial pollution accompanied the decline of industrialization at the end of the 1970s. At this time, certain French academics began investigating the impact of the organization of economic production on health, in connection with the protection of labour and the environment. They denounced the "invisibility" of the harmful effects that industrial production had historically been causing for workers' health and the environment, and the underlying alliance sealed by capital, science, and public powers. The iconic figure of this movement is the toxicologist Henri Pézerat: the scientist started enquiring into occupational diseases affecting workers exposed to toxic substances, such as asbestos, vinyl chloride, lead, arsenic. He stressed the tight link between occupational diseases and pollution caused by industrial production: workers are in fact the first link in a chain of physical and chemical processes triggered by the ejection of toxic substances into the environment (Thébaud-Mony 2014).

To this scholarship, relayed by unions, we owe the first public mobilizations for the social and judicial "visibility" of *occupational diseases related to exposure to toxic substances*: the most important mobilization was led by the Jussieu Faculty in Paris, where the whole community of students and professors was exposed to asbestos despite the risks as assessed by scientists. This mobilization was picked up by the workers of the Amisol plant in Clermont-Ferrand, who had been exposed to the killer substance up to the closure of the industrial site in the mid-1970s (Thébaud-Mony 2008). The mobilization continued into the twenty-first century: in 2006, for instance, the miners of the biggest French goldmine (Salsigne) denounced the increase in carcinogenic diseases among the former workers and neighbours of the site: despite public surveillance, the mine had been ejecting arsenic into the environment for more than a century. In each of the above cases, social mobilization was accompanied by criminal lawsuits. However, no criminal trial has yet started: in the *Salsigne* case, investigations are still on going today, while in the *Jussieu* and *Amisol* cases pre-trial chambers systematically dismissed the charges of involuntary injuries and manslaughter pronounced against public and private directors. But the criminal courts' approach is now in fact changing: in 2015 and in 2016 the Cour de Cassation annulled the dismissal of charges and ordered the

---

[6]Association Robin des Bois (2016) has recently mapped all the kinds of widespread and historical pollution present across the whole territory.

prosecution of the investigations in both the *Jussieu* and *Amisol* cases.[7] The perception of impunity still prevails in society, however, despite the victims having been granted civil compensation either through social security (economic damages) or the judicial commission for the compensation of offences' victims (moral damages).[8]

The French regulatory powers essentially ignored social claims about historical pollution until certain "scandals" were covered extensively by the media: in the first few years of the twenty-first century, the *Metaleurop Nord* case had an impact in France comparable to that of the *Love Canal* case in the USA (Mazade 2013, 2010). Located in the town of Noyelles-Godault, in the northern region of the country, this plant had been producing lead and zinc since the end of the nineteenth century. The plant was shut down in 2003 due to the bankruptcy of the operator. Despite surveillance carried out by the administration during the previous thirty years, historical pollution by lead, cadmium, and zinc affecting an area of 45 km^2 around the plant was discovered at the moment the site was shut down. Groundwater was also highly polluted with arsenic and zinc. Surveys determined that, depending on the distance from the site, between 30 and 20% of children living in the neighbourhood had lead poisoning.[9] With the *Metaleurop* case, then, historical pollution finally broke into public consciousness, in the form of a widespread environmental and health disaster along with the impunity of a "rogue boss"—*Patron voyou*, in the words used by the French media. Common and environmental regulations appeared powerless to impose any kind of liability for the disaster: the attempt to assign regulatory liability for remediation to Metaleurop's parent corporation failed[10] and the victims' claims for criminal justice were systematically dismissed.[11] The result was that compensation of the victims and the rehabilitation of the site were covered by public funds.[12]

Metal Blanc was another case of historical pollution that had major media resonance in France. This concerned an industry which recycled toxic heavy metals

[7]See Cour de Cassation, Chambre Criminelle, 14 April 2015, case no. 14-85.334 (*Jussieu Faculty* case) and 6 June 2016 (*Amisol* case, not yet published).

[8]This is the Commission d'indemnisation des victimes d'infractions (CIVI), which sits before every Tribunal de Grand Instance.

[9]Lead interferes with a variety of body processes and is toxic to many organs and tissues including the heart, bones, intestines, kidneys, and reproductive and nervous systems. It interferes with the development of the nervous system and is therefore particularly toxic to children, causing potentially permanent learning and behaviour disorders.

[10]Appeal Court of Douai, 2 October 2003 (Lienhard 2003, p. 2571).

[11]Cour de Cassation, Chambre criminelle, 30 October 2007, case no. 06-89.365.

[12]The clean-up of the areas inside the site was undertaken by a private company which took the site on with the aim of creating a centre for waste treatment and recycling: a substantial part of the brownfield is now a laboratory where scientists study the consequences of pollution, especially soil pollution (Aligon and Douay 2011). The *Metaleurop* scandal would stimulate the legislature to pass important amendments aiming to strengthen the prevention and compensation for damage in the field of "technological risks" (*Loi n 2003-699 du 30 juillet 2003 relative à la prévention des risques technologiques et naturels et à la réparation des dommages*). See below, Sect. 4.1.1.

from car batteries and which since 1968 had operated on the brownfield site of an old foundry close to the small village of Bourg Fidèle (Champagne). In 1996, the operator obtained a permit to double production in exchange for a substantial reduction in polluting emissions in the air and watercourses. The operator doubled its production but did not comply with the administrative prescriptions, therefore aggravating the heavy metal pollution of the area around the village. In 1998, the inhabitants filed criminal lawsuits for the offence of risks caused to other persons. Criminal trials were held and the company was acquitted, until the Cour de Cassation quashed the decisions in 2010. Contrary to the precedents, which in the *Metaleurop* case saw the dismissal of criminal proceedings, the Chambre Criminelle of the Cour de Cassation stated in *Metal Blanc* that the repeated violation of remediation orders by the operator of a hazardous facility may characterize the offence of risks caused to other persons provided by Article 223-1 of the Criminal Code. The corporation was consequently condemned. The penalty applied to the company was certainly ridiculous (a fine of 75,000 Euros). However, this decision remains a milestone since, for the first time, it permitted the characterization of a failure to remediate historical pollution as a criminal offence against persons.[13]

In recent years, we have seen the emergence of similar mobilizations in other fields, the most important of which is certainly *agriculture*. France has developed intensive and animal husbandry since the 1950s: these activities have been—and still are—an important source of pollution in the country. They involve increased use of fertilizers, plant growth regulators, and pesticides: these substances accumulate in ground and surface water and may have harmful effects on wildlife or human health which are remote both temporally and spatially from the polluting substances actually released. In this field too, the recourse to justice is an important part of the social mobilization. Claims before labour judges have multiplied in the most recent years, with the aim of recognizing as an occupational disease the cancer of farmers who have managed and spread pesticides. Relying on a scientific study assessing the carcinogenicity of certain pesticides (Guyton et al. 2015, p. 491), in 2016 the Appeal Court of Rennes described the leukaemia affecting a farmer working for a food cooperative (Nutréa-Triskalia) as an occupational disease.[14] In the same way, the Administrative Courts stated that the government had to reimburse the local administration for the remediation costs for coastal fisheries degradation (green algae) in the region of Brétagne due to an excessive concentration of nitrogen and phosphorus fertilizers in the fields.[15] The French Republic

---

[13]Cour de Cassation, Chambre criminelle, 21 September 2010, case no. 09-86258. For more details on the characterization of the offence of "risks caused to persons" in cases of historical pollution, see Sect. 7.2 below. The assignment of criminal liability for industrial historical pollution has contributed to this phenomenon becoming a public policy priority, as highlighted by Decocq (2010) in the report of the working group on brownfields and historical pollution instituted by the Lille Council.

[14]See the dossier on the website of the NGO *Générations* Futures (www.generations-futures.fr).

[15]Administrative Appeal Court of Nantes, 26 December 2014.

was also condemned by the European Court of Justice for not having taken effective measures to protect water from such nitrate pollution[16]: a special plan for preventing future nitrate contamination was finally adopted in 2015, but no measures taken for remediating historical pollution.

Another field where we witness growing societal mobilization on the harmful effects of historical pollution, especially with regard to soil and groundwater contamination, is that of *military activities* (Association Robin des Bois 2016; Bausinger et al. 2007). During the First and Second World Wars, millions of bombs and missiles were launched across huge areas of the country, especially in the north and the east. At least a quarter of them did not explode and have been releasing lead, arsenic, mercury, and antimony into the soil ever since. Moreover, from the 1970s, the French Army dissolved several military camps that were highly polluted by explosives and other chemical substances. No legal, regulatory, or judicial measures have yet been taken in this field.

These "new fields" confirm the centrality of social and judicial activism in making historical pollution "visible" despite the regulatory failures and data gaps that result from the existing institutional arrangements (Lascoumes 2011): these "arrangements" may perhaps explain the very slender recognition of the concept of historical pollution within the French legal system.

3 In Search of a Legal Definition of Historical Pollution

Our textual research carried out in the main French Codes and regulations shows that the adjective "historical" is never in fact associated with the terms "pollution" or "contamination".[17] However, it is possible to find some elements contributing to the definition of the legal meaning and scope of the concept of historical pollution in scholars' opinions, soft law, and case law which have been especially developed in France in the field of contaminated land.

The notion of historical pollution has above all interested the experts on environmental risk insurance. Callewaert defines pollution as "historical" where such pollution

> is known and stems from activities practised in the past that have now proved to be harmful. (Callewaert 2006, p. 518)

The notion of historical pollution is here adopted in order to gauge whether the related risks can be insured. For the majority of scholars who have raised this question, the answer is in general positive provided that the harmful consequences

[16]European Court of Justice, 4 September 2014, Commission v. France, case C-237/12.

[17]The textual research was carried out on the Environment, Criminal, Civil, Health, and Rural Codes published on the public database *Legifrance.fr*. The adjective "historical" is associated with "emissions" but only when it is a matter of issuing additional quotas for greenhouse gas emissions (Art. R 229-11 of the Environment Code).

resulting from past activities *have not yet occurred*. In other words, risks related to historical pollution can be insured when unknown elements concern either the costs of restoring a polluted site, or the immediate threats to health or to the groundwater and the neighbouring land (Alexandre 2012, p. 68). In this case (the *clause de reprise du passé inconnu*), the purchaser of a site can be insured against the economic consequences of historical pollution, provided they were *unknown* when the purchase was signed.[18]

Other elements for the definition of the concept of historical pollution under French law can be found in a circular published by the Ministry of the Environment on 26 May 2011, on the chain of liabilities applying in case of an industrial site closure.[19] The concept of historical pollution is here recalled in order to delimit its scope to the cases of plants' operators who are *known* but *no longer able to comply* with the obligations of clean-up. Historical pollution is here defined as follows:

> pollution which is not attributable to a clearly identifiable activity or to an operator.

This circular gives two examples of historical pollution: the presence of pollutant substances derived from very ancient activities and the landfill of areas by clinkers (*mâchefers*). With regard to these cases, the circular emphasizes the necessity that we

> not overestimate the risks associated with this type of pollution: the feedback clearly shows that sites with very old pollution that pose problems are rare. Moreover, in the case of embankments, for example, the areas are rarely exposed and pathways that can be active are often nonexistent. The difficulties associated with this type of pollution generally involve issues of changes in use, falling outside of this circular.

Besides the astonishing tendency of the French administration to minimize the environmental and health risks arising from historical pollution, it seems from this circular that the element distinguishing the so-called historical pollution is not so much the passage of time between the ejections of pollutant substances into the environment and the damages, as *the assignment of the damages to a clearly identifiable operator*. As discussed below, in so far as the operator of an industrial facility is identifiable, the concept of historical pollution is in fact almost useless, at least from the administration's perspective: in fact, under French law the "last operator" of an industrial facility is generally considered liable for the pollution caused by previous operators. The approach of the government is slightly binary: if the operator is identifiable, he or she will be liable for any pollution, current or past. If the operator is not identifiable, the pollution is "historical" and it will be the community which takes charge.

---

[18]See below, Sect. 6.1.

[19]Ministry of the Environment, *Circulaire du 26/05/11 relative à la cessation d'activité d'une installation classée chaîne de responsabilités—défaillance des responsables*. Under French law, a Circular is an order addressed to the administrative authorities.

However, this approach has to be nuanced when taking into consideration the case law. In a case held in 2013 before the Administrative Appeal Court of Lyon, the concept of historical pollution was raised by the defendant in order to delimit his remediation liability vis-à-vis that of the previous operator of a contaminated site. The goal is evident: challenging the legitimacy of the administrative notice that imposed upon the last operator the duty of remediation of the groundwater pollution detected in the site. According to the defendant, the groundwater pollution was firstly due to deposits of some 5000 tonnes of cracking residues of trichloroethane produced up until 1972 *by the previous operator of the site*. Moreover, this historical pollution had already been the object of an administrative order (directed to the previous operator of the site) which had already been quashed by a court. Secondly, the historical pollution was localized in a parcel of the site which was not included in the purchase agreement signed between the defendant and the previous operator. The administrative judges rejected both grounds of the appeal. On the first ground, they recognized that the defendant had been "the recipient of a remedial order of *historical pollution*". However, albeit that the *object* of the remedial order was the same, "the actions imposed on the defendant differ from those that had been imposed on the previous operator". On the second ground, the Court pointed out that although the historical pollution was localized in a parcel of the site excluded from the purchase agreement, there was nevertheless "continuity" between the operators of the site. In short: on the one side, the current operator may be liable for historical pollution if it has not been fully remediated by the ancient operator; on the other hand, private agreements cannot be opposed as a defence against remediation liability.[20]

The analysis of these sources does not allow us to identify a clear and stable legal definition of historical pollution. However, some key elements stem from this analysis that do allow us to define the *scope* and the *function* that the concept of historical pollution may have under French law:

- The main scope of application of the concept of historical pollution is defined by *public regulations on contaminated sites*, i.e. by law and regulations on the remediation of pollution (mainly of soil and water) inherited by productive activities.
- In this context, and regardless of the terminology used, the concept of historical pollution is mainly used in practice as a defence against the (vertical) liability that public powers may impose as regards remediating contaminated sites, and in order to punish and compensate in cases of threats to the environment and to health caused by a failure to remediate.

The analysis allows to consider the "Contaminated Land Regime" as the legal background of historical pollution in the French legal system: this is the reason why we consider it useful for our discussion to detail it.

[20] See Sect. 5.3 below.

4 The Legal Background of Historical Pollution: The French "Contaminated Land Regime"

In France, the regulatory and legal framework applying to contaminated land looks more like a "patchwork" than a comprehensive system. In this regard, it may be recalled that contaminated land was first regulated in France through circulars issued by the Ministry of the Environment and addressed to the local authorities in charge of the protection of the environment (Préfets). Several circulars, issued from 1993 to 2011, set out the national policy on contaminated sites (the so-called National Methodology on Contaminated Sites). This policy relies on three main pillars.

(a) The provision of public information about contaminated sites through two national databases: the BASOL database, which contains information on all contaminated sites identified by the environmental authorities, including the interested parties, site location, type of contamination, remediation actions;[21] and the BASIAS database, which keeps track of past industrial activities including information on former industrial and pollution-related sites. This database also contains the remediated sites which were previously listed under BASOL.[22]
(b) The implementation of a risk-based remediation policy. This approach, *limited to the prevention of risks of "the future use" of the site and not extended to its full remediation*, may be seen as a compromise between the reduction of remediation costs and the neutralization of the major risks stemming from contamination. However, damage to the environment is not addressed per se and "residual contamination" is allowed to persist in the site and expose third parties to unknown risks.
(c) The definition of the "chain of responsibilities" for the remediation of the contaminated sites: under the French law, *the last operator* of a contaminated site is considered liable for the remediation of the site. If it is impossible to identify the last operator or he is insolvent, the contaminated site is considered as an "orphan" and the remediation is guaranteed by the State through a public environment agency (ADEME).

The above-mentioned general principles set out in the national policy on contaminated sites have been implemented mainly through uncoordinated regulations —often derived from EU law—adopted in the field of Environmental law over time (Sect. 4.1). These regulations are to be coordinated with general regimes provided under Private (Sect. 4.2) and Criminal law (Sect. 4.3).

[21]Website of the Ministère de l'Environnement, de l'Energie et de la Mer, BASOL database, http://basol.developpement-durable.gouv.fr/.

[22]Website of the Ministère de l'Environnement, de l'Energie et de la Mer, BASIAS database, http://www.georisques.gouv.fr/dossiers/basias.

4.1 Environmental Law

Set out in the French Environment Code of 2000, Environmental Law is the main piece of the "patchwork" of legal tools regulating contaminated land. Historical pollution is never addressed therein directly or per se; however, it may be tackled by the "authorization regime" on hazardous facilities (Sect. 4.1.1) as well as by the Environmental Liability Regime (Sect. 4.1.2).

4.1.1 The "Authorization Regime" on Hazardous Facilities

The origins of the "Authorization Regime" on hazardous facilities in France are very old, dating back to the beginning of the nineteenth century, and the adoption of the imperial decree "on factories and workshops shedding an unsafe or uncomfortable smell" (1810).[23] This regulation introduced prior administrative controls on the workshops and factories which could create nuisances for neighbours. As some historians argue (Fressoz 2012; Massard-Guilbaud 2010), one of the goals of this regulation was to shelter the growing national industry from the grip of the police: during the *Ancien régime* the police in fact maintained constant surveillance on the artisanal and nascent industrial production by injunctions, fines, and prohibitions (Napoli 2003). Under the new regime, judges could only impose compensation for tort damages, without challenging the administrative authorization issued for the benefit of the facility.[24]

First amended in 1917, this regulation underwent deep reform in 1976.[25] It is to this reform, whose main goal was the *prevention* of industrial pollution, that we owe the key notion of "Classified installations for the protection of the environment" (hereafter ICPE). Pursuant to Article L 511-1 of the Environment Code, ICPE may be defined as follows:

[23]*Décret impérial du 15 octobre 1810 relatif aux Manufactures et Ateliers qui répandent une odeur insalubre ou incommode.*

[24]Fressoz (2012) describes this process as a reorganization of "environmental illegalisms" (Foucault). In other worlds, for the benefit of industry, the 1810 decree would have transformed the threats against the environment, from illegalism against "common goods" (environment and public health) into illegalism against "goods" (property). The historian Geneviève Massard-Guilbaud (2010) related an example that clearly describes this assumption: whereas under the *Ancien régime*, a law-breaker who burnt kelp outside of the prescribed period was liable to a 300 lb fine and corporal punishment in the event of reoffending, under the 1810 regime the owner of a soda plant with the relevant authorization, which discharged production waste into a river, runs no risk of criminal charges. The 1810 decree paved the way for the "commodification" of the environment, which permitted the accomplishment in France of the "Great transformation" described by Karl Polanyi (1944).

[25]Law no. 1976-663 of 19 July 1976 (*Loi n 76-663 du 19 juillet 1976 relative aux installations classées pour la protection de l'environnement*).

> Factories, workshops, depots, work sites and, in general, all facilities operated or owned by any public or private person or entity, which might present hazards or drawbacks for the convenience of the neighborhood, or for public health and safety, or for agriculture, or for the protection of nature and the environment, or for the conservation of sites and monuments or elements of the archaeological heritage.[26]

Nomenclature set out in a Conseil d'Etat decree specifies the facilities addressed by this regulation and the different regimes applicable. Depending on the level of the hazard for the environment and human health, ICPE may be subject to a simple *declaration* to the administrative authority, a *registration*, or a prior *authorization*. The authorization sets thresholds, monitoring requirements, and general operational conditions. Additional conditions are provided for quarries,[27] handling and storage of hazardous substances,[28] waste facilities,[29] nuclear plants,[30] as well as for the hazardous installations set out in the EU Industrial Emissions Directive (hereinafter IED Directive)[31] and in the so-called Seveso Directives.[32] The ICPE are under the surveillance of the administration throughout their operation. The administration must periodically review the conditions attached to the authorization and ensure that they are complied with. The ICPE inspection guarantees regular controls. At any time during and after the operation of the facility, the Préfet can order the operator to take remedial actions aimed at preventing health or environmental risks. A similar regime is provided for facilities, ICPE or not, creating threats to water.[33] In brief: the "authorization regime" is the main tool which is provided under French law for the prevention and remediation of soil and water pollution.

The ICPE regime has been repeatedly amended in the last fifteen years. In 2003, the conditions and the scope of remedial actions were broadened: the level of remediation that is to be guaranteed in case of closure of the site is now fixed at the moment the authorization is delivered.[34] This regime was also amended in 2012 in

---

[26]English translation published on the website *Legifrance.fr*.

[27]Arts. L 514-4 *et seq.* of the Environment Code.

[28]Art. L 515-7 of the Environment Code.

[29]Art. L 515-14 of the Environment Code.

[30]Art. L 515-43 of the Environment Code.

[31]Arts. L 515-28 *et seq.* of the Environment Code. EU Directive 2010/75/EU on Industrial Emissions (IED Directive) replaced EU Directive 2008/1/CE on Integrated Pollution Prevention Control (IPPC Directive).

[32]Arts. L 515-32 *et seq.* of the Environment Code. See Directives 82/501/EEC ("*Seveso I*"), 96/82/EC ("*Seveso II*") and 2012/18/EU ("*Seveso III*").

[33]Article L 241-1 of the Environment Code establishes an authorization regime for all installations not appearing in the ICPE nomenclature and resulting in the drawing of surface or underground water ("Water regime").

[34]Law no. 2003-699 of 30 July 2003 (*Loi n°2003-699 du 30 juillet 2003 relative à la prévention des risques technologiques et naturels et à la réparation des dommages*). This rule applies only to authorizations delivered after 2003.

order to implement the IED directive[35] and to strengthen enforcement measures, including criminal sanctions.[36] Finally, the ICPE regulation was amended in 2014 in order to harmonize the different regimes (ICPE, waste and water) and to clarify the hierarchy of persons liable for the remediation of polluted sites.[37]

4.1.2 The Environmental Liability Regime

In addition to the above-mentioned regimes, and whether or not they apply, the environmental liability regime may also be considered—at least in principle—as a part of the general framework applying to contaminated land (Cans 2009).

In 2008, France implemented EU Directive 2004/35/EC on environmental liability with regard to the prevention and remedying of environmental damage.[38] Based on the environmental principles set out in Articles 3 and 4 of the French Environmental Charter[39]—the "preventive" and "polluter-pays" principles—the environmental liability regime establishes

> the conditions to prevent and remedy, pursuant to the polluter-pays principle and at a reasonable cost for society, environmental damages caused by an operator's activity. An operator is a natural or moral person, public or private, who exercises or effectively controls a lucrative or non-lucrative activity.[40]

The Préfet is also the competent authority for the implementation of this regime, which applies only where serious and measurable deterioration directly or indirectly occurs to the environment, and affects the following: (a) land where contamination creates a significant risk of harming human health; (b) water; (c) species and natural habitats protected under European Birds and Habitats Directives; or (d) ecological services.[41]

---

[35]Law no. 2013-619 of 16 July 2013 (*Loi n°2013-619 du 16 juillet 2013 portant diverses dispositions d'adaptation au droit de l'Union européenne dans le domaine du développement durable*).

[36]Decree no. 2012-34 of 11 January 2012 (*Ordonnance n°2012-34 du 11 janvier 2012 portant simplification, réforme et harmonisation des dispositions de police administrative et de police judiciaire du code de l'environnement*).

[37]Law no. 2014-366 of 24 March 2014 (*Loi n°2014-366 du 24 mars 2014 pour l'accès au logement et un urbanisme rénové*). See below, Sect. 5.2.

[38]Law no. 2008-757 of 1 August 2008 (*Loi n°2008-757 du 1er août 2008 relative à la responsabilité environnementale et à diverses dispositions d'adaptation au droit communautaire dans le domaine de l'environnement*).

[39]The Environmental Charter was integrated in French Constitutional Law in 2005 (*Loi constitutionnelle n°2005-205 du 1er mars 2005 relative à la Charte de l'environnement*). The Charter recognizes fundamental rights and obligations on the protection of the environment.

[40]Art. L 160-1 of the Environment Code.

[41]Art. L 161-1 of the Environment Code.

Operators[42] are required to take necessary prevention or remedial measures. Pursuant to EU law, two different regimes are established: (a) a *strict liability regime*, for operators carrying out dangerous or potential dangerous activities (waste, GMOs, and other activities listed in the IPPC-IED Directive);[43] and (b) a *fault-based regime* for operators who carry out other economic activities. The latter will apply only in case species and natural habitats are damaged.

When there is an imminent threat of environmental damage—for example, because an accident has occurred—the operator has the obligation to take measures to prevent or minimize the damage: preventive measures should cover not only measures to prevent the damage from happening, but also actions that prevent or minimize any damage from becoming significant.[44] In case of environmental damage, remedial measures shall be taken by the operator.[45] When the damaged resource is *soil*, remedial measures aim to remove all risks of health damage according to the use of the site at the moment the damage occurred. When the damaged resources are water or biodiversity, three kinds of remedial measures are provided: primary, complementary, and compensatory remediation: all these measures purport to restore the environment to its baseline conditions.[46]

Like the EU Directive, the French Environmental Liability regime carries some exceptions and defences that may restrict its scope in the case of historical pollution. According to national legislation, this regime does not apply in case of "diffuse pollution";[47] when more than thirty years have passed since the emission;[48] when the emission occurred before 30 April 2007 or results from an activity that was definitively shut down before 30 April 2007.[49]

4.2 Private Law

The regulatory regime provided by Environmental law is to be coordinated with the general regime provided by Private law: pursuant to the French Civil Code, any natural or legal person, whether or not operating hazardous facilities submitted to

[42]The subjective scope of the Environmental Liability regime is broader than the "authorization regime."

[43]Art. R 162-1 of the Environment Code.

[44]See Arts. L 162-3 *et seq.* of the Environment Code.

[45]See Arts. L 162-6 *et seq.* of the Environment Code.

[46]For more details on primary, complementary, and compensatory remedial measures, see annexe II of EU Directive 2004/35/CE.

[47]Art. L 161-2, n. 7 of the Environment Code. This article specifies that in case of diffuse pollution, the administrative authority can establish a causal link between the damage and the activities of several operators.

[48]Art. L 161-4 of the Environment Code.

[49]Art. L 161-4 of the Environment Code.

administrative control, may be held liable for *environmental and health damages caused to third parties.*

The legal grounds for a party being held liable under Private Law are Articles 1382–1384 of the French Civil Code. These articles provide that a person is liable for any damage caused to another person as a result of his fault or negligence:

> Article 1382: Every act whatever of man that causes damage to another, obliges him by whose fault it occurred to repair.
>
> Article 1383: We are responsible not only for the damage occasioned by our own act, but also by our own negligence or imprudence.[50]

In addition, under Article 1384 of the Civil Code, a person will be held liable for any damage caused by anything over which that person has control (*garde de la chose*):

> We are responsible not only for the damage caused by our own act, but also for that which is caused by the acts of persons for whom we are responsible, or by things that are in our custody.

The concept of "custody" is linked to the power of use, direction, and control (*usage, direction et contrôle*). The owner of an asset is deemed to have control over that asset unless he can demonstrate that the control has been transferred to another person.

In practice, however, in the field of industrial contamination, third parties tend to use the strict liability regime provided by the so-called abnormal neighbourhood nuisance theory (*trouble anormal de voisinage*). Under this regime, the "holder" of a hazardous facility is liable for repairing damages to third parties even in the absence of any fault. The claimant has only to prove the damage and, when the origin of the pollution is hard to determine, it is the defendant who has to prove that the damage could not have been caused by its facility and prove that other causes are likely. Besides awarding compensation to the claimant, the civil judge may also order remedying measures to the defendant.[51]

4.3 Criminal Law

Criminal law is the last, but not least, piece of the regulatory and legal framework of contaminated land. Several offences may apply in this field, such as misdemeanours (Sect. 4.3.1), water pollution (Sect. 4.3.2), and offences against persons (Sect. 4.3.3).

---

[50]English translation published on the website *Legifrance.fr*.

[51]Art. L 544 of the Civil Code.

4.3.1 Misdemeanours

When the operator of a facility submitted to the authorization regime (ICPE, water, or waste) fails to comply with legislative and administrative prescriptions, several misdemeanours apply under French law. Fines (up to 100,000 Euros) and imprisonment (up to two years) may apply when: (a) the facility is operated without the required authorization, registration, or declaration;[52] (b) the operator fails to comply with the obligation to inform the Préfet of substantial modification of technical and financial capacities;[53] (c) the operator fails to comply with the Préfet's notice to remediate and clean up the site;[54] (d) the operator fails to comply with the Préfet's notice to definitively close down a classified facility that has not operated for three consecutive years.[55] In all the above cases, criminal courts may also order, as complementary sanctions, the closure of the installation and remediation measures.[56]

When misdemeanours cause a threat to the health or the security of any person, or substantial damage to the environment, more severe penalties may apply: for example, the offence of failure to comply with the Préfet's order to remediate pollution may be punished by up to five years' imprisonment and a fine up to 300,000 Euros.[57] In this case, a concurrence of *misdemeanours* with the offence of water pollution and the offences against persons may arise.[58]

4.3.2 The Offence of Water Pollution

When a person, whether operator of an ICPE or not, fails to comply with the above-mentioned prescription and its action seriously degrades water, the offence of "Water Pollution" may also apply. This offence is defined by Article L 216-6 of the Environment Code as:

> The act of disposing of, discharging or letting flow into surface, underground or seawater within the limits of territorial boundaries, directly or indirectly, one or more substances of any kind whose actions or reactions cause, even if only temporarily, harmful effects on

---

[52]Art. L 173-1 of the Environment Code.

[53]Art. L 173-3 of the Environment Code.

[54]Art. L 173-2 of the Environment Code.

[55]Art. L 514-11 of the Environment Code.

[56]Pursuant to Art. L 173-7 of the Environment Code, when the above-mentioned offences are committed by a natural person, complementary sanctions may also include: (a) the publication of the decision; (b) the confiscation of the proceeds of the crime; (c) the prohibition from exercising professional activity related to the offence. Complementary sanctions applicable to legal persons are mentioned below in Sect. 8.

[57]Art. L 173-3 of the Environment Code. In this article, amended in 2012, the echo of Directive 2008/99/EC on the protection of the environment through criminal law is pretty clear.

[58]Environmental offences and offences against persons may apply also to the owner of the site in case the pollution is not caused by an ICPE.

> health, fauna and flora, […] or significant modifications to the normal regimen of water supply or limitations in the use of bathing waters.

The offence of water pollution includes the

> discharge or abandonment of waste in large quantities in surface or underground waters or in seawater within the boundaries of territorial limits, on beaches or in coastal areas.

The offence of water pollution is punishable by two years' imprisonment and a fine up to 150,000 Euros. The French legislator increased the penalties in 2012 in order to implement the EU Directive 2008/99/EC on the protection of the environment through criminal law.[59]

4.3.3 Offences Against Persons

Finally, common offences against persons may also apply when the pollution creates a significant risk of causing death or serious injury to any person. These are the main offences applying under the French Criminal law: (a) Contamination of drinking water; (b) risk caused to other persons; (c) involuntary injury and manslaughter.[60]

(a) The offence of *Contamination of drinking water* is defined by Article L 1324-4 of the *Code de la santé* as follows:

> The act of damaging public facilities destined to receive drinking water or the act of contaminating with substances susceptible to harm health springs, fountains, wells, tanks, pipes, aqueducts, reservoirs supplying public alimentation, is punishable with three years of imprisonment and a fine of 450,000 Euros.

(b) The offence of *Risk caused to other persons* is defined by Article L 223-1 of the Criminal Code as follows:

> The direct exposure of another person to an immediate risk of death or injury likely to cause mutilation or permanent disability by the manifestly deliberate violation of a specific obligation of safety or prudence imposed by any statute or regulation.

---

[59] *Ordonnance n°2012-34 du 11 janvier 2012 portant simplification, réforme et harmonisation des dispositions de police administrative et de police judiciaire du code de l'environnement.*

[60] The French Criminal Code provides also for the offence of *poisoning* (Art. L 221-5). Defined as "the act of endangering the life of another through the use or administration of substances by whose very nature may cause death," this offence is punishable by up to thirty years of criminal detention, in addition to life imprisonment when the offence is committed under certain circumstances. Despite its potential, poisoning is unlikely to be applicable to acts of pollution, which are in most cases involuntary. The Cour de Cassation in fact affirmed that "criminal poisoning can only be characterized if the author acted with the intention to cause death, a factor shared by poisoning and other crimes that intentionally threaten lives" (Cour de Cassation, Chambre Criminelle, 18 June 2003, case no. 02-85.199).

Specific to French criminal law, this offence is particularly suitable to punish acts of pollution causing risks to human health. However, the penalties provided (one year's imprisonment and a fine of €15,000) weaken its deterrent effect.

(c) When pollution causes serious injury to the physical or psychological integrity of a person or the death of any persons, the offence of involuntary injury (Arts. 222-7 *et seq.* of the *Code pénal*) or involuntary manslaughter (Art. 221-6 of the *Code pénal*) may in principle apply. However, and whatever the potential qualification, the challenge shared by all these offences resides in the need to prove a definite causal link between the act of polluting and the threat to life or health: these offences are therefore unlikely to apply in the environmental field where causation between the degradation of the environment and harm to human wellbeing is still not subject to scientific consensus (Saint-Pau 2013).

With the French general legal framework presented, the discussion on historical pollution may now be completed through the analysis of the different liability regimes—regulatory, civil, and criminal—applying to concrete cases of historical pollution as well as to the hypothetical scenarios envisaged during the research. The analysis will consequently follow the three trajectories, different but complementary, already traced in this paragraph: remedying, compensating, and punishing historical pollution.

5 Remedying Historical Pollution

The main regulatory tool provided by the French legal system in order to deal with the remediation of the contaminated site is the "authorization regime" provided by Environmental Law for hazardous facilities: this regime provides the main clean-up obligations and enforcement measures applying in case of pollution of soils and water (Sect. 5.1). As anticipated,[61] in 2014, the regimes applying to the different hazardous facilities (ICPE, water, and waste) were harmonized and the hierarchy of persons liable for remediation specified (Sect. 5.2). The description of the relevant provisions and regulatory practices on remediation has to be finally enriched by reference to case law: administrative courts have indeed specified the assignment of regulatory liabilities in case of transactions involving contaminated lands, which is a topical question for our discussion on the assignment of liability for historical pollution (Sect. 5.3).

5.1 Remedial Obligations and Enforcement Measures

As discussed above, the explicit obligation to remediate pollution produced by an industrial facility was first introduced in 1976 under the ICPE regulation: pursuant

[61]See above, Sect. 4.1.1.

to the existing Article L 512-6-1 of the Environment Code, when a facility subject to an authorization regime is definitively shut down, the operator has the legal obligation to clean up the site and leave it in a state which does not represent a nuisance or a danger to environment or human health.[62] After the 2003 amendment to the ICPE regime, remedial obligations may be imposed even before the facility's shutdown: pursuant to the new Article L 512-20 of the Environment Code, "complementary remedial measures" may be imposed by the Préfet at any time during the operation, especially when the operator declares modifications or extensions of the facility or when an accident or an incident occurs on the site. Identical remedial obligations are provided under the water[63] and waste regime:[64] remedial obligation may be imposed on the operator when the operation is shut down and at any moment if a threat to the environment or human health is proven.[65] The content of the remedial obligation is not duly defined by law: it is generally set out by the local administrative authorities on a case-by-case basis.[66]

If the operator of the contaminated site does not voluntarily comply with the remedial obligations, the Préfet may enforce them. In this case, the Préfet will first of all issue a notice (*mise en demeure*) at the destination to the non-compliant operator. If he or she does not comply with that notice, the Préfet may (a) require that the operator deposits with a public accountant an amount of money corresponding to the cost of the remediation to be carried out. This sum will be returned to the operator gradually as the required clean-up measures are performed (*consignation*). (b) Carry out the clean-up of the site at the expense of the operator (*travaux d'office*). (c) Suspend the operation of the ICPE until the clean-up has been carried out (*suspension du fonctionnement de l'installation*). (d) Impose an administrative fine of 15,000 Euros and a daily *astreinte* of 1500 Euros.[67]

---

[62]The Conseil d'Etat held that the clean-up obligation arising from the 1976 regime applies also to the operator of sites which, albeit already having closed, would have been submitted to the ICPE Regime (Conseil d'Etat, 20 March 1991, *Rodanet*, application no. 83.776; Conseil d'Etat, 16 November 1998, *Compagnie des bases lubrifiantes*, application no. 182816).

[63]See Arts. L 214-1 *et seq.* of the Environment Code. Under the water regime, in case the operator is not identifiable, the owner of the contaminated land may also be liable for remediation.

[64]See Art. L 541-3 of the Environment Code. As we will discuss below, under the waste regime, the persons liable for remediation are the "waste producer" and the "waste holder".

[65]There are no regulatory or legal value limits for contaminants in soil or groundwater. However, as we underlined above, several technical guidelines have been issued by the Ministry of the Environment in the context of the "National Methodology".

[66]An exception was introduced in 2012 in the *Ordonnance 2012-7 du 5 janvier 2012* transposing Directive IED: according to European legislation the installations classified as IED installations may present a "baseline report" either at the moment of the issue of the authorization or at the moment of the first control provided during the operation. The baseline report fixes the t_0 of the operation and the condition for the rehabilitation of the site (see Arts. L 515-28 *et seq.* of the Environment Code).

[67]Art. L 171-8 of the Environment Code.

Enforcement measures shall be applied by the Préfet within a period of thirty years from the closure of the site[68] and may be challenged before administrative courts. As discussed below, administrative measures and fines do not prevent the application of criminal sanctions.[69]

5.2 *The Hierarchy of Persons Liable for Remediation*

After being addressed by soft law regulation and very often incoherent case law (Boivin and Defradas 2013), the hierarchy of persons liable for remediation has been finally specified by the 2014 legislature. These are: the last operator (Sect. 5.2.1), the site owner (Sect. 5.2.2) and, as a last resort, a Public Fund (Sect. 5.2.3).

5.2.1 The Last Operator

Under the French system, the *last operator* of a hazardous facility (one subject to the ICPE, water, or waste regime) is primarily liable for the remediation of pollution. The *last operator* is considered to be the natural or legal person who holds the administrative authorization to operate the facility (*de jure operator*).[70] In accordance with case law, even an irregular operator (de facto *operator*) of a hazardous facility may be considered liable for remediation: pursuant to the principle *nemo auditur propriam turpitudinem allegans*, an irregular operator of a hazardous facility cannot rely on the absence of the permit to escape their obligations.[71] When both *de jure* and de facto operators manage the same facility, administrative judges tend to impose remedial liability on the *de jure* operator.[72]

When the operator is a legal person, remedial liability cannot be placed on the company's shareholders. This is a corollary of the principle of legal personality, which is considered a fundamental principle of French Corporate Law. However, the case law has developed three exceptions to this principle (Boivin and Defradas 2013, p. 110). When the operating company does not comply with the remedial obligations, the shareholders may be liable in the following cases: (1) when they actually control the facility (*Théorie de l'immixtion*); (2) when the operating company was created just to deal with the closure of the facility (*Théorie de la fictivité*); (3) when the operating company was dissolved with the aim of escaping the remedial obligation (*Theorie de la fraude des actionnaires*). With the aim of

[68]Conseil d'Etat, 8 July 2005, *Alusuisse-Lonz-France*, application no. 247.976.

[69]See below, Sect. 7.

[70]Conseil d'Etat, 29 March 2010, *Communauté de communes de Fécamp*, application no. 318.886.

[71]Conseil d'Etat, 21 February 1997, *Wattelez*, application no. 16.078.

[72]Conseil d'Etat, 6 December 2012, *Sté Arcelomittal France*, application no. 333.977.

tempering the effects of the principle of legal personality and allowing the piercing of the corporate veil in case of fraud, an amendment adopted in 2010 provided the possibility of imposing remedial liability upon the parent company when it has deliberately contributed to the bankruptcy of the subsidiary operating company.[73]

As anticipated, in order to encourage the rehabilitation of brownfields, the 2014 regulation on contaminated sites allows the transfer of the remedial obligation from the last operator to a third party, such as private redevelopment companies.[74] The transfer of the remedial obligation is taken over under the control of the Préfet who defines the conditions of the transfer and authorizes it. A decree adopted in 2015 subsequently specified the terms and conditions of this new proceeding: the third party will have to provide financial guarantees matching the cost of the rehabilitation. Should the third party default, the administration will use this amount to complete the rehabilitation of the site. Moreover, the last operator remains financially liable for the remediation of the polluted site should the above-mentioned guarantees turn out for any reason to be impossible to levy.[75]

5.2.2 The Site Owner

The question of whether the owner of a site where a hazardous facility is operated may be considered liable for clean-up has been raised in administrative case law. After fifteen years of conflicting judicial decisions, on the one hand the Conseil d'Etat specified in 2005 that neither the owner nor the holder of an ICPE site can be assigned remedial obligations.[76] On the other hand, the site owner's liability was implemented under the waste and water law regimes. Article L 541-3 of the Environment Code imposes the obligation to remove the waste illegally left on a site on the "waste producer" and the "waste holder".[77] Administrative judges have progressively taken the "waste holder" to be the owner of the site where the waste was illegally collected. Superior jurisdictions later specified the conditions for engaging the site owner's liability: firstly, other "waste holders" are not to be identified;[78] secondly, the site owner must have shown some kind of negligence vis-à-vis the collection of waste on his parcel.[79]

---

[73]Art. L 512-17 of the Environment Code.

[74]Art. L 512-21 of the Environment Code.

[75]See *Décret n°2015-1004 du 18 août 2015 portant application de l'article L 512-21 du Code de l'environnement.*

[76]Conseil d'Etat, 8 July 2005, *Alusuisse-Lonz-France*, application no. 247.976.

[77]This article implements the waste regime provided for in Directive 2008/98/EC of 19 November 2008 on waste.

[78]Conseil d'Etat, 1 March 2013, applications nos. 348.912/354.188.

[79]Conseil d'Etat, 26 July 2011, *Wattelez II*, applications nos. 328.651. The Cour de Cassation (11 July 2012, case no. 11-10.478) confirmed the principle of the subsidiary liability of the owner of a site where waste has been illegally collected. However, its position appears more severe to the extent it allocates to the site owner the burden of proof of its diligence.

After having been extended to the water law regime,[80] the principle of the subsidiary remediation liability of the "negligent owner" has been generalized as a fundamental principle of the new contaminated site regime. Pursuant to the new Article L 556-3 of the Environment Code, the owner of a site subject to the ICPE, environmental liability, waste, or water regime may be liable for remediation when he or she "was negligent or involved in the pollution".[81]

Consequently, in **Hypothetical Scenario 1**,[82] the owner of a contaminated site may be subject to the legal obligation to remediate the contamination of its parcel when the following conditions are fulfilled: (a) the contamination arises from a hazardous facility subject to one of the above regimes; (b) the facility has been definitively shut down; (c) the last operator cannot be found or is insolvent; and (d) the owner has been negligent or complacent vis-à-vis the contamination. If the above conditions are fulfilled, Agent A may incur regulatory liability for the remediation of the pollution caused by the operator (and criminal liability in case of failure to comply).

The hierarchy of the persons who may be liable for the remediation of a contaminated site is now expressly established under Article L 556-3 of the Environment Code:

1. Firstly, the "last operator" (*de jure* or de facto) of a hazardous facility subject to one of the above regimes.
2. A third party to whom the operator properly transferred the remedial obligation.
3. In case of waste pollution, the producer or the holder of the waste.
4. The owner of the site who was negligent or otherwise involved in the contamination.

5.2.3 Public Fund

Finally, when any of the liable persons cannot be found or is insolvent, the site is defined as an "orphan site": in this case, the remediation of the pollution and the rehabilitation of the site are to be carried out by the French Environment and Energy Management Agency (ADEME).[83] After the rehabilitation, the contaminated site becomes public property.

---

[80]Art. L 214-3-1 of the Environment Code.

[81]Our translation of the French (*s'il est démontré qu'il a fait preuve de négligence ou qu'il n'est pas étranger à cette pollution*).

[82]Hypothetical Scenario 1: Agent A owns land and contaminates it over a long period of time. No damage nor threats to human health occur as consequences of the pollution.

[83]Art. L 556-3 of the Environment Code. See also the Ministry of the Environment Circulars of 26 May 2011 and 8 February 2007.

5.3 Acquisition of Contaminated Sites and Regulatory Liability for Remediation

In France, the topical question as to whether the last operator may be liable for the remediation of pollution caused by operators who previously managed the same contaminated site has been raised by case law. This question has been raised in the very common cases of acquisition of productive sites where hazardous facilities, subject to the ICPE or other authorization regime, have been operated.

In the mid-eighties, administrative courts established some criteria in the field of ICPE that may be considered applicable to all cases where a facility subject to an authorization regime is transferred. According to this case law, in order to establish which operator is liable for "historical pollution," it is necessary to distinguish between two situations: (a) the "substitution" of different operators within the same activity; (b) the "succession" of operators performing different activities on the same site.[84]

(a) There is "substitution of operators within the same activity" when a new operator asks for a transfer of the authorization to his name. In this case, the old operator "disappears" in the eyes of the administration and the new operator will consequently be liable for the remediation of the pollution caused both by its own and past activities. In other worlds, *in case of substitution of operators within the same activity, the new operator will be liable for remediating historical pollution*. However, this principle is subject to some exceptions: when the new operator's substitution is irregular (and *a fortiori* when there is no substitution at all), both the new and the previous operators may be liable for the clean-up of the site: while the new (de facto) operator may be also liable for the remediation of the pollution caused by the old (*de jure*) operator, the latter may also be liable for the remediation of the pollution caused by the new operator.[85]
(b) In accordance with case law, when a new operator undertakes a different activity than that of the old operator on the same site, there is not—by definition—substitution in the operation but rather a "succession of different operators on the site". In accordance with this principle, each operator will be liable for the remediation of the pollution caused by its own activity. In the case of succession of different activities within the same site, the public authority shall

[84]This distinction was proposed for the first time by the Conseil d'Etat in *Société des Produits Chimiques Ugine-Kuhlmann, 11 April 1986. See recently* Conseil d'Etat, 8 July 2005, *Alusuisse Lonza France*, application no. 247.976 and Conseil d'Etat, 23 March 2011, *SA Progalva*, application no. 325.618.

[85]Boivin and Defradas (2013, p. 116) highlight that administrative judges tend to limit the principle of common liability sometimes to a charge against the de facto operator (see, for example, Conseil d'Etat, 21 February 1997, *Wattelez*, application no. 160.787), and sometimes to the charge against the *de jure* operator (see, for example, Conseil d'Etat, 6 December 2012, *Arcelormittal France*, application no. 333.977).

consequently address specific remedial prescriptions to each operator, in strict accordance with the "polluter-pays" principle. However, as Boivin and Defradas (2013, p. 113) highlight, *it is in practice very hard or even impossible to "split" the pollution caused by each operator and to individualize remediation prescriptions*. Consequently, the trend among national public authorities is to impose obligation for remediation of historical pollution on the last operator, even in cases of succession.[86]

According to these principles, it is now possible to summarize how regulatory liability for remediating historical pollution may interfere with the most common schemes of acquisition of hazardous facilities and contaminated sites.

(a) In case of the mere *acquisition of shares* in a company operating a hazardous facility, the buyer will be liable for remediating historical pollution: the mere transfer of shares does not in fact involve any change of operator.
(b) In case of *merger or asset acquisition* of a company operating a hazardous facility, the buyer will in general be liable for remediating historical pollution (both in the respect of the *de jure* and de facto operator), except if he or she does not operate the facility or operates a different activity than that exercised by the merged or sold company within the same site.

Although liabilities in mergers or asset acquisition are often limited by warranties or indemnities issued to the benefit of the buyer, such private agreements are not enforceable against the Préfet in case of regulatory liability for remediating historical pollution. The buyer shall consequently comply with the remedial obligation. Under certain conditions, however, the buyer will be entitled to bring a claim against the seller to obtain compensation.

6 Compensating Historical Pollution: Contractual and Tort Law Liability

As discussed above, historical pollution can be also tackled under French law through Private law remedies, such as contractual (Sect. 6.1) and tort liability (Sect. 6.2).

6.1 Contractual Liability

French law provides specific legal warranties that may apply to the case of transfer of contaminated sites. Pursuant to Article L 541-20 of the Environment Code, when

[86]Cour Administrative d'Appel of Douai, 2 March 2006, *Societé BM Chimie*.

an ICPE has been operated on a parcel of land, the seller shall disclose specific information on the operation, on the risks and other main drawbacks thereof, as well as on the eventual storage of chemical or radioactive substances. If the seller does not comply with this legal obligation of disclosure, and the pollution makes the land unfit for the purpose specified in the purchase agreement, the buyer may request, within two years of the discovery of the pollution, either the cancellation of the sale or the reduction of the transfer price. The buyer may also demand the remediation of the pollution at the seller's expense, when the cost of this rehabilitation does not seem disproportionate compared to the transfer price.[87] For this action to be likely to succeed in the case of contaminated sites, three conditions need to be met: (a) the pollution must be anterior to the sale; (b) it must render the parcel unfit for the use which the purchaser intended; and (c) it must not have been revealed by the normal processes of investigation.

Hidden by the course of time and geological accumulations, it is axiomatic that such underlying conditions are met in cases of historical pollution: the seller's contractual liability may consequently be invoked. However, case law shows that the longstanding character of the contamination may also contribute to exempting the seller from contractual liability: the contamination may be so ancient that the seller may be considered as being *unaware of the vice* and, consequently, as *acting in good faith*. As the Cour de Cassation pointed out in a case concerning the transfer of a site affected by historical pollution:

> The contentious elements concerned the operation of the facility from 1903 to 1913, and it seems that they had been all but forgotten when EDF [the actual seller] purchased the land thirty-three years later. Forty-nine years after this acquisition, at the time of the contentious transfer, there was no evidence that EDF had been aware of the elements constituting the vice. In so far as the buyer was unable to establish elements that could implicate EDF, the case is to be necessarily dismissed.[88]

In accordance with these principles, we can now adapt **Hypothetical Scenario 2**[89] to French legal practice. In this scenario, Agent A will not incur any liability for the sole fact of purchasing contaminated land. However, if an ICPE had been operated on the parcel purchased, and the buyer replacing the former operator performs the same activity as the seller, Agent B may incur regulatory liability for the remediation of historical pollution (and criminal liability in case of failure to comply). However, Agent B will be entitled to bring a civil claim against Agent A if the latter did not comply with the legal obligation to disclose any significant risk or inconveniences arising from the operation, and historical pollution is identified

[87]See Art. L 514-20 of the Environment Code, as amended by Law no. 2014-366 of 24 March 2014.

[88]Cour de Cassation, 3rd Chambre civile, 4 November 2004, case n 03-12.499. This is the only case listed in the website *Legifrance.fr* in which the term "pollution historique" is used by civil judges in the sense accepted by this research (accessed in June 2016).

[89]Hypothetical Scenario 2: Agent A owns land and contaminates it. The land is subsequently purchased by Agent B. No damage nor threats to human health occur as consequences of the pollution.

which makes the land unsuitable for its intended use. In this case, and within a period of two years following the discovery of the contamination, Agent B may claim a partial refund of the transfer price, a cancellation of the sale, or the remediation of the land at the seller's expense if the cost of such remediation is not disproportionate to the purchase price.

6.2 Tort Law Liability

By providing that the authorization to operate a hazardous facility "is granted subject to the rights of third parties," Article L 514-19 of the Environment Code enshrines the principle that the operator is not absolved of Tort Law Liability.

According to the major case law, ground or water pollution can be classified as an "abnormal neighbourhood nuisance" (*troubles anormaux du voisinage*). This theory, whose origins date back to ancient case law, provides for liability for anyone who makes nuisances in excess of the normal neighbourhood inconveniences. After conflicting decisions, the Cour de Cassation specified that this is a case of *strict liability*: therefore, there is no requirement to prove fault.[90] It has therefore been very easy for civil judges to rely on the abnormal neighbourhood nuisance theory to impose civil liability for industrial pollution, whether or not caused by an ICPE operator (Boivin 2010, pp. 169–170).

The historical nature of the pollution does not seem sufficient to prevent the civil liability of the current operator or site owner: neither ignorance of the historical pollution nor the fact that it arises from a previous activity can in fact block strict liability. To impose civil liability for historical pollution, it is sufficient to prove the damage. Even the principle of "individual preoccupation" or "anteriority" provided for by Article L112-16 of the *Code de la Construction* seems unable to block civil liability for historical pollution. According to this principle—which was initially intended to prevent farmers from being sued by the owners of country homes and was subsequently extended to all productive activity—the person who settles in the neighbourhood of an existing productive activity cannot claim any damages under the abnormal neighbour nuisance theory. However, the principle of individual preoccupation applies only provided the pollutant activity is led "in conformity with the legislative and regulatory measures in force" and "at the same conditions".[91] This means that the principle of preoccupation does not prevent the civil judge from taking into account whether the site was operated in compliance with the applicable regulations and whether relevant changes occurred in the operation, leading to a significant increase in nuisance or to the creation of new ones. The legislation is so

---

[90]The Cour de Cassation recognized that the strict liability does not constitute a disproportionate breach of property law as protected by Art. 1 of the First Additional Protocol to the European Convention for the Protection of Human Rights. See Cour de Cassation, 3rd Chambre civile, 23 October 2003, case no. 02-16.303.

[91]Art. L 112-16 *Code de la construction et de l'habitation*.

unclear and the qualitative and quantitative traceability of an industrial activity is so hard to ensure, that is rather hard for an operator to rely on the principle of preoccupation to escape strict civil liability for historical pollution (Boivin 2010, p. 183).

Under the French Civil Code, those entitled to file proceedings and claim for damages are those harmed in some way by the pollution found on a site, including recognized environmental protection NGOs. Such tort actions must be made within five years of the discovery of the damage.

Whether liable or not under Tort Law for damages caused by historical pollution, the operator of a contaminated site who does not comply with the remedial obligations may also incur criminal liability.

7 Punishing Historical Pollution

Criminal law plays a central role in dealing with historical pollution: failing to comply with a notice imposing remediation of historical pollution may constitute a misdemeanour and, in case of damage to elements of the environment, an environmental offence (Sect. 7.1). When the historical contamination also creates a threat to human health, common offences against persons may also apply (Sect. 7.2).

7.1 Misdemeanours and Environmental Offences

Since criminal liability is clearly shaped by regulatory liability, criminal courts—as administrative courts—generally consider the *last operator* criminally liable for historical pollution. In so far as he is supposed to have "substituted" the old operators in the full knowledge of the existence of historical pollution, and being in any case obliged to remediate it, criminal courts may impose criminal liability upon the last operator for failure to comply with legislative or administrative prescriptions concerning the remediation of historical pollution. If he fails to comply with the remediation obligation, the last operator may thus incur both regulatory *and* criminal liability for historical pollution. This approach likely moves towards vicarious liability. In accordance with case law, the last operator in fact may not only incur criminal liability if he fails to comply with historical pollution remediation orders *which he is the recipient of*; he may incur criminal liability even if he fails to comply with the *remedial orders addressed to the old operator*—in this case, under the purchase agreement the last operator would have agreed to implement the remediation measures imposed by the Préfet on the old operator, who has henceforth "disappeared".[92] While private agreements cannot be opposed by the last operator as a defence against regulatory or criminal liability for historical pollution, courts rely on private agreements as a foundation for the last operator's liability.

On this basis, criminal liability for historical pollution may also be imposed on the de facto operator, i.e. the legal or natural person who operates the facility without formally replacing the old operator. In a case decided in 2015, the Criminal Chamber of the Cour de Cassation imposed criminal liability for (operation without authorization) and environmental offences (water pollution) committed by the previous (*de jure*) operator of the facility, on the legal person performing an ICPE on the basis of a contract for service.[93] Criminal judges went very far in this case: they described the de facto operator as *the co-perpetrator of the offences committed by the formal operator of the facility*. The arguments used by the Court are interesting for our discussion. The Court did not deny that the *actus reus* of the offences was entirely realized by the last operator: on the one hand, it operated the facility without the required authorizations; on the other hand, it discharged into the water table 240,000 m^3 of waste which polluted a watercourse adjacent to the facility. However, the court considered the de facto operator of the facility liable for both offences in so far as he

> left to pursue a classified facility without the required authorization [...] (and) continued deliberately to operate the facility being aware that the operation led to the ejection of (pollutant) substances into the watercourse.

According to the Court, the de facto operator had all the technical and juridical means to require the *de jure* operator to obtain the necessary authorizations and prevent and remediate the pollution; but he failed to do so, and, consequently, may be considered the co-perpetrator of the offences. This decision is both surprising and interesting. *Surprising* because it restricts the scope of the defence of due diligence against criminal liability under environmental law: in order to avoid criminal liability, the operator shall not only prove that he did everything possible to prevent the offence; he shall also prove that he used his leverage to avoid or minimize the environmental offences of other actors with whom he has relationships, especially contractual, regardless of whether he materially contributes to the offence. But this decision is *interesting*, too, because it echoes the concepts of "due diligence" and "sphere of influence" which have been introduced into international discourse and soft law in order to encourage corporations to take into account their impact—and

---

[92]Cour de Cassation, Chambre criminelle, 1 April 2008, case no. 07-86096.

[93]Cour de Cassation, Chambre criminelle, 13 January 2015, case no. 13-88183. As Boivin and Defradas (2013, p. 225) highlight, criminal courts assign to the company directors a presumption of negligence in case of failure to comply with environment, hygiene, or security legislation.

leverage-based responsibility.[94] In accordance with these concepts, corporations not only have *negative responsibility* couched in terms of avoiding, preventing, minimizing, or eliminating harmful social and environmental effects; they also have *positive responsibility* in terms of supporting, promoting, improving, fulfilling, or otherwise contributing to the protection of society and the environment . As Pereira (2014, p. 66) has highlighted concerning human rights violations, the concepts of "due diligence" and "sphere of influence" may have an impact on the legal concept of complicity: a term which under national (and international) law designates the joint commission (co-perpetration) or the facilitation of the offence (aiding and abetting), *complicity* may be extended to corporations' failure to comply with the positive responsibility to protect society and the environment. Even though the approach of the French criminal courts fragrantly violates the *nullum crimen sine lege* principle, it confirms the trend of judicial hybridization of criminal liability with concepts issuing from corporate social responsibility in order to strengthen the legal protection of the environment (d'Ambrosio and Tricot 2016).

In accordance with this case law, we may conclude that in **Hypothetical Scenario 3**,[95] Agent B is likely to incur criminal liability both for aggravated regulatory offences and water pollution.

7.2 Offences Against Persons

In the case that the pollution causes risks to human health, the offence of "risks caused to other persons" (*mise en danger d'autrui*) may also apply. Pursuant to Article 223-1 of the Criminal Code:

> The direct exposure of another person to an immediate risk of death or injury likely to cause mutilation or permanent disability by the manifestly deliberate violation of a specific obligation of safety or prudence imposed by any statute or regulation is punished by one year's imprisonment and a fine of €15,000.

We have not found any case law where this offence has been applied to a hypothesis of historical pollution. However, we can affirm that the offence of risks caused to other persons may apply to cases of historical pollution. The Chambre criminelle of the Cour de Cassation stated in the above-mentioned *Metal Blanc* case that the *mens rea* of the offence requires that the ICPE operator fails to comply with several remedial orders. According to the Court:

---

[94]See Global Compact and ISO 26000.

[95]Hypothetical Scenario 3: Agent A owns land and contaminates it. The polluted land is subsequently purchased by Agent B. In the meantime, land contamination results in contamination of groundwater which is regularly used in agriculture.

> The multiplicity of failures to comply and the necessity for the administration to apply administrative measures to enforce the administrative prescription underlying the authorization, characterize the manifestly deliberate violation of the prescriptions. Consequently, the accused corporation is to be found guilty.[96]

In so far as remediation orders may—and in most cases do—apply to historical pollution, the failure to remediate in accordance with such orders may be considered as a manifestly deliberate violation of a specific obligation concerning safety and prudence.

The offence may also be committed in case the last operator fails to remediate historical pollution. Like the "original" ejection of polluting substances, the failure to comply with the remediation order may expose persons to the "immediate risk of death or injury likely to cause mutilation or permanent disability". The proof of this element is usually given, on a case-by-case basis, through scientific expertise.[97] However, the expertise is not necessary when the exposure's impact on human health is already known. This "presumption" can nevertheless be reversed when the absence of an actual measurable risk on the public has been established. For instance, the operators of a refinery that caused sulphur dioxide air pollution above the population alert level were found not guilty on the basis of a toxicological study that found no measurable impact of the incident on the population.[98] It follows that the term "immediate risk" characterizing the *actus reus* of the offence relates less to the temporal lag between the acts of pollution and the harmful effects, than to the scientific consensus that the action (or the inaction) of the operator created a risk to human health. It is, however, still to be established whether one year's imprisonment and 15,000 Euros fine (75,000 Euros for corporations) are penalties proportional to the offence, especially when it is committed by economic actors who are fully aware of the health and safety risks caused by their actions or inactions (Rousseau 2013).

In **Hypothetical Scenario 4**,[99] when failing to comply with remedial orders, Agent B may be liable for both aggravated *misdemeanours* and environmental offences. Agent B may be also liable for the offences of "risk caused to another person" and "contamination of drinking water".

As mentioned above, when pollution (historical or not) becomes a real threat to human life or health, the offences of involuntary bodily harm and manslaughter may in principle apply. However, technical obstacles concerning proof of causation prevent the application of these offences to cases of contamination. In accordance with French legal literature and practice, criminal causation has to be established under the *equivalency of conditions* theory: a relation of necessity (even indirect

---

[96]Cour de Cassation, Chambre criminelle, 21 September 2010, case no. 09-86.258. See also Cour de Cassation, Chambre criminelle, 5 April 2011, case no. 09-83.277.

[97]Cour de Cassation, Chambre criminelle, 6 October 2009, case no. 09-81.037.

[98]Cour de Cassation, Chambre criminelle, 4 October 2005, case no. 04-87654.

[99]Hypothetical Scenario 4: Agent A owns land and contaminates it. The polluted land is subsequently purchased by Agent B. In the meantime, land contamination results in contamination of groundwater which is regularly used in human nutrition.

and/or partial) must therefore exist between the conduct and the result (Saint-Pau 2012, p. 679). French criminal courts are very strict about the establishment of legal causation and refuse to rely on any form of scientific assertions (population-based, probabilistic, etc.). This means that the act of polluting, while positively proven, is not on its own enough to conclude the aforementioned offences, as long as uncertainty exists with regard to the actual consequences affecting the life and health of identified individuals. Hence, in the current state of scientific knowledge, involuntary offences against persons are unlikely to be applied in the field of anthropogenic pollution (Saint-Pau 2013).

This is not the case of the **Hypothetical Scenario 5**,[100] where there is evidence of a causal link between the pollution and the death or injury of a number of people: in the case Agent B is likely to be condemned for involuntary bodily harm or manslaughter under French law.

8 Corporate Criminal Liability

Corporate criminal liability was first introduced in the French legal order by the Criminal Code of 1994. Initially applicable to a limited number of offences, corporate criminal liability has been extended to all offences as from 31 December 2005.[101]

Consequently, a legal entity may be criminally liable for pollution-related offences committed on its behalf both by its legal representatives and by employees acting on behalf of the company through an express power of attorney (*délégation de pouvoir*). Article 121-2 of the Criminal Code applies a regime of liability to legal entities based on criteria of *organic identification* or of liability by *ricochet*. There are three main consequences to this regime of liability. First, the elements constituting the offence must be considered in relation to the physical person acting on behalf of the legal entity and in relation to the legal entity itself.[102] Second, the existence of a distinct fault committed by the legal entity is not required.[103] Third, while the existence of adequate compliance procedures and control systems may be taken into account by the courts in considering the context of the offence, they do

[100]Hypothetical Scenario 5: Agent A owns land and contaminates it. The polluted land is subsequently purchased by Agent B. There is evidence of a causal link between the pollution and the death or injury of a number of people.

[101]As of 31 December 2005, the Perben II Law of 9 March 2004, concerning amendments in line with changes in criminality, extended the criminal liability of legal entities to all offences, with the only exception being deliberate exclusion.

[102]See, in particular, Cour de Cassation, Chambre Criminelle, 4 December 2011, case no. 01-80.455, with regard to the offence of pollution provided for in Art. L 432-2 of the Environment Code.

[103]Cour de Cassation, Chambre Criminelle, 26 June 2001, case no. 00-83.466.

not of themselves constitute a defence or a mitigating factor in determining the penalty.

The maximum fine applicable to a corporate entity is five times the fine applicable to natural persons. For example, in case of failure to comply with the Préfet's order, a legal entity may be fined up to 375,000 Euros, as compared with a fine of up to 75,000 Euros for individuals. Where expressly provided by law, the following additional penalties may be imposed on legal entities: (1) dissolution, where the corporate entity was created to commit a felony, or, where the felony or misdemeanour carries a sentence of imprisonment of three years or more, where the corporate entity was diverted from its objectives in order to commit the crime; (2) prohibition from exercising, directly or indirectly, one or more social or professional activities, either permanently or for a maximum period of five years; (3) placement under judicial supervision for a maximum period of five years; (4) permanent closure or closure for up to five years of one or more of the premises of the company that were used to commit the offence in question; (5) disqualification from public tenders, either permanently or for a maximum period of five years; (6) disqualification from public tenders, either permanently or for a maximum period of five years; (7) prohibition on making a public appeal for funds, either permanently or for a maximum period of five years; (8) prohibition on drawing checks; (9) confiscation of the object which was used or intended to be used for the commission of the offence, or of the assets which are the product of it; and (10) the publication of the judgement.[104]

9 Concluding Remarks

To conclude the discussion in this chapter, it must first be pointed out that, in the French legal system, remediation, compensation, and punishment of historical pollution are mainly informed by Environmental Law, and more precisely by the "authorization regime" provided for hazardous facilities. This regime governs the operation of a large range of industrial facilities that may represent a nuisance or a danger to the environment and to human health. Hazardous facilities subject to this regime are operated under the control of the administrative authority, which may at any moment impose remedial actions when the operation causes threats to the environment or human health. Remedial actions are generally imposed on the *last operator* of the hazardous facility or on the third party who has been assigned the rehabilitation of the contaminated site. In case of the "organization" of the irresponsibility of the last operator, the operator's parent company or its shareholders may be also liable for remediation. A "negligent owner" of the site may have subsidiary liability for remediation where he was involved in the contamination and the operator "disappears". In case the persons liable for remediation do not comply

[104] Art. 131-39 of the Criminal Code.

with the remedial obligation, the administrative authority may enforce remediation and apply fines.

In accordance with regulatory and judicial practice, the last operator is generally liable for historical pollution, i.e. for the remediation of pollution caused by the old operators of a contaminated site. Historical pollution may be used as a defence against regulatory liability only in case the last operator performs a different activity on the contaminated site than that performed by previous operators. In this case, remediation actions are to be imposed in accordance with the polluter-pays principle. However, in practice it will be hard to "split" the different contaminations, and the result is that even in case of "succession" of different operators on the same site, regulatory liability for historical pollution is likely to be imposed on the last operator. The last operator will be also liable for tort law damages claimed by third parties under the strict liability regime applied by civil judges. Warranties against historical pollution, generally agreed to the benefit of the last operator, cannot be used as a defence either against regulatory liability or tort law liability. The last operator may only claim damages against the seller if he did not comply with the obligations of disclosure pertaining to the contamination of the site.

The regime of regulatory liability for historical pollution shapes criminal liability. Many misdemeanours are in fact provided under environmental law in case of failure to comply with legislative or administrative prescriptions: the result is that the last operator—whether a natural or legal person—when not complying with a remediation order (concerning historical pollution) incurs criminal liability for historical pollution. Misdemeanours may be aggravated when historical pollution causes risks for or damage to the environment or human health. In this case, environmental offences or offences to persons may apply as well. The risk of applying vicarious liability in the field of historical pollution is consequently high. The case law is too recent for a definitive assessment of the matter. However, a trend appears: that of increasingly shaping criminal liability according to regulatory liability, and hybridizing the respective criteria for assigning personal liabilities. If this trend is confirmed, we can conclude that in France the assignment of liability for historical pollution may be as effective as the risk of vicarious liability is high.

References

Alexandre, A. (2012). *Risques environnementaux. Approches juridique et assurantielle Europe et Amérique du Nord*. Bruxelles: Bruylant.

Association Robin des Bois. (2016). *Atlas. La France Toxique*. Arthaud: Paris.

Aligon, D., & Douay, F. (2011). *Site Atelier Metaleurop. Synthèse des travaux de recherche réalisés autour de l'ancienne fonderie de Noyelle-Godaut.* Rapport ADEME.

Asselaim, J. -C. (2011). *Histoire économique de la France du XVIIIe siècle à nos jours. 2. Depuis 1918*. Paris: Seuil.

Bausinger, T., Bonnaire, E., & Preub, J. (2007). Exposure assessment of a burning ground for chemical ammunition on the Great War battlefields of Verdun. *Science of the Total Environment, 382,* 259–271.

Boivin, J.-P. (2010). *Pratique du contentieux des installations classées et des carrières*. Paris: Éditions Le Moniteur.

Boivin, J.-P., & Defradas, F. (2013). *Sites et sols pollués: outils juridiques, méthodologiques et financiers*. Paris: Éditions Le Moniteur.

Callewaert, V. (2006) Les assurances contre les atteintes à l'environnement et ses alternatives: une réponse satisfaisante? In G. Viney, & B. Dubuisson (Eds.), *Les responsabilités environnementales dans l'espace européen*. Bruxelles-Paris: Bruylant-LGDJ.

Cans, C. (Eds.) (2009). *La responsabilité environnementale: Prévention, imputation, réparation*. Paris: Dalloz.

Cour de Cassation (2014). *Le temps*.

Crutzen, P. J. (2002). Geology of mankind. *Nature, 415*(6867), 23.

d'Ambrosio, L., & Tricot, J., (2016). Panorama en droit pénal. In K. Martin-Chenut, & R. De Quenaudon (Eds.), *La RSE saisi par le droit. Perspectives interne et internationale*. Paris: Pedone.

d'Ambrosio, L. (2016). L'affaire Eternit: quelles leçons? In K. Martin-Chenut & R. De Quenaudon (Eds.), *Développement durable: Mutations ou métamorphoses de la responsabilité?*. Paris: Pedone.

Decocq, O. (2010). Mission d'information et d'évaluation friches industrielles et pollution historique. *Rapport Friches industrielles et pollutions historiques*.

Fressoz, J.-B. (2012). *L'Apocalypse Joyeuse*. Paris: Seuil.

Guyton, K., et al. (2015). Carcinogenicity of tetrachlorvinphos, parathion, malathion, diazinon, and glyphosate. *The lancet Oncology, 16*, 490–491.

Laurian, L. (2008). Environnemental justice in France. *Journal of Environmental Planning and Management, 51*, 55–79.

Lascoumes, P. (2011). *Une démocratie corruptible. Arrangements, favoritisme et conflits d'intérêts*. Paris: Seuil.

Lienhard, A. (2003). Affaire Metaleurope: quand l'intégration des filiales d'un groupe confine-t-elle à la fictivité? *Recueil Dalloz*, 2571.

Massard-Guilbaud, G. (2010). *Histoire de la pollution industrielle en France, 1789–1914*. Paris: Éditions EHESS.

Mazade, O. (2013). «Patron voyou»: de la désignation publique à la sanction juridique. *Le cas Metaleurop. Champ pénal, 10*, 2013.

Mazade, O. (2010). *La reconversion des hommes et des territoires. Le cas Metaleurop*. Paris: L'Harmattan.

Napoli, P. (2003). *La naissance de la police moderne. Pouvoir, normes, société*. Paris: La Découverte.

Ost, F. (1999). *Le temps du droit*. Paris: Odile Jacob.

Pereira, B. (2014). Iso 26000, due diligence, sphère d'influence et droits de l'homme. *Revue de l'organisation responsable, 9*, 60–75.

Polanyi, K. (1944). *La Grande transformation*. Paris: Gallimard.

Rousseau, F. (2013). Essai d'une reconfiguration des infractions pénales contre les personnes à l'aune des «scandales sanitaires». *Droit pénal*, étude n°11.

Saint-Pau, J.-C. (2013). La répression de la destruction de biens environnementaux. In A. Gogorza & R. Ollard (Eds.), *La protection pénale de l'environnement*. Paris: Éditions Cujas.

Saint-Pau, J. -C. (2012). Les causalités dans la théorie de l'infraction. In *Mélanges en l'honneur du professeur Jacques-Henry Robert*. Paris: LexisNexis.

Thébaud-Mony, A. (2014). *La science asservie. Santé publique: les collusions mortifères entre industriels et chercheurs*. Paris: La Découverte.

Thébaud-Mony, A. (2008). *Travailler peut nuire gravement à votre santé*. Paris: La Découverte.

Viñuales, J. (2016). *Law and Anthropocene*, C-EENRG Working Papers, 2016-4 (pp. 1–72). Cambridge Centre for Environment, Energy and Natural Resource Governance, University of Cambridge.

Chapter 13
Historical Pollution and Human Rights Violations: Is There a Role for Criminal Law?

Donato Vozza

Contents

D. Vozza (✉)
Seconda Università di Napoli, Naples, Italy
e-mail: donato.vozza@gmail.com

F. Centonze and S. Manacorda (eds.), *Historical Pollution*,
DOI 10.1007/978-3-319-56937-6_13

1 The Relationship Between Historical Pollution and Human Rights Violations

Writing a study on the relationship between historical pollution and human rights violations from an international perspective is a particularly complex task for at least two reasons: first of all, a legal concept of "historical pollution" has not been developed in the case law of international and regional human rights bodies; furthermore, there is no research on this specific subject. Nevertheless, it must be highlighted that several cases decided by regional human rights bodies fall under the general definition of historical pollution elaborated for the purposes of this research.[1] In addition, the vast legal literature on the "greening" of human rights provides a theoretical basis for starting comprehensive research on the topic (Desgagné 1995; Shelton 1991, 2002b, 2006, 2007, 2008; Bosselmann 2001; Boyle and Anderson 1996; Jarvis and Sherlock 1999; DeMerieux 2001; Loucaides 2004; San José 2005; Fitzmaurice and Marshall 2007; Pedersen 2008, 2010; Francioni 2010; Anton and Shelton 2011; Boyle 2007, 2012; Zarsky 2012; Boer 2015). In particular, such a connection may be read in the light of a twofold right-based approach to environmental protection (Shelton 2007, p. 10).

As a first approach, environmental protection, as recognized by Principle 1 of the Declaration of the United Nations Conference on the Human Environment (known also as the Stockholm Declaration), is a "precondition for the enjoyment of other human rights" enshrined in current international and regional laws (Shelton 2007, p. 10; Boyle 2012, p. 617). Accordingly, people who have suffered serious physical or psychological problems or damage to their houses or property through environmental contamination may bring actions before human rights bodies with the goal of ascertaining the infringements of their rights (Shelton 2006, p. 131). Indeed, because the full enjoyment of fundamental rights, such as the rights to life, health, and property, may be limited by historical soil and groundwater contamination (Shelton 2007, p. 10), States should ensure, under human rights law, that the general condition of the environment does not deteriorate to the point of damaging individuals' rights (Shelton 2006, pp. 129 ff.). Therefore, human rights may be breached in the event that a State fails to take appropriate action to prevent or resolve cases of historical pollution that produce adverse effects on human health and property.

According to the second approach, procedural human rights serve to better protect the environment (Shelton 2007, pp. 10–11). Such an approach to environmental protection is based, *inter alia*, on Principle 10 of the Rio Declaration on Environment and Development, which declares that access to information, public participation in decision-making, and access to justice should be guaranteed in

[1]Concerning the definition of "historical pollution" see in this volume Chaps. 2 and 3.

environmental matters, and on Principle 17 of the same text, according to which competent public authorities must undertake an environmental impact assessment (EIA) for "proposed activities that are likely to have a significant adverse impact on the environment". From such a perspective, the protection of the environment and people from historical pollution should also be based on the effective safeguarding of procedural rights. Consequently, human rights violations may result from the State's failure to recognize these typologies of rights in scenarios of environmental contamination.

The pages that follow will pay attention to the provisions contained in supranational human rights law applicable to cases of historical pollution (Sect. 2) and to the main decisions and reports delivered by regional courts and bodies for the protection of human rights in cases of historical pollution (Sect. 3). The main goal of this work is to determine whether, and to what extent, an obligation arises from human rights to prevent and punish cases of death or personal injury in scenarios of historical contamination (Sects. 4, 5 and 6).

2 The Legal Framework: Preliminary Remarks

Many international treaties protect human rights from any type of violation, including historical pollution caused by the negligent actions of private or public companies that in the past have released toxic substances into the air, soil, or groundwater, and by persistent omissions on the part of public authorities in the safeguarding of human rights. Before dealing with the case law of the international and regional human rights bodies on historical pollution, attention must therefore be paid to the description of the substantive and procedural rights applied by these bodies (UNEP 2014, pp. 2 ff.; Shelton 2007, pp. 12 ff.).

2.1 Substantive Human Rights and Historical Pollution

Authoritative scholars agree that one of the main consequences arising from the fact that many agreements on the protection of human rights were adopted "before environmental protection became a matter of international concern" is that there are currently few explicit provisions relating to the safeguarding of the environment as a right in the existing supranational legal framework (Shelton 2007, p. 12). Taking into account these first legal data, the question arises as to whether and to what extent the existing human rights enshrined in international law may effectively be used to protect humans from the adverse effects of historical pollution. A logical second question is whether certain explicit provisions relating to the protection of the environment have been interpreted and applied by international and regional bodies in cases of historical pollution.

At the international level,[2] specific provisions on the right to health can be found in certain relevant agreements, such as the International Covenant on Economic, Social, and Cultural Rights (IESCR), which in Article 11 recognizes "the right of everyone to an adequate standard of living for himself and his family, including adequate food, clothing and housing", and in Article 12, Paragraph 1, protects "the right of everyone to the enjoyment of the highest attainable standard of physical and mental health", from which derives the obligation for the State to adopt the necessary actions for "the improvement of all aspects of environmental and industrial hygiene" (Shelton 2007, p. 12). According to General Comment No. 15 of the Committee on Economic, Social, and Cultural Rights (CESCR), Article 11 and 12 of the IESCR also represent the legal basis of the right to water, which "is indispensable for leading a life in human dignity. It is a prerequisite for the realization of other human rights".[3] These provisions, which impose upon States the duty to protect the purity of water for domestic use, have recently been invoked by the European Center for Constitutional and Human Rights, Multiwatch, Derechos Humanos sin Fronteras, CooperAcción, and inhabitants of Espinar (in Cusco, Peru), in a legal report submitted to the UN Special Rapporteur on the right to water and to the UN Working Group on transnational corporations with reference to a case of historical contamination. Their submission complained of a serious and manifest violation of the right to water of the inhabitants of Espinar as a result of historical groundwater pollution caused by a copper mine.[4]

At the regional level, at least three recent human rights treaties expressly recognize the protection of the right to the environment in addition to other rights, such as civil and political rights (i.e. the rights to life and property) and cultural and social rights (i.e. the rights to health, water, food, and culture): Article 24 of the African Charter on Human and Peoples' Rights (ACHPR) recognizes the right of peoples "to a general satisfactory environment favourable to their development"; Article 11 of the Additional Protocol to the American Convention on Human Rights in the Area of Economic, Social, and Cultural Rights, and Article 28 of the Arab Charter on Human Rights both protect the "right to a healthy environment"; last, but not least, Article 37 of the European Union Charter of Fundamental Rights (EUCFR) recognizes "environmental protection".

---

[2]For international and regional legal materials and cases on human rights and environment see, amongst others, UNEP 2014 and Council of Europe (2012).

[3]See United Nations, General Comment 15, 2003, "The Right to Water," The Economic and Social Council, E/C.12/2002/11.

[4]The UN Special Rapporteur on the right to water and the UN Working Group on transnational corporations, which were called upon to pronounce on the legal report concerning this case of historical pollution, are both part of the Special Procedures of the Human Rights Council—namely, independent experts designated by the Human Rights Council to study and report back respectively on the right to safe drinking water and on the issue of human rights and business in order to disseminate and implement the UN Guiding Principles on Business and Human Rights (see Roos 2013; Jagers 2011).

Despite the fact that many international agreements—such as the European Convention on Human Rights (ECHR)—do not explicitly encompass the right to a healthy environment, in the interpretation of regional bodies, environmental issues may be subsumed under the right to life, the right to health, the right to respect for private and family life, freedom of expression, freedom of assembly and association, the right to effective remedy, and the right to property, among others. These cases have been discussed before the European Court of Human Rights (ECtHR), the Inter-American Commission on Human Rights (IACommHR), and the African Commission on Human and Peoples' Rights (ACommHPR). The enjoyment of human rights presupposes a high-quality environment: it is unthinkable that people can live and survive without persistent problems or obstruction in an area that is heavily contaminated with toxic substances in the air, the soil, or the groundwater due to past manufacturing processes. International and regional human rights bodies have been called upon to decide on cases where applicants have complained of violations of their human rights due to contamination of the environment caused by industrial activities that began in the past and are still ongoing, or due to dangerous waste buried in old industrial areas or brownfield sites. In many cases, governments have been held responsible for their failure to take appropriate measures to protect victims from long-term pollution (see Sect. 3).

2.2 *Procedural Human Rights and Historical Pollution*

Public participation in decision-making processes on environmental issues plays a central role in the prevention and management of cases involving historical pollution. In recent decades, many States and the international community in general have become aware of the great potential of involving citizens in decision-making processes on environmental matters. That said, special consideration shall be given to international and regional instruments regulating these rights (Shelton 2006, pp. 132 ff.).

According to Principle 10 of the Rio Declaration on Environment and Development, several global and regional instruments recognize procedural rights, among them the UNECE Convention on Access to Information, Public Participation in Decision-making, and Access to Justice in Environmental Matters, also known as the Aarhus Convention, the most important legal tool adopted for strengthening procedural rights in environmental matters (Shelton 2007, pp. 15–16).[5] The Aarhus Convention should be considered as a treaty on human rights for

[5]The *UNECE Convention on Access to Information, Public Participation in Decision-Making, and Access to Justice in Environmental Matters* was adopted in Aarhus, Denmark, on 25 June 1998 and entered into force in 2001. Several instruments have been adopted in the context of the European Union for implementing this Convention. See Council Decision 2005/370/EC of 17 February 2005 on the conclusion, on behalf of the European Community, of the Convention on access to information, public participation in decision-making and access to justice in

three main reasons (Boyle 2012, p. 622). First, it recognizes procedural rights as functional in the protection of human rights. Second, it directly addresses individuals as rights holders. Third, the procedural rights of the Aarhus Convention have been incorporated into the case law developed by regional Courts and Commissions. That said, the right to access to information retained by public competent authorities on the state of environmental elements and the living conditions of people affected by environmental degradation is the first procedural right regulated in this agreement (Art. 2(3)). The State must ensure the accessibility and transparency of environmental information upon application of individuals, or *ex officio* when human rights violations may be involved. Furthermore, the agreement ensures public participation in decision-making processes relating to specific industrial activities and general policies on environmental matters. Under Article 6, the public concerned must be informed of the decision-making process relating to the authorization of the industrial activities listed in Annex I to the Aarhus Convention, which relates, *inter alia*, to the energy sector, the production and processing of metals, mineral industry, chemical industry, and waste management. In addition, this discipline is applicable to decisions on activities not listed in Annex I which may have a significant influence on the environment. The public must be made aware of the decision-making process on environmental matters in an appropriate way and be allowed to present opinions that public bodies must assess when they adopt a measure. The public concerned must be informed of the final measures taken. Lastly, Article 9 provides that access to justice must be granted if a request for information has been rejected in an unreasonable manner.

A further strategic instrument to prevent cases of historical pollution that can affect people's fundamental rights is the EIA of certain private and public activities. In accordance with Principle 17 of the Rio Declaration on Environment and Development, international legal instruments such as the Convention on Environmental Impact Assessment in a Transboundary Context and Directive 2011/92/EU on the assessment of the effects of certain public and private projects on the environment oblige States to have recourse to EIAs when adverse long-term effects on the environment are foreseeable.

Although substantive and procedural rights in environmental matters must be distinguished (UNEP 2014), it should be stressed that there is a close connection

---

(Footnote 5 continued)

environmental matters; Directive 2003/4/EC of the European Parliament and of the Council of 28 January 2003 on public access to environmental information and repealing Council Directive 90/313/EEC; Directive 2003/35/EC of the European Parliament and of the Council of 26 May 2003 providing for public participation in respect of the drawing up of certain plans and programmes relating to the environment and amending with regard to public participation and access to justice Council Directives 85/337/EEC and 96/61/EC; Regulation (EC) No. 1049/2001 of the European Parliament and of the Council of 30 May 2001 regarding public access to European Parliament, Council, and Commission documents; Regulation (EC) No. 1367/2006 of the European Parliament and of the Council of 6 September 2006 on the application of the provisions of the Aarhus Convention on access to information, public participation in decision-making, and access to justice in environmental matters to community institutions and bodies.

between them. This link is particularly evident in the light of the jurisprudence of regional human rights bodies. For instance, the ECtHR has recognized that infringements of the right to life (Art. 2) and of the right to private and family life (Art. 8) may arise when public authorities fail to inform people living in a contaminated area about the risks arising from their exposure to environmental pollution.[6] Furthermore, the IACommHR, in a chapter of its *1997 Report on the Situation of Human Rights in Ecuador*, concluded that the protection of the right to life and the physical integrity of the inhabitants of the Oriente—a region of Ecuador—affected by oil exploitation activities could be better achieved through safeguarding the effective exercise of procedural rights in environmental matters, namely the right to information (Art. 13), the right to participate in decision-making processes (Art. 23), and the right to an effective remedy (Art. 25).

In conclusion, the absence of recognition of procedural rights to the public in cases of historical pollution may entail an infringement of substantive fundamental rights.

3 The Jurisprudence of Regional Human Rights Bodies Relating to Historical Pollution

In the context of a wide variety of decisions in environmental matters (UNEP 2014, pp. 49 ff.; Council of Europe 2012), attention will be paid to a select group of judgments and reports delivered by regional human rights bodies in response to actions brought by victims of human rights, often with the support of non-governmental human rights organizations (NGOs),[7] against those states considered responsible for not having taken appropriate measures to protect the fundamental rights of victims from the adverse effects of historical pollution. More specifically, the focus will be on three groups of cases decided by the ACommHPR, the IACommHR, and the ECtHR.

A tiny number of cases decided by these bodies relate to the phenomena of historical pollution in a narrow sense, namely cases of contamination "caused by

[6]In the *Taşkın* case, the Strasbourg Court has interpreted the right to private and family life in the field of environmental protection (Art. 8) on the basis of the procedural rights recognized in the UNECE Aarhus Convention on Access to Information, Public Participation in Decision-making, and Access to Justice in Environmental Matters. By interpreting a human right in the light of the Aarhus Convention, the European Court indirectly binds States—such as Turkey—that have not ratified such an instrument to comply with the procedural rights in environmental matters. On this point, see Arato (2012, p. 371). Moreover, the infringement of the duty to have recourse to an EIA may constitute the legal basis for ascertaining the violation of certain substantive rights. On the duties of EIA see the *Giacomelli* case decided by the ECtHR. For a comment, see Krstic and Cuckovic (2015, pp. 176, 179), Kravchenko and Bonine (2012, p. 274). See, amongst others, IACommHR, *Community San Mateo de Huanchor and its members v. Peru Report* (Para. 66).

[7]On the role of the NGOs in the development of a "greening international jurisprudence," see Zengerling (2013).

past (often industrial) activities on a site" (Brandon 2013, p. 199). In these cases, environmental damage usually becomes apparent after a long latency period following the release of toxic substances into soil or groundwater. Likewise, health problems may develop slowly over a long period of time. Another feature of these cases is that polluting behaviours were often expressly authorized by a competent authority according to the domestic law at the time of the industrial releases, because they were not considered likely to cause environmental damage and human health problems on the basis of extant scientific information at the time when the activity was performed.

A second group of cases to be examined falls within a different paradigm concerning human rights violations that arise from past exposure of workers to toxic substances—for example, asbestos or nuclear radiation. In effect, past human exposure to toxic substances and historical pollution can be considered to share a number of corresponding issues: firstly, the damage to the environment and to human health is the result of the use of toxic substances in the context of past industrial activities; secondly, the damage to the environment and to health occurs after a latency period. Both paradigms give rise to "legacy problems", with the consequence that the legal issues that arise in these cases are similar. From this perspective, *Howald Moor and others v. Switzerland* and *Brincat and others v. Malta*, both decided by the ECtHR, deserve closer attention.

The third type of cases worthy of attention can be subsumed under the definition of historical pollution in a broad sense. These are cases of environmental pollution produced by industrial activity over time, often decades, resulting in damage to human health. This paradigm is very different from that of historical pollution in a strict sense, since the industrial activities producing environmental pollution are still ongoing. Therefore, historical pollution is often the outcome of a complex combination of past and new contamination. Cases such as *SERAC and CESR v. Nigeria*, *Dubetska and Others v. Ukraine*, *Fadeyeva v. Russia*, and *Smaltini v. Italy* decided by the ECtHR can be considered as examples of historical pollution in a broad sense.

3.1 The Decision of the African Commission on Human and Peoples' Rights in the Ogoni *Case*

One leading case is the landmark decision of the ACommHPR in *Social and Economic Rights Action Center (SERAC) and the Center for Economic and Social Rights (CESR) v. Nigeria* concerning Communication No. 155/1996 (Skogly 1997; Shelton 2002a; Wheeler et al. 2002; Coomans 2003; Nwobike 2005; Eweje 2006): this has been the first judgment of a human rights body on a case concerning the violation of the right to environment enshrined expressly in a regional instrument (Van der Linde and Louw 2003; Shelton 2007, p. 29) that has contributed to the development of the protection of human rights in the African continent (Coomans 2003; Nwobike 2005). Moreover, the African Commission, in this decision, has articulated for the first time the "duties of governments in Africa to monitor and control the activities of multinational corporations" (Shelton 2007, p. 29; Eaton 1997).[8]

In addition, this decision concerns an important case of historical pollution (in a broad sense), as is evident from the following analysis of the facts. In March 1996, two NGOs—namely SERAC and CESR—presented a petition to the African Commission in which they complained about the violation of human rights of people living in the Ogoni region of Nigeria, a large area polluted by oil extraction activities carried out over the course of many decades by petroleum companies. The communication alleged that these activities caused serious environmental and health problems and that the Nigerian Government failed to prevent this environmental degradation. Additionally, the petitioners affirmed that the government committed human rights abuses through military operations against people protesting about environmental degradation. Consequently, the communication alleged violations of the right to freedom from discrimination (Art. 2), the right to life (Art. 4), the right to property (Art. 14), the right to health (Art. 16), the right to housing (Art. 18(1)), the right to free disposal of wealth and natural resources (Art. 21), and the right to a general satisfactory environment (Art. 24).

After having declared the petition admissible, the ACommHPR first acknowledged four types of duties for a State in order to comply with the human rights established in the convention, namely "the duty to respect, protect, promote, and fulfil" them. With reference to the environmental matter, the Commission highlighted that:

> Government compliance with the spirit of Articles 16 and 24 of the African Charter must also include ordering or at least permitting independent scientific monitoring of threatened environments, requiring and publicising environmental and social impact studies prior to any major industrial development, undertaking appropriate monitoring and providing information to those communities exposed to hazardous materials and activities and providing meaningful opportunities for individuals to be heard and to participate in the development decisions affecting their communities. (Para. 53)

In line with these principles, the Commission examined, *inter alia*, the conduct of the Government of Nigeria in relation to the rights to health and to a general satisfactory environment. Having pointed out that, according to the European and American jurisprudence on human rights, public authorities should take measures to protect the rights of people from acts that may be perpetrated by private parties, it concluded that the Nigerian government had not protected the local people since it had "given the green light to private actors" (Para. 58) to destroy the environment

[8]Before the decision of the ACommHPR in the *Ogoni* case, Eaton (1997, pp. 264–265) highlighted that "foreign petroleum companies have pervaded the Niger Delta for over 38 years, their noxious products and wastes now polluting virtually all corners of the region".

and to damage the health of local communities exposed to the above serious contamination.

The government was required to cease all abuse of Ogoni communities, to prosecute public officials involved in human rights violations, to inform these communities about environmental and human health risks, to involve citizens in decision-making processes in environmental matters, and last, but not least, to:

> ensure adequate compensation to victims of the human rights violations, including relief and resettlement assistance to victims of government-sponsored raids, and undertaking a comprehensive clean-up of lands and rivers damaged by oil operations.

Despite the political importance of the decision, it should, however, be pointed out that the decision in the Ogoni case has not received adequate legal implementation.

3.2 *The Reports of the Inter-American Commission on Human Rights*

In the American system of human rights, the number of cases that victims and NGOs have submitted to the Inter-American Commission on Human Rights (IACommHR) with the aim of ascertaining the responsibilities of States for the infringements of fundamental rights abuses has progressively increased (Anicama 2008, p. 14). Initially, regional bodies especially emphasized the duty of the State to protect the right to land of the indigenous people recognized by Article 21 of the American Convention on Human Rights (ACHR) (Cuneo 2005), while latterly they also have recognized the importance of other fundamental rights—including the rights of the child—with reference to environmental matters (Scott 2000; Spieler 2010).

In such a context, the case of historical pollution addressed in the *1997 Report on the Situation of Human Rights in Ecuador* prepared by the IACommHR is very interesting. However, it is worth highlighting that other petitions have been held admissible by the IACommHR with regard to multiple violations of the rights of victims due to industrial emissions and state omissions. In some of these cases, moreover—such as *La Oroya Community v. Peru*[9] and *Community San Mateo de Huanchor and its members v. Peru*—the IACommHR has imposed on Peru the adoption of precautionary measures (health assistance and care programmes, environmental impact assessment studies, clean-up programmes), with the goal of protecting human rights in the context of complex scenarios of environmental degradation due to industrial activities.

---

[9]For a comment, see Spieler 2010.

3.2.1 The 1997 Report on the Situation of Human Rights in Ecuador

In Ecuador, contamination caused by oil exploitation activities during the second half of the last century is a serious historical pollution problem. In 1990, this issue was brought to the attention of the IACommHR by inhabitants and indigenous residents in the Oriente, a region of Ecuador. The applicants complained that oil exploitation activities carried out by state-owned and private oil companies in the Oriente since 1960 had contaminated water, soil, and air, to the detriment of health and environment. The case was handled by the IACommHR within a general country report on the situation of human rights in Ecuador of 1997, in particular in Chapter VIII, which concerned "The human rights situation of the inhabitants of the interior of Ecuador affected by development activities":

> *The exploitation of oil resources in the Oriente since the 1960 s* […] *has had a profound impact on the region and its people.* […] These inhabitants of oil development sectors have been unanimous in claiming that the operations generally, and the improper handling and disposal of toxic wastes in particular, have jeopardized their lives and health. They claim that oil exploitation activities taking place in or near their communities have contaminated the water they use for drinking, cooking and bathing, the soil they cultivate to produce their food, and the air they breathe. (emphasis added)

The inhabitants therefore argued that the Ecuador had failed to regulate and monitor oil exploitation activities. In this respect, the Commission highlighted that the government had undertaken certain legislative measures to address the activities of the oil sector and to prevent the violation of individual rights of people living in the Oriente. Among the various actions undertaken by the Ecuadorian government, one of the more interesting was a series of agreements it concluded with the oil company Texaco between 1994 and 1995 following its decision in the spring of 1992 to withdraw from its activities in the Oriente,[10] the aim of which was the clean-up of contaminated sites.[11]

Nevertheless, the Commission argued that the practical measures adopted by Ecuador to develop domestic legislation were few. From the Commission's point of view, although each country has the independence to exploit—including with foreign investment—resources that can be found in its own territory, it must, however, protect the environment in a sustainable manner. Moreover, the Commission ruled that the absence of regulation or supervision of environmental standards may cause violations of human rights protected by the ACHR. In this respect, it pointed out that "where environmental contamination and degradation pose a persistent threat to human life and health, the foregoing rights are

[10]The case concerning the responsibilities of Texaco for damages caused to the environment and human health in Ecuador has seen further developments before the US and Ecuador Tribunals. See Patel (2012).

[11]IACommHR stressed that "[o]ne of the Government's most visible activities with respect to the effects of oil development has been its effort to ensure that *Texaco finance and implement a plan to clean up areas that were contaminated during the company's twenty-plus years of operation in the Oriente*" (emphasis added).

implicated". Therefore, Ecuador was required to take certain positive measures to ensure that oil companies complied with the obligations of the ACHR and to safeguard the physical integrity and the life of the local people. In addition, the Commission invited Ecuador to take all necessary measures to prevent environment pollution and damage to human health through the recognition of procedural rights, namely the right to information (Art. 13), the right to participate in decision-making processes (Art. 23), and the right to an effective remedy (Art. 25) (Shelton 2007, p. 31). Lastly, the Commission invited both the State and corporations to take measures to decontaminate polluted soil and groundwater:

> As the Commission observed at the conclusion of its observation in loco: "Decontamination is needed to correct mistakes that ought never to have happened." *Both the State and the companies* conducting oil exploitation activities are responsible for such anomalies, and both should be responsible for correcting them. It is the duty of the State to ensure that they are corrected. (emphasis added)

As highlighted in the literature, "this is the first time that an Inter-American supervisory organ explicitly referred to responsibilities held by both companies and states" (Anicama 2008, p. 26).

3.3 The Case Law of the European Court of Human Rights

As mentioned above, the ECtHR cannot receive complaints that only pursue the aim of safeguarding environmental good quality, because the ECHR does not expressly recognize such a right. However, according to the *Manual on Human Rights and Environment* drafted and published by the Council of Europe in 2012, "environmental factors may affect individual Convention rights" (Council of Europe 2012, p. 7): in these cases, the Strasbourg Court can decide on applications filed by the victims of environmental pollution. In a such context, European human rights bodies have considered certain cases of historical pollution, especially when they have assessed the alleged violations of the right to life (Art. 2 ECHR), the right to private and family life (Art. 8 ECHR), and the right to property (Art. 1 of Protocol No. 1 to ECHR) in the context of industrial activities (Xenos 2007; Shelton 2015).[12] Moreover, there are other cases of historical pollution that have been decided by the Strasbourg Court with reference to the right to a fair trial (Art. 6).

[12]In asserting the right to life in the context of industry, the European Court has often used the definition of dangerous activities established in the Convention on Civil Liability for Damage Resulting from Activities Dangerous to the Environment (Lugano, 21 June 1993) and in the Convention on the Protection of the Environment through Criminal Law (Strasbourg, 4 November 1998). In the leading case *Öneryıldız v. Turkey*, the European Court used these international instruments to define, under ECHR Art. 2, the waste recovery and disposal operations as dangerous activities (Paras. 59–60). Although these legal instruments adopted by the Council of Europe are not binding, the European Court considers them relevant in interpreting the fundamental rights laid down in the European Convention on Human Rights.

3.3.1 Human Health Problems Caused by the Failure of Public Authorities to Clean up a Contaminated Site

Two typologies of cases in which the applicants have alleged human rights violations as a result of environmental contamination caused by wastes can be found in the case law of the ECtHR.

In cases such as *Giacomelli v. Italy* the Strasbourg Court has found a violation of the right to respect private and family life because the applicant was adversely affected by ongoing pollution caused by a waste disposal plant that performed its activities without an environmental impact assessment. The latter decision shows how human rights may compel States to prevent long-term pollution.

On the other hand, *Locascia and others v. Italy* is a noteworthy case of historical pollution currently pending before the ECtHR. The applicants complained of a breach of the right to life (Art. 2) and the right to respect for private and family life (Art. 8) caused by the operations of a waste management facility situated in an area known as "Lo Uttaro" (Caserta, Italy), and by the failure of the competent public authorities to clean up and reclaim the contaminated area after the closure of the plant. The waste disposal facility was managed by a company in the 1980s and 1990s, after which it was closed. In 2001, a report produced by the Carabinieri and quoted in the application highlighted "the absolute unfitness of the 'Lo Uttaro' area to host a waste disposal plant in the light of a situation of serious environmental degradation and pollution" (Para. 5). On 11 November 2006, when a serious waste crisis occurred in the Campania region, the plant was reopened and, consequently, individuals living near the polluted area filed an application to the Tribunal of Naples seeking to suspend the waste disposal activities. On 7 November 2007, local public authorities ordered the interruption of the waste disposal operation. The criminal justice authorities subsequently ordered the seizure of the plant and carried out criminal investigations relating to the waste cycle in Campania. On 4 August 2009, the public authorities signed an agreement to secure and reclaim the contaminated site, but this programme was not concretely implemented. Therefore, historical contamination had a negative impact on the community. In this regard, in 2007 the World Health Organization (WHO) published the results of an epidemiological study on the health impact of the waste cycle in Campania, pointing out that the death rate from cancer in the provinces of Caserta and Naples was higher than the regional average by at least 16% for men and 11% for women, "with peaks of 19.3% for men and 18.2% for women in relation to stomach cancer" (Paras. 39–40). Moreover, a high number of birth defects were registered in the South of Caserta (Para. 40). In conclusion, the WHO study's findings

> *indicate the presence of an area characterised by elevated cancer mortality rates* and by elevated occurrence of birth defects, corresponding with that area where most waste disposal sites are concentrated … *these preliminary findings are consistent with a possible*

> *contributory role of waste-related exposures in determining ill health in the area over time.*[13] (Para. 41; emphasis added)

After exhausting any internal remedies, the applicants complained about the violation of the right to life and the right to private and family life due to the fact that the State had not adopted all the measures necessary to protect their fundamental rights. In detail:

> *Invoking Articles 2 and 8 of the Convention, the applicants complain about the danger to their health and the interference with their private life and home caused by the operation of the plant and by the failure of the authorities to secure, clean-up and reclaim the area after the closure of the plant.* (Para. 44; emphasis added)

The expected judgment of the Strasbourg Court could define the duties of States under human rights law in a complex scenario of historical pollution.

3.3.2 Deaths and Damages to Human Health as a Result of Past Exposure to Asbestos and Nuclear Testing

In the case law of the ECtHR, there are a number of judgments in which the Court has held certain States responsible for the violations of the right to life (Art. 2), the right to respect for private and family life (Art. 8), the right to access to court, and the right to a fair and public hearing within a reasonable time (Art. 6, Para. 1) in relation to cases concerning people whose work in the past brought them into contact with materials made with asbestos, or who shared a home with workers exposed to asbestos (e.g. women who contracted asbestos-related illnesses because they cleaned the clothes of their husbands employed in asbestos factories[14]).

In *Brincat and others v. Malta*, a group of applicants employed by a public repair yard from 1968 to 2003 argued that they had been constantly and intensely exposed to asbestos in the course of their professional activities. From the applicants' point of view, their prolonged contact with asbestos resulted in harm to their private and family life, and even death in the case of one of them, Mr. Attard. In 2009 the applicants sought compensation by bringing constitutional redress proceedings, but they were dismissed. Having exhausted all domestic legal remedies, the applicants filed a claim before the ECtHR, invoking a violation of the right to life and the right to private and family life, since the State had failed to adopt the measures necessary to protect them.

---

[13]Such a conclusion was shared by a study of the Superior Institute of Health on the impact on human health of waste disposal in Campania (Paras. 42–43).

[14]See, for example, the *Janssen* and *Klasen* cases where the applicants complained about the length of the proceedings before the German Courts for claimed compensation (Art. 6, Para. 1, ECHR) related to cases of asbestos-related illness they contracted through cleaning the clothes of their husbands, who worked in asbestos factories. In both the above cases, the ECtHR held that there had been a violation of Art. 6, Para. 1. See also the *Bullerwell* case.

In point of admissibility, the ECtHR affirmed that Article 2 of the ECHR is not only applicable to cases of deaths caused by unjustified use of force by public agents. Each state also has a duty to protect the right to life in the framework of "industrial activities which by their nature are dangerous", such as the management of collection sites and waste storage, nuclear testing, toxic emissions of factories, and the exposure to asbestos of workers employed in a company owned or controlled by the government (Para. 81). By applying these principles to the present case, the ECtHR assessed the complaints lodged by victims, highlighting that on the basis of the medical examination, the death of the worker was caused by a malignant mesothelioma, "known to be a rare cancer associated with asbestos exposure" (Para. 83). As regards the position of the other applicants with respiratory problems caused by their past exposure to asbestos, the ECtHR applied Article 8 of the ECHR. That said, the ECtHR was deemed to be competent *ratione temporis* to rule on the present case, taking into account that the ECHR came into force in Malta on 23 January 1967, after the occurrence of the facts that were at the origin of the alleged violations (Para. 86).

In deciding on the merits of the case, the ECtHR argued that the Maltese State had implicitly admitted that it had become aware of the hazards of asbestos around 1987 and during this period it had adopted the relevant measures. However, according to the Court, "the Maltese Government knew or ought to have known of the dangers arising from exposure to asbestos at least as from the early 1970s" (Para. 106), for at least two reasons: firstly, hundreds of scientific papers produced by scientific researchers from the 1930s had demonstrated the harmful effects of asbestos on human health; secondly, in the 1970s there were asbestos lawsuits in Malta. Nevertheless, from the mid-1980s to the early 2000s, the Maltese Government had failed to adequately protect the health of workers exposed to asbestos.[15] Likewise, no information on the risks to health was communicated to the applicants, with the consequence that Malta was responsible for the infringements of Article 2 and 8 of the ECHR.

The applicants also complained that the Government had failed to prosecute those responsible for the death of Mr. Attard. As regards this allegation, the ECtHR observed that—"unlike in medical negligence cases, where a civil remedy may suffice"—each State has a duty under Article 2 of the ECHR to investigate cases relating to deaths deriving from dangerous activities. Although Mr. Attard died in the context of dangerous activities that took place in the past, the Court noted that

> *the present case concerns a death that did not result from one particular instance but rather from circumstances which were spread over a number of decades*, and which did not ensue from uncertain conditions—indeed, the State's responsibility has been established by the domestic courts in similar circumstances even where there was no investigation. [...] There is no doubt that during that period information about asbestos related consequences was publicly available. (Para. 124; emphasis added)

[15]The ECtHR argued that "the only practical measure that appears to have been taken by the State, as the employer, was to distribute masks, on unspecified dates and at unspecified intervals (if distributed repeatedly at all)" (Para. 112).

From the moment that the cause of the death of Mr. Attard was not confined to the knowledge of public officials, but became publicly available, the Court established that the domestic judges were under no obligation to conduct an investigation *ex officio*. It followed that there had been a breach of the procedural obligations under Article 2 of the Convention.

The second decision worthy of comment is *Howald Moor and others v. Switzerland*. In this case, the applicants were the relatives of Mr. Moor, a mechanic who died in 2005 from a malignant pleural mesothelioma due to his workplace exposure to asbestos from 1965 to 1978. They brought proceedings before the Strasbourg Court because they considered their right of access to a Court under Article 6, Paragraph 1 of the ECHR to have been infringed. The request for damages that they had filed before domestic Courts was considered time-barred by virtue of domestic law, which prescribed a limitation period of ten years for such actions, measured from the date of the last exposure of the workers to asbestos and not from the date at which the damage became apparent. From the point of Mr. Moor's final exposure to asbestos in 1978, the limitation period had expired in 1989. In ruling on this complaint, the ECtHR held that the right to a fair trial guaranteed by Article 6, Paragraph 1 guarantees an effective judicial remedy to victims to protect their civil rights. However, the right of access to a court may be limited: although each State enjoys a certain margin of appreciation in this regard, limitations to the right of access can be considered compatible with Article 6, Paragraph 1 only if they are prescribed by law and proportionate. Moreover, the ECtHR stated that the aim of limitation periods in personal injury cases is, *inter alia*, to safeguard legal certainty and

> to protect potential defendants from stale claims which might be difficult to counter and prevent the injustice which might arise if courts were required to decide upon events which took place in the distant past on the basis of evidence which might have become unreliable and incomplete because of the passage of time. (Para. 72)

In applying these principles to the present case, the ECtHR noted that this specific case concerned a complex issue relating to the starting point (*dies a quo*) of the limitation period of the claims for damages brought by asbestos victims before domestic tribunals. From the Court's point of view, given that the latency period of diseases related to asbestos exposure could last several decades, any action for damages was a priori destined to fail in the event that the absolute limit of 10 years began from the date on which the person was exposed to asbestos. Actions for damages were excluded before asbestos victims could have become tangibly aware of the violation of their rights. In conclusion, even if the Swiss legal rule on the statute of limitation pursued a legitimate aim, it was disproportionate. As the Strasbourg Court stated:

> the systematic application of the rule to persons suffering from diseases which could not be diagnosed until many years after the triggering events deprived those persons of the chance to assert their rights before the courts. (Para. 77)

Consequently, the Court concluded that

> where it was scientifically proven that a person could not know that he or she was suffering from a certain disease, that fact should be taken into consideration in calculating the limitation period. (Para. 78)

Taking into account the circumstances of the case, the ECtHR stated that the legal rule on the statute of limitation in question had restricted the right of access of the applicants to the point of causing an infringement of Article 6, Paragraph 1 of the ECHR.

Another group of cases decided by the ECtHR concerns people who were diagnosed with leukaemia, multiple myeloma, or other types of cancers many years —often decades—after their exposure to nuclear tests organized by the British military in the Pacific Islands in the 1950s. Moreover, the ECtHR has been seised of a case in which the applicant, who had contracted leukaemia, was the daughter of a person exposed to radiation at Christmas Island in the 1950s. In *LCB v. UK*, the applicant submitted that the State had not provided her family with information on the level of her father's exposure to radiation. However, the ECtHR found that there was no violation of Article 2 of the ECHR, given the lack of "a causal link between the exposure of a father to radiation and leukaemia in a child subsequently conceived" (Para. 39). Moreover, it "could not reasonably hold, therefore, that, in the late 1960s, the UK authorities could or should, on the basis of this unsubstantiated link, have taken action in respect of the applicant" (Para. 39). It followed that there was no conventional violation of the right to life.

3.3.3 Deaths and Damages to Human Health as a Result of Environmental Pollution Caused by Long-Term Industrial Emissions

Last, the focus will be on three cases of historical contamination in the broad sense: *Fadeyeva v. Russia*, *Dubetska and Others v. Ukraine*, and *Smaltini v. Italy*.

The first case to discuss is *Fadeyeva v. Russia.*[16] In 1982, the applicant, Mrs. Fadeyeva, moved to the city of Cherepovets near the Severstal steel plant, the biggest iron smelter in the Russian State, employing tens of thousands of people. This plant was built during the Soviet era and was, at the time of the facts, public property. Despite the fact that the public authorities delimited a "sanitary security zone" with the purpose of isolating the plant from the residential area, the applicant nevertheless suffered damage from pollution. In 1974, the local authorities planned the transfer of the local residents to a safer area, but no measures were adopted in practice.

In 1993 the industrial unit was privatized, while the flats of the plant remained in the hands of the local public authorities. During the 1990s, the Russian government

[16]See also the *Ledyaeva* case.

adopted two programmes to protect the environment in the vicinity of the plant, including a special programme financed by the World Bank (Para. 15). Nonetheless, the facility continued to cause environmental pollution. In 1995 the applicant brought proceedings before the Russian courts in order to obtain her resettlement. In 1996, a domestic court stated that the applicant had the right to be resettled, and consequently it invited the competent public bodies to include her on a "priority waiting list" for a new apartment. The applicant lodged an appeal, but it was rejected; she was then added to the waiting list for new accommodation. In 1999, the applicant filed a new appeal, but this too was rejected. In 2002 the applicant was diagnosed with various illnesses of the nervous system.

Consequently, the applicant submitted a complaint to the Strasbourg Court contending that the state had failed to protect her right to respect for private and family life against environmental contamination deriving from the plant's industrial activities. In deciding on the complaint, the Court observed that Article 8 of the ECHR does not guarantee environmental protection: however, a violation of such a right may arise when adverse effects of environmental pollution attain a minimum level of severity, whose valuation is relative and depends on all the circumstances of the case, such as "the intensity and duration of the nuisance, and its physical or mental effects". Moreover, "the general context of the environment should also be taken into account", even if "there would be no arguable claim under Article 8 if the detriment complained of was negligible in comparison with the environmental hazards inherent to life in every modern city" (Para. 69).

Applying these principles to the present case, the Court noted, first of all, that the medical certificate presented by the applicant did not establish a causal connexion between pollution and the damage suffered by applicant. Given that the ECHR entered into force in Russia on 5 May 1998, only the period subsequent to that date was taken into consideration for the purposes of assessing the applicant's complaints (Para. 82). In this respect the Court pointed out that since 1998, pollution had surpassed a series of limits established by domestic environmental rules (Para. 87). By combining such a presumption with other indirect evidence, it concluded that environmental degradation had had an adverse effect on the health of the applicant, to the point of reaching a level of severity that fell within the scope of application of the right in question.

That said, the Court highlighted that the plant was not state-owned or otherwise controlled by the state during the period under assessment, having been denationalized in the early 1990s. Nevertheless, it argued that the state continued to exercise control over industrial activities by imposing conditions on the managers of the plant. Consequently, in the Court's view, the Russian State had failed to regulate industry. Furthermore, the Court took into consideration the fact that environmental degradation was the result of a situation that was "long-standing and well known". In detail:

> The Court observes in this respect that the Severstal steel plant was built by and initially belonged to the State. The plant malfunctioned from the start, releasing gas fumes and

> odours, contaminating the area, and causing health problems and nuisance to many people in Cherepovets […]. Following the plant's privatisation in 1993, the State continued to exercise control over the plant's industrial activities by imposing certain operating conditions on the plant's owner and supervising their implementation. The plant was subjected to numerous inspections by the Federal Environmental Agency and administrative penalties were imposed on the plant's owner and management […]. *The environmental situation complained of was not the result of a sudden and unexpected turn of events, but was, on the contrary, long-standing and well known* […]. As in López Ostra […], in the present case the municipal authorities were aware of the continuing environmental problems and applied certain sanctions in order to improve the situation. (Para. 90; emphasis added)

Additionally, the European Court stressed that environmental pollution could not be attributed to a large number of smaller sources because the main cause of the pollution in the city was due to the activity of the plant. Since the public authorities in this case were without doubt in a position to assess the environmental harm and take suitable measures to reduce the pollution, the European Court considered that there was a connection between the contamination and the state's obligation to take positive obligations to protect human rights in question.

Lastly, the Court had to check whether there was a "fair balance" between the interests of the applicant and those of the society as a whole, as required by Paragraph 2 of Article 8. In this regard, the Court considered that the activities of the facility in question contributed to the economy of the region in accordance with a "legitimate aim" under Paragraph 2 of Article 8. In determining whether, in pursuing this objective, the public authorities had balanced in a fair way the interests of the applicant and those of the community as a whole, the Court pointed out that the State had failed, on the one hand, to give a solution to the applicant, and on the other, to reduce the harmful emissions. Although the Strasbourg Court pointed out that the State had a wide margin of appreciation in its choice of measures to protect fundamental rights, it failed to operate a fair balance of the interests at stake. Therefore, the Court concluded for the violation of the conventional right because the prolonged exposure of Mrs. Fadeyeva to environmental pollution had caused her numerous drawbacks and suffering. Consequently, the Court recognized a sum in compensation for non-pecuniary damage suffered by the victim only for the period after the ECHR entered into force in Russia. With regard to the request of pecuniary damage, the European judges concluded that the applicant proved not to have suffered economic loss due to industrial emissions.

In conclusion, the Strasbourg Court has held a State responsible for the infringement of the right to private and family life even where the domestic authorities have not directly interfered in the private life of the applicant, but have failed to fulfil positive obligations laid down in the Convention. Therefore, the responsibility of the State within the meaning of the Convention also derives from its failure to comply with the legislation on industrial emissions, in regulating private industry, or in adopting measures aimed at reducing emissions. Furthermore, the privatization of a public enterprise that has caused the pollution of the environment does not exclude the responsibility of a State where, especially in cases of long-term pollution, it fails to adopt all the appropriate measures to protect conventional rights.

In *Dubetska and Others v. Ukraine*,[17] the applicants lived in a village situated in the Chervonograd coal-mining basin. They complained that the chronic pollution caused by two state-owned companies over four decades of coal-mining activities had had a negative impact on the environment, health, and housing. According to an environmental assessment commissioned by the State Committee for Geology and Mineral Resource Utilisation in 1998,

> *All the coal-mining industry operational in the region for over forty years has been negatively affecting the environment*: spoil heaps from the mines and the coal-processing factory have been created, from which dust with a high concentration of toxic components spreads into the atmosphere and the soil … systems of water drainage of the mines … and cesspools … of the coal-processing factory are sources of pollution of surface and underground waters … Rocks from the spoil heaps contain a variety of toxic heavy metals, leaching of which results in pollution of soils, surface and underground waters … Very serious polluters … are cesspools of mining waters and factory tailing ponds …, which in the event of the slightest disturbance of the hydro-insulation cause pollution of surface and ground waters … The general area of soil subsidence is about 70 square kilometres […] All this inevitably affected the health of people living in the area. (Para. 15; emphasis added)

Other reports highlighted that people living in this village were exposed to high risks of developing cancer and respiratory diseases due to high levels of soil and groundwater contamination. The applicants therefore complained that the authorities had failed to protect their right to private and family life (Art. 8). In ruling on that complaint, the ECtHR first of all declared the application partly admissible. It considered itself "not competent *ratione temporis* to examine the State actions or omissions in addressing the applicants' situation prior to the date of the entry of the Convention into force with respect to Ukraine (11 September 1997)", but only the "'applicants' complaints, which relate to the period after this date" (Para. 82). That said, the ECtHR ruled that there was a breach of the right to respect for private and family life for the reason that the Ukrainian Government had not adopted effective measures to protect the applicants from pollution. The pollution reduction plan adopted by the public authorities remained unrealized, and the judicial authorities had failed to resettle the applicants into a clean area. In conclusion, despite the introduction by the Ukrainian State of criminal legislation into the environmental field, according to the Court "for more than twelve years the State authorities have not been able to put in place an effective solution for the applicants' personal situation". Thus, the European Court held that there was a breach of Article 8.

Lastly, attention should be paid to *Smaltini v. Italy*, a case held inadmissible by the ECtHR. Although the Strasbourg Court rejected the victims' complaints, such a judgment relates to a case of ongoing pollution or—as described in the previous pages—of historical pollution in the broad sense. Moreover, it is the first judgment against Italy for contamination caused in the city of Taranto by ILVA, one of Europe's largest steel factories. Despite the negative result of this application, the Court did assess alleged infringements of the right to life under its procedural aspect. In particular, this is one of the few cases in which the Strasbourg Court has

[17]For a general comment, see Fitzmaurice (2011).

decided on a criminal law issue in environmental matters. Moreover, it must be noted that new applications have recently been brought to the ECtHR by other victims of industrial pollution in the Taranto area.[18]

In the light of the above, it is necessary to frame the facts. In September 2006, Mrs. Smaltini—resident in the city of Taranto, a short distance from ILVA—discovered she had contracted acute myeloid leukaemia. She filed a complaint with the public prosecutor's office of Taranto against a manager of the company, arguing that her leukaemia was a result of environmental degradation caused by pollution released from the ILVA plant. The applicant argued that certain managers of ILVA had already been sentenced for having authorized illegal emissions that had, over the course of years, caused a significant increase in the number of deaths from different forms of cancer in the area of Taranto (Para. 8). Other documents were attached in support of the complaint, including a hospital certificate, a report on the emission of carcinogenic substances produced by the plant, and a report of the Superior Health Institute in Taranto. In 2007, the Taranto public prosecutor's office, having concluded that a causal link between ILVA's emissions and pathology contracted by the applicant was not established, requested the dismissal of the criminal case. The claimant filed an opposition to the archiving of the criminal proceedings, highlighting that the causal link between the harmful emissions and the development of cancer were confirmed by research by the Italian Association against Leukaemia, Lymphoma and Myeloma. Moreover, she alleged that the causal link between the substances produced by the ILVA, such as dioxins, PM10, and PCBs, and the development of cancers and leukaemia in the area of Taranto was confirmed by studies developed by the International Agency for Research on Cancer, based in Lyon. In 2008, the judge for preliminary investigations refused the request of the public prosecutor to archive the criminal proceedings. He ordered haematological expertise in order to establish the relationship between the pathology and the pollution produced by ILVA. The testimony of the two experts, a coroner and a haematologist, highlighted that the origin of acute myeloid leukaemia was unknown: it could arise at all ages and in all environments, in Italy or elsewhere in the world, without distinction of sex. Although in the scientific world suspicion had emerged that the pollutants examined could cause illness, this information was not scientifically proven. Moreover, the experts argued that there was no greater incidence of acute myeloid leukaemia in people living in the region of Taranto compared to those living in other Italian regions. On the basis of the objective scientific evidence, therefore, the consultants excluded the possibility that ILVA's polluting emissions caused leukaemia. The prosecutor again demanded the dismissal of the criminal case, but the claimant filed a further opposition requesting the admission of new evidence. In 2009, the judge for preliminary investigations

[18]See the *Cordella* case.

archived the criminal proceedings, since he considered adding further evidence to be superfluous. However, the applicant brought an application before the ECtHR complaining of an infringement of Article 2 of the ECHR.

In ruling on this complaint, the European Court argued that the applicant did not complain that the national authorities had failed to put in place the necessary legal or administrative measures to protect her life (Para. 50). On the contrary, she alleged only that in the criminal proceedings against the ILVA manager, the national judges had failed to establish the existence of a causal link between ILVA's emissions and the applicant's disease. With this, the European Court first of all reiterated that Article 2 of the ECHR obliges States to protect the right to life through criminal law; serious violations of such a right should therefore be repressed and punished (Para. 52). Moreover, it underlined that Article 2 requires States to proceed with an official, independent, and impartial investigation to ensure the prosecution of serious violations of the right to life caused by dangerous activities, to the extent that the results of the investigation justify this repression (Para. 53).

Consequently, in assessing whether the applicant's complaint was scrupulously examined by the national courts (Para. 54), the Strasbourg Court noted that according to the national reports there was no higher incidence of leukaemia in the area of Taranto compared to other Italian regions. Moreover, the area of Taranto did not figure prominently among territorial areas where women were mainly affected by this disease (Para. 57). In addition, further epidemiological studies published in 2009 and 2012 did not prove the existence of a causal relationship between the ILVA's polluting emissions and leukaemia in the province of Taranto (Para. 58). Lastly, taking into consideration that the claimant benefited from contradictory additional investigations to ascertain the existence of a causal link, but without success, the European Court concluded that "without prejudice to the results of scientific studies to come", the dismissal of the criminal proceedings due to a lack of scientific evidence of the causal link necessary to punish the person charged with a criminal offence did not lead to the infringement of the right to life of the victim (Para. 60). The application was therefore declared manifestly unfounded.

4 The Duty to Protect Human Rights and the Environment from the Harmful Effects of Historical Pollution

In the light of the above, state parties to the regional conventions on human rights have a number of duties to fulfil in order to protect human rights and the environment from the negative effects of different forms of historical pollution. This section focuses on the obligations borne by States under human rights law.

4.1 Human Rights at Stake

The typologies of human rights at stake with reference to cases of historical pollution are various: they include civil, political, and social rights, depending on the conventional system of reference. In the European human rights system, the ECtHR has not obliged states to directly protect soil, air, or groundwater from pollution, but it does require them to defend existing human rights from the adverse consequences of environmental contamination. In the American human rights system, the IACommHR has highlighted that each State must defend, *inter alia*, the right to land of indigenous people and the rights of children to be protected from contamination caused by industrial activities over time. In the African human rights system, the ACommHPR has underlined the necessity to protect the right to a general satisfactory environment expressly enshrined in Article 24 of the ACHPR, as well as other human rights (life, property, housing, etc.), with the goal of preventing the infringement of these rights resulting from ongoing human activities. Furthermore, all these organs have underlined the relevance of procedural rights in environmental matters as means by which each State may redress cases of historical contamination.

4.2 The Scope Ratione Personae *of Human Rights*

As regards the scope *ratione personae* of these rights, the subjects that must respect human rights are first of all States themselves.[19] As shown above, regional human rights bodies have developed several theories on the State's duty to protect human rights. The ACommHPR has stated that each State has four duties for each right, namely the "duty to respect, protect, promote and fulfil" it, representing a mix of positive and negative obligations. Likewise, the ECtHR has developed a theory of obligations that distinguishes between positive and negative obligations and, with reference to positive obligations, between substantive and procedural obligations (Sudre 1995; Akandji-Kombe 2007, pp. 5 ff.).

Nevertheless, certain obligations based on human rights are designed to indirectly bind private actors, such as companies involved in dangerous activities. In this connection, it must be underlined that the ECtHR has developed a jurisprudence concerning the duty on the part of the State to protect human rights in the context of industry—both public and private—through the regulation of dangerous activities (Xenos 2007). For instance, in *Fadeyeva v. Russia*, Russia has been considered responsible by the European Court for having failed to regulate and monitor a dangerous activity after its privatization, with a consequent violation of

[19]According to Boyle (2007, p. 486), "the starting point for any discussion of human rights and the environment is that a failure by the state to regulate or control environmental nuisances or to protect the environment may interfere with individual rights".

the right to life. In its *1997 Report on Ecuador*, the IACommHR has directly requested public bodies and companies to clean up the contaminated area in the Oriente region.

4.3 The Scope Ratione Temporis *of Human Rights*

With regard to the *temporal scope* of the rights in discussion, the most interesting indications come from the European human rights jurisprudence. In cases of damages to health or life resulting from ongoing pollution, the Court has operated a distinction between facts that have occurred before and after the entry into force of the Convention in the State sued by the victim, establishing that only in relation to the latter may it assess the infringement of human rights alleged by the victims. A paradigmatic example is the valuation made by the Court in the *Dubetska and Others v. Ukraine* case, where the victims complained of damage to their health and homes due to historical contamination produced by mining exploitation activities for more than 40 years (from 1960 to 2003), part of which occurred before the entry into force of the ECHR in Ukraine in 1997. In this case, the Court limited its assessment only to the facts that occurred after the Ukrainian State had formally completed its process of accession to the European conventional system.

4.4 The Content of Human Rights: Duties to Prevent, Compensate, and Clean up

Each State must satisfy diverse obligations, in the framework of a specific system of protection of human rights, in order to protect human beings from the negative consequences of historical pollution. Effectively, public bodies are required to adopt key measures to prevent and manage phenomena of historical pollution that may damage the environment and human well-being.[20] Several obligations can be distinguished among them, and these are detailed here:

- *The obligation to prevent the infringements of human rights in the context of industry*, by regulating dangerous activities that may adversely affect human rights (i.e. the right to life, the right to respect for private and family life, the right to property).
- *The obligation to prevent and handle complex phenomena of historical pollution by recognizing the procedural human rights of the people concerned.* In detail, each State must provide *information* on the risks to human health and

[20]However, it should be noted that the ECtHR recognizes a wide margin of appreciation on the part of States in their choice of measures for the protection of persons: what matters is that the measures taken in practice are effective at protecting individuals.

environment to those communities living near industrial units or contaminated sites. In this connection, the ECtHR has increasingly emphasized the importance of the precautionary principle in the protection of the fundamental rights enshrined in the convention.[21] Moreover, *people who may be affected by environmental pollution should be involved in the decision-making process.*

- *The obligation to protect victims of environmental degradation and historical pollution with healthcare programmes and other social and urban policies* (e.g. the resettlement of victims in non-polluted residential areas in order to prevent their further exposure to the effects of historical contamination).
- *The obligation to protect the environment through the cleaning up of contaminated sites to prevent these negative situations from interfering with the enjoyment of fundamental rights.* For instance, in the *Ogoni* case, the ACommHR obliged Nigeria to "undertake a comprehensive clean-up of lands and rivers damaged by oil operations". In its 1997 Report on Ecuador, the IACommHR required Ecuador to clean up contaminated land in the Oriente. Even if a such duty cannot be derived from the ECHR, interesting proceedings are pending before the ECtHR, in which the applicants have claimed an infringement of their physical integrity as a consequence of the failure of public bodies to clean up a contaminated area near their house.
- The obligation to guarantee that victims have the right to effective remedies to obtain, *inter alia*, compensation for the damage that they have suffered as a result of historical pollution. In the Ogoni case, the ACommHR asked Nigeria to ensure effective remedies and adequate compensation for the victims of the human rights violations. Likewise, the ECtHR has required many States to satisfy the victims of historical pollution with compensation for pecuniary and non-pecuniary damages.

4.5 Scientific Knowledge and Legal Responses

Human health problems deriving from exposure to toxic substances spread through air, soil, or groundwater by past industrial activities may become apparent only after a long time. In these cases, the Strasbourg Court assesses the scientific knowledge (*scientific time*) available to the Government at the time of the toxic emissions (*industrial time*) in order to establish whether a State has fulfilled its positive obligations to protect the right to life (Art. 8) and the right to private life (Art. 8) of the victims (*juridical time*).

A paradigmatic case is that of *Brincat and Others v. Malta*, in which the ECtHR condemned Malta because a group of workers was left for long a time without adequate safeguards against the dangers of asbestos, although the Government

[21]On the concept of risk in the light of the jurisprudence of the ECtHR, see Hilson (2009), Green (2013). Moreover, see the *Guerra* case.

"knew or ought to have known" the relevant dangers at least by the early 1970s. In the light of the above, if a Government knows, on the basis of scientific evidence, the risks to life for individuals exposed to industrial emissions, this entails its primary duty to adopt a legal framework to protect them. If it fails in its duty, there may be a breach of the right to life or the right to respect for private and family life.

Moreover, according to the case law of the European Court, each State should consider the features of the latent damages in fixing the limitation period. In particular, scientific data shows that certain diseases may occur many years after the victim's exposure to toxic substances such as asbestos (*scientific time*), and, consequently, the starting date for calculating the limitation period for civil claims should start from when these damages become apparent (*juridical time*). A different solution may give rise to a breach of the right enshrined in Article 6, Paragraph 1 of the ECHR. Lastly, it is also under discussion whether the principles relating to the statute of limitations established in the *Howald Moor* judgment may also be extended to criminal proceedings.

5 The Protection of Victims Through Criminal Law

The relationship between human rights and criminal law can be considered "paradoxical" (Nicosia 2006, pp. 255 ff.; Tulkens 2011; Viganò 2011; Manacorda 2012; Pulitanò 2013; Tulkens and van de Kerchove 2014). On the one hand, certain human rights, such as the principle of legality, the principle of non-retroactivity, and the principle of the presumption of innocence have the purpose of limiting the recourse to criminal punishment.[22] On the other hand, with the development of the case law of regional human rights bodies, "the scenario is reversed: the criminal law is called into play to protect human rights", and "this trend is itself the result of the development and extension of the theory of positive obligations" (Tulkens 2011, p. 582).

Effectively, the Inter-American Court of Human Rights and the ECtHR require contracting States to punish and prosecute those responsible for the infringement of certain fundamental rights, such as the right to life, the prohibition of torture, serious ill treatment, and slavery, and the right to liberty protected by the Human Rights Conventions. In this regard, scholars have focused on the State's duty to investigate and punish deaths resulting from military use of weapons, inhuman or degrading treatment, torture, forced disappearances, and trafficking in human beings (Mowbray 2002; Basch 2007; De Vero 2007; Chevalier-Watts 2010; Lacroix 2011; Viganò 2011; Satzger 2012, 143 ff.; Manacorda 2012; Giudicelli-Delage et al. 2013), while a small number of studies have explored the specific issue of whether States have a duty to punish and prosecute infringements

[22]As emphasized by Tulkens (2011, p. 579), this is the "defensive role of human rights," a role of "neutralizing" the criminal law.

of fundamental rights in the event of deaths due to environmental damage (Scalia 2015).

The question at stake is whether the criminal law may have a role in protecting human rights in cases of historical pollution. In that regard, attention will be paid to three different scenarios that have emerged in the case law of the regional human rights bodies.[23] In the first scenario, public agents reacted violently against minority local communities who exercised their right to protest in opposition to environmental degradation due to industrial activities carried out over the course of several decades. In the second scenario, certain people died as a result of a methane explosion caused by the decomposition of waste that had accumulated and not been properly handled in a public landfill over a long period. In the third scenario, people died as a result of their past or long-term exposure to toxic substances.

5.1 Human Rights Abuses in Historical Pollution Scenarios

With reference to the first scenario, it should be pointed out that Nigeria was held responsible by the African Commission in the *Ogoni* case because the Nigerian military authorities responded with massive violence to public protests organized by Ogoni communities in opposition to the environmental degradation caused by oil companies. In that situation, the Commission appealed to the Nigerian Government to "ensure protection of the environment, health and livelihood of the people of Ogoniland" by "stopping all attacks on Ogoni communities and leaders by the Rivers State Internal Securities Task Force and permitting citizens and independent investigators free access to the territory" and "conducting an investigation into the human rights violations described above and prosecuting officials of the security forces, NNPC and relevant agencies involved in human rights violations" (ACHPR, 155/96 Social and Economic Rights Action Center (SERAC) and Center for Economic and Social Rights (CESR), Nigeria).

5.2 Loss of Life in the Context of Dangerous Activities

The second case worthy of attention is that of *Öneryıldız v. Turkey*, decided by the ECtHR (Xenos 2007). Although the decision concerns an accident caused by a

[23]On closer examination, only a small number of cases concern historical pollution in the narrow sense.

methane explosion in a rubbish tip, it serves as a benchmark to carry out a more in-depth reflection on the obligations of States to prosecute and punish authorities that fail to prevent a breach of the right to life from the negative effects of historical pollution. The principles established in this case have been used by the European Court in decisions concerning cases of historical pollution in the broad sense—as in *Smaltini v. Italy*—and cases of past exposure of workers to asbestos—as in *Brincat and others v. Malta*. To highlight the different approaches followed by the European Court in cases of disaster and historical pollution, it will be necessary to compare the conclusions adopted by the European Court in these cases.

In the case of *Öneryıldız v. Turkey*, Mr. Öneryıldız, a Turkish national, lived with close relatives in an Istanbul slum that surrounded a rubbish tip managed since the 1970s by the public authorities. Although the authorities knew that the landfill posed serious risks to the slum's inhabitants, no measures were taken to avoid an explosion of methane generated by the decomposition of waste. In 1993, an explosion of waste destroyed a number of houses, including that of the applicant, who lost nine relatives. Consequently, the criminal and administrative authorities carried out investigations. At the end of the criminal proceedings, the mayors of Ümraniye and Istanbul were held responsible for "negligence in the performance of their duties" under Article 230, Paragraph 1 of the Turkish Criminal Code, and were given suspended fines by the Istanbul Criminal Court. Moreover, the applicant brought an action before the Istanbul Administrative Courts to obtain compensation for damages arising from the death of his family and the loss of his home. At the end of the process, the applicant and his surviving sons obtained compensation of around €2077 for non-pecuniary damage and €208 for pecuniary damage. However, the victims did not receive these payments. After all internal remedies had been exhausted, the applicant brought an application before the ECtHR complaining of a violation of Article 2 (the right to life), Article 6, Paragraph 1 (the right to trial within a reasonable time), Article 8 (the right to respect for private and family life), Article 13 (the right to an effective remedy), and Article 1 of Protocol No. 1 (the right to property). On 18 June 2002, the European Court held Turkey responsible for the violations of Article 2 and Article 1 of Protocol No. 1 of the ECHR. At the request of the Turkish government, the case was remitted to the Grand Chamber, which sentenced the Turkish State for violation of Articles 2 and 1 of Protocol No. 1 and Article 13 of the Convention as regards the complaint under Article 2 and Article 1 of Protocol No. 1.

In deciding on these complaints, the ECtHR found a violation of Article 2 in its substantive aspect. From this point of view, Article 2 obliges contracting States to take all appropriate measures to prevent violations of the right to life as a result of dangerous activities. Consequently, the European Court recognized a violation of the right to life, in its substantive aspect, because despite knowing the potential risk of death linked to a possible methane explosion in the landfill site, the public authorities had not taken the necessary measures to protect the right to life of the inhabitants of the slum. From a substantive point of view, the State has an obligation to provide a legal response in cases of alleged violations of the right to life.

In this regard, the European Court established that "the obligations deriving from Article 2 do not end there" (Para. 91):

> Where lives have been lost in circumstances potentially engaging the responsibility of the State, that provision entails a duty for the State to ensure, by all means at its disposal, an adequate response—judicial or otherwise—so that the legislative and administrative framework set up to protect the right to life is properly implemented and any breaches of that right are *repressed and punished*. (Para. 91; emphasis added)

From a procedural perspective, the European Court distinguished between infringements of the right to life caused intentionally and infringements of those rights caused unintentionally. When the death is a result of unintentional conduct, a positive obligation to put in place an "effective judicial system" within the meaning of Article 2, in its procedural aspect, may be satisfied if "civil, administrative or even disciplinary remedies were available to the victims" (Para. 92). In this case, the European Court decided to apply the principles developed in relation to the use of lethal force. According to this principle, "an official investigation is justified not only because any allegations of such an offence normally give rise to criminal liability, but also because often, in practice, the true circumstances of the death are, or may be, largely confined within the knowledge of State officials or authorities" (Para. 93). Such considerations have been considered valid under certain conditions in connection with cases of deaths or injuries that have occurred in the context of dangerous industrial activities. Moreover, the European Court stated that:

> Where it is established that the negligence attributable to State officials or bodies on that account goes beyond an error of judgment or carelessness, in that the authorities in question, fully realising the likely consequences and disregarding the powers vested in them, failed to take measures that were necessary and sufficient to avert the risks inherent in a dangerous activity […], the fact that those responsible for endangering life have not been charged with a criminal offence or prosecuted may amount to a violation of Article 2, irrespective of any other types of remedy which individuals may exercise on their own initiative. (Para. 93)

Therefore, the European Court has established that a criminal proceeding is not necessary if the death is a result of an "error of judgment or carelessness" on the part of public officials. Nevertheless, it should be stressed that the line that distinguishes the concept of "negligence" from that of "carelessness" is very thin. Although in this paragraph the European Court has used certain dogmatic concepts, its reasoning seems to be more pragmatic than dogmatic. Thus, the principle stated by the European Court can be read clearly only with reference to this specific case.

Moreover, criminal proceedings must meet a number of requirements to satisfy the standards required by Article 2 in its procedural aspect: they must be official, independent, impartial, prompt, effective, and capable of ensuring the application of the criminal penalties. A criminal investigation must have the purpose of ascertaining the facts leading to death or injury in the context of industry, and identifying the responsible authorities. Lastly, the European Court has established that although Article 2 does not provide for a "right for an applicant to have third parties prosecuted or sentenced for a criminal offence", violations of the right to life should not go unpunished. In the Court's view, "this is essential for maintaining public confidence and ensuring adherence to the rule of law and for preventing any appearance of tolerance of or collusion in unlawful acts" (Para. 96). First of all, the European Court has established that it is not necessary to punish and prosecute environmental pollution behaviour, but only human rights violations that result from a lack of protection by public bodies of the environmental law. Consequently, the State has an obligation to protect the right to life by introducing criminal offences against the persons responsible.

By applying these principles to the present case, the European Court considered that the investigating authority had acted with exemplary promptness and had established with diligence the circumstances which led to the death of the inhabitants of the slums. On 4 April 1996, the Criminal Court of Istanbul sentenced the two mayors to a suspended fine of TRL610,000 (approximately €9.70) "for negligent omissions in the performance of their duties within the meaning of Article 230 sec. 1 of the Criminal Code" (Para. 116). In this regard, the European Court reiterated that it is not for the Court to assess whether or not these people are responsible. However, it observed that in this case the only purpose of the criminal proceedings "was to establish whether the authorities could be held liable for 'negligence in the performance of their duties' under Article 230 of the Criminal Code, which provision does not in any way relate to life-endangering acts or to the protection of the right to life within the meaning of Article 2" (Para. 116). It therefore concluded that there had been an infringement of Article 2 of the ECHR in its procedural aspect, on the grounds that the right to life had not received adequate protection by law in connection with the fatal accident caused by a dangerous activity.[24]

5.3 *Deaths and Personal Injuries as a Result of Historical Pollution*

Despite the fact that the ECtHR has progressively emphasized the protection of the right to life through criminal law in cases of disasters (Lacroix 2011), it does not seem that this result can also be referred to cases of historical pollution. Only a partial answer can be given to the question of whether there the contracting State has a duty to protect the right to life under Article 2 of the ECHR through criminal law with reference to cases of deaths caused by historical pollution. There are three main reasons for this. Firstly, the ECtHR has been asked to rule on alleged infringements of the right to life in its procedural aspect only in a limited number of

---

[24] Among other things, the European Court concluded that there had been a violation of the right to an effective remedy, with reference to Art. 2, since the applicant had not been paid the damages awarded in the decision of the Administrative Court of Istanbul. In this regard, the Court has reiterated that "The timely payment of a final award of compensation for anguish suffered must be considered an essential element of a remedy under Article 13 for a bereaved spouse and parent" (Para. 152).

cases, such as *Brincat* and *Smaltini*, that relate to environmental pollution due to past industrial activities. Secondly, only the decision delivered by the ECtHR in the *Smaltini* case centres on a complex situation of environmental contamination caused by toxic substances emitted over time by an industrial plant. In different terms, the *Brincat* case concerns the death of a person who contracted a malignant pleural mesothelioma due to his past exposure to asbestos. Thirdly, although in both the above-mentioned cases the applicants had complained about the procedural violation of the right to life because they believed that the judicial authorities had not carried out criminal investigations that fulfilled the minimum standards required by Article 2 of the ECHR, the ECtHR ruled that Malta and Italy could not be considered responsible for having breached the fundamental right to life of the applicants. In fact, in the two cases in question the European Court recalled the obligations of the State to prosecute the people responsible for deaths in the context of dangerous activities, but did not consider these obligations as having been breached. In the *Brincat* case, the ECtHR established that no investigation *ex officio* was necessary on the grounds that the applicant's death was not confined within the knowledge of state officials. In *Smaltini v. Italy*, the Strasbourg Court established that the decision of the public authorities to discontinue the criminal proceedings was not in breach of Article 2, because the applicants had not demonstrated, in the light of the scientific data available at the time of the events, a causal link between the toxic emissions and his diseases.

Therefore, the ECtHR seems to take different approaches in cases of disaster and historical pollution. Although the ECtHR has emphasized the duty of States to prosecute those persons who are responsible for disasters which give rise to harmful effects for human health, up to now it seems to have adopted a more cautious approach in relation to liability for deaths resulting from historical pollution.

For several reasons, it is difficult to ascertain a violation of the right to life in its procedural aspect in relation to cases of historical pollution. First of all, the European Court has drawn up a general principle according to which it cannot rule on the alleged violations of fundamental rights if these occurred before the entry into force of the ECHR in domestic systems. Secondly, although there is a general trend to develop private duties under human rights law (Hessbruegge 2005), the ECtHR decides on the responsibilities of States regarding the violation of fundamental rights. Therefore, the question is whether a State may be held responsible by the ECtHR for having filed a criminal case for deaths or injuries caused by toxic substances released in the past into the air, soil, and groundwater. Several critical questions arise in connection with the use of criminal responses in relation to cases of deaths or injuries deriving from historical pollution. For instance, the standard of proof beyond reasonable doubt concerning the causal relationship between environmental pollution and long-term personal injury is difficult to achieve. Thus, the ECtHR may rarely deem a violation of the right to life under the procedural profile where national courts have archived a criminal case in the absence of proof of the causal relationship between industrial emissions and the damage suffered by the victims on the basis of the scientific knowledge available at the time of the facts. Furthermore, in cases of historical pollution in a narrow sense, the event (damage to environmental and human health) was often not foreseeable when the conduct took place. Historical pollution is characterized by chronological distance between the pollution event and the physical harm, so it is often difficult to prove *mens rea*. In this regard, it should be pointed out that from the European Court's view, there is no obligation on the part of the State to prosecute offences of negligence: in fact, "if the infringement of the right to life is not intentional, civil, administrative or even disciplinary remedies may be a sufficient response". Given that injuries or deaths in cases of historical pollution derive mainly from unintentional conducts, other remedies can be adopted by domestic Courts.

The satisfaction of the victims can also take place by means of compensation or administrative and disciplinary measures.[25] However, administrative or disciplinary sanctions may constitute criminal measures under the European Convention. In line with the *Engel* criteria outlined by the European Court,[26] Article 6 (right to a fair trial), Article 7 (no punishment without law),[27] and Article 4, Protocol No. 7 (*ne bis in idem*) of the ECHR should be applied to these penalties. By applying the principle of non-retroactivity, the principle of legality, the principle of personal responsibility, and the presumption of innocence[28] in administrative penalty proceedings, it is once again difficult for domestic Courts to identify and punish (legal or physical) persons who have caused historical pollution in a narrow sense.

6 The Legal Responses to Historical Pollution in the Light of Human Rights: Strengths and Weaknesses

Undoubtedly, historical pollution may pose significant risks to life, human health, and property. In these cases, responsibility cannot lie only with the corporations concerned, but, according to the case law of the regional human rights bodies, States too have a primary duty to protect the environment and human rights from historical pollution, by adopting the appropriate legislative and administrative frameworks. In addition, certain regional bodies, such as the ECtHR, oblige contracting States to protect the right to life in the context of industry through criminal law and effective criminal investigations. On this point, the ECtHR has emphasized the duties of the State to prosecute the culprits of industrial, environmental, and technological disasters in which people lose their lives. Nevertheless, it has up to the present adopted a more cautious approach in assessing the responsibility of States for failing to investigate deaths caused by historical pollution or past exposure to toxic substances. In these latter cases, characterized by damages that become apparent a long time after polluting activities were carried out, criminal

[25]About the victim-centric perspective adopted by the ECtHR, see Valentini (2012).

[26]See Delmas-Marty 1987; Esposito (2008, pp. 307 ff.), Mazzacuva (2013).

[27]See Bernardi (2001), Manes (2012).

[28]See Panebianco (2014).

proceedings are often inadequate to meet the victims' real needs, which can be better satisfied with civil and administrative remedies. However, only future developments of the case law of the human rights bodies will be able to show whether, and if so to what extent, the criminal law can play a role in strengthening the protection of human rights in cases of historical pollution.

It should be noted that such an approach to the protection of the environment from historical pollution has many limits. These rights constitute ties only for States that adhere to regional human rights systems, with the result that many industrialized or developing countries have no obligation to observe the indications of these supranational organs. In addition, the force of these rights is not the same in all the conventional systems, depending on the binding nature of the judgment and decisions delivered by regional bodies: from this point of view, the European conventional system may be considered as the more incisive in the implementation of the protection of human rights. Nevertheless, the "anthropocentric" approach developed by the ECtHR does not allow for the protection of the environment regardless of the existence of victims. In the Court's view, there is no violation of the Convention if there is a serious degradation of the environment that does not interfere with the enjoyment of human rights. From this perspective, several authors envisage the need to recognize an autonomous value in the right to a good environment.

Recognition at the international and regional levels of the right to a sound environment could constitute a further way to impose upon public authorities the duty to clean land contaminated by past industrial activities. However, this is not enough. On the one hand, it would be appropriate to consider the opportunity to strengthen the protection of human rights and the environment through international instruments aimed at regulating the responsibilities of transnational corporations with regard to human rights (Chopra 1994; Eaton 1997; Hillemanns 2003; Morgera 2006; Schmidt 2005; Shinsato 2005).[29] On the other hand, only the development of an international policy on land contaminated by past industrial activities can lead to results that are more incisive in the protection of human rights.

References

Akandji-Kombe, J. F. (2007). *Positive obligations under the European Convention on Human Rights. A guide to the implementation of the European Convention on Human Rights.* Human rights handbook, No. 7. Strasbourg: Council of Europe.

Anicama, C. (2008). State responsibilities to regulate and adjudicate corporate activities under the Inter-American human rights system. Report on the American convention on human rights.

[29]In this context, it would be necessary to give greater force to the existing international and regional instruments: see, for example, *The United Nations Code of Conduct for Transnational Corporations*; *The OECD Guidelines for Multinational Enterprises*; *Agenda 21 of the United Nations Conference on Environment in Development*, *UN Norms on the Responsibilities of Transnational Corporations and Other Business Enterprises with Regard to Human Rights.*

Available at http://www.reports-and-materials.org/sites/default/files/reports-and-materials/State-Responsibilities-under-Inter-American-System-Apr-2008.pdf

Anton, D. K., & Shelton, D. L. (2011). *Environmental protection and human rights*. Cambridge: Cambridge University Press.

Arato, J. (2012). Constitutional transformation in the ECtHR: Strasbourg's expansive recourse to external rules of international law. *Brooklyn Journal of International Law, 37*(2), 349–387.

Basch, F. F. (2007). The doctrine of the Inter-American Court of Human Rights regarding states' duty to punish human rights violations and its dangers. *American University International Law Review, 23*(1), 195–229.

Bernardi, A. (2001). Nessuna pena senza legge (art. 7). In S. Bartole, B. Conforti, & G. Raimondi (Eds.), *Commentario alla Convenzione europea per la tutela dei diritti dell'uomo e delle libertà fondamentali* (pp. 249–306). Padova: Cedam.

Boer, B. (2015). *Environmental law dimensions of human rights* (Vol. 21). Oxford: Oxford University Press.

Bosselmann, K. (2001). Human rights and the environment: Redefining fundamental principles? In B. Gleeson & N. Low (Eds.), *Governance for the environment: Global problems, ethics and democracy* (pp. 118–134). London: Palgrave.

Boyle, A. (2007). Human rights or environmental rights: A reassessment. *Fordham Envtl. L. Rev., 18*, 471–511.

Boyle, A. (2012). Human rights and the environment: Where next? *European Journal of International Law, 23*(3), 613–642.

Boyle, A. E., & Anderson, M. R. (1996). *Human rights approaches to environmental protection*. Oxford: Clarendon Press.

Brandon, E. (2013). *Global approaches to site contamination law*. Dordrecht: Springer Science & Business Media.

Chevalier-Watts, J. (2010). Effective investigations under Article 2 of the European Convention On Human Rights: Securing the right to life or an onerous burden on a state? *European Journal of International Law, 21*(3), 701–721.

Chopra, S. K. (1994). Multinational corporations in the aftermath of Bhopal: The need for a new comprehensive global regime for transnational corporate activity. *Valparaiso University Law Review, 29*(1), 235–284.

Coomans, F. (2003). The Ogoni case before the African Commission on human and peoples' rights. *International and Comparative Law Quarterly, 52*(03), 749–760.

Council of Europe. (2012). *Manual on human rights and the environment: Principles emerging from the case-law of the European Court of Human Rights*. Strasbourg: Council of Europe Publishing.

Cuneo, I. M. (2005). The rights of indigenous peoples and the Inter-American Human Rights System. *Arizona Journal of International and Comparative Law, 22*, 53–63.

De Vero, G. (2007). La giurisprudenza della Corte di Strasburgo. In G. De Vero & G. Panebianco (Eds.), *Delitti e pene nella giurisprudenza delle Corti europee* (pp. 11–58). Torino: Giappichelli.

Delmas-Marty, M. (Dir.) (1987). La "matière pénale" au sens de la Convention européenne des droits de l'homme, flou du droit pénal. *Revue de science criminelle et de droit pénal comparé*, 819–862.

DeMerieux, M. (2001). Deriving environmental rights from the European Convention for the Protection of Human Rights and Fundamental Freedoms. *Oxford Journal of legal studies, 21* (3), 521–561.

Desgagné, R. (1995). Integrating environmental values into the European Convention on Human Rights. *American Journal of International Law, 89*, 263–294.

Eaton, J. P. (1997). The Nigerian tragedy, environmental regulation of transnational corporations, and the human right to a healthy environment. *Boston University International Law Journal, 15*, 261–297.

Esposito, A. (2008). *Il diritto penale "flessibile"*. Giappichelli: Quando i diritti umani incontrano i sistemi penali. Torino.

Eweje, G. (2006). Environmental costs and responsibilities resulting from oil exploitation in developing countries: The case of the Niger Delta of Nigeria. *Journal of Business Ethics, 69*(1), 27–56.
Fitzmaurice, M. (2011). The European court of human rights, environmental damage and the applicability of Article 8 of the European Convention on Human Rights and Fundamental Freedoms. *Environmental Law Review, 13*(2), 107–114.
Fitzmaurice, M., & Marshall, J. (2007). The Human Right to a Clean Environment-Phantom or reality? The European Court of Human Rights and English courts perspective on balancing rights in environmental cases. *Nordic Journal of International Law, 76*(2), 103–151.
Francioni, F. (2010). International human rights in an environmental horizon. *European Journal of International Law, 21*(1), 41–55.
Giudicelli-Delage, G., Manacorda, S., & Tricot, J. (2013). *"Devoir de punir?" Le système pénal et la protection internationale du droit à la vie*. Paris: Société de législation comparée.
Green, A. (2013). Rights, risk and the value of life: A critical analysis of the right to life under the European Convention on Human Rights. *Journal of Law and Medicine, 20,* 877–887.
Hessbruegge, J. (2005). Human rights violations arising from conduct of non-state actors. *Buffalo Human Rights Law Review, 11,* 21–88.
Hillemanns, C. F. (2003). UN norms on the responsibilities of transnational corporations and other business enterprises with regard to human rights. *German Law Journal, 4*(10), 1065–1080.
Hilson, C. (2009). Risk and the European Convention on Human Rights: Towards a new approach. *Cambridge Yearbook of European Legal Studies, 11*(1), 353–398.
Jagers, N. (2011). UN Guiding Principles on business and human rights: Making headway towards real corporate accountability. *Netherlands Quarterly of Human Rights, 29*(2), 159–163.
Jarvis, F., & Sherlock, A. (1999). The European Convention on Human Rights and the environment. *European Law Review, 24*, HR-15, 15–29.
Kravchenko, S., & Bonine, J. E. (2012). Interpretation of human rights for the protection of the environment in the European Court of Human Rights. *Pacific McGeorge Global Business & Development Law Journal, 25,* 245–287.
Krstic, I., & Cuckovic, B. (2015). Procedural aspects of Article 8 of the ECHR in environmental cases-The greening of human rights law. *Annals of the Faculty of Law in Belgrade*, 170–189.
Lacroix, C. (2011). L'influence de la jurisprudence de la Cour européenne des droits de l'Homme. Le droit au procès pénal en cas de catastrophes. *RISEO, 3,* 173–186.
Loucaides, L. (2004). Environmental protection through the jurisprudence of the European convention on human rights. *The British Year Book of International Law, 75*(1), 249–267.
Manacorda, S. (2012). "Dovere di punire"? Gli obblighi di tutela penale nell'era della internazionalizzazione del diritto. *Rivista italiana di diritto e procedura penale, 55*(4), 1364–1401.
Manes, V. (2012). Art. 7 §§ I–XV. In S. Bartole, P. De Sena, & V. Zagrebelsky (Eds.), *Commentario breve alla Convenzione europea per la salvaguardia dei diritti dell'uomo e delle libertà fondamentali* (pp. 258–288). Padova: Cedam.
Mazzacuva, F. (2013). La materia penale e il "doppio binario" della Corte europea: le garanzie al di là delle apparenze. *Rivista italiana di diritto e procedura penale*, 1899–1940.
Morgera, E. (2006). The UN and corporate environmental responsibility: Between international regulation and partnerships. *Review of European Community & International Environmental Law, 15*(1), 93–109.
Mowbray, A. (2002). Duties of investigation under the European Convention on Human Rights. *International and Comparative Law Quarterly, 51*(02), 437–448.
Nicosia, E. (2006). *Convenzione europea dei diritti dell'uomo e diritto penale*. Torino: Giappichelli.
Nwobike, J. C. (2005). The African commission on human and peoples' rights and the demystification of second and third generation rights under the African Charter: Social and Economic Rights Action Center (SERAC) and the Center for Economic and Social Rights (CESR) v Nigeria. *African Journal of Legal Studies, 1*(2), 129–146.

Panebianco, G. (2014). The *Nulla Poena Sine Culpa* Principle in European Courts Case Law. The perspective of the Italian criminal law. In S. Ruggeri (Ed.), *Human Rights in European Criminal Law* (pp. 47–78). Switzerland: Springer International Publishing.

Patel, S. (2012). Delayed justice: A case study of Texaco and the Republic of Ecuador's operations, harms, and possible redress in the Ecuadorian Amazon. *Tulane Environmental Law Journal, 26,* 71.

Pedersen, O. W. (2008). European environmental human rights and environmental rights: A long time coming? *Georgetown International Environmental Law Review, 21*(1), 73–111.

Pedersen, O. W. (2010). The ties that bind: The environment, the European convention on human rights and the rule of law. *European Public Law, 16*(4), 571–595.

Pulitanò, D. (2013). Diritti umani e diritto penale. *Rivista italiana di diritto e procedura penale,* 1613–1636.

Roos, S. R. (2013). UN guiding principles on business and human rights. In S.O. Idowu, N. Capaldi, L. Zu, & A. Das Gupta (Eds.), *Encyclopedia of Corporate Social Responsibility* (pp. 2592–2601). Heidelberg, New York: Springer.

San José, D. I. G. (2005). *Environmental protection and the European convention on human rights* (Vol. 88). Strasbourg: Council of Europe Publishing.

Satzger, H. (2012). *International and European criminal law.* München: Beck.

Scalia, V. (2015). *The European Court of Human Rights and environmental crime.* Catania: Study in the framework of the EFFACE research project, University of Catania.

Schmidt, T. M. (2005). Transnational corporate responsibility for international environmental and human rights violations: Will the United Nations' norms provided the required means? *California Western International Law Journal, 36,* 217–247.

Scott, I. K. (2000). The Inter-American System of human rights: An effective means of environmental protection. *Virginia Environmental Law Journal, 19,* 197.

Shelton, D. (1991). Human rights, environmental rights, and the right to environment. *Stanford Journal of International Law, 28*(1), 103–138.

Shelton, D. (2002a). Decision regarding communication 155/96 (Social and Economic Rights Action Center/Center for Economic and Social Rights v. Nigeria). Case No. ACHPR/COMM/A044/1. *American Journal of International Law, 96*(4), 937–942.

Shelton, D. (2002b). Human rights and the environment: Jurisprudence of human rights bodies. *Environmental Policy and Law, 32*(3–4), 158–162.

Shelton, D. (2006). Human rights and the environment: What specific environmental rights have been recognized? *Denver Journal of International Law and Policy, 35*(1), 129–171.

Shelton, D. (2007). Human rights, health and environmental protection: Linkages in law and practice. *Human Rights and International Legal Discourse, 1,* 9–59.

Shelton, D. (2008). Human rights and the environment: Problems and possibilities. *Environmental Policy and Law, 38*(1–2), 41.

Shelton, D. (2015). Legitimate and necessary: Adjudicating human rights violations related to activities causing environmental harm or risk. *Journal of Human Rights and the Environment, 6*(2), 139–155.

Shinsato, A. L. (2005). Increasing the accountability of transnational corporations for environmental harms: The petroleum industry in Nigeria. *Northwestern Journal of Human Rights, 4* (1), 186–209.

Skogly, S. I. (1997). Complexities in human rights protection: Actors and rights involved in the Ogoni conflict in Nigeria. *Netherlands Quarterly of Human Rights, 15*(1), 47–60.

Spieler, P. (2010). The La Oroya case: The relationship between environmental degradation and human rights violations. *Human Rights Brief, 18,* 19–79.

Sudre, F. (1995). Les obligations positives dans la jurisprudence européenne des droits de l'homme. *Revue trimestrielle des droits de l'homme, 23,* 363–384.

Tulkens, F. (2011). The paradoxical relationship between criminal law and human rights. *Journal of International Criminal Justice, 9*(3), 577–595.

Tulkens, F., & van de Kerchove, M. (2014). Criminal law and human rights. In S. Body-Gendrot, M. Hough, K. Kerezsi, R. Lévy, & S. Snacken (Eds.), *The Routledge handbook of European criminology* (pp. 91–106). New York: Routledge.

UNEP Compendium on Human Rights and the Environment. Selected international legal materials and cases (UNEP/CIEL). (2014). Nairobi: UNON Publishing Services Section.
Valentini, V. (2012). *Diritto penale intertemporale. Logiche continentali ed ermeneutica europea.* Milano: Giuffré.
Van der Linde, M., & Louw, L. (2003). Considering the Interpretation and Implementation of Article 24 of the African Charter on Human and Peoples' Rights in Light of the SERAC Communication. *African Human Rights Law Journal, 3*, 167–171.
Viganò, F. (2011). Obblighi convenzionali di tutela penale. In V. Manes–G. Zagrebelsky (Eds.), *La Convenzione europea dei diritti dell'uomo nell'ordinamento penale italiano* (pp. 243–298). Milano: Giuffré Editore.
Wheeler, D., Fabig, H., & Boele, R. (2002). Paradoxes and dilemmas for stakeholder responsive firms in the extractive sector: Lessons from the case of Shell and the Ogoni. *Journal of Business Ethics, 39*(3), 297–318.
Xenos, D. (2007). Asserting the right to life (Article 2, ECHR) in the context of industry. *German Law Journal, 8*(3), 231–254.
Zarsky, L. (2012). *Human rights and the environment: Conflicts and norms in a globalizing world.* London: Earthscan.
Zengerling, C. (2013). *Greening international jurisprudence: Environmental NGOs before international courts, tribunals, and compliance committees.* Leiden: Martinus Nijhoff Publishers.

Cases

ACommHR, Communication 155/96, The Social and Economic Rights Action Center and the Center for Economic, and Social Rights/Nigeria. Case No. ACHPR/COMM/A044/1 (ACommHR 27 May 2002).
ECtHR, *Brincat and others v. Malta*, Appl. Nos. 60908/11, 62110/11, 62129/11, 62312/11 and 62338/11, 24 July 2014.
ECtHR, *Bullerwell v. United Kingdom*, Appl. No. 48013/99, 12 December 2002.
ECtHR, *Cordella and Others v. Italy*, Appl. Nos. 54414/13 and 54264/15, Communicated Case, 27 April 2016.
ECtHR, *Dubetska and others v. Ukraine*, Appl. No. 30499/03, 10 February 2011.
ECtHR, *Fadeyeva v. Russia*, Appl. No. 55723/00, 9 June 2005, 30 November 2005.
ECtHR, *Giacomelli v. Italy,* Appl. No. 59909/00, 2 November 2006.
ECtHR, *Guerra and others v. Italy*, Appl. no. 116/1996/735/932, 19 February 1998.
ECtHR, *Howald Moor and Others v. Switzerland*, Appl. No. 52067/10 and 41072/11, 11 March 2014.
ECtHR, *Janssen v. Germany*, Appl. no. 23959/94, 20 December 2001.
ECtHR, *Klasen v. Germany*, Appl. No. 75204/01, 5 October 2006.
ECtHR, *LCB v. United Kingdom*, Appl. No. 14/1997/798/1001, 9 June 1988.
ECtHR, *Ledyaeva and Others v. Russia*, Appl. Nos. 53157/99, 53247/99, 53695/00, 56850/00, 26 October 2006.
ECtHR, *Locascia and others v. Italy*, Appl. No. 35648/10, Communicated case, 5 March 2013.
ECtHR, *Öneryıldız v. Turkey* [GC], Appl. No. 48939/99, 30 November 2004.
ECtHR, *Smaltini v. Italy*, Appl. No. 43961/09, 24 March 2015.
ECtHR, *Taşkın and others v. Turkey,* Appl. No. 46117/99, 10 November 2004.
IACommHR, *Community of La Oroya v. Peru*, Admissibility Report (Report No. 76/09—Petition 1473-06), 5 August 2009.
IACommHR, *Community San Mateo de Huanchor and its members v. Peru Report*, Admissibility Report (Report No. 69/04, Petition 504/03, 15 October 2004).
IACommHR, *Report on the Situation of Human Rights in Ecuador, OEA/*Ser.L/V/II.96, doc. 10 rev. 1 (1997).

Chapter 14
Historical Pollution and Long-Term Liability: A Global Challenge Needing an International Approach?

Donato Vozza

Contents

D. Vozza (✉)
Seconda Università di Napoli, Naples, Italy
e-mail: donato.vozza@gmail.com

F. Centonze and S. Manacorda (eds.), *Historical Pollution*,
DOI 10.1007/978-3-319-56937-6_14

1 Historical Pollution: A Local Problem in Global Perspective

From the second half of the last century, many developed countries—notably the United States, Japan, and various Western European states—began to take the phenomenon of historical pollution seriously, especially after the discovery of mercury in the water ecosystem of Minamata Bay (Japan) (Upham 1979; Funabashi 2006; Ministry of the Environment, Japan 2011; Coulter 2014, p. 12), the identification of contamination at the former Fischer landfill site (Austria), and severe contamination in the residential area of Love Canal (United States) (Brandon 2013, p. 2).

Such cases led to the introduction of national remediation programmes for contaminated land and environmental damage (Brandon 2013, pp. 2–3). In the USA, for instance, the incident at Love Canal led in 1980 to the Comprehensive Environmental Response, Compensation, and Liability Act (known also as CERCLA or "Superfund Act"), which provides for liability to remediate contamination from historical pollution (Mamutse and Fogleman 2013, p. 587).[1]

In recent decades, the contamination of land and groundwater caused by past activities has emerged as a problem with global scope and considerable nuance. Cases have occurred in Eastern Europe, Asia, South America, and Africa (Brandon 2013, pp. 17–20). In Eastern Europe, Caucasus, and Central Asia, for instance, "historic soil damage is often treated as an independent issue, largely due to the legacy of heavy pollution from state-owned enterprises during the Soviet era and the liability challenges that emerged in connection with their bankruptcy or privatisation" (OECD Report 2012, p. 15; Earnhart 2004); in South America, there is a problem of diffuse mercury pollution deriving from past gold mining (Pestana and Formoso 2003); in South Africa, groundwater pollution caused by acid mine drainage "is a historical problem which is aggravated by present mining practices" (Feris 2012, p. 4); in West/Central Africa, the oil industry "has been polluting Ogoniland in the southeast region of Nigeria for the past fifty years" (Percival et al. 2012, p. 770); and in Tanzania there is the well-known case of the DDT stockpile in the small town of Old Korogwe, "which was leaching into the soil and the adjacent Pangani River, a source of water and food for the town" (Percival et al. 2012, p. 770).

[1]Between 1960 and 1980 cases of historical contamination emerged in all the countries of Western Europe. In the early 1980s, severe arsenic contamination of groundwater was found at an abandoned factory at Bocholt in Belgium. In 1980, chemical waste was discovered in the subsoil of a new housing development at Lekkerkerk in the Netherlands. Severe contamination of soil and groundwater as a consequence of industrial and mining activities was found in the Lower Swansea valley and at the village of Shipham. The discovery of these and other cases led all the States to introduce national remediation programmes (see Prokop et al. 2000).

The phenomena are unquestionably significant; what, then, are the solutions that have been adopted at international level? Although the answer is complex, we can identify several broad approaches to addressing historical pollution.

First, we see a process of legal transplantation of the most advanced domestic models on environmental liability to countries that have only recently become faced with historical pollution. For example, the CERCLA, one of the better legal instruments introduced for preventing and remedying contamination from historical pollution, has in part influenced the EU law on environmental liability (Vanheusden 2007, p. 568; Percival et al. 2012),[2] as well as the current debate on the issue concerning the remediation of environmental damage in important countries such as China (Jeffery and Zhao 2012; Zhao 2013). Such a process, by which countries "imitate" the American model, could be described as a "horizontal circulation" or "legal transplantation" (Watson 1974; Sacco 1991; Mattei 2008) of the CERCLA. Nevertheless, it should be emphasized that there is a manifest note of caution in the way CERCLA is circulating from the USA to other legal systems. Thus, although the laws of many countries are gradually moving towards the US model, no country has yet decided to adopt an environmental programme equivalent to the CERCLA (Percival et al. 2012, p. 771).

Second, it should be noted that certain new international agreements have been adopted with the aim of preventing future historical pollution. For example, recognizing the lessons of Minamata Disease, a global agreement known as the Minamata Convention on Mercury (Kumamoto 10 October 2013) has been adopted to reduce the use of mercury (Coulter 2014). Moreover, given that persistent organic pollutants (POPs) have toxic properties, resist degradation, and accumulate in ecosystems, a Convention on Persistent Organic Pollutants for reducing and eliminating the use and production of these substances was adopted in Stockholm on 22 May 2001. Conversely, there are still no sufficient international and regional instruments on protecting soil against contamination nor on the management of contaminated land. As regards the liability regime for historic environmental harms, it must be emphasized that the main indications here come from soft law instruments, such as the OECD's *Liability for Environmental Damage in Eastern Europe, Caucasus and Central Asia (EECCA): Implementation of Good International Practices* (OECD 2012). In the EU framework, the harmonization of domestic laws on environmental liability is working through the European Liability Directive, but as noted above, this regional law cannot be applied to historic environmental damage.

[2]With regard to EU countries, Vanheusden (2007) has noted that "*Until recently, the historical contamination of land has not been the subject of effective, formalized legal attention.* Member States—or regions within a Member State—have only begun to introduce legislation on the remediation of soil contamination within the last ten to twenty years, often inspired by the American approach" (emphasis added).

Third, according to an "oblique" dynamic (Alogna 2016, p. 20), international and regional lending institutions such as the World Bank have developed many projects intended to remedy groundwater and soil contamination, by imposing specific conditions on the countries concerned to allocate the necessary funds for clean-up programmes (Brandon 2013, p. 292). In this regard, it should be highlighted that in its 2012 report *Getting to Green—A Sourcebook of Pollution Management Policy Tools for Growth and Competitiveness*, the World Bank develops a specific approach for addressing historical contamination, by providing —first of all—a definition of "legacy pollution."

Having made these three observations from a general perspective, the attention of this chapter now shifts to the content of the international instruments and global initiatives for remedying environmental damage and protecting human health from historical pollution. After studying the relationship between the general principles of environmental law and the concept of historical pollution, the chapter presents a careful examination of the international measures adopted for preventing and remedying historic contamination.

2 International Principles of Environmental Law

The first objective of this chapter is to identify the relationship between key international principles in environmental matters, namely the polluter-pays principle and the precautionary principle (De Sadeleer 2002; Pedersen 2010; Sands et al. 2012, pp. 187 ff.; Renna 2012), and the concept of historical pollution. As will be described in the pages that follow, the application of the polluter-pays principle to phenomena of historical pollution is problematic (Shelbourn 1994), and in many cases frankly impossible (Munir 2013, p. 9). In other words, the costs of remediation of contaminated sites by past activities are often sustained by the collective because, for instance, the operator cannot be identified (i.e., cases of orphan sites) or cannot be held responsible under a liability regime that accepts the principle of no retroactivity as a limit to the application of the polluter-pays principle. However, the principles that govern environmental matters may be viewed in both a retrospective and a prospective sense. Accordingly, the polluter-pays principle should not only be read as a guiding principle for remedying environmental damage, but also as a principle that can help develop measures for preventing historical contamination (e.g., eco-taxes). Likewise, the principles of precaution and prevention represent not only two key tools currently employed for avoiding future cases of historical pollution, but also play an important role in guiding operators in procedures for decontaminating polluted land.

2.1 *Historical Pollution as a Limit to the Application of the Polluter-Pays Principle?*

According to the polluter-pays principle, a juridical or physical person that causes damage to soil, air, water, flora, and fauna should pay the costs for cleaning the degraded environment. This principle, which has economic origins, was first recognized by the Organization for Economic Cooperation and Development (OECD) in a non-binding document drafted in 1972.[3] Over the following decades, the principle was recognized in several international agreements on the environment, as well as in EU Law, but the various formulations are not uniform, leading to difficulties in clearly identifying its boundaries (Larson 2005, pp. 547 ff.). It has been argued, for instance, that the principle has different meanings and rationales (Luppi et al. 2012, pp. 17–20), including a pedagogical function consisting in encouraging producers to respect the environment in the context of dangerous activities.

As a principle of environmental policy recognized by several international agreements, the polluter-pays principle has contributed to the convergence of environmental responsibility systems in numerous developed countries, especially in the USA and the European Union (Larson 2005, p. 564). In these countries, the polluter-pays principle has the following two meanings: in a positive meaning, a firm that causes environmental damage should pay the decontamination costs ("those who pollute must pay"),[4] while in a negative meaning the community, in principle, should not bear the costs of environmental damage ("those who do not pollute do not pay"). Subverting the original logic of the principle, certain emerging countries (e.g., India) have elaborated a new model of government liability for environmental harm that can be defined as a "government-pays regime," namely a system on the basis of which public authorities are jointly-and-severally or primarily responsible for the harms caused by private activities (Luppi et al. 2012).

Such a principle, in its traditional formulation, may appropriately be applied only if the contaminator can be identified, with the consequence that it does not operate in cases of orphan sites[5] where the real contaminator who should pay the costs cannot be found (Munir 2013, p. 19). However, even where the polluters can be identified, it is difficult to establish whether they can be deemed responsible under the polluter-pays principle for pollution due to activities carried out before that environmental law was adopted (ibid.).

The answer to such a question can be sought only in the laws adopted at internal, European, and international level in accordance with the principle in question. In this connection, the relationship between this principle and the liability regime applicable to the phenomenon of historical pollution may vary from one system to

[3]Recommendation on Guiding Principles concerning International Economic Aspects of Environmental Policies adopted by the OECD Council on 26 May 1972, C(72) 128.

[4]As evidenced by the doctrine, the polluter-pays principle requires the polluter to bear the primary obligation for the remediation of sites that they have contaminated. See Brandon 2013, p. 82.

[5]On this issue, see Larson 2013, pp. 991 ff.

another (Stallworthy 1996; Mamutse and Fogleman 2013). On the one hand, several countries—such as the USA—"have enacted liability regimes to remediate land contaminated by historic pollution" (Mamutse and Fogleman 2013, p. 581). This regime has been criticized for imposing retroactive liability, often borne by the current owners of contaminated land (ibid.). Other countries, on the other hand, do not have a regime of responsibility to remediate land and groundwater contamination from historical pollution. Even in EU law, the Environmental Liability Directive (henceforth ELD), based on the polluter-pays principle, is not applicable to historic damage to the environment, namely damage where "more than 30 years have passed since the emission, event or incident, resulting in the damage, occurred," and to damage caused before 30 April 2007 (Art. 17). In this event, the costs of restoring sites contaminated from historical pollution will be borne by public authorities (de Sadeleer 2015, p. 234). Moreover, although this Directive establishes the principle of non-retroactivity of responsibility, it should be noted that, since it is a minimum, the Member States are free to adopt a different rule at domestic level which recognizes the retroactive effect of the liability (Bergkamp 2005, pp. 391–392). However, such a rule aimed to extend responsibility to cases of historical pollution would not constitute implementation of EU law.

Having outlined the general limits to the application of the polluter-pays principle in cases of historical pollution under certain liability regimes, attention turns to the principles of prevention and precaution.

2.2 *The Precautionary Principle: A Complex Tool for Preventing Long-Term Risks*

In his *Risk Society: Towards a New Modernity*, Ulrich Beck distinguished between traditional risks and the risks of modernity,[6] pointing out that the latter, characterized by scientific uncertainty, have local and global consequences (Beck 1992). The international and European responses to this lack of scientific certainty are based on the precautionary principle (de Sadeleer 2010, p. 185; Girela 2006, p. 273): this principle imposes limits on the use of new technologies that, in a condition of scientific uncertainty, could cause grave or irreparable damage to the environment and human health (Freestone and Hey 1996; Boutillon 2002, p. 430).

From the latter half of the 1980s onwards, many international instruments have incorporated the precautionary principle with regard to water pollution, waste management, and other fields of environmental law. In EU law, the principle is recognized in Article 191(2) TFEU and in other areas of health and safety. Taking into account the multiple formulations of the principle, the following discussion addresses its binding nature at international level and its relation with scientific and

[6]On the characteristics of the new generation of risks see de Sadeleer 2010, p. 185.

economic data (Hickey and Walker 1994, p. 424). As regards its binding nature at international level, since the principle is contained in a number of international instruments ranging from non-binding to binding texts, it is unclear whether it may be described as an international recommendation, approach, standard, or obligation (Hickey and Walker 1994/1995, p. 437). With regard to its relation with scientific and economic data, the degree of scientific uncertainty that requires the adoption of appropriate measures for avoiding damage to human health or the environment is not clear, since it depends on several factors, including the capabilities and socio-economic conditions of the State in question. On this point, scholars have distinguished between at least two versions of the precautionary principle, depending on whether the measures to be taken for the protection of human health and the environment are strong or weak (Sunstein 2003, p. 1014). These two versions of the principle have different impacts on the development of domestic economies, however.

Having thus summarily presented the precautionary principle, it must be clarified that this principle, along with the principle of prevention, is currently used for managing long-term risks. In particular, in the context of a lack of scientific certainty, the precautionary principle aims to prevent pollution events that develop after a long period of time and whose adverse effects on human health and the environment only become apparent decades later (Lee 2003, p. 86), while the principle of prevention is used for avoiding undeniable risks to humans and environment. By virtue of these principles, recent decades have seen the adoption of global and regional initiatives and agreements for preventing the release of toxic substances into the environment (e.g., POPs), for regulating dangerous activities (e.g., waste management), and for conserving environmental compartments (especially soil and groundwater). All the instruments described below are intended to avoid cases of historical pollution occurring in the future. Nevertheless, it should be noted that the precautionary principle also plays a significant role with reference to the remediation and management of sites contaminated by past activities (Brandon 2013, p. 71).

3 Supranational Legal Frameworks: Preliminary Remarks

In outlining the current international law framework with respect to the multifaceted phenomenon of historical pollution, it is necessary to distinguish between international instruments adopted for preventing risks to the environment in a long-term perspective, and those instruments developed for remedying damage to the environment caused by past activities.

Section 4 discusses the international and regional instruments adopted for preventing long-term risks of damage to the environment and human health, according to the principles of prevention and precaution. International instruments on persistent organic pollutants and other substances with toxic properties are discussed in Sect. 4.1; international measures regulating hazardous activities that have long-term harmful consequences for the environment and human health are discussed in Sect. 4.2; measures for the protection of soil and groundwater from industrial contamination are discussed in Sect. 4.3.

Section 5 provides a brief overview of the international instruments which, in line with the polluter-pays principle, regulate civil liability for damage caused by dangerous activities, and assesses to what extent they contain elements for remedying historic damage to the environment or damage caused by past activities. Particular attention will be devoted to the OECD's *Liability for Environmental Damage in EECCA: Implementation of Good International Practices* (OECD 2012), which contains interesting information about the liability regimes adopted by countries with reference to historic damage (Sect. 5.1); as well as the ELD, as interpreted by the EU Court of Justice in relation to cases of historic damage and gradual and diffuse pollution (Sects. 5.1 and 5.2), and other EU Directives aimed at imposing obligations on operators after the closure of a site (Sect. 5.3). The focus then turns to an old proposal of the Commission to introduce a directive on the protection of soil in the EU framework (Sect. 6.1), and a number of global initiatives undertaken by international and regional lending institutions, networks, and NGOs for the effective development of policies on contaminated sites (Sect. 6.2).

A final question that needs to be answered is whether international agreements and EU law provide for criminal-law responses to past polluting activities that cause current harm to the environment and human health (Sect. 7).

4 Instruments for Preventing Environmental Pollution in a Long-Term Perspective

In line with these preliminary remarks, attention now turns to international and regional instruments adopted for preventing damage to the environment and human health in a long-term perspective. First, there will be a discussion of international instruments on persistent organic pollutants and other substances with permanent toxic properties. In this connection, it will be shown that many international and EU instruments that restrict the use, disposal, storage, and transport of persistent organic pollutants and heavy metals have been adopted to prevent cases of historical pollution, but this strategic approach needs to be further developed in the years to come. Secondly, the chapter will focus on international and regional instruments regulating dangerous activities from which long-term harmful consequences for the environment and human health can derive. In this regard, it must be noted that in the EU context a strategy has been developed based on the protection of the environment through the supranational regulation of present-day dangerous activities, including Directives and Regulations on the prevention and control of industrial emissions (the EU IPPC Directive), on the assessment of the effects of public and private projects on the environment (the EU EIA Directive), and on the participation of organizations in a community eco-management and audit scheme (the EU EMAS Regulation)—these being three different general instruments that seek in a synergetic way to prevent the contamination of soil and water. In addition, a specific legal framework on the management of waste, especially that deriving from extractive industries, may be considered an important instrument for protecting the environment. Lastly, it will be highlighted that one of the main limitations of the current global strategies for the prevention of historical contamination is the absence of an international agreement on soil protection.

4.1 Reducing and Eliminating the Use of POPs and Other Toxic Substances

In the most developed countries, the main cause of historical pollution is the widespread use of persistent organic pollutants (Fiedler 2002) and other toxic substances in the context of industrial activities carried out over the last century.[7] Consequently, the most efficient strategy for protecting the environment from the adverse effects of historical contamination is, on the one hand, to prevent new releases of POPs and heavy metals into the environment and, on the other, to decontaminate polluted soil and groundwater.[8]

With reference to the first strategy, many binding agreements have been adopted in the last decade for reducing the use, disposal, transport, and storage of persistent organic pollutants or heavy metals in accordance with the precautionary and prevention principles. The most important instrument in force for preventing contamination in accord with the precautionary principle is the Stockholm Convention on Persistent Organic Pollutants (Lallas 2001). Because these pollutants "possess toxic properties, resist degradation, bioaccumulate and are transported, through air, water and migratory species, across international boundaries and deposited far from

[7]See Forgotson 1964, Milhollin 1979, pp. 1 ff., and Martin 1985, pp. 299–300.

[8]Such a strategy has been emphasized by the European Union Commission in the Communication from the Commission to the Council, the European Parliament and the Economic and Social Committee—Community strategy for Dioxins, Furans and Polychlorinated Biphenyls (2001/C 322/02) (COM(2001) 593 final), according to which "PCBs, and that is the main difference with dioxins, are intentionally produced chemicals, that were manufactured for decades before the ban in marketing and use was adopted in 1985 due to their reproductive toxicity and bio-accumulative effects. The main part of these products, which are persistent and bioaccumulable in fat of biota, is now spread in soils, sediments and in the whole aquatic environment ('historical pollution')." Consequently, "[t]he most efficient way to reduce the levels [of PCBs] in the foodchain is to reduce the contamination in the environment. This should be done by: (1) avoiding 'new releases' in the environment; (2) addressing 'historical pollution.'" Moreover, see Communication from the Commission on the Results of the Risk Evaluation and the Risk Reduction Strategies for the Substances: Cadmium and Cadmium Oxide—(2008/C 149/03).

their place of release, where they accumulate in terrestrial and aquatic ecosystems," they produce an adverse impact on the environment and on human health in a long-term perspective. After several amendments, the Stockholm Convention now regulates tens of POPs divided into three annexes, and the Commission is considering a proposal to extend these lists to additional toxic substances. Annex A lists the POPs whose production and use shall be eliminated, while Annex B lists those whose production shall be restricted, and Annex C specifies POPs from unintentional production which must be reduced or eliminated. This Convention also provides that States must develop strategies to identify and decontaminate sites where there is a presence of these toxic substances. In similar terms to the Stockholm Convention on POPs, there is the Protocol to the 1979 Convention on Long-Range Transboundary Air Pollution on Persistent Organic Pollutants (CLRTAP), which was adopted in Aarhus, 24 June 1998, to prevent or minimize emissions of persistent organic pollutants.[9] On the basis of this Protocol, each State must adopt measures to eliminate the use of substances included in a first annex and restrict the use of substances listed in a second annex.

Other relevant instruments adopted at international level for minimizing the use and production of polluting chemicals are: (a) the Strategic Approach to International Chemicals Management (SAICM), based on five pillars, including the need to develop adequate information on polluting chemicals and their sound management; (b) the Rotterdam Convention on the Prior Informed Consent Procedure for Certain Hazardous Chemicals and Pesticides in International Trade (Rotterdam, 10 September 1998), according to which every chemical product included in a specific list indicated in the agreement must receive prior consent of the importer before being exported; (c) the Protocol on Pollutant Release and Transfer Registers to the Aarhus Convention (Kiev, 21 May 2003), which aims to ensure public access to information through the provision of a register, namely the Pollutant Release and Transfer Registers (PRTR). Lastly, a number of international agreements and protocols, such as the Aarhus Protocol on Heavy Metals to the 1979 Convention on Long-range Transboundary Air Pollution (Aarhus, 24 June 1998) and the Minamata Convention on Mercury, have been adopted with the aim of reducing the use of heavy metals in the context of industrial and commercial activities. At regional level, many of the above-mentioned international instruments have been adopted in the EU framework.[10]

---

[9]Art. 1, Para. 7, of the Convention establishes that POPs are organic substances that: (a) possess toxic characteristics; (b) are persistent; (c) bioaccumulate; (d) are prone to long-range transboundary atmospheric transport and deposition; and are likely to cause significant adverse human health or environmental effects near to and distant from their sources.

[10]The Rotterdam Convention on the Prior Informed Consent Procedure for Certain Hazardous Chemicals and Pesticides in International Trade was transposed into the European Union with Decision 2003/106/EC. However, in Case C-94/03 the Court of Justice of the European Union annulled the decision, arguing that it would have had to be based exclusively on the rules of the EC Treaty concerning the common commercial policy and not on the article concerning the environmental policy. By Decision 2006/730/EC, the Rotterdam Convention was again transposed. Regulation (EC) no. 1907/2006 of the European Parliament and of the Council of 18 December

It is clear, then, that many international and European instruments that restrict the use of persistent organic pollutants and heavy metals have been adopted to prevent cases of historical pollution. However, this strategic approach stands in need of further development in the future.

4.2 Regulating Dangerous Activities: The Management of Hazardous Waste in EU Law

Historical contamination of soil and groundwater is often a consequence of industrial releases resulting from past activities carried out under a lack of appropriate regulations.[11] For preventing further cases of this type of pollution, a plurality of instruments regulating dangerous activities or industrial emissions have been approved within the EU framework, notably Directive 2010/75/EU of the European Parliament and of the Council of 24 November 2010 on Industrial Emissions (Integrated Pollution Prevention and Control—IPPC),[12] Directive 2011/92/EU of the European Parliament and of the Council of 13 December 2011 on the Assessment of the Effects of Certain Public and Private Projects on the Environment (EIA) , as emended by Directive 2014/52/EU, and Regulation No. 1221/2009 of the European Parliament and of the Council of 25 November 2009 on the Voluntary Participation by Organisations in a Community Eco-management and Audit Scheme (EMAS), as emended. Again, these three different general instruments seek to work synergistically to prevent the contamination of soil and water.

(Footnote 10 continued)

2006 Concerning the Registration, Evaluation, Authorisation and Restriction of Chemicals (REACH) implements the Rotterdam Convention (Recital no. 2). Furthermore, the European Union has adopted the following international instruments: Regulation (EU) no. 649/2012 of the European Parliament and of the Council of 4 July 2012 Concerning the Export and Import of Hazardous Chemicals; Council Decision 259/2004/EC of 19 February 2004 Concerning the Conclusion, on Behalf of the European Community, of the Protocol to the 1979 Convention on Long-Range Transboundary Air Pollution on Persistent Organic Pollutants; Council Decision 2006/61/EC of 2 December 2005 on the Conclusion, on Behalf of the European Community, of the UNECE Protocol on Pollutant Release and Transfer Registers (PRTR); the Council Decision Concerning the Conclusion, on Behalf of the European Community, of the Stockholm Convention on Persistent Organic Pollutants was adopted on 14 October 2004 and emended by several Commission regulations.

[11]For instance, chemical manufacture (e.g., petroleum and petrochemical industries, pharmaceutical manufacture, pesticide manufacture, etc.); metal and mineral extraction, refining, storage, and use (e.g., metal treatment, mining industries, asbestos products manufacture, hydrocarbon exploration, etc.); waste recycling, treatment, and disposal; storage and use of explosives; production and distribution of GM products, etc.

[12]Recital No. 24 of the IPPC Directive establishes that "[i]n order to ensure that the operation of an installation does not deteriorate the quality of soil and groundwater, it is necessary to establish, through a baseline report, the state of soil and groundwater contamination."

However, it should be noted that there is a specific legal framework in the EU system on the management of waste, which can be considered as strategic in preventing damage to the environment from a long-term perspective. The reference instrument is Directive 2008/98/EC of the European Parliament and of the Council of 19 November 2008 on Waste, which defines certain key concepts such as those of waste, recovery, and disposal, and establishes the requirements for a form of waste management that is in compliance with respect for the environment and human health (Art. 13). The Directive provides that each State has the duty to prepare waste management plans (Art. 28) and waste prevention programmes (Art. 29) and that the entities and undertakings who engage in activities concerning waste treatment operations are subject to periodic inspection. On the basis of Article 36, Member States must adopt effective, proportionate, and dissuasive sanctions for violation of the Directive's provisions. Another pillar of the European law on waste is Council Directive 1999/31/EC of 26 April 1999 on the Landfill of Waste (as emended), which has been adopted with the aim to prevent long-term harmful environmental effects that may be caused by waste.[13] This Directive regulates a number of things, including the application for and granting of authorization to carry out the management of waste, control and surveillance procedures, and closure procedures.

In addition to these general directives concerning waste, other directives regulate the management of waste produced by industry. The paradigmatic example is the EU legal framework on waste from the titanium dioxide industry. Refuse from the titanium dioxide industry has prejudicial risks for human health and the environment, and for this reason the European Union adopted three directives[14] which require Member States to take all necessary actions to make sure that such waste is dealt with without causing problems for human health and the environment. The second example is Directive 2006/21/EC of the European Parliament and of the Council of 15 March 2006 on the Management of Waste from Extractive Industries and Amending Directive 2004/35/EC, intended to prevent pollution caused by waste from extractive industries.[15] Physical or legal persons responsible for the management of waste must prevent or reduce the impact of these operations on the

---

[13]In particular, Recital No. 12 of the Directive clarifies that it has the aim "to indicate clearly the requirements with which landfill sites must comply as regards location, conditioning, management, control, closure and *preventive and protective measures to be taken against any threat to the environment* in the short as well as *in the long-term perspective, and more especially against the pollution of groundwater by leachate infiltration into the soil*" (emphasis added).

[14]See: Council Directive 78/176/EEC of 20 February 1978 on Waste from the Titanium Dioxide Industry; Council Directive 82/883/EEC of 3 December 1982 on Procedures for the Surveillance and Monitoring of Environments Concerned by Waste from the Titanium Dioxide Industry; Council Directive 92/112/EEC of 15 December 1992 on Procedures for Harmonizing the Programmes for the Reduction and Eventual Elimination of Pollution Caused by Waste from the Titanium Dioxide Industry.

[15]Recital No. 18 of the Preamble states that the aim of the Directive is "to indicate clearly the requirements with which waste facilities servicing the extractive industries should comply as regards location, management, control, closure and *preventive and protective measures to be taken*

basis of the best available techniques. In addition, operators should adopt a waste management plan, whose objective is also to prevent negative long-term effects on the environment.[16] The waste management plan must contain certain mandatory elements, including a description of the negative impact of the deposit of waste on the environment and on human health, the proposal of a plan for closure and post-closure procedures, and a survey of the soil and groundwater which will be affected by the landfill of waste. Moreover, a financial guarantee must be provided to the public authorities by operators who deal with the depositing of extractive waste, so that "there are funds readily available at any given time for the rehabilitation of the land affected by the waste facility" (Art. 14(1)(b)). In this case too, effective, proportionate, and dissuasive sanctions must be applied by States in case of breaches of the national provisions adopted to implement this Directive.

On the basis of this brief analysis, then, it can be seen that the second strategy adopted in the European context to prevent historical pollution is based on the adoption of laws aimed at regulating dangerous activities, including waste management activities.

4.3 Protecting Soil and Groundwater Against Contamination

As highlighted by researchers, historical contamination can be found in soil and groundwater, but not in the air, since "contamination disappears due to dilution within a relatively short timeframe" (Swartjes 2011, p. 1025); however, the air can constitute a means of transport of POPs from which historical soil and groundwater contamination could derive, and thus the international protection of the air against industrial emissions must be considered a central tool for preventing future damage to human health and the environment.[17]

That said, it should be stressed that, although a global approach to soil protection is necessary to prevent historical contamination of land, current international and regional agreements especially govern certain types of soil damage, such as

(Footnote 15 continued)

against any threat to the environment in the short and long-term perspectives, and more especially against the pollution of groundwater by leachate infiltration into the soil" *(emphasis added).*

[16]Art. 5(2)(c)(ii) of this Directive states that the waste management plan "prevents or at least minimises any long-term negative effects, for example attributable to migration of airborne or aquatic pollutants from the waste facility."

[17]At international level see, among other instruments to protect air from pollution, the Protocol to the 1979 Convention on Long-Range Transboundary Air Pollution on Persistent Organic Pollutants (CLRTAP).

desertification[18] and soil degradation[19] (Hannam and Boer 2002, pp. 57–58; Brandon 2013, p. 55; Wyatt 2008, p. 192), with the consequence that only a few instruments and global initiatives aim at protecting the soil against contamination. In EU law, the Draft EU Framework Directive for Soil Protection presented by the European Commission in 2006 was recently withdrawn,[20] while the legal framework on the protection of groundwater is more advanced.[21] On the basis of this brief survey, it is possible to claim that one of the main limitations of the current global strategies for the prevention of historical pollution is the absence of a general international agreement concerning the protection of soil against contamination.[22]

5 Environmental Liability for Historic Damage in International Perspective

Following on from these preliminary remarks, the next pages provide a brief overview of the international instruments regulating civil liability for damage caused by dangerous activities, in order to assess whether and to what extent they contain elements relevant to remedying damage caused to the environment by past activities. Not all typologies of conventions on civil liability will be taken into consideration[23]; rather, the discussion below reviews general conventions on liability for environmental damage,[24] devoting particular attention to the international perspective, and notably the OECD's *Liability for Environmental Damage in EECCA: Implementation of Good International Practices* (OECD 2012), which contains interesting suggestions about the predisposition of liability regimes to

[18]See, for example, United Nations Convention to Combat Desertification (UNCCD).

[19]See, for example, European Union Protocol on the Implementation of the Alpine Convention of 1991 in the Field of Soil Conservation (1998).

[20]European Commission, Proposal for a Directive of the European Parliament and of the Council Establishing a Framework for the Protection of Soil, COM (2006) 232 final.

[21]At EU level, see, for example, Directive 2000/60/EC of the European Parliament and of the Council of 23 October 2000 Establishing a Framework for Community Action in the Field of Water Policy and Directive 2006/118/EC of the European Parliament and of the Council of 12 December 2006 on the Protection of Groundwater Against Pollution and Deterioration.

[22]Currently, the possibility of introducing an international agreement for the protection of soil is under discussion by authoritative scholars: see Brandon 2013, pp. 47–48, and Boer and Hannam 2015.

[23]Without any pretensions to being exhaustive, we may distinguish between the following international agreements: conventions on civil liability for environmental damage; conventions on civil liability for damage caused by oil pollution; conventions on civil liability for damage caused by the transport of dangerous goods; conventions on civil liability for nuclear risks; conventions on liability for damage resulting from the movement of hazardous wastes; conventions on civil liability and compensation for damage caused by industrial accidents; conventions on civil liability for damage caused by space objects.

[24]See, among others, Bowman and Boyle 2002, and de La Fayette 2010.

address historical damage. Within the EU framework, specific attention will be paid to the ELD and to the jurisprudence of the European Court of Justice on cases of historical and diffuse pollution. The discussion will review a range of topics, namely the legal framework for environmental liability (definition of environmental damage, requirements for damage remediation and compensation, and liability regime—i.e., strict or fault-based liability, and joint or proportional liability), the assessment of environmental damage, and the connection between liability and financial or insurance guarantee. And last but not least, it provides an overview of the main provisions contained in EU law which impose procedures on operators after the closure of a site, exhibiting the new strategies developed for gradually extending the responsibility of the operators for past activities.

5.1 International Instruments on Liability for Damage Caused by Dangerous Activities

Before addressing the legal framework relating to civil liability for damage caused by past activities, a distinction first needs to be made between international liability and domestic liability for damage to the environment. In this respect, even though there are several agreements which establish international liability for trans-boundary damage, namely "border-crossing damage via land, or air" (Xue 2003), attention here will focus on international agreements and initiatives which establish national liability for environmental damage.

Furthermore, a distinction must also be made between civil liability for sudden and accidental pollution and civil liability for gradual and historical pollution. Generally, international agreements or soft law that require States to adopt a civil liability regime for environmental damage do not refer to gradual or historical pollution, but to sudden or accidental pollution. Therefore, only certain provisions or recommendations incorporated in international and regional instruments of hard or soft law are relevant to research on the provisions relating to liability for damage caused by past activities.

That said, special attention will be paid to certain provisions of the Convention on Civil Liability for Damage Resulting from Activities Dangerous to the Environment (1993), the Resolution on Responsibility and Liability under International Law for Environmental Damage (1997), the Guidelines for the Development of Domestic Legislation on Liability, Response Action and Compensation for Damage Caused by Activities Dangerous to the Environment (2010), and the OECD's 2012 report *Liability for Environmental Damage in EECCA: Implementation of Good International Practices.*

First of all, the 1993 Convention on Civil Liability for Damage resulting from Activities Dangerous to the Environment, better known as the Lugano Convention, is an instrument—currently not in force—adopted in the context of the Council of Europe in order to ensure the prevention and remedying of environmental damage caused by dangerous activities according to the polluter-pays principle. The civil liability regime established by this Convention concerns certain dangerous activities, namely operations involving the dangerous substances expressly indicated in the Convention,[25] GMOs, and the operation of sites for the treatment or recycling of waste as set out in a non-exhaustive list in Annex II to the Convention and for the permanent deposit of such waste. Moreover, it covers a wide range of damages, such as death or personal injury, property damage, damage to the environment (natural resources such as water, air, soil, fauna, and flora), and the costs of preventive measures. As clarified by Paragraph 34 of the *Explanatory Report*, the Convention should be applied only in cases of damage caused by dangerous activities, but not in cases of deaths or personal injuries arising from non-dangerous activities.

It is important to distinguish between the responsibility regimes for damage caused by the permanent deposit of waste (Art. 7) and those resulting from other dangerous activities listed in the Convention, including the treatment of waste (Art. 6). On this point, Paragraph 34 of the Explanatory Report to the Convention clarifies that the choice to make a distinction between the liability regime for the treatment of waste and that for the permanent deposit of waste derives from the need to take into account the different natures of the damage resulting from these operations.

As regards as the liability regime for damage resulting from waste treatment operations and other dangerous activities, Article 6 of the Lugano Convention distinguishes between situations depending on whether the damage is the consequence of an incident that consists in a single, continuous, or series of occurrences. Where damage is a consequence of a single-occurrence incident, the operator in control of the dangerous activity at the moment in which the incident took place is to be held responsible for the damage to the environment or human health. If the damage is a result of a plurality of events caused by several operators, the latter are jointly and severally liable unless one of the operators proves that he caused only part of the damage. Finally, if the damage occurred after the cessation of all activities on a site, the last operator who carried out the activity is responsible, unless he can prove that all or part of the damage was caused by a supervening event.

The liability regime to be applied to damage caused by the operator of a site responsible for the permanent deposit of a waste is exceptional, because it is not based on the general rule that liability is based on the time when the incident took place, but—as recognized by Article 7 and specified by the Explanatory Report to the Convention—"on the time when the damage becomes known." The latter

---

[25]Dangerous substances are: (a) substances which have properties listed in Annex I, Part A to the Convention (i.e., toxic, very toxic, harmful, carcinogenic, dangerous for the environment, etc.), which cause a "significant risk for man, the environment or property"; (b) substances listed in Annex I, part B, which are dangerous over certain concentrations (Art. 2, Para. 2). Annex I refers to the EU instruments on the approximation of domestic law relating to the classification, packaging, and labelling of dangerous preparations.

provision is not to be applied to the treatment of waste on a site which falls under the liability regime in Article 6, with the consequence that where several operations of a different nature—such as treatment of waste under Article 6 and permanent deposit of waste under Article 7—are carried out on the same site by the same operator or several operators, different regimes of liability must be distinguished (Art. 7, Paras. 2 and 3).

The operator's liability under the Convention is based on a strict responsibility scheme and is subject to a limited number of exemptions. The relevant clauses include those that exclude the liability of the operator for damage caused in compliance with a measure imposed by a public body, or "by pollution at tolerable levels under local relevant circumstances" (Art. 8). Moreover, when a victim has contributed with fault to cause the damage, the compensation may be excluded or reduced (Art. 9). In order to ensure a more effective strict liability regime, there is a special provision on the evidence of causality between the activity and the damage as a result of dangerous activities, according to which "the court shall take due account of the increased danger of causing such damage inherent in the dangerous activity" (Art. 10).[26] Lastly, for guaranteeing effective compensation for damage in connection with environmental or human health caused by a dangerous activity, the Convention obliges States parties to introduce an obligatory financial security system (Art. 12).

The Lugano Convention also recognizes the right of victims to access information held by public authorities (Arts. 14–15) and operators (Art. 16): the operators must provide only the information strictly required by the victims to assess the right to compensation (i.e., the concentration of waste), and may refuse to provide information that would incriminate them (Art. 16, Para. 5).

In relation to the limitation period for proposing action for compensation, Article 17 of the Convention contains two deadlines: firstly, the action for compensation must be made within three years from the time when the applicant "knew or ought reasonably to have known of the damage"; secondly, the legal action cannot be brought before the domestic court more than thirty years after the site was closed.[27] Further provisions govern the jurisdiction and enforcement of decisions given by a court on the matter.

---

[26]However, Para. 63 of the Explanatory Report to the Convention clarifies that this provision "does not create a true presumption of a causal link."

[27]Art. 17 states: "Limitation periods. (1) Actions for compensation under this Convention shall be subject to a limitation period of three years from the date on which the claimant knew or ought reasonably to have known of the damage and of the identity of the operator. The laws of the Parties regulating suspension or interruption of limitation periods shall apply to the limitation period prescribed in this paragraph. (2) However, in no case shall actions be brought after thirty years from the date of the incident which caused the damage. Where the incident consists of a continuous occurrence the thirty years' period shall run from the end of that occurrence. Where the incident consists of a series of occurrences having the same origin the thirty years' period shall run from the date of the last of such occurrences. In respect of a site for the permanent deposit of waste the thirty years' period shall at the latest run from the date on which the site was closed in accordance with the provisions of internal law."

In addition to the Lugano Convention, another instrument that must be mentioned is the 1997 Resolution on Responsibility and Liability under International Law for Environmental Damage, adopted on 4 September 1997 by the Institute of International Law (IIL). Without prejudice to the question of criminal responsibility of natural or juridical persons (Art. 1), the Resolution is a non-binding document that, distinguishing between international responsibility, responsibility for harm alone, and civil liability, promotes the reform of domestic regimes regarding both "responsibility" and "liability" for environmental damage (Brandon 2013, p. 83). With reference to civil liability, the Resolution states that the operator who carries out the dangerous activity should be primarily liable for the damage to the environment on the basis of strict liability, and thus without fault. A special rule recognizes, on the one hand, the necessity to establish a causal link between activity and damage in order to declare the operator responsible for the environmental damage; and, on the other hand, the possibility of providing for a presumption of causality between certain types of activity and cumulative and long-standing damage to the environment.[28] The operator may be exempted from civil liability, but not from international responsibility and responsibility from harm alone, if it has acted in full compliance with domestic rules and government controls (Art. 6). Moreover, apportioned and joint-and-several responsibility must be considered (Art. 11). There are other general recommendations on limits to civil liability, exemptions from civil liability, insurance,[29] preventive measures, response actions, and compensation for damage.

In order to provide guidance to States, on 26 February 2010 the United Nations Environment Programme (UNEP) adopted the Guidelines for the Development of Domestic Legislation on Liability, Response Action and Compensation for Damage Caused by Activities Dangerous to the Environment. The UNEP Guidelines define the concept of "damage" in a broad sense, including environmental damage, loss of life, personal injury, damage to property arising from environmental damage, costs of reinstatement and preventive measures, and "pure economic loss,"[30] while leaving to each domestic system the task of defining the concept of "dangerous activity." The Guidelines provide for a strict liability regime for the operator who caused the accident resulting in damage to the environment or human health; a series of grounds for exclusion of liability (such as damage caused by a third party or by an operator in compliance with an obligation imposed by a public authority);

---

[28]Under Art. 7: "A causal nexus between the activity undertaken and the ensuing damage shall normally be required under environmental regimes. This is without prejudice to the establishment of *presumptions of causality relating to* hazardous activities or *cumulative damage or long-standing damages not attributable to a single entity but to a sector or type of activity*" (emphasis added).

[29]Moreover, Art. 10 establishes that "[f]oreseeability of damage in general terms of risk should not affect the availability of insurance."

[30]Under the Guidelines, the term "pure economic loss" means "loss of income, unaccompanied by personal injury or damage to property, directly deriving from an economic interest in any use of the environment and incurred as a result of environmental damage."

joint-and-several or apportioned liability in cases where the damage was caused by a plurality of operators; and an obligation for the operator to have a financial guarantee to cover the damage.

Although the UNEP soft law instrument refers to "incidents," some Guidelines do address the issue of domestic legislation in cases of historical pollution. Under Guidelines 4(4) and 6(2), domestic legislation may also provide the following additional grounds for excluding or mitigating liability: "(a) That the activity was expressly authorized and fully in conformity with an authorization, given under domestic law, that allows the effect on the environment; (b) That the damage was caused by an activity which was not likely to cause damage according to the state of scientific and technical knowledge at the time that the activity was carried out." These two additional exceptions have different characteristics. The first covers cases of activity authorized under domestic law, while the second relates to situations where an operator carried out an activity in the past that, according to the state of scientific and technical knowledge at that time, was not considered likely to damage the environment or human health, and only later became an acknowledged source of damage. Moreover, Guideline 12 introduces a time limit for long-term liability concerning environmental harms.[31]

To conclude, several instruments have been adopted in the international context to regulate civil liability for environmental damage caused by dangerous activities, but not all of these are mandatory for States. In a narrow or broad sense, they define the concepts of damage, dangerous activities and dangerous substances, grounds for the exclusion of liability, different regimes of liability in the event that the damage is caused by one or many operators, financial guarantees to cover liability, and so on. All such instruments introduce a time limit for assessing the liability of the operator with reference to environmental damage. Nevertheless, certain interesting provisions have been noted, such as Articles 6 and 7 of the Lugano Convention, which regulate the liability regimes for damage occurring after the cessation of dangerous activities on a site. These indications may be taken into consideration for understanding how issues of civil liability for damage caused by past activities may be addressed.

[31]Guideline 12 states: "Time limits for presentation of claims. (1) Domestic law should establish that claims for compensation are inadmissible unless they are brought within a certain period of time from the date the claimant knew or ought to have known of the damage and the identity of the operator. In addition, claims should be inadmissible unless they are brought within a certain period of time following the occurrence of the damage. (2) Where the damage-causing incident is a series of occurrences having the same origin, the time limits established under the present guideline should run from the last of such occurrences. Where the damage-causing incident consists of a continuous occurrence, such time limits should run from the end of that continuous occurrence."

5.1.1 OECD Recommendations on Liability for Environmental Damage in Eastern Europe, Caucasus, and Central Asia

One instrument deserves particular emphasis as regards civil liability for damage caused by dangerous activities, namely the OECD's 2012 *Liability for Environmental Damage in EECCA: Implementation of Good International Practices*, which makes explicit reference to the liability for historical (residual) pollution. The OECD has elaborated four recommendations with the aim of assisting the countries of Eastern Europe, Caucasus, and Central Asia (EECCA) to reform "their environmental liability systems, particularly to be able to tackle the need for huge investments to clean up past pollution" (OECD Report 2012, p. 10). With these recommendations, the OECD has encouraged these countries to reform their systems of environmental liability, taking as a reference the most advanced regimes used in the United States (namely, CERCLA) and the European Union. In fact, the OECD's report undertakes comparative research on the different systems of environmental liability, covering three topics: the legal framework for environmental liability (definition of environmental damage, requirements for damage remediation and compensation, regime of liability—i.e., strict or fault-based liability, and joint or proportional liability); the assessment of environmental damage; and the relationship between environmental responsibility and financial guarantee (insurance, self-insurance, bank guarantee, or collective financial fund).

As already mentioned, the OECD's report is also a key reference as regards historical pollution. Specifically, each part of the report presents certain recommendations that States should take into account in setting out a legal framework concerning the liability for historic environmental damage. Firstly, as regards the liability regime for historical environmental damage, the OECD Report contains a discussion *inter alia* on "*Special Liability Regimes for Historic Damage*," coming to the following conclusions:

> There is no fundamental legal distinction between liability for new as opposed to historic (residual) environmental damage. The principal difference is whether a responsible party can be identified and how remediation is paid for. (OECD Report 2012, pp. 15 ff.; emphasis added)

Secondly, the 2012 OECD report shows that, taking into account the various rules adopted in different countries, the assessment of the "remediation standard for new and historic damage" may be different (OECD Report 2012, p. 22).

Thirdly, according to Monti (2002), the 2012 OECD report shows how difficult it is to insure for damage subsequent to "gradual pollution events": in effect, insurance is not capable of achieving its purpose as regards this kind of pollution "which develops slowly and secretly over a long period of time and whose damaging effects show only in the long run" (Monti 2002, p. 11). The reason for this is that such damage manifests the two characteristics of factual uncertainty and long-term effects (Monti 2002, p. 11). Consequently, for cases of past chronic pollution and diffuse pollution, it is worth considering the proposal to introduce a no-fault compensation fund (Monti 2002, pp. 18–19).

In conclusion, the 2012 OECD report provides important indications about an appropriate liability regime for historical contamination. First, it shows that historical pollution is a world-wide phenomenon, but which each country tends to try and solve by itself. Secondly, it has highlighted that the main difference between the liability regimes for new as opposed to historic damage lies in the fact that, in the latter cases, the liable operator usually cannot be found (e.g., for reasons of bankruptcy of the polluter, etc.), with the consequence that responsibility for remedial measures may be transferred to the landowner, even if not the real polluter. However, such a form of liability is problematic, both because it has retroactive effects and because it may affect a third party in good faith. It is therefore necessary to find a "fair" solution which affirms the landowner's liability for the assumption and cost of the clean-up procedures. In this regard, a promising proposal could be to introduce a register of contaminated lands, so that a third party may determine whether a site is contaminated. Only in this case could it be called upon to clean up an area contaminated by past activity; otherwise, the public authority should bear the costs of the decontamination. Thirdly, this report is significant in that it canvasses the possibility of introducing a compensation fund at domestic level for damage resulting from historical and diffuse pollution.

5.2 *Directive 2004/35/EC on Environmental Liability and Historical Pollution*

One of the main pillars of the EU legal framework on environmental matters is Directive 2004/35/EC of the European Parliament and of the Council of 21 April 2004 on Environmental Liability with Regard to the Prevention and Remedying of Environmental Damage (European Liability Directive, ELD). This Directive, adopted after in-depth debate,[32] is based on the polluter-pays principle and aims both to avoid damage to the environment (water, soil, flora, and fauna) and to remediate contaminated land.[33] As highlighted by an authoritative scholar, "the ELD has a misleading name: it is not a civil liability regime" (Bergkamp 2005,

[32]Adopted after a long discussion generated by the Green Paper on Remedying Environmental Damage (COM (93) 47 final, Brussels, 14 May 1993), the White Paper on Environmental Liability (European Commission, COM (2000) 66 final, Brussels, 9 February 2000), and the Legislative Proposal for a Directive (COM (2002) 17 final, Brussels, 23 January 2002).

[33]The ELD does not apply to environmental damage regulated by the international agreements incorporated in Annex IV. See International Convention of 27 November 1992 on Civil Liability for Oil Pollution Damage; International Convention of 27 November 1992 on the Establishment of an International Fund for Compensation for Oil Pollution Damage; International Convention of 23 March 2001 on Civil Liability for Bunker Oil Pollution Damage; International Convention of 3 May 1996 on Liability and Compensation for Damage in Connection with the Carriage of Hazardous and Noxious Substances by Sea; Convention of 10 October 1989 on Civil Liability for Damage Caused During Carriage of Dangerous Goods by Road, Rail and Inland Navigation Vessels.

p. 389), but rather introduces at EU level an administrative liability regime for preventing and restoring environmental damage and a special procedure for cooperation between States.

The ELD must be applied to damage caused by an "operator."[34] As Fogleman (2015, p. 10) highlights:

> The ELD imposes liability for preventing and remediating land damage on an operator whose activities cause the contamination. Member States may impose more stringent provisions that include imposing liability on the owner of contaminated land.[35]

The ELD recognizes two liability regimes for environmental damage. On the one hand, the Directive provides for strict responsibility for environmental harm caused by activities listed in the Directive on integrated pollution prevention and control and in other EU Directives, incorporated in Annex III. On the other hand, the Directive provides fault liability for damage to protected species and natural habitats caused by activities other than those indicated in Annex III. Notably, the ELD concerns not only avoiding imminent harm to the environment, but also remedying damage that has already occurred, in particular by establishing a series of procedures that have to be followed (including at the request of concerned people)[36] by operators and public authorities to ensure that harmful contaminants are removed from land. According to the polluter-pays principle, the operator must bear the costs of preventive and remedial actions, unless it can prove that the harmful effects were caused, for instance, by a third party or for other reasons established by the Directive.

The main provisions used to assess whether and to what extent the ELD can be used for remedying historic damage are contained in Articles 17, 8(4), 4(5), and Recital 13. First, Article 17 establishes that the ELD cannot be applied retroactively[37] or to "damage, if more than 30 years have passed since the emission, event or incident, resulting in the damage, occurred." Consequently, this directive cannot be applied to historical damage.[38] Secondly, Article 8(4) lays down that an operator

---

[34]Article 2(6) of the ELD establishes that an "operator" is "any natural or legal, private or public person who operates or controls the occupational activity or, where this is provided for in national legislation, to whom decisive economic power over the technical functioning of such an activity has been delegated, including the holder of a permit or authorisation for such an activity or the person registering or notifying such an activity."

[35]On the issue see Fogleman 2015, pp. 9 ff.

[36]The ELD recognizes, *inter alia*, the right of people affected by environmental damage to request that competent public authorities take necessary measures. See Art. 12.

[37]Art. 17 of the ELD also establishes that this Directive shall not apply to: " —damage caused by an emission, event or incident that took place before the date referred to in Article 19(1), — damage caused by an emission, event or incident which takes place subsequent to the date referred to in Article 19(1) when it derives from a specific activity that took place and finished before the said date.".

[38]de Sadeleer 2015, p. 234) has highlighted that "*[t]he ELD does not apply retroactively to historic pollution* or damage caused before 30 April 2007. This means that the costs of cleaning up and restoring contaminated sites before this date will eventually be borne by public authorities"

may be exonerated from environmental liability under domestic law when the damage has been caused by an industrial emission or activity expressly authorized by a public authority on the basis of the conditions expressed in the Directive, or one that "was not considered likely to cause environmental damage according to the state of scientific and technical knowledge at the time when the emission was released or the activity took place." Thirdly, Article 4(5) and Recital 13 preclude applying the ELD to damage caused by diffuse pollution if it is not possible to establish a causal link between the operator's activity and the environmental damage.[39]

Thus, there are limits to the application of the ELD to historic damage or to damage caused by diffuse and widespread pollution. Nevertheless, the Directive does not prohibit Member States maintaining stricter provisions on preventing and remedying environmental damage.[40] In other words, each State may decide to allow retroactive effect to the environmental liability regime, but such a choice cannot be considered an implementation of EU Law. In two important judgments, the EU Court of Justice has interpreted the polluter-pays principle—recognized in Article 174(2) EC and in Directive 2004/35/EC—in relation to cases of historic damage and gradual and diffuse pollution.

5.2.1 The Causal Link Between Operators' Activities and Diffuse Pollution

In *Raffinerie Mediterranee (ERG) SpA and others*,[41] the Grand Chamber of the EU Court of Justice interpreted the polluter-pays principle in relation to a case of diffuse pollution caused by a plurality of operators in the petrochemical sector who succeeded one another on the same territory over the course of five decades. The European decision originated from a referral by an Italian Court (TAR Sicily) for a preliminary ruling in the context of an administrative judgment between petrochemical companies and public authorities concerning the remediation of historical

(Footnote 38 continued)

(emphasis added). On the temporal provisions of the ELD see Fogleman 2014, pp. 144 ff. For critical discussion, see Brandon 2013, p. 203.

[39]Recital No. 13 of the ELD establishes that "Not all forms of environmental damage can be remedied by means of the liability mechanism. For the latter to be effective, there need to be one or more identifiable polluters, the damage should be concrete and quantifiable, and a causal link should be established between the damage and the identified polluter(s). Liability is therefore not a suitable instrument for dealing with pollution of a widespread, diffuse character, where it is impossible to link the negative environmental effects with acts or failure to act of certain individual actors."

[40]On the differences between the ELD and national laws on contaminated land adopted in certain EU member States with reference to historical pollution, see Fogleman 2015, p. 11.

[41]Case C-378/08, Raffinerie Mediterranee (ERG) SpA, Polimeri Europa SpA and Syndial SpA v. Ministero dello Sviluppo economico and Others, 9 March 2010 (ECLI:EU:C:2010:126). For comment, see Bergkamp 2010, pp. 355 ff., Bertolini 2010, pp. 1607 ff.

contamination at Roadstead of Augusta.[42] The Italian administrative authorities imposed measures on the companies of Roadstead for remedying environmental damage, on the basis that their activities are situated close to the contaminated area, without establishing a causal relationship between the activities and the harm and "without any distinction being made between past and present pollution" (Para. 27). Since the Italian Court considered it impossible to determine the degree of responsibility of the operators that had operated in the area (Para. 28), it suspended the main process and proposed a referral to the Court of Justice for a preliminary ruling, specifically concerning whether the polluter-pays principle could preclude such a practice (Para. 34).

In ruling on this point, the Court of Justice first stated that Directive 2004/35, Article 17, "applies to damage caused by an emission, event or incident which took place after 30 April 2007 where such damage derives either from activities carried out after that date or activities which were carried out but had not finished before that date" (Para. 41). However, the European judges argued that it is for the national court to establish whether Directive 2004/35 is *ratione temporis* applicable to the main proceedings concerning a case of past and diffuse pollution (Paras. 38–47). Where there are no temporal or material conditions for the application of the Directive, the referring judge should apply national law; conversely, if those requirements are met, the referring Court should take into account the guidance provided by the EU Court of Justice.

Consequently, the Court of Justice recalled that it is necessary to establish a causal link in order to apply the Directive to cases of diffuse pollution (Art. 4(5) and Art. 11(2) of Directive 2004/35, Paras. 52–54).[43] Nevertheless, Directive 2004/35 does not specify the standard for assessing a causal connection between the releases of pollutants and environmental damage, so that the task of defining this standard falls within the competence of the Member States. In line with these indications given by the Court of Justice, the competent administrative authority may impose corrective actions on an operator on the basis that it is carrying out its activity in a polluted zone, but these measures may be considered in line with the polluter-pays principle only if the competent authorities "have plausible evidence capable of justifying its presumption, such as the fact that the operator's installation is located

[42]As underlined by the EU Court of Justice, this is a zone "affected by recurring incidents of environmental pollution, dating back as far as the 1960s, when the Augusta-Priolo-Melilli site was established as a hub for the petroleum industry. Since that time, numerous undertakings operating in the hydrocarbon and the petrochemical sectors have been set up and/or succeeded one another in the region" (Para. 19).

[43]According to the EU Court of Justice, "*Such a causal link could easily be established where the competent authority is confronted with pollution which is confined to a particular area and period of time and is attributable to a limited number of operators*. On the other hand, that is not the case with diffuse pollution phenomena and, therefore, the legislature of the European Union considered that, in the case of such pollution, a liability mechanism is not an appropriate instrument where such a causal link cannot be established. Consequently, Article 4(5) of Directive 2004/35 provides that the directive is to apply to that kind of pollution only where it is possible to establish a causal link between the damage and the activities of individual operators" (Para. 54; emphasis added).

close to the pollution found and that there is a correlation between the pollutants identified and the substances used by the operator in connection with his activities" (Para. 57).

From the European Court's point of view, if the referring Court considers Directive 2004/35 applicable, it is necessary to distinguish whether the damage was caused by the activities listed in Annex III of the Directive, or by activities that are not indicated in this Annex (Para. 61). This distinction is relevant because the liability for environmental damage caused by the activities listed in Annex III is based on a strict liability scheme, while the liability for environmental damage caused by other activities is based on fault or negligence (Para. 61). That said, the Court has, first of all, emphasized that operators active in the chemical sector—such as those involved in the main proceeding—are included in Annex III of ELD, with the result that they are subject to the strict liability regime (Paras. 62–63). Nevertheless, the Court of Justice also pointed out that, under Article 11(2) of Directive 2004/35, read in conjunction with Recital 13 in the Preamble, it is for the domestic authorities to determine the typology of operator and to apply the corresponding liability regime. In both of the above cases of responsibility, the referring Court—according to the indications given by the European Court of Justice—should establish a causal relationship between the polluting activities and the damage: in the absence of this element, the administrative authority cannot ask the operators to adopt appropriate remedying measures.

5.2.2 The Exemption of Innocent Landowners from Liability for Historic Environmental Damage

A significant decision relating to a case of environmental damage caused by past activities was delivered by the Court of Justice on 4 March 2015 in the *Fipa Group* case.[44] The origin of the facts dates back to the 1960s–1980s, when two companies belonging to the Montedison SpA group managed an industrial production site for insecticides and herbicides in the province of Massa Carrara (Tuscany). As a result of these activities, there was contamination of land by chemicals including dichloroethane and ammonia. In the mid-1990s, the remediation of this land was carried out and completed by the operator. Between 2006 and 2008, this land was sold on to three companies, namely Fipa Group Srl, Tws Automation Srl, and Ivan Srl; and between 2007 and 2011 the competent administrative authority ordered these companies to execute new specific remediation measures, which were appealed before the Regional Administrative Court of Tuscany.

With three separate judgments, the domestic Court annulled the acts prescribed by the public authorities on the grounds that the latter could not impose the

[44]Case C-534/13, Ministero dell'Ambiente e della Tutela del Territorio e del Mare and Others v Fipa Group Srl, Tws Automation Srl, Ivan Srl, 4 March 2015 (EU:C:2015:140). For comment, see Pozzo et al. 2015, pp. 1071 ff., and Varvaštian 2015, pp. 270 ff.

measures in question on operators that had no direct responsibility for the historic damage to the environment. The public authorities appealed against those judgments before the Council of State, which decided to suspend the proceedings and proposed a referral for a preliminary ruling to the Court of Justice on the question:

> whether the EU principles of environmental law, as set out in Article 191(2) TFEU and in Directive 2004/35, in particular the "polluter pays" principle, must be interpreted as precluding national legislation such as that at issue in the main proceedings, which, in cases where it is impossible to identify the polluter of a plot of land or to have that person adopt remedial measures, does not permit the competent authority to require the owner of the land (who is not responsible for the pollution) to adopt preventive and remedial measures, that person being required merely to reimburse the costs relating to the measures undertaken by the competent authority within the limit of the market value of the site, determined after those measures have been carried out. (Para. 38)

In ruling on this point, the Court of Justice has established certain important principles applicable to the liability of the current landowners of a site contaminated by past activities. First, the Court argued that it is a task of the referring Court to determine whether Directive 2004/35 must be applied to the facts in the main proceedings under Article 17 thereof (de Sadeleer 2015, p. 234): that said, the European Court also emphasized that in this case the applicability of the ELD is doubtful, because the polluting activities ceased in the late 1980s and the contaminated area was partially remedied in the mid-1990s,[45] before the entry into force of the Directive on 30 April 2007. Moreover, in the European Court's view, if the referring Court concludes that Directive 2004/35 may be applied to the facts of the main proceedings, it must be ascertained whether the current owners of the contaminated soil can be considered an "operator" under the ELD. In this regard, despite the ELD distinguishing between a subjective and objective regime of environmental liability, in both the above cases it is necessary to ascertain a causal link between the operators' behaviour and the relevant damage,[46] otherwise the situation is governed by national law. That said, the Court of Justice argued that in this case there seems to be no causal link between the activities carried out by the three companies and the contamination in question. Nevertheless, each State may adopt "more stringent measures in relation to the prevention and remedying of environmental damage, including the identification of additional responsible parties, provided that such measures are compatible with the Treaties" (Para. 61). Consequently, a domestic law that, when polluters cannot be identified, requires

[45]According to the EU Court of Justice, "given that, according to the facts described in the documents before the Court, the historic environmental damage at issue in the main proceedings stems from economic activities undertaken by former owners of the land currently held by Fipa Group, Tws Automation and Ivan, respectively, it is unlikely that Directive 2004/35 is applicable *ratione temporis* in the main proceedings" (Para. 43).

[46]As summarized by de Sadeleer (2015, p. 237), "*mere owners of historically contaminated occupational sites, who are not responsible for the environmental damage, must not take on the burden of remedying pollution to which they have not contributed*" (emphasis added).

innocent landowners to pay only the costs of measures undertaken within the limits of the market value of the site can be considered in line with EU law (Para 63).

5.3 Closure and Post-Closure Care Requirements for Landfill Sites in EU Law

Particularly interesting is a group of provisions contained in EU law aimed at imposing closure and post-closure procedures for dangerous sites, on the basis of which operators are responsible for adverse effects that may derive from a site after its closure.

One of the main examples is Directive 2006/21/EC of the European Parliament and of the Council of 15 March 2006 on the Management of Waste from Extractive Industries and Amending Directive 2004/35/EC. This Directive fixes the general principles and procedures on the basis of which an operator must handle waste from extractive industries in respect of "long-term perspectives, and more especially against the pollution of groundwater by leachate infiltration into the soil" (Recital 18). In fact, Member States shall ensure that each operator adopt a waste management plan, whose objective is to prevent or at least minimize "any long-term negative effects, for example attributable to migration of airborne or aquatic pollutants from the waste facility" (Art. 5(1)(c)(ii)). The Directive pays specific attention to closure and after-closure procedures for waste facilities: notably, Article 12(4) states that "the operator shall be responsible for the maintenance, monitoring, control and corrective measures in the after-closure phase for as long as may be required by the competent authority."

A further example of a liability regime established within EU law concerning environmental damage that may occur in a long-term perspective is Directive 2009/31/EC on the Geological Storage of Carbon Dioxide, which aims to safeguard the environment against negative effects resulting from the storage of CO_2 (De Figueiredo et al. 2005; Ingelson et al. 2010; Haan-Kamminga 2011). Under this Directive, an operator has to take preventive and remedying measures even after the closure of a site. However, at least twenty years after the closure of the storage site, liability for preventing and remedying damage to the environment and human health shall be transferred, according to the requirements laid down in the above Directive, from the operators to the competent public authorities (Arts. 17–20).

In view of the above, EU policy can be seen to be moving towards the management of long-term liability for uncertain damage to the environment and human health caused by past hazardous activities. On the basis of the above-mentioned Directive, the operator remains responsible for monitoring after the closure of the site, and is responsible for corrective measures on the basis of a post-closure plan, until that liability is transferred to the public authority.

6 Global Initiatives on the Management of Contaminated Land and Brownfield Redevelopments

A comparative survey conducted by Brandon has shown that countries have developed two generations of laws on contaminated land: the first generation of laws follows the "command and control" approach to remediation, while the second generation is based on brownfield redevelopments and voluntary remediation (Brandon 2013, p. 3). Despite these developments, a global approach to site contamination law does not yet exist (Brandon 2013), and the reluctance of many countries to participate in an international response to site contamination, and therefore on historic contamination, stems mainly from the fact that they consider it a local problem. In the EU framework, the first generation of laws on environmental liability concerning new contamination was developed with the adoption of the ELD, while a policy on brownfield redevelopments for old contamination is blocked because of strong opposition from a number of Member States (Vanheusden 2007, p. 563). Nevertheless, a number of global initiatives have been undertaken by international and regional lending institutions, networks, and NGOs, aimed at the effective development of policies on contaminated sites.

6.1 Brownfield Redevelopment in the EU Framework

Brownfield redevelopment has become one of the main problems of the European Union (Vanheusden 2003, 2007, 2009; Petersen 2008). According to Vanheusden (2007, p. 561):

> *The European Union is faced with both very old and recent soil contamination.* Some of the old soil contamination dates back to the accelerated industrial development in the beginning of the nineteenth century. With its more than two hundred years of industrialization, Europe faces a soil contamination problem resulting from the use and presence of dangerous substances in many production processes.

The European Union, unlike the United States, lacks a legal and strategic approach to brownfield redevelopment. At present, there is no Europe-wide inventory on contaminated land, just as there is no EU Directive on soil. In this respect, it must be underlined that in the context of the sixth environment action programme, entitled *Environment 2010: Our Future, Our Choice*, the EU Commission planned the development of a thematic strategy for soil protection.[47] In developing that strategy, in April 2002 the EU Commission published a first communication entitled *Towards a Thematic Strategy for Soil Protection* in which

---

[47]For detail, see Communication from the Commission to the Council, the European Parliament, the Economic and Social Committee and the Committee of the Regions on the Sixth Environment Action Programme of the European Community, "Environment 2010: Our future, Our choice" [COM (2001) 31 final].

it placed emphasis on the need to develop a strategy on the main causes of soil degradation and contamination. Subsequently, the Commission developed a strategy on soil protection through three different initiatives, namely the Communication from the European Commission to other Institutions Concerning "Thematic Strategy for Soil Protection" (2006), the Proposal for a Directive of the European Parliament and of the Council Establishing a Framework for the Protection of Soil and Amending Directive 2004/35/EC (2006), and an Impact Assessment (2006).

Following its 2006 Communication, the European Commission published its Proposal for a Directive of the European Parliament and of the Council Establishing a Framework for the Protection of Soil and amending Directive 2004/35/EC. In line with Recital 21 of the Preamble of the proposed Directive:

> Earlier industrialisation and poor or inappropriate management practices have left a legacy of hundreds of thousands of contaminated sites in the Community which call for *a common strategy to manage historical contamination of soil in order to prevent and mitigate harmful effects on human health and the environment.* (Recital 21)

The proposal was approved through a resolution of the European Parliament on 13 November 2007.[48] However, the text was amended by the parliamentary body at different points, and the Directive failed to receive approval from the European Council. A group of Member States opposed the proposal for various reasons, citing the need to respect the principle of subsidiarity and arguing that the topic of soil protection is a local or national problem. Moreover, they argued that a Directive was unnecessary given the existence of national and regional measures and that it would have meant new costs for those countries that already have a specific and effective policy on soil (Brandon 2013, p. 51). After lengthy discussion, in 2014 the European Commission decided to withdraw the proposal.

Central to the proposed Directive were the provisions on the contamination of soil (Recitals 21–29). The Directive required Member States to take progressive steps to prevent soil contamination (Art. 9), to adopt an inventory of contaminated sites, to identify those sites where hazardous activities had taken place in the past, and to remediate contaminated sites listed in national inventories.[49] Particularly significant was Article 12, in its original version, which called on Member States to introduce a rule which required private parties involved in the sale of a site to

---

[48]European Parliament Legislative Resolution of 14 November 2007 on the Proposal for a Directive of the European Parliament and of the Council Establishing a Framework for the Protection of Soil and Amending Directive 2004/35/EC [COM(2006)0232-C6-0307/2006-2006/0086(COD)].

[49]Art. 13 provided for the following: "Remediation. (1) Member States shall ensure that the contaminated sites listed in their inventories are remediated. (2) Remediation shall consist of actions on the soil aimed at the removal, control, containment or reduction of contaminants so that the contaminated site, taking account of its current use and approved future use, no longer poses any significant risk to human health or the environment. (3) Member States shall set up appropriate mechanisms to fund the remediation of the contaminated sites for which, subject to the polluter pays principle, the person responsible for the pollution cannot be identified or cannot be held liable under Community or national legislation or may not be made to bear the costs of remediation."

submit a report on the state of the soil to the competent authority.[50] Moreover, the Directive established that Member States had to establish appropriate mechanisms to fund the remediation of contaminated sites where the real polluter could not be identified. Lastly, Article 22 of the draft Directive provided that Member States had to adopt sanctions with reference to transgressions of national provisions adopted in the implementation of the Directive.

In sum, then, except for some significant interventions by the Court of Justice,[51] it may be concluded that the EU currently lacks an incisive tool for addressing historical pollution. Nevertheless, the Decision No 1386/2013/EU of the European Parliament and of the Council of 20 November 2013 on a General Union Environment Action Programme to 2020 "Living well, within the limits of our planet" (7th EAP) "shall ensure that by 2020: (e) land is managed sustainably in the Union, soil is adequately protected and the remediation of contaminated sites is well underway."

[50]Art. 12 stated the following: "Soil status report. (1) Where a site is to be sold on which a potentially polluting activity listed in Annex II is taking place, or for which the official records, such as national registers, show that it has taken place, Member States shall ensure that the owner of that site or the prospective buyer makes a soil status report available to the competent authority referred to in Article 11 and to the other party in the transaction. (2) The soil status report shall be issued by an authorised body or person appointed by the Member State. It shall include at least the following details: (a) the background history of the site, as available from official records; (b) a chemical analysis determining the concentration levels of the dangerous substances in the soil, limited to those substances that are linked to the potentially polluting activity on the site; (c) the concentration levels at which there are sufficient reasons to believe that the dangerous substances concerned pose a significant risk to human health or to the environment. (3) Member States shall establish the methodology necessary for determining the concentration levels referred to in paragraph 2(b). (4) The information contained in the soil status report shall be used by the competent authorities for the purposes of identifying contaminated sites in accordance with Article 10(1)."

[51]Before the approval of Directive 2008/98/EC of the European Parliament and of the Council of 19 November 2008 on waste and repealing certain Directives, which excludes unexcavated contaminated soil from the scope of the Directive, many scholars considered the ruling of the European Court of Justice in *Van de Walle and others* to be important for the development of an EU policy on brownfield redevelopment (Vanheusden 2007, p. 566; McIntyre 2005). In that decision, the Court of Justice ruled that hydrocarbons that have been spilled accidentally into the soil can be considered waste under the Directive 75/442/EEC. According to the Court of Justice, if the hydrocarbons which are the source of the pollution were not to be regarded as "waste" for the fact that they have been spilled unintentionally, their holder would be subtracted from the obligations laid down in this Directive, in contrast with the prohibition of abandonment and uncontrolled discharge of waste. Moreover, the Court of Justice argued that the soil contaminated by a spill of hydrocarbons must be classified as waste: "in that case, the hydrocarbons cannot be separated from the land which they have contaminated and cannot be recovered or disposed of unless that land is also subject to the necessary decontamination" (Para. 52). In conclusion, the Court of Justice ruled that "the petroleum undertaking which supplied the service station can be considered to be the holder of that waste within the meaning of Article 1(c) of Directive 75/442 only if the leak from the service station's storage facilities which gave rise to the waste can be attributed to the conduct of that undertaking" (Para. 61).

6.2 The Management of Contaminated Land in the Instruments of Lending Institutions, Networks, and NGOs

As already mentioned, a number of global initiatives have been undertaken by international and regional lending institutions, networks, and NGOs concerning the development of policies on contaminated sites. Of notable interest are, first of all, the many projects developed by international banks with the goal of remediating groundwater and soil contamination around the world (Brandon 2013, p. 292). These institutions have imposed loan conditions that countries involved in their projects allocate the necessary funds to carry out clean-up programmes (Brandon 2013, p. 292). In this regard, the *World Bank Group Environmental, Health, and Safety (EHS) Guidelines* "provides a summary of management approaches for land contamination due to anthropogenic releases of hazardous materials, wastes, or oil, including naturally occurring substances." These management approaches, that refer to areas contaminated by "historic or current site activities," require that four steps are followed: "(1) Risk screening; (2) Interim risk management; (3) Detailed quantitative risk assessment; and (4) Permanent risk reduction measures."

Moreover, the World Bank Group has also adopted *A Sourcebook of Pollution Management Policy Tools for Growth and Competitiveness*, in which a specific approach for addressing historical contamination has been developed. This sourcebook is relevant, first of all, because it offers a definition of the notion of "legacy pollution."

> Legacy pollution refers to any pollution that remains from past activities where there is no immediately responsible party who can be held liable for the pollution and compelled to carry out remediation. […] Legacy pollution sites are characterized by the existence of potentially harmful materials, which are known to be toxic to humans or destructive to the environment in proximity to communities or sensitive ecosystems that are being impacted or are at significant risk from the pollution. In serious cases, where toxic substances such as heavy metals and organic chemicals are involved, there is the potential to cause severe health impacts to human populations and to result in significant and long-term damage to the environment. (World Bank Group 2012, pp. 57–58)

The World Bank's description is essential in order to understand the concept of historical pollution. In addition, the *Sourcebook* shows that the management of land contaminated by past activities is a complex problem, because "the key challenges that govern the responses are usually institutional and financial" (World Bank Group 2012, p. 60). From this point of view, emphasis is laid on public–private partnerships as a means to address historical pollution, these being a framework through which public parties (local and national governments), private parties (polluters and current landowners), communities affected by historical pollution, institutions, experts, and academics may participate in the management of contaminated land (World Bank Group 2012, pp. 60–61). Moreover, the *Sourcebook* shows how providing economic support for addressing historic pollution problems represents one of the main challenges to which public authorities are called upon to give a response. As argued in the *Sourcebook*, these responses to legacy pollution could have positive effects, principally regarding the health of people living near contaminated areas,[52] but they are also expensive. Noting that "hazardous and legacy waste problems are usually relatively localized—if severe—in their impacts," the *Sourcebook* encourages governments to adopt national programmes as used in other countries (e.g., the Superfund adopted in the USA).

Last but not least, it should be noted that another key role in the development of strategies regarding contaminated land management is played by a number of European and international networks (International Committee on Contaminated Land, Common Forum on Contaminated Land in Europe, NICOLE, Cabernet, and so on).[53] This shows the importance of flexible tools such as these in addressing historical pollution.

7 Criminal Liability and Historical Pollution: A Complex Relationship

A final question that arises when discussing legal solutions to cases of historical pollution is whether there are criminal responses to be found within the relevant international agreements. In this regard, it must be noted that despite noteworthy international legal developments in environmental matters (DiMento 2010), responses to environmental damage remain focused on non-criminal solutions. In effect, international law contains only a very limited number of provisions requiring States to introduce criminal sanctions or other punitive sanctions for the violation of environmental rules[54] (Mégret 2011, p. 198). Although a series of arguments militate for the criminalization of environmental damage at international level (Mégret 2011; Neyret 2014, 2015), there is still strong resistance by States to the adoption of an international agreement on the protection of environment through

[52]However, the indications of the *Sourcebook* cannot be applied in cases of diffuse pollution. See World Bank Group (2012, pp. 66–67).

[53]For recent comment on the issue, see Azam 2016.

[54]For example, Art. 55 of Protocol Additional to the Geneva Conventions of 12 August 1949, and Relating to the Protection of Victims of International Armed Conflicts (Protocol I), 8 June 1977 states that "Protection of the natural environment—(1) *Care shall be taken in warfare to protect the natural environment against* widespread, *long-term* and severe *damage*. This protection includes a prohibition of the use of methods or means of warfare which are intended or may be expected to cause such damage to the natural environment and thereby to prejudice the health or survival of the population. (2) Attacks against the natural environment by way of reprisals are prohibited" (*emphasis added*). Article 9(5) of the Basel Convention on the Control of Transboundary Movements of Hazardous Wastes and their Disposal establishes that "Each Party shall introduce appropriate national/domestic legislation to prevent and punish illegal traffic." Art. VII (2) of the Convention on the Prevention of Marine Pollution by dumping of Wastes and other Matter of 29 December 1972 states that "Each Party shall take in its territory appropriate measures to prevent and punish conduct in contravention of the provisions of this Convention."

criminal law. The reasons for this, according to Mégret, are fourfold at least—namely, limitations due to the nature of the system (international law), the nature of the interests protected (the environment), the discipline (international environmental law), and the nature of the project (criminal law) (Mégret 2011, pp. 204 ff.). Therefore, *de iure condito*, criminal liability of industrial operators for historical contamination is not regulated in any international agreement.

Indeed, there are additional reasons for believing that historical pollution, in its narrowest definition, will not be considered from a criminal perspective at international level in the near future. Firstly, according to the principle of non-retroactivity, historical environmental damage resulting from past polluting activities carried out before the relevant environmental regulations were adopted cannot be criminalized. Secondly, such criminalization would run counter to the principle of culpability, since a person would be liable for conduct committed without fault at the time of the facts. Thirdly, there are many practical obstacles to criminalizing such acts. As already noted in relation to liability regimes for historical pollution, responsibility for historical damage is difficult to ascertain given the long elapsed time between conduct and event: it thus becomes difficult to establish a causal link between the operator's activities and the actual damage, and so difficult to establish guilt. This is true *a fortiori* in criminal proceedings, where such conditions would make it difficult to assign responsibility for the crime beyond any reasonable doubt. Fourthly, there is a political-criminal reason: in many countries, the main responses to this phenomenon consist in concrete actions to protect human health and the environment, including the reclamation of contaminated soils. In this regard, it may be observed that an administrative response, according to which the polluter is called upon to clean up contaminated land and to meet the associated costs, may be more effective than a criminal penalty. Furthermore, such an approach is based on the principles of *extrema ratio* and subsidiarity, and so administrative and civil responses would be preferable for protecting the victims and the environment (Faure and Svatikova 2012). As highlighted in the World Bank report examined in the preceding pages, scenarios of historical pollution are complex, since there are a range of interests to reconcile, namely the protection of the victims versus the protection of the fundamental rights of the accused, the protection of the health of local communities affected by contamination versus the preservation of jobs and the development of the economy, and so on. Reconciliation procedures thus seem to be the most appropriate to solve problems arising from historical pollution.

Unlike in the international context, specific criminal law rules and penalties have been laid down for the protection of the environment in the European context.[55] Within EU law, it is possible to distinguish between two typologies of rules.

---

[55]First, on 4 November 1998 the member States of the Council of Europe adopted the Convention on the Protection of the Environment through Criminal Law. However, the Convention never entered into force. In EU law, regulations for environmental crimes have been introduced via two Directives, namely Directive 2008/99/EC of the European Parliament and of the Council of 19 November 2008 on the Protection of the Environment through Criminal Law and Directive

On the one hand, Directive 2008/99/EC on the Protection of the Environment through Criminal Law requires Member States to apply effective, proportionate, and dissuasive sanctions again "unlawful" polluting operations "committed with at least serious negligence" and which cause or may cause damage to the environment and to human health (Siracusa 2008; Plantamura 2009; Vagliasindi 2012; Faure 2010; Öberg 2011; Heine 2011; Klip 2015, p. 238; Mitsilegas et al. 2016). In particular, Article 3 of Directive 2008/99/EC requires States to adopt rules for punishing nine broad types of conduct, among them:

(a) "the discharge, emission or introduction of a quantity of materials or ionising radiation into air, soil or water, which causes or is likely to cause death or serious injury to any person or substantial damage to the quality of air, the quality of soil or the quality of water, or to animals or plants;
(b) the collection, transport, recovery or disposal of waste, including the supervision of such operations and the aftercare of disposal sites, and including action taken as a dealer or a broker (waste management), which causes or is likely to cause death or serious injury to any person or substantial damage to the quality of air, the quality of soil or the quality of water, or to animals or plants;
(c) the operation of a plant in which a dangerous activity is carried out or in which dangerous substances or preparations are stored or used and which, outside the plant, causes or is likely to cause death or serious injury to any person or substantial damage to the quality of air, the quality of soil or the quality of water, or to animals or plants; [...]"

However, offences adopted in domestic law in compliance with such clauses of Article 3 of Directive 2008/99/EC cannot be applied retroactively to criminal proceedings in which parties are prosecuted for death or serious injury to a person or substantial damage to the environment as a consequence of polluting activities that took place before that Directive came into force in domestic systems. This is a consequence of the application of the principle of non-retroactivity as enshrined in Article 49(1) CFREU. In such a case, each Member State can only apply its national law.

Although there is no doubt that historical contamination is generally a consequence of past polluting activities committed by juridical persons, this does not change the result: although under Articles 6 and 7 juridical persons should also be punished for all offences established in Article 3, such a provision cannot be applied to operations committed before the entry into force of the Directive.

On the contrary, the national rules implementing this Directive can be applied in criminal proceedings concerning cases of contamination occurring as a consequence of operations carried out after its entry into force.[56] In this respect, it must

(Footnote 55 continued)

2009/123/EC of the European Parliament and of the Council of 21 October 2009 Amending Directive 2005/35/EC on Ship-source Pollution and on the Introduction of penalties for Infringements.

[56]As regards the rights of victims in criminal proceedings in environmental matters, see Cardwell et al. 2011 and Rafaraci 2015.

be noted that the offences established in Article 3 also require Member States to punish operations which are "likely to cause" damage to the environment or human health.

On the other hand, several non-criminal Directives on waste require Member States to adopt effective, proportionate, and dissuasive sanctions for infringements of rules concerning measures to be adopted after the closure of a waste site. These can be considered as rules aimed at strengthening the protection of the environment after all activities on a site have ceased, by considering operators responsible for the protection of site in a long-term perspective, until responsibility should be transferred to public authorities. Lastly, a rule which is of particular interest, but which never entered into force, is Article 22 of the Draft EU Framework Directive for Soil Protection (2006), recently withdrawn by the European Commission, according to which "Member States shall lay down the rules on penalties applicable to infringements of the national provisions adopted pursuant to this Directive and shall take all measures necessary to ensure that they are implemented."

8 New International Answers for a Common Problem?

The reluctance of many countries to develop an international approach to the management of historical damage to the environment and human health (i.e., harms caused by past activities) turns on the various reasons illustrated in the previous pages. In summary, these reasons can be subdivided into the categories of political, practical, economic, cultural, and juridical reasons. The political reasons that hinder the adoption of an international instrument to address historical pollution stem from the fact that this type of contamination mainly produces negative effects within the territory of a single state. Being seen as a local problem, it is treated as issue that must be addressed by local governments (World Bank Group 2012, p. 11). In addition, different countries have different perceptions of the importance of environmental protection: while some countries have developed an approach that is more advanced concerning the protection of ecosystems, others—especially developing economies—prioritize industrialization at the expense of environmental protection. In addition, direct interventions to address historical contamination also have significant economic costs, and often there may be no private interest in supporting such efforts. States may even entirely lack the financial capacity to bear the costs of recovery, especially where interventions are complex and expensive. Finally, countries also follow different legal approaches in addressing the issue of past contamination.

Having highlighted the limits of international intervention in such matters, however, it is important to stress that the contamination of the soil is a common problem for humanity: what is really at stake here are the rights of new and future generations who will be exposed to these forms of pollution. In addition, it has to be emphasized that an international approach to the problem will allow Member States to share the best strategies for dealing with such complex and nuanced phenomena.

From this point of view, the OECD's recommendations as elaborated in *Liability for Environmental Damage in EECCA: Implementation of Good International Practices* (OECD 2012), and the *Sourcebook of Pollution Management Policy Tools for Growth and Competitiveness* drafted by the World Bank, remain stand-out examples of ways to approach the issue. These could indeed form the basis of a more ambitious project to develop international *guidelines* for addressing historical pollution.

References

Alogna, I. (2016). The circulation of the model of sustainable development: Tracing the path in a comparative law perspective. In V. Mauerhofer (Ed.), *Legal aspects of sustainable development. Horizontal and sectorial policy issues* (pp. 13–33). Switzerland: Springer International Publishing.

Azam, M. M. (2016). Soil contamination and remediation measures: Revisiting the relevant laws and institutions. In H. Hasegawa, I. M. D. Mofizur Rahman & M. A. Rahman (Eds.), *Environmental remediation technologies for metal-contaminated soils* (pp. 99–124). Tokyo: Springer Japan.

Beck, U. (1992). *Risk society: Towards a new modernity*. London: Sage Publications.

Bergkamp, L. (2005). Implementation of the environmental liability directive in EU member states. *ERA Forum, 6*(3), 389–400.

Bergkamp, L. (2010). Comment on Case C-378/08, 9 March 2010; Joined Cases C-379/08 and 380/08, 9 March 2010; Joined Cases C-478/08 and C-479/08, 9 March 2010. *Journal for European Environmental & Planning Law, 7*(3), 355–361.

Bertolini, P. (2010). Il Principio "chi inquina paga" e la responsabilità per danno ambientale nella sentenza della Corte di giustizia 9 marzo 2010—procedimento C-378/08. *Rivista italiana di diritto pubblico comunitario, 6,* 1607–1632.

Boer, B., & Hannam, I. (2015). Developing a global soil regime. *International Journal of Rural Law and Policy, 1,* 1–13.

Boutillon, S. (2002). The precautionary principle: Development of an international standard. *Michigan Journal of International Law, 23*(2), 429–470.

Bowman, M., & Boyle, A. E. (Eds.). (2002). *Environmental damage in international and comparative law: Problems of definition and valuation*. Oxford, New York: Oxford University Press.

Brandon, E. (2013). *Global approaches to site contamination law*. Dordrecht: Springer.

Cardwell, P. J., French, D., & Hall, M. (2011). Tackling environmental crime in the European Union: The case of the missing victim? *Environmental Law & Management, 23*(3), 113–121.

Coulter, M. (2014). The Minamata convention on mercury: Past, present, and future environmental health. *Sustainable Development Law & Policy, 14*(3), 12–48.

De Figueiredo, M. A., Reiner, D. M., & Herzog, H. J. (2005). Framing the long-term *in situ* liability issue for geologic carbon storage in the United States. *Mitigation and Adaptation Strategies for Global Change, 10*(4), 647–657.

de La Fayette, L. A. (2010). International liability for damage to the environment. In M. Fitzmaurice, D. M. Ong, & P. Merkouris (Eds.), *Research handbook on international environmental law* (pp. 320–360). Cheltenham: Edward Elgar Publishing.

de Sadeleer, N. (2002). *Environmental principles. From political slogans to legal rules*. Oxford: Oxford University Press.

de Sadeleer, N. (2010). The principles of prevention and precaution in international law: Two heads of the same coin? In M. Fitzmaurice, D. M. Ong, & P. Merkouris (Eds.), *Research handbook on international environmental law* (pp. 182–199). Cheltenham: Edward Elgar Publishing.

de Sadeleer, N. (2015). Preliminary reference on environmental liability and the polluter pays principle: Case C-534/13, Fipa. *Review of European, Comparative & International Environmental Law, 24*(2), 232–237.

DiMento, J. F. C. (2010). *The global environment and international law*. Austin: University of Texas Press.

Earnhart, D. (2004). Liability for past environmental contamination and privatization. *Environmental & Resource Economics, 29*(1), 97–122.

Faure, M. (2010). Effective, proportional and dissuasive penalties in the implementation of the environmental crime and shipsource pollution directives: Questions and challenges. *European Energy and Environmental Law Review, 19*(6), 256–278.

Faure, M., & Svatikova, K. (2012). Criminal or administrative law to protect the environment? Evidence from Western Europe. *Journal of Environmental Law, 24*(2), 253–286.

Feris, L. (2012). The public trust doctrine and liability for historic water pollution in South Africa. *Law, Environment and Development Journal, 8*(1), 1. [iv].

Fiedler, H. (2002). *Persistent organic pollutants*. Berlin, Heidelberg: Springer.

Fogleman, V. (2015). Landowners' liability for remediating contaminated land in the EU: EU or national law? Part I: EU law. *Environmental Liability—Law, Policy and Practice, 23*(1), 6–22.

Fogleman, V. (2014). The temporal provisions of the environmental liability directive: The start date, direct effect and retrospectivity. *Environmental Liability—Law, Policy and Practice, 22* (4), 144–155.

Forgotson, E. H. (1964). Liability for long-term latent effects of toxic agents. *American Bar Association Journal, 50*(2), 142–145.

Freestone, D., & Hey, E. (1996). *The precautionary principle and international law: The challenge of implementation*. The Hague, London, Boston: Kluwer Law International.

Funabashi, H. (2006). Minamata disease and environmental governance. *International Journal of Japanese Sociology, 15*(1), 7–25.

Girela, M. A. R. (2006). Risk and reason in the European Union Law. *European Food and Feed Law Review, 5*, 270–285.

Haan-Kamminga, A. (2011). Long-term liability for geological carbon storage in the European Union. *Journal of Energy & Natural Resources Law, 29*(3), 309–331.

Hannam, I., & Boer, B. (2002). *Legal and institutional frameworks for sustainable soils: A preliminary report. IUCN environmental policy and law paper no. 45*. IUCN, Gland, Switzerland and Cambridge, UK in collaboration with IUCN Environmental Law Centre, Bonn, Germany.

Heine, G. (2011). Recenti sviluppi e principali questioni del diritto penale dell'ambiente in Europa. *Rivista trimestrale diritto penale dell'economia, 24*(1–2), 105–122.

Hickey Jr., J. E., & Walker, V. R. (1994/1995). Refining the precautionary principle in international environmental law. *Virginia Environmental Law Journal*, 14(3), 423–454.

Ingelson, A., Kleffner, A., & Nielson, N. (2010). Long-term liability for carbon capture and storage in depleted North American oil and gas reservoirs: A comparative analysis. *Energy Law Journal, 31*(2), 431–470.

Jeffery, M. I., & Zhao, X. (2012). Developing a national contaminated land liability scheme in China: The comprehensive environmental response, compensation, and liability act revisited. *Journal of Energy & Natural Resources Law, 30*(4), 423–465.

Klip, A. (2015). *European criminal law. An integrative approach* (3rd ed.). Cambridge, Antwerp, Portland: Intersentia.

Lallas, P. L. (2001). The Stockholm Convention on persistent organic pollutants. *American Journal of International Law, 95*(3), 692–708.

Larson, E. T. (2005). Why environmental liability regimes in the United States, the European Community, and Japan have grown synonymous with the polluter-pays principle. *Vand. J. Transnat'l L., 38*(2), 541–576.

Larson, R. B. (2013). *Orphaned pollution. Arizona State Law Journal, 45*(3), 991–1032.

Lee, M. (2003). The precautionary principle in the court of first instance. *King's Law Journal, 14*(1), 86–91.

Luppi, B., Parisi, F., & Rajagopalan, S. (2012). The rise and fall of the polluter-pays Principle in developing countries. *International Review of Law and Economics, 32*(1), 135–144.

Mamutse, B., & Fogleman, V. M. (2013). Environmental claims and insolvent companies: The contrasting approaches of the United Kingdom and the United States. *British Journal of American Legal Studies, 2*(2), 579–633.

Martin, T. J. (1985). Long-term liability for hazardous waste induced injury in Missouri: Latent harm sufferers beware. *Wash. UJ Urb. & Contemp. L., 28,* 299–343.

Mattei, U. (2008). Circolazione dei modelli giuridici. In *Enciclopedia del diritto, Annali* I. Milano: Giuffrè.

McIntyre, O. (2005). The all-consuming definition of 'waste' and the end of the 'contaminated land' debate? *Journal of Environmental Law, 17*(1), 109–127.

Mégret, F. (2011). The problem of an international criminal law of the environment. *Columbia Journal of Environmental Law, 36*(2), 195–254.

Milhollin, G. (1979). Long-term liability for environmental harm. *University of Pittsburgh Law Review, 41*(1), 1–26.

Ministry of the Environment, Japan. (2011). *Lessons from Minamata disease and mercury management in Japan*, January 2011, Available at https://www.env.go.jp/en/focus/docs/files/20110101-39.pdf

Mitsilegas, V., Fitzmaurice, M., & Fasoli, E. (2016). The relationship between EU criminal law and environmental law. In V. Mitsilegas, M. Bergström, & T. Konstadinides (Eds.), *Research handbook on EU Criminal Law* (pp. 272–293). Cheltenham, Northampton: Edward Elgar Publishing.

Monti, A. (2002). *Environmental risks and insurance. A comparative analysis of the role of insurance in the management of environment-related risks.* Paris: OECD Report.

Munir, M. (2013). History and evolution of the polluter pays principle: How an economic idea became a legal principle?. *Available at SSRN 2322485.*

Neyret, L. (2014). Pour la reconnaissance du crime d'écocide. *Revue juridique de l'environnement, 39*(HS01), 177–193.

Neyret, L. (Ed.). (2015). *Des écocrimes à l'écocide. Le droit pénal au secours de l'environnement.* Paris: Bruylant.

Öberg, J. (2011). Criminal sanctions in the field of EU Environmental Law. *New Journal of European Criminal Law, 2*(4), 402–425.

OECD Report. (2012). *Liability for environmental damage in Eastern Europe, Caucasus and Central Asia (EECCA): Implementation of good international practices.* Paris: OECD Publishing.

Pedersen, O. W. (2010). Environmental principles and environmental justice. *Environmental Law Review, 12*(1), 26–49.

Percival, R. V., Cooper, K. H., & Gravens, M. M. (2012). CERCLA in a global context. *Southwestern Law Review, 41*(4), 727–772.

Pestana, M. H. D., & Formoso, M. L. L. (2003). Mercury contamination in Lavras do Sul, south Brazil: A legacy from past and recent gold mining. *Science of the Total Environment, 307*(1–3), 125–140.

Petersen, M. (2008). European soil protection law after the setback of December 2007–existing law and outlook. *European Energy and Environmental Law Review, 17*(3), 146–155.

Plantamura, V. (2009). Una nuova frontiera europea per il diritto penale. *Diritto penale e processo, 15*(7), 911–921.

Pozzo, B., Vanheusden, B., Bergkamp, L., & Brans, E. (2015). The remediation of contaminated sites and the problem of assessing the liability of the innocent landowner: A comparative law perspective. *European Review of Private Law, 6*(23), 1071–1119.

Prokop, G., Schamann, M., & Edelgaard, I. (2000). *Management of contaminated sites in Western Europe (Topic report No 13/1999).* Luxembourg: Office for Official Publications of the European Communities.

Rafaraci, T. (2015). New perspectives for the protection of the victims in the EU. In S. Ruggeri (Ed.), *Human rights in European criminal law. New developments in European legislation and case law after the Lisbon Treaty* (pp. 215–225). Switzerland: Springer International Publishing.

Renna, M. (2012). I principi in materia di tutela dell'ambiente. *Rivista quadrimestrale di Diritto dell'Ambiente*, 2012(Settembre), 62-83.

Sacco, R. (1991). Legal formants: A dynamic approach to comparative law (Installment II of II). *The American Journal of Comparative Law, 39*(2), 343–401.

Sands, P., Peel, J., Fabra, A., & MacKenzie, R. (2012). *Principles of international environmental law* (3rd ed.). Cambridge: Cambridge University Press.

Shelbourn, C. (1994). Historic pollution-Does the polluter pay?. *Journal of Planning and Environmental Law*, 703–709.

Siracusa, L. (2008). La competenza comunitaria in ambito penale al primo banco di prova: la direttiva europea sulla tutela penale dell'ambiente. *Rivista trimestrale di diritto penale dell'economia, 21*(4), 863–901.

Stallworthy, M. (1996). Historic pollution, the restoration of contaminated land and the regulatory potential of the regime under the Environment Act 1995. *Oil & Gas Law & Taxation Review, 10,* 184–192.

Sunstein, C. R. (2003). Beyond the precautionary principle. *University of Pennsylvania Law Review, 151*(3), 1003–1058.

Swartjes, F. A. (2011). *Dealing with contaminated sites. From theory towards practical application*. Springer, Netherlands: Dordrecht.

Upham, F. K. (1979). After Minamata: Current prospects and problems in Japanese environmental litigation. *Ecology Law Quarterly, 8*(2), 213–268.

Vagliasindi, G. M. (2012). The European harmonisation in the sector of protection of the environment through criminal law: The results achieved and further needs for intervention. *The New Journal of European Criminal Law, 3*(3–4), 320–331.

Vanheusden, B. (2003). Towards a legal framework in the EU for brownfield redevelopment. *European Energy and Environmental Law Review, 12*(6), 178–186.

Vanheusden, B. (2007). Brownfield redevelopment in the European Union. *Boston College Environmental Affairs Law Review, 34*(3), 559–575.

Vanheusden, B. (2009). Recent developments in European policy regarding brownfield remediation. *Environmental Practice, 11*(04), 256–262.

Varvaštian, S. (2015). Environmental liability under scrutiny: The margins of applying the EU 'polluter pays' principle against the owners of the polluted land who did not contribute to the pollution. Case C-534/13 Ministero dell'Ambiente e della Tutela del Territorio e del Mare and Others v Fipa Group Srl and Others [2015] (ECJ, 4 March 2015) (Fipa Group and Others). *Environmental Law Review, 17*(4), 270–276.

Watson, A. (1974). *Legal transplants: An approach to comparative law*. Edinburgh: Scottish Academic Press.

World Bank Group (2012). *Getting to green—A sourcebook of pollution management policy tools for growth and competitiveness*. Washington, DC.

Wyatt, A. M. (2008). The dirt on international environmental law regarding soils: Is the existing regime adequate? *Duke Environmental Law & Policy Forum, 19,* 165–207.

Xue, H. (2003). *Transboundary damage in international law* (Vol. 27). Cambridge: Cambridge University Press.

Zhao, X. (2013). *Developing an appropriate contaminated land regime in China. Lessons learned from the US and UK*. Berlin, Heidelberg: Springer.

Index

F. Centonze and S. Manacorda (eds.), *Historical Pollution*,
DOI 10.1007/978-3-319-56937-6

MIX
Papier aus verantwortungsvollen Quellen
Paper from responsible sources
FSC® C105338

If you have any concerns about our products,
you can contact us on
ProductSafety@springernature.com

In case Publisher is established outside the EU,
the EU authorized representative is:
Springer Nature Customer Service Center GmbH
Europaplatz 3, 69115 Heidelberg, Germany

Printed by Libri Plureos GmbH
in Hamburg, Germany